Marine Biotechnology

Volume 1
Pharmaceutical and Bioactive
Natural Products

Marine Biotechnology

Contributors

Ami Ben-Amotz, National Institute of Oceanography, Israel Oceanographic and Limnological Research, Tel-Shikmona, Haifa 31080, Israel

Matthew W. Bernart, College of Pharmacy, Oregon State University, Corvallis, Oregon 97331

Mary A. Bober, Marine Science Institute, and Department of Biological Sciences, Unviersity of California, Santa Barbara, California 93106

Bruce F. Bowden, Department of Chemistry and Biochemistry, James Cook University, Townsville, Queensland 4811, Australia

Martha Cohen-Parsons, Department of Chemistry, University of Illinois, Urbana, Illinois 61801

Brent R. Copp, Department of Medicinal Chemistry, University of Utah, Salt Lake City, Utah 84112

Phil Crews, Department of Chemistry and Biochemistry, and Institute of Marine Sciences, University of California, Santa Cruz, California 95064

Marianne S. de Carvalho, Marine Science Institute, and Department of Biological Sciences, University of California, Santa Barbara, California 93106

D. John Faulkner, Scripps Institution of Oceanography, University of California at San Diego, La Jolla, California 92093-0212

William Fenical, Marine Research Division, Scripps Institution of Oceanography, University of California at San Diego, La Jolla, California 92093-0236

Mark P. Foster, Department of Medicinal Chemistry, University of Utah, Salt Lake City, Utah, 84112

William H. Gerwick, College of Pharmacy, Oregon State University, Corvallis, Orgeon 97331

Lisa M. Hunter, Department of Chemistry and Biochemistry, and Institute of Marine Sciences, University of California, Santa Cruz, California 95064

Chris M. Ireland, Department of Medicinal Chemistry, University of Utah, Salt Lake City, Utah, 84112

Robert S. Jacobs, Marine Science Institute, and Department of Biological Sciences, University of California, Santa Barbara, California 93106

Peer B. Jacobson, Marine Science Institute, and Department of Biological Sciences, University of California, Santa Barbara, California 93106

Paul R. Jensen, Marine Research Division, Scripps Institution of Oceanography, University of California at San Diego, La Jolla, California 92093-0236

Leonard A. McDonald, Department of Medicinal Chemistry, University of Utah, Salt Lake City, Utah, 84112

Koji Nakanishi, Suntory Institute for Bioorganic Research, Shimamoto, Mishima-gun, Osaka, Japan, and Department of Chemistry, Columbia University, New York, New York 10027

Yoko Naya, Suntory Institute for Bioorganic Research, Shimamoto, Mishima-gun, Osaka, Japan

Isabel Pinto, Marine Science Institute, and Department of Biological Sciences, University of California, Santa Barbara, California 93106

Derek C. Radisky, Department of Medicinal Chemistry, University of Utah, Salt Lake City, Utah 84112

Donald W. Renn, FMC Corporation, Rockland, Maine 04841

Kenneth L. Rinehart, Department of Chemistry, University of Illinois, Urbana, Illinois 61801

Mukesh K. Sahni, Department of Chemistry, William Paterson College, Wayne, New Jersey 07470

Francis J. Schmitz, Department of Chemistry and Biochemistry, University of Oklahoma, Norman, Oklahoma 73019

Gurdial M. Sharma, Department of Chemistry, William Paterson College, Wayne, New Jersey 07470

Lois S. Shield, Department of Chemistry, University of Illinois, Urbana, Illinois 61801

Yuzuru Shimizu, Department of Pharmacognosy and Environmental Sciences, College of Pharmacy, The University of Rhode Island, Kingston, Rhode Island 02881

J. Christopher Swersey, Department of Medicinal Chemistry, University of Utah, Salt Lake City, Utah 84112

Kazuo Tachibana, Suntory Institute for Bioorganic Research, Shimamoto, Mishima-gun, Osaka, Japan; *current address*: Department of Chemistry, University of Tokyo, Hongo, Bunkyo, Japan

Stephen I. Toth, Department of Chemistry and Biochemistry, University of Oklahoma, Norman, Oklahoma 73019

Allen B. Williams, Marine Science Institute, and Department of Biological Sciences, University of California, Santa Barbara, California 93106

Preface

Biotechnology may be defined as the application of scientific and engineering principles to the processing of materials by biological agents to provide goods and services (Bull *et al.*, 1982, p. 21) or as any technique that uses living organisms (or parts of organisms) to make or modify products, to improve plants or animals, or to develop microorganisms for specific use (OTC, 1988). In line with these broad definitions we can consider marine biotechnology as the use of marine organisms or their constituents for useful purposes in a controlled fashion. This series will explore a range of scientific advances in support of marine biotechnology. It will provide information on advances in three categories: (1) basic knowledge, (2) applied research and development, and (3) commercial and institutional issues. We hope the presentation of the topics will generate interest and interaction among readers in the academic world, government, and industry. This first volume examines chemical and biological properties of some natural products that are useful or potentially useful in research and in the chemical and pharmaceutical industries. One chapter describes a system for producing such substances on a large scale.

Biotechnology incorporates molecular biology in order to go beyond traditional biochemical technology such as the production of antibiotic drugs from bacterial cultures in bioreactors. Development of the technology for production of antibiotics in this way resulted from fundamental advances in chemistry, pharmacology, microbiology, and biochemical engineering. It is likely that molecular biology will be used to improve the efficiency of this technology. One of the objectives of this series is to emphasize the importance of interdisciplinary science and to encourage improvement of the environment for it in academic institutions. Another objective is to demonstrate the rich array of materials and processes in the marine world that have no terrestrial counterparts and to suggest that the workings of this diverse world should be explored aggressively. The majority of plant and invertebrate phyla on earth are either exclusively or predominantly marine. Many of these marine organisms are so poorly understood that they have yet to be fully described and named. Developing a thorough understanding of the ocean's biological materials and processes will provide new information for technological development. The authors who have contributed to this volume make this point eloquently.

We acknowledge the enthusiasm and cooperation with which the advisory board and authors have endorsed this series and its first volume.

David H. Attaway
Oskar R. Zaborsky

REFERENCES

Bull, A. T., Holt, G., and Lilly, M. D., 1982, *Biotechnology: International Trends and Perspectives*, Organisatsion for Economic Co-operation and Development, Paris.
OTC (Office of Technology Assessment), 1988, *New Developments in Biotechnology 4: U.S. Investment in Biotechnology*, Office of Technology Assessment, Congress of the United States, Washington, D.C.

Contents

Chapter 3

Pharmacological Studies of Novel Marine Metabolites 77

Robert S. Jacobs, Mary A. Bober, Isabel Pinto, Allen B. Williams,
Peer B. Jacobson, and Marianne S. de Carvalho

Chapter 4

Eicosanoids and Related Compounds from Marine Algae 101

William H. Gerwick and Matthew W. Bernart

Chapter 5

Marine Proteins in Clinical Chemistry 153

Gurdial M. Sharma and Mukesh K. Sahni

Chapter 6

Medical and Biotechnological Applications of Marine Macroalgal Polysaccharides 181

Donald W. Renn

Chapter 7

Antitumor and Cytotoxic Compounds from Marine Organisms 197

Francis J. Schmitz, Bruce F. Bowden, and Stephen I. Toth

Chapter 8

Antiviral Substances 309

Kenneth L. Rinehart, Lois S. Shield, and Martha Cohen-Parsons

Chapter 9

The Search for Antiparasitic Agents from Marine Animals 343

Phil Crews and Lisa M. Hunter

Chapter 10

Dinoflagellates as Sources of Bioactive Molecules 391

Yuzuru Shimizu

Chapter 11

Production of β-Carotene and Vitamins by the Halotolerant Alga *Dunaliella*

Chapter 12

Marine Microorganisms: A New Biomedical Resource

Chapter 13

Academic Chemistry and the Discovery of Bioactive Marine Natural Products 459

D. John Faulkner

Biomedical Potential of Marine Natural Products

Chris M. Ireland, Brent R. Copp, Mark P. Foster, Leonard A. McDonald, Derek C. Radisky, and J. Christopher Swersey

1. INTRODUCTION

Marine natural products, the secondary or nonprimary metabolites produced by organisms that live in the sea, have received increasing attention from chemists and pharmacologists during the last two decades. Interest on the part of chemists has been twofold: natural products chemists have probed marine organisms as sources of new and unusual organic molecules, while synthetic chemists have followed by targeting these novel structures for development of new analogs and new synthetic methodologies and strategies (Albizati *et al.*, 1990). The rationale for investigating the chemistry of marine organisms has changed over the past several decades. Early investigations were largely of a "phytochemical" nature, reporting detailed metabolite profiles similar to those reported for terrestrial plants in previous decades. However, analogous to investigations of terrestrial plants, more recent studies of marine organisms have focused on their potential applications, particularly to the treatment of human disease and control of agricultural

Chris M. Ireland, Brent R. Copp, Mark P. Foster, Leonard A. McDonald, Derek C. Radisky, and J. Christopher Swersey • Department of Medicinal Chemistry, University of Utah, Salt Lake City, Utah 84112.

Marine Biotechnology, Volume 1: Pharmaceutical and Bioactive Natural Products, edited by David H. Attaway and Oskar R. Zaborsky. Plenum Press, New York, 1993.

pests (Fautin, 1988). Pharmacological evaluations of marine natural products have likewise undergone an evolution over the past two decades: beginning with the early investigations of toxins, followed by studies of cytotoxic and antitumor activity, to the present day, where a myriad of activities based on whole-animal models and receptor-binding assays are being pursued. The intent of this chapter is to look back at the evolution of biomedically oriented natural product studies of marine organisms, to chronicle the key developments, discoveries, and advances in the level of sophistication that have fueled further interest in this field, and finally to look forward at the future biomedical potential of marine natural products.

This chapter will not provide a comprehensive review of marine natural products; there are already several treatises that give such coverage (Faulkner, 1991 and earlier reviews cited therein). Additionally, several chapters in this volume provide detailed discussions of individual groups of metabolites or approaches to the treatment of specific diseases. Instead, the look back will highlight selected examples where marine natural products have been subjected to pharmacological investigations beyond the preliminary screening stage. Particular attention will be given to those metabolites that are being or have been developed as pharmaceuticals, and compounds that have made significant contributions to our understanding of cellular processes at the biochemical level. It is difficult to distinguish between these two categories because the difference between an effective drug and a molecular probe lies, more often than not, in the degree of toxicity associated with the compound.

Foregoing an in-depth review, it is still useful to look at trends as the field has developed. In the decade from 1977 to 1987 approximately 2500 new metabolites were reported from a wide variety of marine organisms ranging from prokaryotic microbes and soft-bodied invertebrates to vertebrate fish. These metabolites are equally diverse in their biosynthetic origins, spanning mevalonate, polyketide, and amino acid metabolic pathways (Ireland *et al.*, 1988, 1989). An analysis of the phyletic distribution of these compounds (Fig. 1) shows that the majority (93%) are confined to four groups (macroalgae, coelenterates, echinoderms, and sponges), largely a reflection of the abundance and ease of collection of these organisms. Comparing these results with those of an earlier survey for the period of 1977–1985 (Ireland *et al.*, 1988), and also the results for the last 2 years of the present survey (Fig. 2), several trends are observed. While in each breakdown the same four groups dominate (accounting for greater than 92% of the compounds in each case), their relative contributions change. The contribution from macroalgae decreases significantly, whereas sponges become the dominant source of compounds. The increase in studies of sponges explains, in large part, the increasing diversity of structural types reported from marine organisms, because sponges have a wider range of biosynthetic capabilities than any other group of marine invertebrates. The number of compounds reported from echinoderms has also increased dramatically due largely to recent comprehensive investigations of

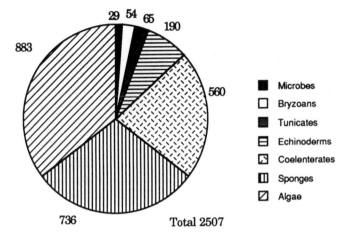

Figure 1. Phyletic distribution of newly reported secondary metabolites for 1977–1987.

saponins produced by starfish (Bruno *et al.*, 1990, and references therein). Unfortunately, these compounds tend to be generally toxic to erythrocytes. Although the contributions from ascidians and microbes remain modest, there is increasing interest in these groups. Interest in the former reflects recognition that in spite of their small size and general difficulty of collection, ascidians produce an array of very potent natural products; didemnin B (1) the only marine natural

(1)

product currently in human clinical trials, is from an ascidian (Dorr *et al.*, 1986, 1988; Stewart *et al.*, 1986). Microbes are being recognized for the nearly unlimited potential of culturable marine Actinomycetes, bacteria, and cyanobacteria as

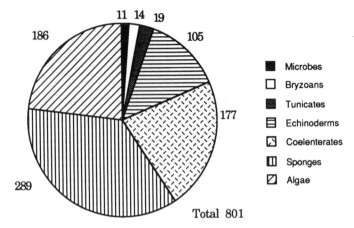

11 14 19
186   105

186

289

177

Total 801

■ Microbes
□ Bryzoans
▨ Tunicates
⊟ Echinoderms
◨ Coelenterates
Ⅲ Sponges
▨ Algae

Figure 2. Phyletic distribution of newly reported metabolites for 1986–1987.

renewable sources of pharmaceuticals (see Chapter 12 by Fenical and Jensen). It is likely that in the next decade these two groups will assume important roles in marine biomedical studies. It is also important to point out that the number of compounds reported annually is increasing steadily, indicating that marine organisms will continue to be significant sources of natural products. While the above analysis can only be retrospective and is inherently flawed because it partially reflects the specific interests of investigators in the field, it also serves to indicate which groups of marine organisms are currently receiving the greatest attention from the perspective of drug development, and also provides some insights into which groups will play important roles in future biomedical studies of marine natural products.

Looking back over the past several decades, we can see that biomedical investigations of marine natural products have focused mainly on a few areas: CNS membrane-active toxins and ion channel effectors, anticancer and antiviral agents, tumor promoters, anti-inflammatory agents, and metabolites that control microfilament-mediated processes. The important developments within these areas will be highlighted in the following sections.

2. MARINE NATURAL PRODUCTS THAT ACT AT MEMBRANE RECEPTORS

Perhaps the most important molecules (from a cellular physiology and pharmacology standpoint) derived from marine sources to date are the very potent and specific sodium (Na$^+$) channel blockers tetrodotoxin (**2**) and saxitoxin (**3**).

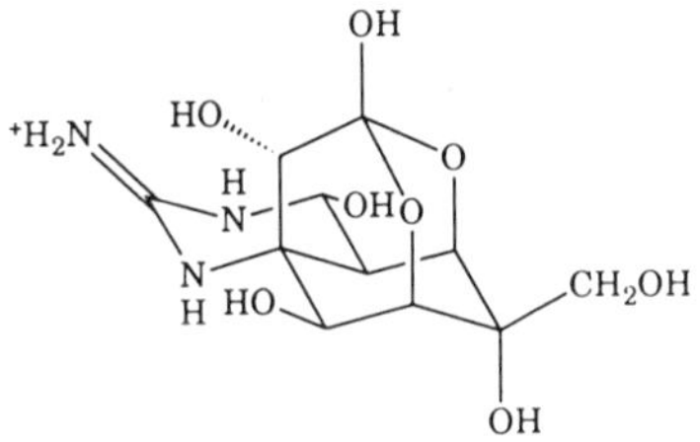

(2) 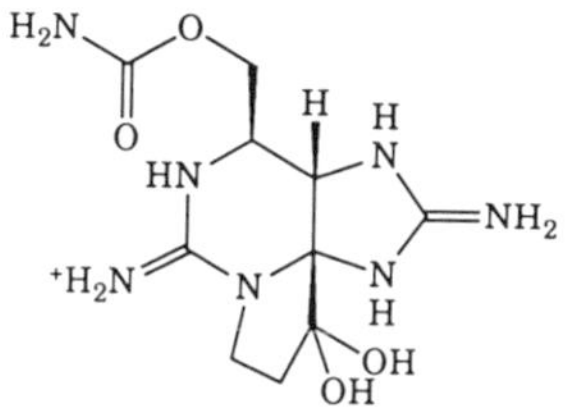   (3)

Pharmacological studies with tetrodotoxin and saxitoxin played the major role in developing the concept of Na^+ channels in particular and membrane channels in general. These compounds are small, structurally unrelated organic cations that act as Na^+ channel occluders, with no effect on the gating mechanism of the blocked channel. They interact only on the external surface of Na^+ channels, interrupting passive inward flux of Na^+ ions. Numerous reviews on tetrodotoxin and saxitoxin have been published, covering their history, chemistry, physiology, and pharmacology and developments pertaining to the characterization of Na^+ channels (Catterall, 1980; Hall *et al.*, 1990; Kao, 1966; Narahashi, 1974, 1988; Kao, 1986).

Although these compounds have been important for conceptualizing the existence of ion channels, the mechanism by which they act as occluders remains in question. Two models have been proposed to explain the action of these toxins. The simplest occluder model entails binding of the inhibitor to the pore, blocking ion transport; the second model involves binding of the toxin to a receptor on the exterior of the protein, causing a distinct change in channel permeability. In fact, stoichiometry and binding studies suggest that tetrodotoxin and saxitoxin bind at or near the external mouth of the channel, and their blocking action is inhibited by a variety of monovalent and divalent cations. Recent evidence has put into question the simple "plug" model: the structural dependence for activity of the toxins appears to be more complicated than necessary for simple pore blockage, and physiological behavior extends beyond simple occlusion (Kao, 1986; Strichartz *et al.*, 1986; Strichartz and Castle, 1990).

These toxins have proven extremely useful and popular chemical tools for neurophysiology and neuropharmacology studies. Numerous experiments have been conducted using tetrodotoxin and saxitoxin to study the biophysics of Na^+ channel action (e.g., Cai and Jordan, 1990), Na^+ channel expression and assembly during cellular development (e.g., Wollner *et al.*, 1988), and the potential antiarrhythmic properties of tetrodotoxin against occlusion-induced arrhythmias (Abraham *et al.*, 1989). Binding studies with ^{3}H-labeled toxins were used to determine the density of Na^+ channels in various tissue membranes (Narahashi, 1988), and the toxins have also proven applicable in monoclonal antibody studies of lymphocyte activation (e.g., Pinchuk and Pinchuk, 1990).

The potency and selectivity of tetrodotoxin and saxitoxin for Na^+ channels has led to their common use in voltage clamp analyses in which it is desirable to separate the transient current (carried mostly by Na^+ ions) and the steady-state current (carried by K^+) ions (Narahashi, 1974). Tetrodotoxin and saxitoxin will undoubtedly continue to be important fundamental tools for the study of Na^+ channels and excitatory phenomena.

The marine dinoflagellate *Ptychodiscus brevis* (formerly *Gymnodinium breve*), collected from red tides in the Gulf of Mexico, produces a range of potent, high-molecular- weight polyether neurotoxins known as the brevetoxins (**4,5**) (Lin

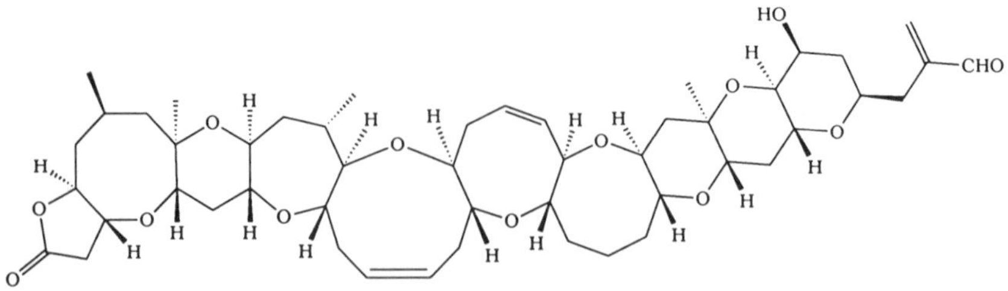

(**4**)

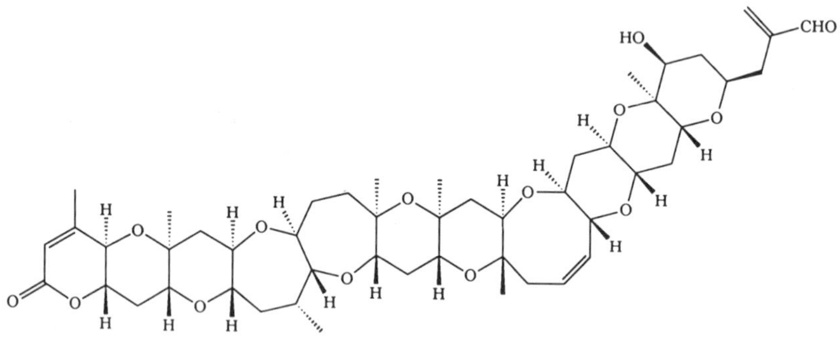

. (**5**)

et al., 1981; Shimizu *et al.*, 1968a, b; van Duyne, 1990). There are many different physiological effects elicited by the brevetoxins, including ichthyotoxicity, depression of respiratory and cardiac function, and dose-related muscular contractions leading to twitching, spasms, and eventual death (Baden, 1989). All these effects are consistent with sodium channel activation. Preliminary studies indicated that the brevetoxins bind specifically to site 5 of voltage-sensitive sodium channels, resulting in increased cell resting potential (Strichartz and Castle, 1990). The resulting increase in cellular sodium influx causes nerve depolarization and neurotransmitter release, followed by failure of neuromuscular transmission (Edwards *et al.*, 1989; Baden, 1989). Studies with [³H]-brevetoxin-3 established the number of sodium channel binding sites in rat and turtle brain tissue (Edwards *et al.*, 1989). Structure–activity relationship studies have been initiated with the brevetoxins for the purpose of characterizing binding site 5 of the voltage-sensitive sodium channels (Trainer *et al.*, 1990).

The venoms of the predatory cone snails (*Conus* spp.) contain small, pharmacologically active peptides, which are targeted to various ion channels and receptors. Due to the ability of the conotoxins to discriminate between closely related receptor subtypes, they have become valuable tools for research in neuroscience (Olivera *et al.*, 1990a). μ-Conotoxin GIIIA (also known as geographutoxin II or GTX II), isolated from the piscivorous cone snail *Conus geographus*, a 22-residue polycyclic peptide bearing three hydroxyprolines and three disulfide bridges (Olivera *et al.*, 1990a, b), was shown to bind, and hence occlude, site 1 of the sodium channel in skeletal muscle. μ-Conotoxin GIIIA is an important tool for studying skeletal muscular neurotransmission because it has no discernible effect on sodium channels in cardiac, brain, and nervous tissue (Cruz *et al.*, 1989; Hong and Chang, 1989; Strichartz and Castle, 1990).

ω-Conotoxin GVIA (ω-CgTx), a 27-amino acid peptide isolated from *C. geographus* (Yoshikami *et al.*, 1989; Olivera *et al.*, 1990a, b), is a neuron-specific antagonist with high affinity for voltage-sensitive calcium channels (VSCC) which control neurotransmitter release in response to depolarization of nerve termini. ω-Conotoxin GVIA interacted with N- and/or L-type channels, depending upon tissue type, and was also instrumental in characterizing these channel subtypes and their distribution in various tissue (Plummer *et al.*, 1989; Karschin and Lipton, 1989; Yoshikami *et al.*, 1989). In conjunction with specific L-type calcium channel blocking ligands, the relative importance of N- and L-type VSCCs to the release of various neurotransmitters has been assessed in parallel experiments using ω-conotoxin GVIA and nifedipine (Wessler *et al.*, 1990), nilvadipine (Takemura *et al.*, 1989a), and PN 200-110 (Keith *et al.*, 1989; Mangano *et al.*, 1990; Dutar *et al.*, 1989; Herdon and Nahorski, 1989; Fredholm, 1990). ^{125}I-Labeled toxin was used to determine the distribution of calcium channels in rat brain tissue (Takemura *et al.*, 1989b), to identify N- and L-type VSCCs in bovine adrenomedullary plasma membranes (Jan *et al.*, 1990; Ballesta *et al.*, 1989), and to visualize toxin-binding sites in brain tissue of normal and cerebellar mutant mice (Maeda *et al.*, 1989).

Maitotoxin, one of the principal toxins responsible for ingestive ciguatera fish poisoning, is a high-molecular-weight polyether neurotoxin produced by the marine dinoflagellate *Gambierdiscus toxicus* (Santostasi *et al.*, 1990); it causes death in mice (170 ng/kg) when injected intraperitoneally. Numerous pharmacological effects are elicited by maitotoxin in a dose-dependent manner, typically at concentrations in the pico- to nanomolar range. These effects include increased cellular calcium uptake, neurotransmitter/hormone release, phosphoinositide breakdown, contraction of smooth and skeletal muscle, and stimulation effects on the heart (Gusovsky and Daly, 1990). Most effects of maitotoxin appear to be due to either its interaction with extracellular calcium or the enhanced influx of calcium (Gusovsky and Daly, 1990; Ohizumi and Kobayashi, 1990; Wu and Narahashi, 1988). This was supported by a recent study which showed that all muscle-contracting effects of maitotoxin were profoundly suppressed or abolished by calcium-ion entry blockers and polyvalent cations (Ohizumi and Kobayashi, 1990). The potential of maitotoxin as a unique pharmacological tool for studying calcium transport is yet to be fully realized.

The nicotinic/acetylcholine receptor (nAChR) represents a focal point to understanding the process of signal transduction in biological systems. Characterization of this receptor/ion channel is critical to understanding voluntary neuromuscular transmission, as it is present at the postsynaptic endplates of skeletal muscle. The nAChR also plays an important role in signal transduction in the autonomic and central nervous systems. Signal transduction occurs by binding of agonists to their receptors on the α subunits (one receptor per subunit) producing a transient opening of the ion channel, resulting in membrane depolarization (Stroud *et al.*, 1990). Several phylogenetically and structurally diverse marine natural products have proven useful probes for studying the nAChR.

The α-conotoxins (Table I), isolated from cone shells (Olivera *et al.*, 1985), contain from 13 to 15 amino acids and possess a consensus sequence that presumably preserves some structural and thereby functional homology (Pardi *et al.*, 1989). The α-conotoxins (so named because of their pharmacological similarity to the snake α-toxins) were shown to act as antagonists of the nAChR at the endplate region of neuromuscular junctions blocking neuromuscular transmission. This binding was reversible and inhibited by preincubation with other nAChR antagonists, such as α-bungarotoxin (Olivera *et al.*, 1990a). Whole-animal studies showed that the α-conotoxins were selective for nACh receptors at neuromuscular junctions; α-conotoxins-GI and -MI had no effect on blood pressure, heart rate, or responses to vagal and preganglionic stimulation at concentrations which induced paralysis (Marshall and Harvey, 1990). More interestingly, α-conotoxin-SI, which has a proline at position nine instead of a positively charged residue, displayed phylogenetic selectivity toward nACh receptors, indicating that structurally these

Table I.   The Various Conotoxins and Their Linear Peptide Sequences

Name of peptide	Sequence
α-Conotoxin GI	ECCNPACGRHYSC-NH$_2$
α-Conotoxin-GIA	ECCNPACGRHYSCGK-NH$_2$
α-Conotoxin-MI	GRCCHPACKNYSC-NH$_2$
α-Conotoxin-SI	ICCNPACGPKYSC-NH$_2$
μ-Conotoxin-GIIIA	RDCCTPPKKCKDRQCKPQRCCA-NH$_2$
ω-Conotoxin-GVIA	CKSPGSSCSPTSYNCCRSCNPYTKRCY-NH$_2$

P: hydroxyproline.

receptors are a nonhomogeneous family of proteins (Zafaralla *et al.*, 1988). α-Conotoxins have been used in conjunction with monoclonal antibodies and other agonists and antagonists of the nAChR to probe the structure of the agonist receptor sites on the α-subunits (Dowding and Hall, 1987). The results from this investigation supported a model with two binding sites for ACh on the receptor, and indicated that these sites are structurally distinct. The high selectivity of α-conotoxins to nACh receptors at neuromuscular junctions suggests they will continue to be useful probes of this system (Olivera *et al.*, 1988, 1990a; Kobayashi *et al.*, 1989; Wu and Narahashi, 1988).

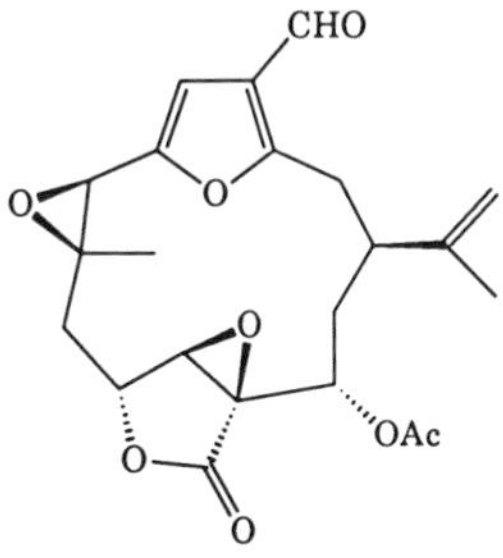

(6a) 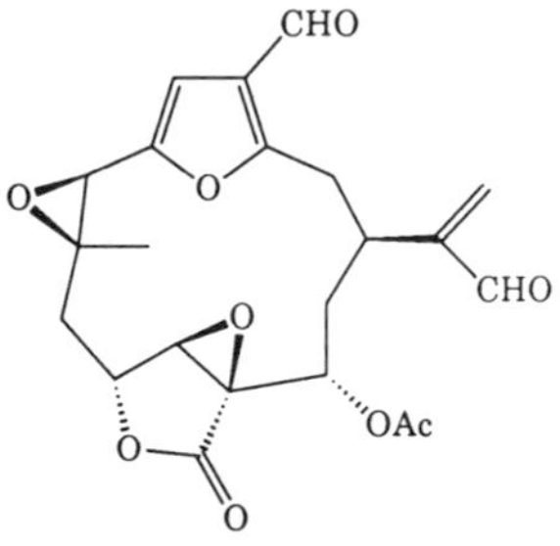   (6b)

Lophotoxin (**6a**) and its analog-1 (**6b**), diterpene lactones isolated from gorgonian corals, *Lophogorgia* spp. (Fenical *et al.*, 1981), are paralytic toxins that produce an irreversible postsynaptic blockade at neuromuscular junctions (Culver and Jacobs, 1981; Culver *et al.*, 1985). Lophotoxin irreversibly inactivated the nAChR by preferential binding to tyrosine-190 (Abramson *et al.*, 1988) at one of the two primary agonist sites (Culver *et al.*, 1984), via Michael addition or Schiff base formation (Culver *et al.*, 1985). In addition, lophotoxin was shown to be a selective, high-affinity antagonist at the neuronal nAChR, blocking nicotinic transmission in autonomic ganglia. Lophotoxin is an important tool for studying this receptor because most α-neurotoxins demonstrate little or no physiological activity at this receptor (Sorenson *et al.*, 1987). The importance and biomedical uses of lophotoxin have been reviewed (Jacobs *et al.*, 1985; Taylor *et al.*, 1988).

Neosurugatoxin (**7**), a reversible nAChR antagonist isolated from the Japanese ivory mollusk, *Babylonia japonica* (Kosuge *et al.*, 1981), was useful for characterizing the ganglionic nAChR into two different subclasses. It is generally accepted that there are two subclasses of nACh receptors in nervous tissue, the so-called high- and low-affinity agonist receptors (Billiar *et al.*, 1988). Neosurugatoxin binds with high selectivity at the high-affinity agonist site, as indicated by the observation that it did not inhibit α-bungarotoxin binding, a known antagonist at the low-affinity site (Billiar *et al.*, 1988; Bourke *et al.*, 1988). The selectivity of neosurugatoxin for specific receptors in nervous tissue illustrated the pharmacological distinction between ganglionic nACh receptors and those at the neuro-

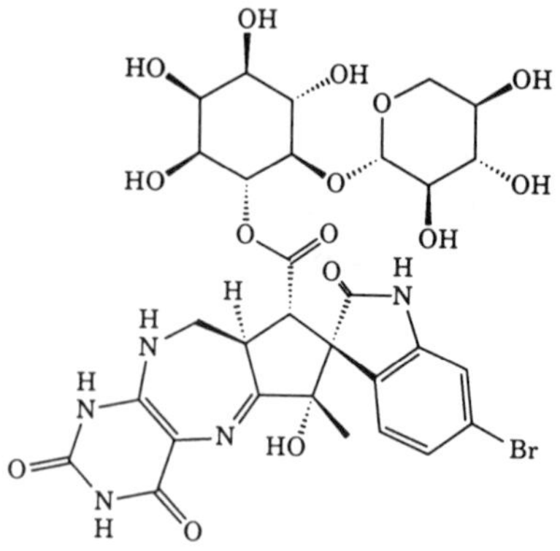

(7)

muscular junction (Wada *et al.*, 1989; Hayashi and Yamada, 1975; Hayashi *et al.*, 1984; Rapier *et al.*, 1990).

Recently, a study was conducted using α-conotoxins-GIA and -MI, lophotoxin and its analog-1, neosurugatoxin, and α-bungarotoxin to probe the structural differences between neuronal nAChR subunit combinations expressed in *Xenopus* oocytes (Luetje *et al.*, 1990). The results indicated that neosurugatoxin and the conotoxins distinguished between muscle and neuronal subunit combinations, whereas lophotoxin and analog-1 distinguished between neuronal subunit combinations on the basis of differing α subunits.

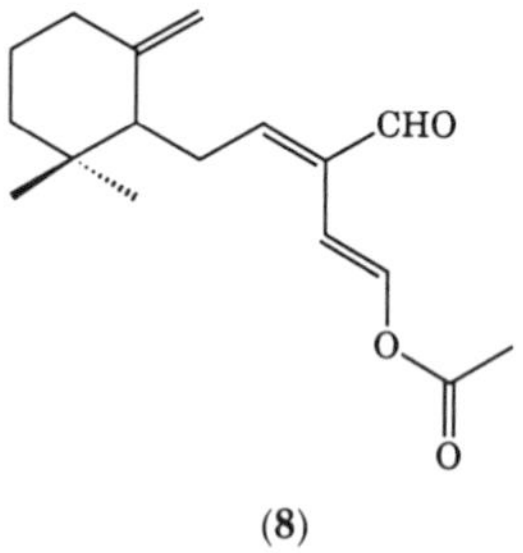

(8)

Onchidal (**8**), an acetate ester isolated from the mollusk *Onchidella binneyi* (Ireland and Faulkner, 1978), was shown to be an irreversible substrate inhibitor of acetylcholinesterase with a novel mechanism of action, with no effect on α-bungarotoxin binding to the nAChR (Abramson *et al.*, 1989b). Irreversible inhibition was prevented by coincubation with agents that block or modify the ACh binding site. Thus, onchidal may prove useful for the design of new anticholinesterase insecticides and in identifying the active site residues that contribute to binding and hydrolysis of ACh.

The importance of adenosine as a second messenger in signal transduction has led to considerable interest in compounds that bind adenosine receptors. A pair of modified nucleosides that act as adenosine analogs have been used to study adenosine receptors in a variety of systems. (5'-deoxy-) 5-Iodotubercidin (**9**),

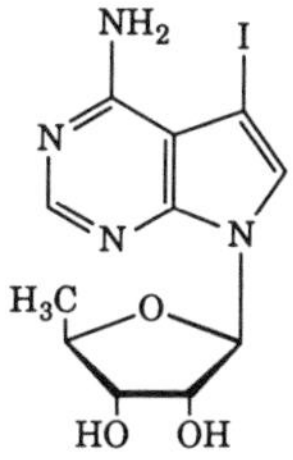

(**9**) 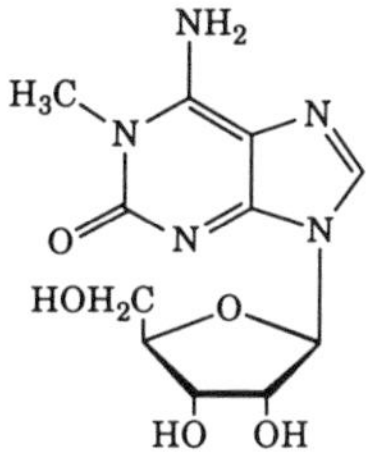   (**10**)

isolated from the red alga *Hypnea valetiae* (Kaslauskas *et al.*, 1983), inhibited adenosine uptake in brain tissue, and is one of the most potent and specific inhibitors of adenosine kinase reported to date (Davies *et al.*, 1984). The specificity and potency of this molecule toward adenosine kinase led to its wide application in studies of purine/nucleotide metabolism and regulation (e.g., Das and Steinberg, 1988; Dawicki *et al.*, 1988; Weinberg *et al.*, 1988; Bontemps and van den Berghe, 1989; Kather, 1990). 1-Methylisoguanosine (**10**), an orally active adenosine analog isolated from the marine sponge *Tedania digitata* (Baird-Lambert *et al.*, 1980), exhibited properties as a muscle relaxant, inducing hypothermia and cardiovascular effects similar to adenosine, and was used to aid in the characterization of these receptors (Williams *et al.*, 1987).

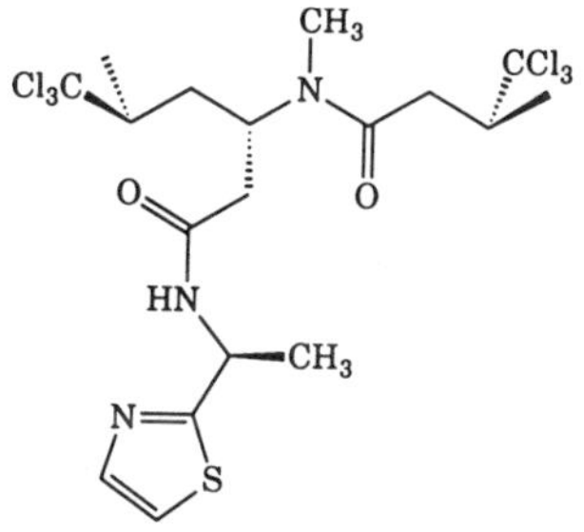

(**11**)

Dysidenin (**11**), a hexachlorinated alkaloid isolated from the sponge *Dysidea herbacea* (Charles *et al.*, 1978, 1980; Biskupiak and Ireland, 1984), inhibits iodide transfer in thyroid cells. This molecule may provide insight into the mechanism of the elusive "iodide pump," as it inhibits iodide transport by a different mechanism than ouabain (Van Sande *et al.*, 1990).

3. ANTITUMOR COMPOUNDS

Spongothymidine (arabinosyl thymine; araT) (12) and spongouridine (arabinosyl uracil; araU) (13) were isolated in the early 1950s (Bergmann and Feeney,

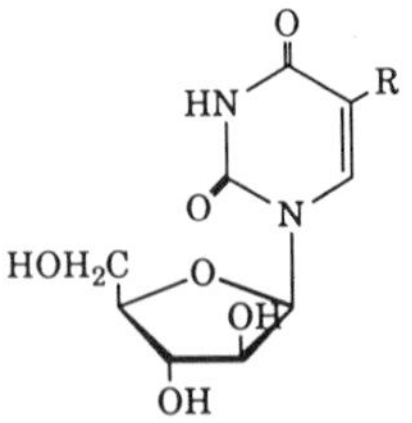

(12) Spongouridine (araU): R = H

(13) Spongothymidine (araT): R = CH$_3$

1951; Bergmann and Burke, 1955) from the Caribbean sponge *Tethya crypta*. This discovery ultimately led to the synthesis of a new class of arabinosyl nucleoside analogs. One such analog, arabinosyl cytosine (ara-C) (14), displayed *in vivo*

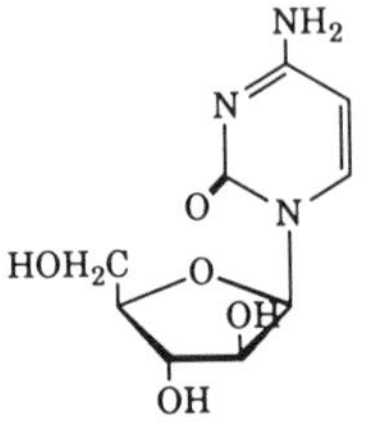

(14)

antileukemic activity (Evans *et al.*, 1961, 1964). Ara-C derived its activity by conversion to arabinosyl cytosine triphosphate, incorporation into cellular DNA, and subsequent inhibition of DNA polymerase (S. Cohen, 1977). It is currently in clinical use for treatment of acute myelocytic leukemia and non-Hodgkin's lymphoma (de Andrea *et al.*, 1990; Minigo *et al.*, 1990; Geller *et al.*, 1990). The arabinosyl nucleosides served a catalytic role in promoting the search for antitumor compounds from marine sources. Over the subsequent four decades, several research programs have focused on this objective. Although many compounds of marine origin have been isolated which possess *in vitro* activity, only didemnin B (1) has been evaluated in clinical trial. A review of cytotoxic compounds is undertaken elsewhere in this volume (Chapter 7).

The didemnins, a series of new depsipeptides isolated from the ascidian *Trididemnum solidum*, showed promising cytotoxic, antitumor, antiviral (Rinehart *et al.*, 1981a–c), and immunosuppressive activities (Montgomery and Zukoski, 1985; Montgomery *et al.*, 1987). Didemnin B, the most potent of this group, exhibited cytotoxicity against L1210 murine leukemia cells *in vitro* at a dose of

0.001 μg/ml (Rinehart *et al.*, 1981b). Didemnin B was also reported active in stem-cell assays against a variety of human tumors, including mesothelioma, sarcoma, hairy cell leukemia, and carcinomas of the breast, ovary, and kidney (Rossof *et al.*, 1983; Jiang *et al.*, 1983). Didemnin B also showed significant *in vivo* activity against intraperitoneally implanted B16 melanoma (T/C 160 @ 1 mg/ kg) and P388 leukemia cells (T/C 199 @ 1 mg/kg) in mice (Rinehart *et al.*, 1981b, 1983) (T/C = mean survival time of test group compared to that of control group, as a percent). At the highest nontoxic dose, didemnin B effected a 45% survival rate with a 2.7-fold life extension of rats challenged with Yoshida ascites tumor cells (Famiani, 1987). These promising *in vivo* results led to preclinical toxicology studies and ultimately to evaluation of didemnin B in phase I clinical trials in the United States as an anticancer agent (Dorr *et al.*, 1986, 1988; Stewart *et al.*, 1986). Preclinical toxicological studies suggested that the major target organs for toxicity were the lymphatics, gastrointestinal tract, liver, and kidney. Didemnin B caused hepatic toxicity with a decrease in clotting factors (Chun *et al.*, 1986; Stewart *et al.*, 1986). Based on these results, it was anticipated that the maximum tolerated dose in humans would be defined by gastrointestinal or hepatotoxicity. Didemnin B completed phase I clinical trials, where maximum tolerable clinical doses were established. It is slated for phase II trials, where it is hoped that objective tumor response will be observed.

Didemnin B was reported to inhibit protein synthesis more than DNA synthesis, and to a lesser extent RNA synthesis, in B16 and L1210 cells (Crampton *et al.*, 1984; Li *et al.*, 1984). Also, didemnin B reportedly suppressed phospho-rylation of a 100-kDa epidermal cytosolic protein along with immunological and inflammatory reactions induced by the tumor promoter 12-*O*-tetradecanoylphorbol-13-acetate (TPA) (Gschwendt *et al.*, 1987). It was further noted that inhibition of protein synthesis was not due to decreased amino acid uptake, since cytotoxicity was not reversed by high concentrations of several amino acids. Furthermore, flow cytometric results indicated that low doses of didemnin B (10 ng/ml) inhibited progression of the cell cycle from G_1 to S phase (Crampton *et al.*, 1984). The exact mechanism(s) of action of didemnin B are unknown.

In vivo evaluation of several marine natural products yielded several promis-ing leads for future development. Mycalamide A (**15**), isolated from a *Mycale* sp. sponge (Perry *et al.*, 1988), exhibited marked *in vitro* cytotoxicity and good *in vivo* activity against P388 leukemia (T/C 156 @ 10 μg/kg), B16 melanoma (T/C 245 @ 30 μg/kg), and M5076 ovarian sarcoma (T/C 233 @ 60 μg/kg) (Burres and Clement, 1989).

Halichondrin B (**16**), one of several polyether macrolides isolated from the marine sponge *Halichondria okadai*, showed excellent activity *in vivo* against P388 leukemia (T/C 323 @ 10 μg/kg) and B16 melanoma (T/C 244 @ 5 μg/kg) (Hirata and Uemura, 1986). These promising results (T/C of 300 is considered a cure in this protocol) have generated much interest in the potential of the

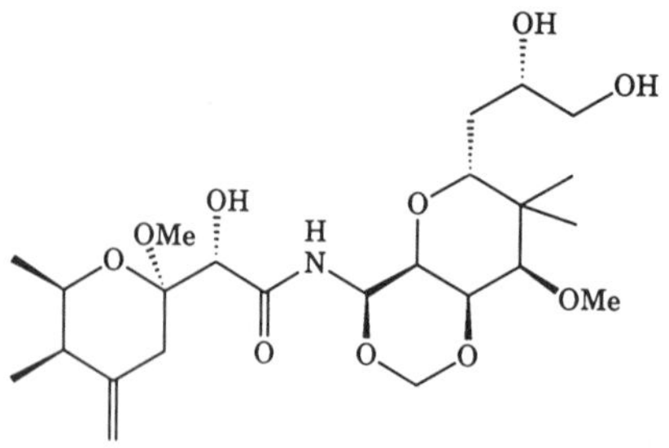

(15)

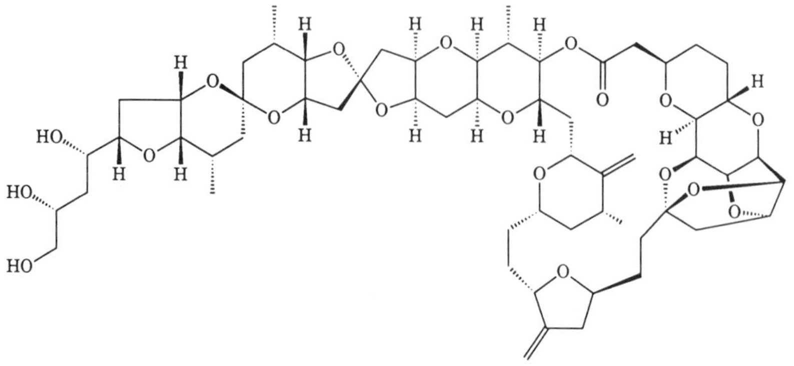

(16)

halichondrins as clinically useful anticancer agents. Unfortunately, their development has been hampered by limited supplies of material.

Dolastatin 10 (**17**), a cytotoxic peptide from the opisthobranch mollusk *Dolabella auricularia*, was reported to show exceptional antineoplastic activity *in vivo* against PS leukemia (T/C 169–202 @ 1–4 µg/kg), and B16 melanoma (T/C 142–238 @ 1.44–11.1 µg/kg), and effected 17–67% curative responses at 3.25–26

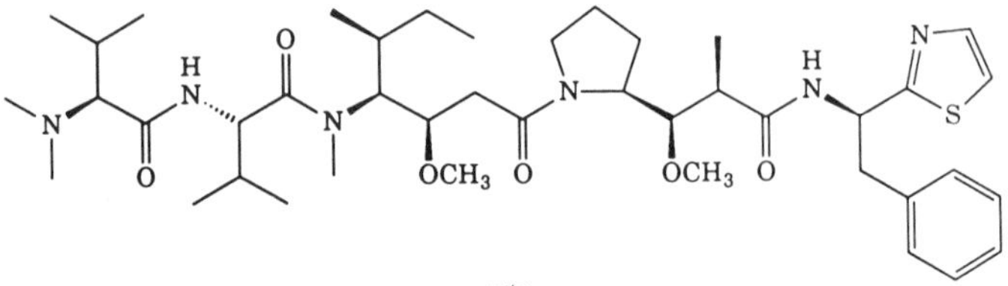

(17)

µg/kg against human melanoma xenographs in nude mice (Pettit *et al.*, 1987). Complete evaluation of dolastatin 10 was hindered by limited supplies until the total synthesis of dolastatin 10 was reported (Pettit *et al.*, 1989). The ability of dolastatin 10 to inhibit the growth of L1210 murine leukemia cells *in vitro*, in nanomolar concentrations, has been attributed to its ability to act as an antimitotic agent, causing cells to become arrested in metaphase. This arrest results from interaction of the compound with tubulin, inhibition of polymerization, and prevention of microtubule assembly (Bai *et al.*, 1990a). Dolastatin 10 noncompetitively inhibited the binding of radiolabeled vincristine to tubulin and inhibited nucleotide exchange on tubulin without displacing nucleotide present in the exchangeable site. This suggested close proximity of the dolastatin 10 binding site to those of the nucleotide and the vinca alkaloids on the same subunit of tubulin (Bai *et al.*, 1990b).

The long-standing reports of *in vivo* antitumor activity in extracts of *Ecteinascidia turbinata* have recently been attributed to the ecteinascidins (**18, 19**)

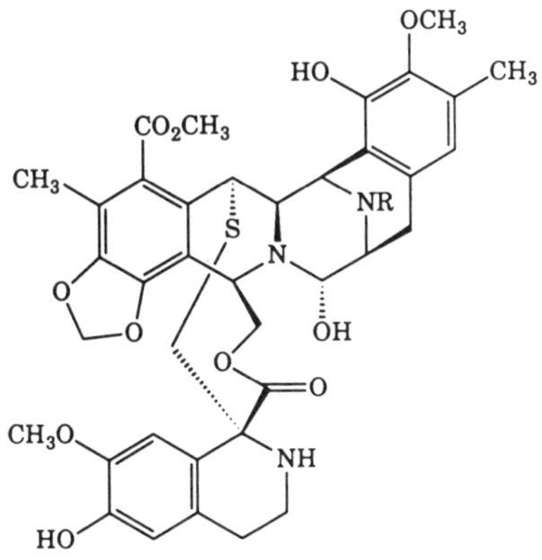

(**18**) Ecteinascidin 743: R = CH$_3$
(**19**) Ecteinascidin 729: R = H

(Reinhart *et al.*, 1990; Wright *et al.*, 1990). Ecteinascidin 743 was reported to effect a half-maximal inhibition of L1210 leukemia cells at 0.5 ng/ml while exhibiting notable activity *in vivo* against P388 leukemia (T/C 167 @ 15 μg/kg). The most active agent, ecteinascidin 729, showed potent activities against both P388 murine leukemia (T/C 214 @ 3.8 μg/kg) and B16 melanoma (T/C 246 @ 10 μg/kg) (Reinhart *et al.*, 1990).

Dercitin (**20**), a new acridine alkaloid isolated from a *Dercitus* sp. sponge (Gunawardana *et al.*, 1988), was found to prolong the life of P388-bearing mice (T/C 170 @ 5 mg/kg). Preliminary data suggested dercitin derives it activity by intercalating DNA (Burres *et al.*, 1989).

Aeroplysinin-1 (**21**), isolated from the sponges *Aplysina aerophoba* (Fattorusso *et al.*, 1970) and *Ianthella ardis* (Fulmor *et al.*, 1970), was recently reported to have significant antileukemic activity in the L5178ly cell/NMRI mouse system (T/C 388 @ 50 mg/kg daily for 5 days) (Kreuter *et al.*, 1989).

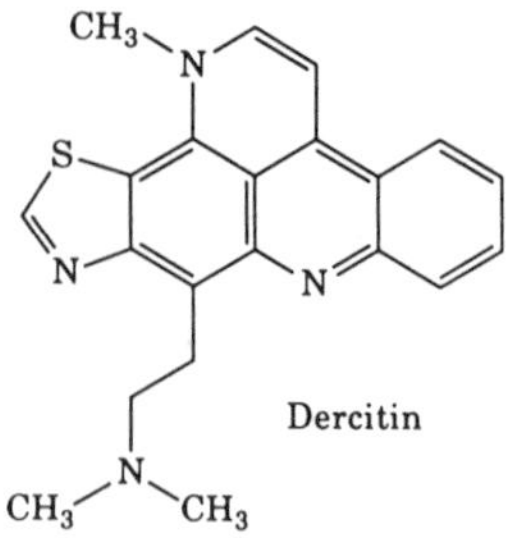

Dercitin

(20)

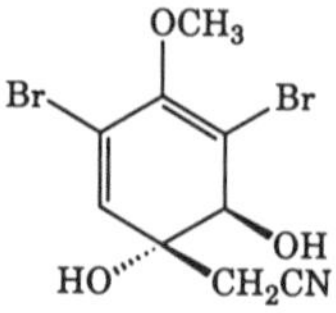

(21)

4. TUMOR PROMOTERS

Protein kinase C (PKC), an enzyme ubiquitous in eukaryotic cells, mediates transduction of extracellular signals into intracellular events, and is the major receptor for tumor-promoting phorbol esters. Structurally, PKC consists of a single polypeptide chain ($\sim$80 kDa) with two functionally different domains: a hydrophobic domain that may function in membrane attachment, and a hydrophilic domain containing the catalytic site. Functionally, PKC is a large family of isozymes [reviewed in Nishizuka (1988)], differing in their distribution within the cell. They have different molecular weights, are nonequivalent to monoclonal antibodies, and have different autophosphorylation sites.

When an extracellular ligand binds a cell surface receptor, secondary messengers such as diacyglycerol are produced. Effector proteins, such as PKC, respond to secondary messengers and activate a variety of biochemical systems within the cell. In the case of PKC, this action involves protein phosphorylation. Formation of a diacylglycerol–PKC complex initiates a protein phosphorylation cascade that, among other functions, controls cell regulation and proliferation. Proliferating and transformed cells display higher PKC activity; unregulated protein phosphorylation has been linked with tumor promotion. Phorbol esters, a class of natural products that promote tumors in animals exposed to carcinogens, activate PKC. TPA mimics the action of DAG, resulting in uncontrolled proliferation and transformation of the initiated cell. Discovery that PKC contains the phorbol ester receptor has greatly aided the study of PKC signal transduction and understanding of the mechanism of phorbol ester tumor promotion. The structure and function of PKC, as well as its activation by phorbol esters, have been the topic of many recent reviews (Farago and Nishizuka, 1990; Bouvier, 1990; Jaken, 1990; Hannun and Bell, 1989; Nishizuka, 1989; Nakadate, 1989).

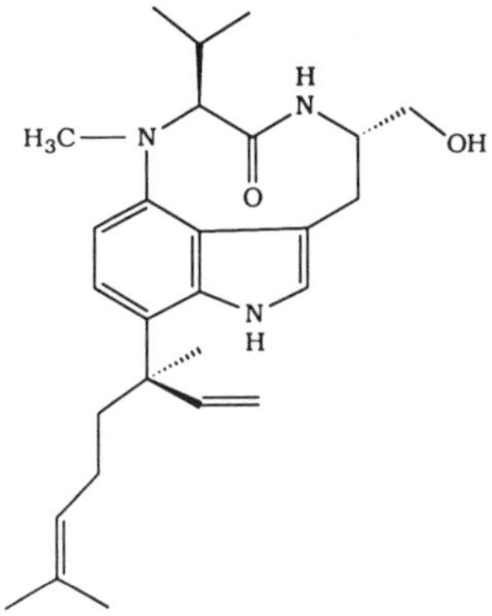

(22)

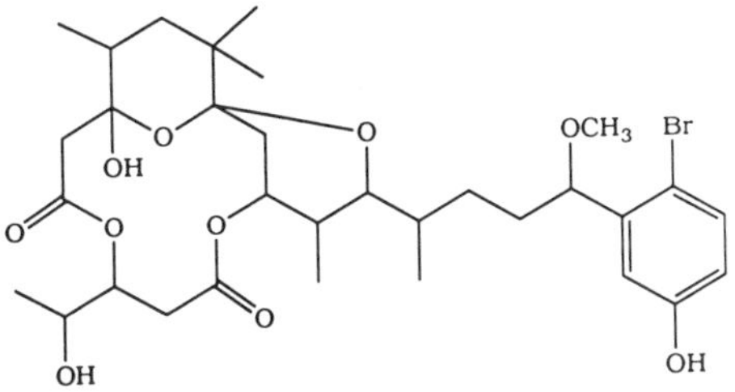

(23)

Further understanding of PKC's function in cellular biochemistry was made possible by the discovery that the marine natural products lyngbyatoxin (teleocidin A-1) (22) and aplysiatoxin (23) bind to the phorbol receptor (Fujiki and Sugimura, 1987). Although lyngbyatoxins, aplysiatoxins, and TPA are structurally quite dissimilar, they compete for the same receptor on PKC, evidence that the tumor-

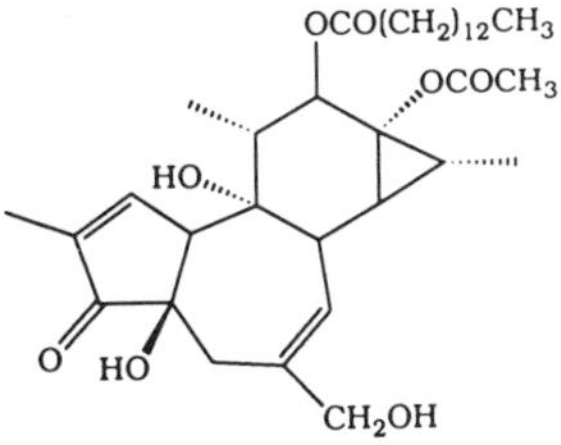

TPA

promoting activities ascribed to TPA are the direct consequence of activating PKC. Insight into the electrostatic structure of the phorbol ester receptor was gained by comparing superimposed computer models of the promoters TPA, teleocidin, and aplysiatoxin (Thomson and Wilkie, 1989; Itai *et al.*, 1988; Irie *et al.*, 1987; Wender *et al.*, 1986; Jeffrey and Liskamp, 1986). Although this approach produced conflicting results, use of these data to design synthetic activators and repressors could provide powerful probes for PKC's mechanism of action. This information may lead to the design of useful tumor-suppressing compounds.

TPA-type tumor promoters are useful tools in studying the many functions of PKC. Recently, teleocidin was used to study the regulation of tumor necrosis factor alpha as influenced by expression of interleukin-1 (IL-1) and IL-2 receptors, and interactions between B-cell lymphokines (Benjamin *et al.*, 1990). Teleocidins and TPA were also shown to simulate directly IL-2 function in cytotoxic T cells (Gavériaux and Loor, 1989), to potentiate the activity of decay-accelerating factor (Bryant *et al.*, 1990), and to stimulate B-cell gamma-interferon production (Benjamin *et al.*, 1986).

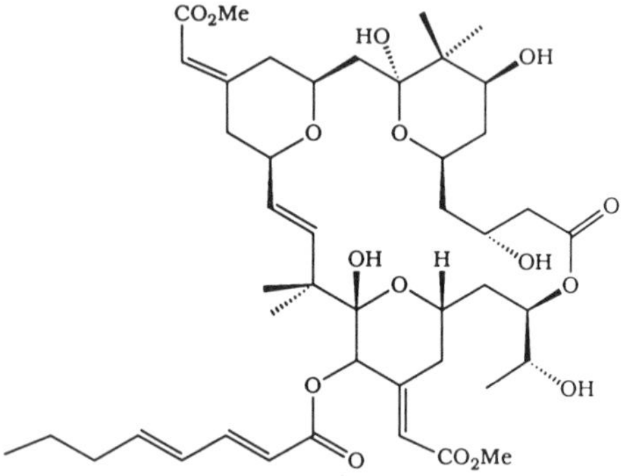

(**24**)

Bryostatin 1 (**24**), isolated from the bryozoan *Bugula neritina* (Pettit *et al.*, 1982), is a cyclic macrolide with an oxygen-rich cavity reminiscent of cyclic ionophore antibiotics. These antibiotics cause tumor destruction at the cellular level by transporting chelated cations such as K^+, Ca^{2+}, or Na^+ across the cell membrane. Preliminary evidence that bryostatin 1 chelated Ag^+ suggested administration to quiescent Swiss 3T3 cells (Berkow and Kraft, 1985) while monitoring the intra- and extracellular ion gradient. Although bryostatin 1 failed to induce any cation flux, it activated DNA synthesis, a response characteristic of phorbol esters. Furthermore, preincubation of 3T3 cells with phorbol 12-myristate 13-acetate (PMA) rendered the cells unresponsive to treatment with bryostatin B; likewise, bryostatin 1 desensitized the cells to TPA. Bryostatin 1 prevented [³H]-phorbol 12,13-dibutyrate binding to Swiss 3T3 cell receptors. What makes bryostatin so valuable is that although administration of TPA-type promoters causes tumors in animals, administration of the bryostatins does not, and indeed they inhibit tumor promotion by phorbol esters (Hennings *et al.*, 1987). The bryostatins show enormous potential for inhibiting PKC tumor promotion, and may have use as antitumor agents.

The cytotoxic macrolides bistratene A [also known as bistramide A (Gouiffès *et al.*, 1988a, b)] (**25**) and bistratene B (Degnan *et al.*, 1989) (**26**), isolated from the

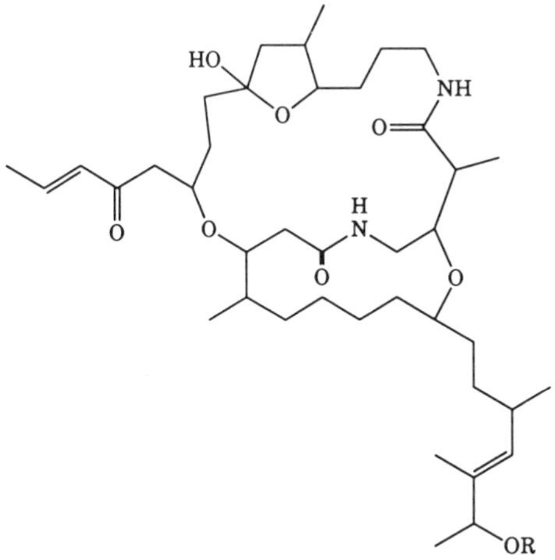

(**25**) Bistratene A: R = H
(**26**) Bistratene B: R = COOH₃

ascidian *Lissoclinum bistratum*, were shown to enhance the phospholipid-dependent activity of type II protein kinase C (Watters *et al.*, 1990). These compounds may also prove useful tools for probing the mechanisms of cell regulation.

Recently, a new group of non-TPA-type tumor promoters (okadaic acid class) was reported. Their mechanism of action is distinct from that of TPA-type promoters in that they do not bind to or activate protein kinase C, although they do modulate protein phosphorylation. While TPA-type promoters have illuminated many aspects of cellular regulation by enhancing protein phosphorylation, promoters of the okadaic acid class have made similar contributions by virtue of their specific inhibition of phosphatases 1 and 2A (PP-1 and PP-2A) and resulting effects on protein dephosphorylation (P. Cohen *et al.*, 1990; Ishihara *et al.*, 1989; Suganuma *et al.*, 1988, 1990; Bialojan and Takai, 1988).

Okadaic acid (**27**), a polyether derivative of a C_{38} fatty acid, initially isolated from the sponges *Halichondria okadai* and *H. melanodocia* (Tachibana *et al.*, 1981), was subsequently shown to be produced by several species of dinoflagellates and to be concentrated in a variety of filter feeders (Fujiki *et al.*, 1988). [³H]-Okadaic acid was reported to bind specifically to particulate and cytosolic

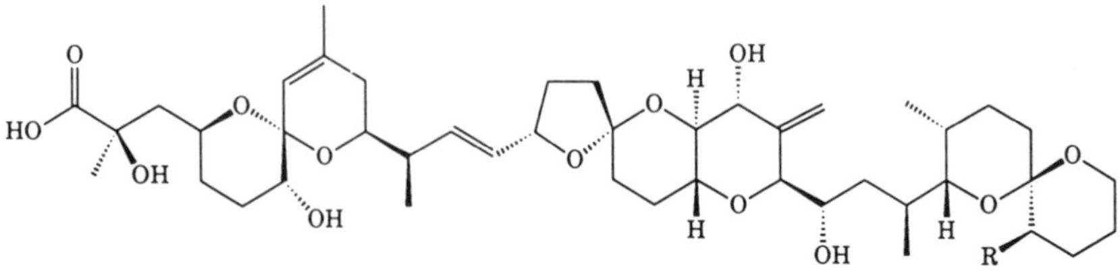

(**27**) Okadaic acid: R = H
(**28**) Dinophysistoxin 1: R = CH₃

fractions of mouse skin, suggesting interaction with both membrane-bound and cytosolic macromolecular receptor(s) (Suganuma *et al.*, 1989). This binding was not inhibited by TPA-type tumor promoters. Okadaic acid inhibited phosphatase activity in a dose-dependent manner in experiments using cytosolic fractions of homogenized mouse brain which retain both kinase and phosphatase activity (Sassa *et al.*, 1989; Suganuma *et al.*, 1989). In experiments with smooth muscle, okadaic acid induced muscle fiber contraction due to enhanced myosin light-chain phosphorylation (Shibata *et al.*, 1982; Takai *et al.*, 1987; Obara *et al.*, 1989; Kodama *et al.*, 1986; Bialojan *et al.*, 1988). This hyperphosphorylation was shown to be due to the inhibitory effects of okadaic acid on protein phosphatases (Takai *et al.*, 1987). Okadaic acid also induced hyperphosphorylation of a 60-kDa protein in primary human fibroblasts (Issinger *et al.*, 1988), which has been shown to be specifically dephosphorylated by PP-1 and PP-2A (Schneider *et al.*, 1989). It was further shown that okadaic acid induced phosphorylation of the 10-kDa elongation factor 2 (EF-2) protein, a substrate of the Ca^{2+}/calmodulin-dependent protein kinase III (Redpath and Proud, 1989). Okadaic acid played a significant role in elucidating cellular processes involving phosphorylation/dephosphorylation (Haystead *et al.*, 1989).

In a surprising development, okadaic acid reverted the phenotype of NIH 3T3 cells transformed by *raf* and *ret*-II oncogenes (Sakai *et al.*, 1989). This result fueled speculation that phosphatases regulate interconversion of normal and malignant cells. Okadaic acid induced *cdc2* kinase activation in *Xenopus* oocytes by inhibiting PP-2A during interphase (Félix *et al.*, 1990; Goris *et al.*, 1989; Kipreos and Wang, 1990). The *cdc2* gene product, a 34-kDa protein kinase, is responsible for activating a major protein phosphorylation cascade that is important in regulating the cell cycle. While the level of *cdc2* kinase is constant, there is a cyclic rise and fall in its activity during the cell cycle, thought to result from a balance between kinase phosphorylation and phosphatase dephosphorylation (Kipreos and Wang, 1990). These findings provided important clarification of the two-stage tumor promotion model (Fujiki and Sugimura, 1987): tumor promotion apparently results from hyperphosphorylation of protein kinase (e.g., PKC) effector proteins which initiate cell proliferation; dephosphorylation of these proteins by PP-1 or PP-2A downregulates cell proliferation. Therefore in normal cells, phosphatases function as tumor suppressors in the context of this model.

There have been many examples of the use of okadaic acid as a probe for identifying physiologically relevant protein phosphatases and for identifying cellular processes that are regulated by phosphorylation/dephosphorylation, such as ion channel function (Kume *et al.*, 1989; Hescheler *et al.*, 1988) and the process of vision (Palczewski *et al.*, 1989). Okadaic acid was also used in combination with inhibitor proteins to develop a method for quantification and identification of phosphatases (P. Cohen, 1989; Takai *et al.*, 1989; MacKintosh and Cohen, 1989; P. Cohen *et al.*, 1989a, b). Dinophysistoxin 1 (35-methylokadaic acid) (**28**) (Fujiki *et al.*, 1988) and acanthifolicin (9,10-episulfide okadaic acid) (**29**) (Schmitz *et al.*, 1981) inhibited PP-1 and PP-2A with similar potencies to okadaic acid (Fujiki *et al.*, 1989b; P. Cohen *et al.*, 1990).

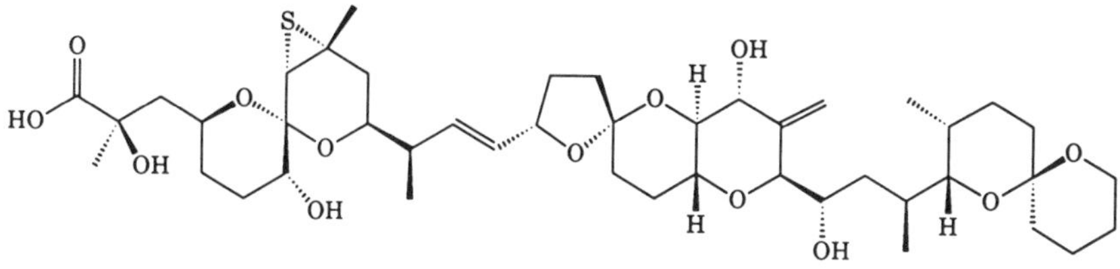

(**29**)

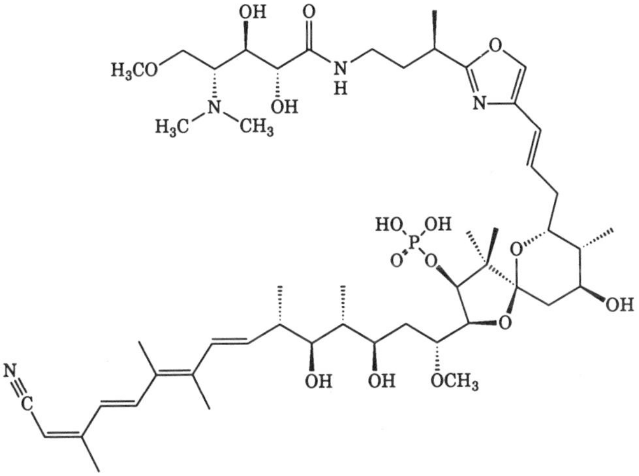

(**30**)

Calyculin A (**30**), isolated from the marine sponge *Discodermia calyx* (Kato *et al.*, 1986), binds to okadaic acid receptors, and was a potent tumor promoter on mouse skin (Suganuma *et al.*, 1990). Calyculin A was reported to be a more potent inhibitor of PP-1 and PP-2A than okadaic acid (Ishihara *et al.*, 1989).

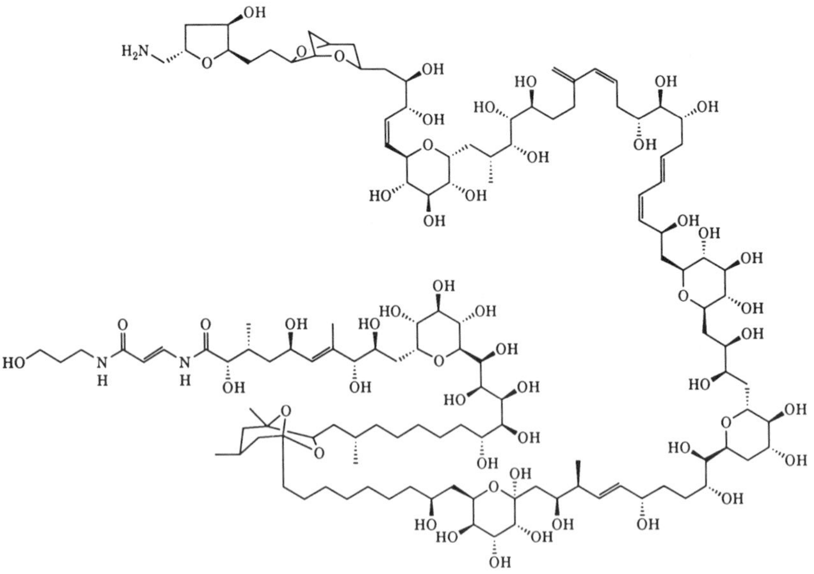

(31)

Palytoxin (**31**), a very potent toxin from the coelenterate *Palythoa toxica* (Moore and Bartolini, 1981; Uemura *et al.*, 1981), was shown to stimulate arachidonic acid metabolism synergistically with TPA-type promoters, suggesting that palytoxin activates an alternative signal transduction pathway (Fujiki *et al.*, 1986; Levine and Fujiki, 1985; Levine *et al.*, 1986). Palytoxin downregulated the epidermal growth factor (EGF) by reducing the number and affinity of EGF binding sites. This downmodulation requires extracellular sodium, and a correlation exists between palytoxin-induced sodium uptake and inhibition of EGF binding. These results suggest that palytoxin activates a sodium pump, and that sodium may act as a second messenger in this signal transduction pathway. Palytoxin should therefore provide an additional useful tool for probing cellular regulation processes (Wattenberg *et al.*, 1987, 1989a,b).

    Sarcophytol A (**32**), an oxygenated cembrane isolated from the soft coral

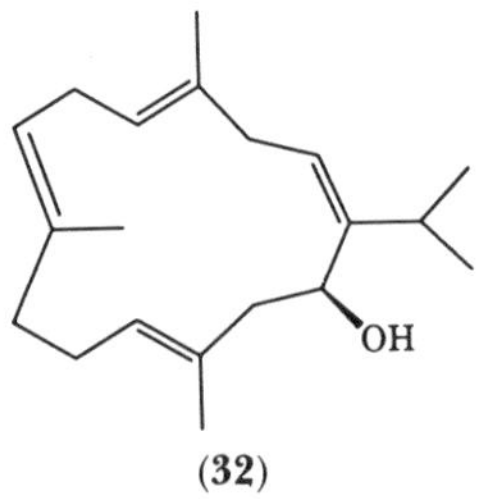

(32)

Sarcophyton glaucum (Kobayashi *et al.*, 1979), was recently shown to be anti-tumorigenic. Sarcophytol A inhibited development of *N*-methyl-*N*-nitrosourea-induced large bowel carcinomas in rats. It also suppressed sodium deoxycholate induction of ornithine decarboxylase, a marker for tumor promotion, in the large bowel mucosa (Narisawa *et al.*, 1989). Sarcophytol A also inhibited tumor promotion by teleocidin in a two-stage carcinogenesis experiment on mouse skin (Fujiki *et al.*, 1989a). Thus sarcophytol may prove to be an important probe for studying the mechanism(s) of carcinogenesis.

5. ANTI-INFLAMMATORY/ANALGESIC COMPOUNDS

Several marine natural products with anti-inflammatory and analgesic properties have found use in studying the roles of arachidonic acid metabolism and calcium mobilization in inflammation. Proinflammatory stimuli induce their effect ultimately through the mobilization/release of calcium ions (Ca^{2+}) from intra- or extracellular stores. Mobilization of calcium is believed to be initiated by binding of an agonist to its receptor. The receptor transduces a signal via a guanine nucleotide-binding protein (G-protein), activating a phospholipase (e.g., PLA_2, PLC) which hydrolyzes membrane phospholipids to produce a series of second messengers (e.g., inositol triphosphate, IP_3, and arachidonic acid, AA). Inositol triphosphate binding to its receptor on the rough endoplasmic reticulum results in Ca^{2+} release from intracellular stores. Extracellular calcium release is believed to depend on release of arachidonic acid. This fatty acid is metabolized via either the cyclooxygenase pathway to prostaglandins, prostacyclins, and thromboxanes, or the lipoxygenase pathway to tetraenoic acids, leukotrienes, and lipoxins. Binding of these metabolites to their receptors on hormone-activated Ca^{2+} channels leads to Ca^{2+} mobilization from extracellular sources. Many agents that mediate inflammation (accompanied by pain) and proliferation exert their physiological effects via modulation of phospholipid metabolism and Ca^{2+} mobilization. Consequently, compounds which inhibit a phospholipase and/or Ca^{2+} mobilization are anti-inflammatory, and thereby analgesic (Wheeler *et al.*, 1988; Mayer and Jacobs, 1988).

Manoalide (**33**), a nonsteroidal sesterterpene isolated from the sponge *Luffariella variabilis* (de Silva and Scheuer, 1980), emerged as a potent tool for studying inflammation. Manoalide irreversibly inhibited PLA_2 (Glaser and Jacobs, 1986; Mayer *et al.*, 1988; Jacobson *et al.*, 1990), inhibiting arachidonic acid release and thus its subsequent metabolism to prostaglandins and leukotrienes. It blocked phorbol ester (e.g., PMA)-induced inflammation, but not arachidonic acid-induced response. This property made manoalide very important for elucidating the role of PLA_2 in arachidonic acid release for eicosanoid

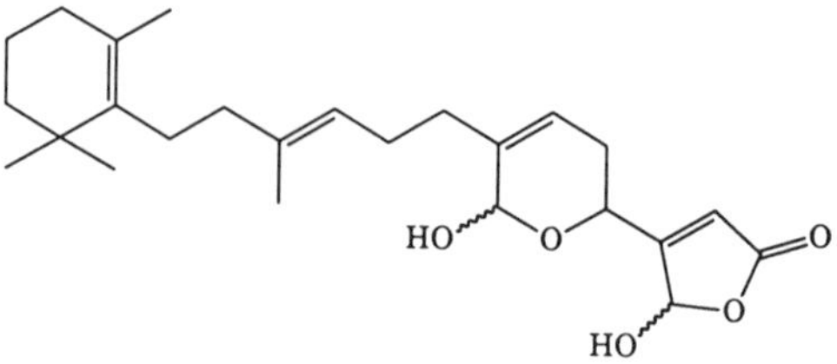

(33)

biosynthesis. Recent studies localized the manoalide binding site on PLA_2 (Glaser *et al.*, 1988), defined the pharmacophore responsible for PLA_2 activation (Glaser *et al.*, 1989), and examined the range of phospholipases inhibited by manoalide (e.g., Lister *et al.*, 1989; Jacobson *et al.*, 1990; Ulevitch *et al.*, 1988).

In addition to inhibiting PLA_2, manoalide inhibited 5-lipoxygenase (de Vries *et al.*, 1988), leading to speculation that its anti-inflammatory activity was due in part to inhibition of leukotriene biosynthesis. However, the most important factor contributing to the anti-inflammatory activity of manoalide was attributed to its inhibitory effect on Ca^{2+} channels (Wheeler *et al.*, 1988). Interestingly, at low concentrations manoalide inhibited calcium channels with no effect on phospho-inositide metabolism. The ability of manoalide to dissect these two components of the inflammation process may prove to be its most useful attribute in studying the role of Ca^{2+} signaling in inflammation and proliferation (e.g., Barzaghi *et al.*, 1989).

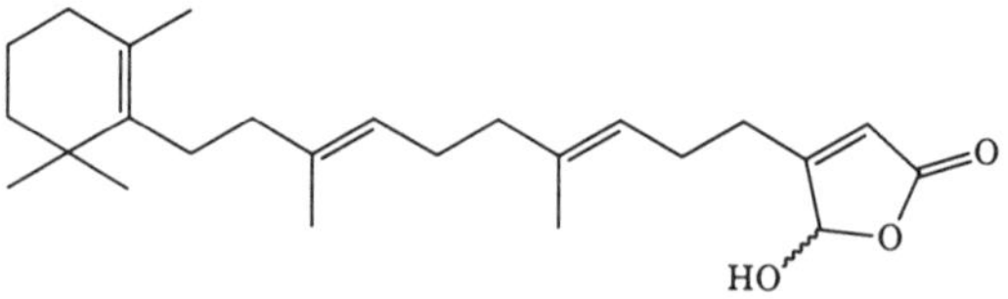

(34)

Luffariellolide (**34**), an analog of manoalide isolated from the same organism, also exhibited anti-inflammatory activity, was slightly less potent than manoalide, and was a partially reversible PLA_2 inhibitor (Albizati *et al.*, 1987).

Pseudopterosins A and E (**35a**, **35b**), members of a family of diterpene ribosides isolated from the gorgonians *Pseudopterogorgia bipinata* and *P. elisabethae*, exhibit potent anti-inflammatory and analgesic activities and act as reversible inhibitors of lipoxygenase and PLA_2 (Luedke, 1990; Look *et al.*, 1986).

While the previous compounds all exhibited anti-inflammatory activity, 15-acetylthioxy-furodysinin lactone (**36**), isolated from a *Dysidea* sp. (Carté *et al.*, 1989), elicited the opposite response. The compound caused intracellular Ca^{2+} mobilization that was blocked by LTB_4 receptor antagonists. 15-Acetylthioxy-furodysinin lactone binds LTB_4 receptors with high affinity and activates the

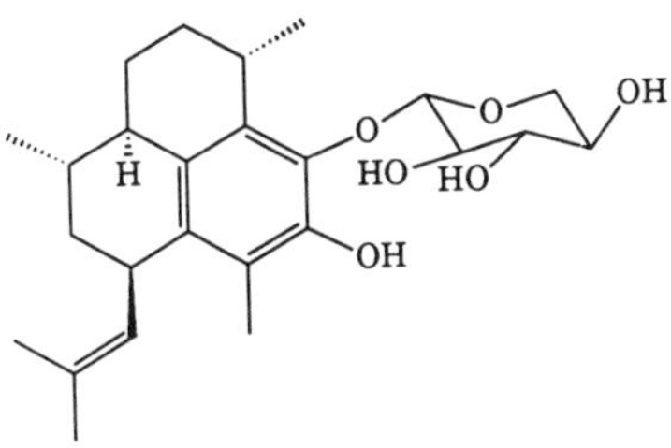

(35a)

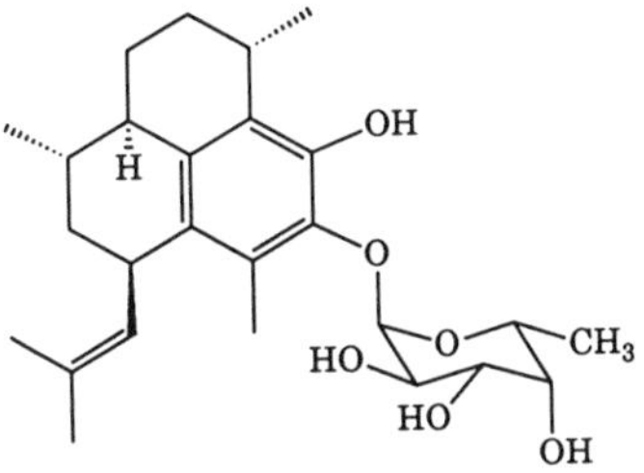

(35b)

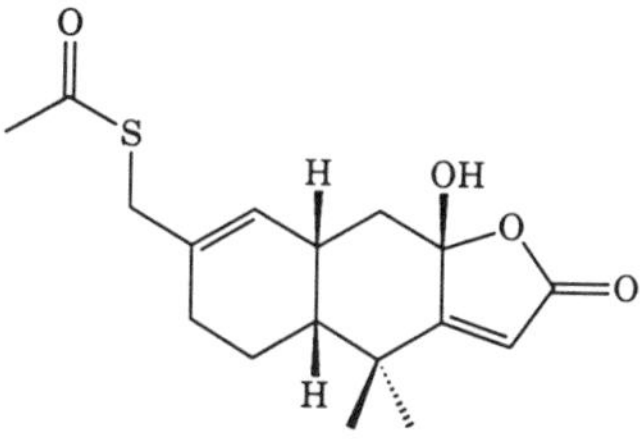

(36)

receptor-mediated signal transduction processes related to LTB$_4$. Thus, it will likely be of use in studying the role of leukotrienes in inflammation (Mong *et al.*, 1990).

6. ANTIVIRAL AGENTS

The search for viral chemotherapeutic agents from marine sources has been disappointing. The only compound reported thus far to show significant therapeutic activity is ara-A, a semisynthetic based on the arabinosyl nucleosides isolated from the sponge *Tethya crypta* (Bergmann and Feeney, 1951; Bergmann and Burke, 1955). A number of marine metabolites have shown very promising *in vitro* activity; however, only the didemnins demonstrated *in vivo* activity.

The didemnins, depsipeptides isolated from the Caribbean ascidian *Trididemnum solidum* (Rinehart *et al.*, 1981a), were reviewed earlier in the antitumor section. In addition to their antitumor activity, they exhibited *in vitro* and *in vivo* antiviral properties. Didemnin B significantly reduced the yields of DNA and RNA viruses by 10^4–10^5 at 0.5 µg/ml, and in *in vivo* tests protected over 70% of a population of mice from lethal vaginal doses of herpes simplex type 2 virus (intravaginal application three times per day) (Rinehart *et al.*, 1983), and a lethal subcutaneous challenge of Rift Valley fever virus with a 90% survival rate at 0.25 mg/kg (Canonico *et al.*, 1982).

Patellazole B (**37**), isolated from the ascidian *Lissoclinum patella* (Zabriskie

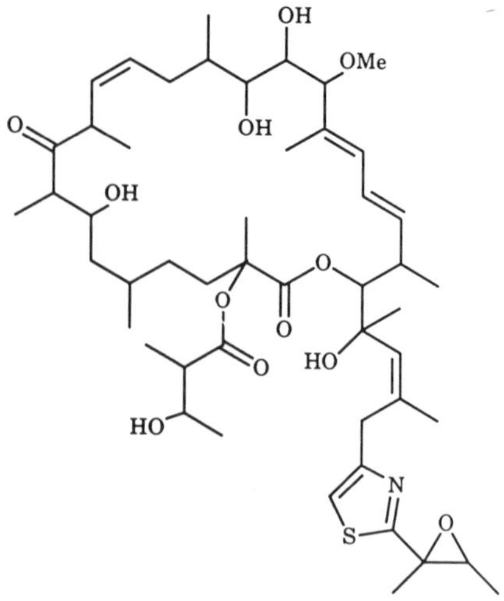

(37)

et al., 1988), exhibited very potent antiviral activity against herpes simplex viruses. In an *in vitro* assay, patellazole B inhibited viral replication of HSV-1 and HSV-2 at 0.5 and 60 ng/ml, respectively. Patellazole B was cytotoxic toward the host Vero cells at concentrations 1000 times greater than that at which it was active against HSV-1 (Ireland and Maiese, 1990, unpublished data).

The eudistomins (**38**), a class of β-carbolines isolated from the Caribbean ascidian *Eudistoma olivacea* were active in shipboard antiviral assays (Rinehart *et al.*, 1981c). In subsequent experiments five of the eudistomins showed enhanced activity after activation with UV-A light (Hudson *et al.*, 1988).

Several derivatives of the sesquiterpene hydroquinone avarol isolated from sponges of the genus *Dysidea* were recently shown to inhibit HIV-1 reverse transcriptase. Avarol F (**39**) and avarone E (**40**), the two most effective derivatives,

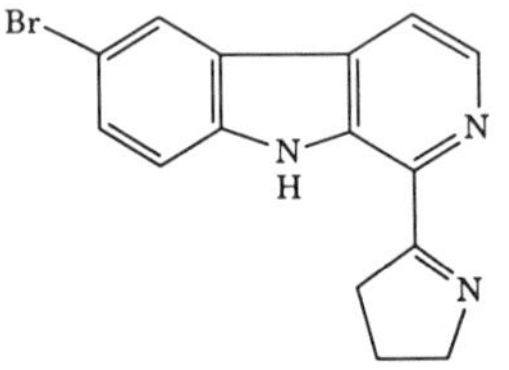

(**38a**) Eudistomin H 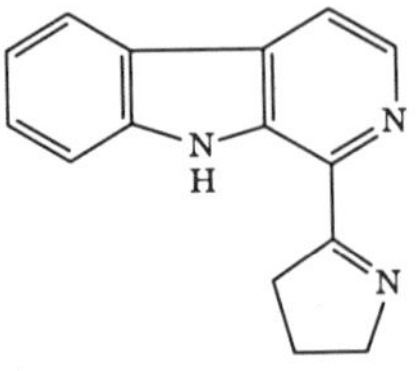   (**38b**) Eudistomin I 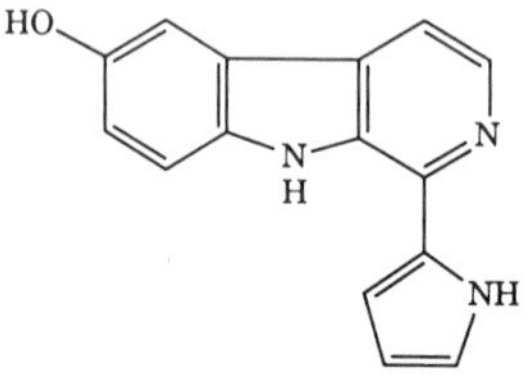   (**38c**) Eudistomin M

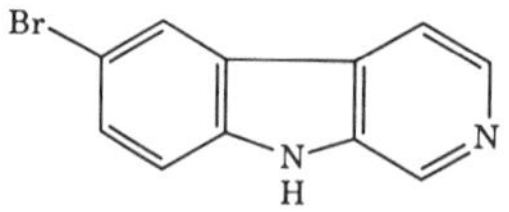

(**38d**) Eudistomin N 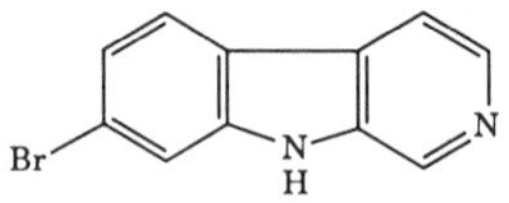   (**38e**) Eudistomin O

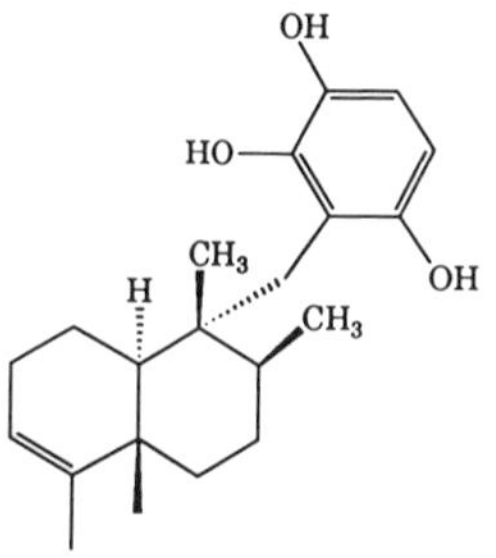

(39) 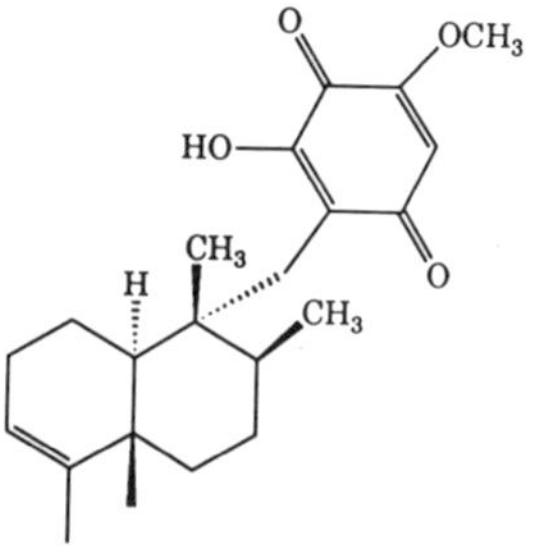   (40)

showed 2- to 10-fold greater DNA polymerase activity than RNase H activity. Further results for one derivative, avarone E, suggested that HIV-1 RT binding occurred at sites different from those of DNA synthesis substrates, dGTP, and primer template (Loya and Hizi, 1990).

A class of sulfonic acid-containing glycolipids that inhibit HIV-1 was isolated from two species of microcultured blue-green algae (Gustafson *et al.*, 1989). The extract of *Lyngbya lagerheimii* contained glycolipids A and B (**41**), whereas the

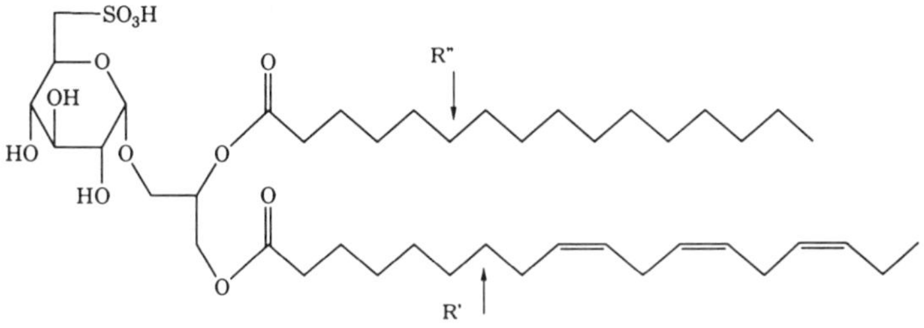

	R'	R''
(41a)	18:3	16:0
(41b)	18:3	16:0
(41c)	18:3	16:0
(41d)	18:3	16:0

extract of *Phormidium tenue* contained C and D. The degree of HIV-1 protection varied substantially between different host cell lines, but seemed to be relatively independent of acyl chain length and degree of unsaturation. These glycolipids are structural components of chloroplast membranes, and although they occur widely in higher plants, algae, and photosynthetic microorganisms, they have not been previously associated with HIV-1 inhibitory activity (Gustafson *et al.*, 1989).

Algal sulfated polysaccharides, including the metabolite fucoidin (fucoidan) of the *Laminaria* sp. kelp (brown algae) were shown to inhibit a variety of DNA and RNA enveloped viruses, including herpes simplex virus and HIV (Baba *et al.*, 1988). Very importantly, fucoidin exhibited antiviral activity at concentrations several orders of magnitude lower than its anticoagulant threshold. *In vitro* studies with peripheral mononuclear cells from AIDS patients suggested fucoidin binds the *env* protein of HIV. This result suggested that HIV invaded target cell cytoplasm slowly and that fucoidin can still react with HIV in the cell membrane (Sugawara *et al.*, 1989).

7. METABOLITES WHICH AFFECT MICROFILAMENT-MEDIATED PROCESSES

Transformation of chemical bond energy into motion plays a key role in many cellular processes. Microfilaments are responsible for this transformation, which occurs in fertilization and early development processes, phagocytosis, protein synthesis, and microorganismal propulsion (Stryer, 1981). Several marine-derived metabolites have contributed to our understanding of these cellular and molecular biochemical processes in unique ways.

The latrunculins, 2-thiazolidinone-containing macrolides isolated from the Red Sea sponge *Latrunculia magnifica* (Kashman *et al.*, 1980), were reported to disrupt the organization of microfilaments but not microtubules in cultured cells (Spector and Shochet, 1983). Latruculin A (**42**) was an order of magnitude more

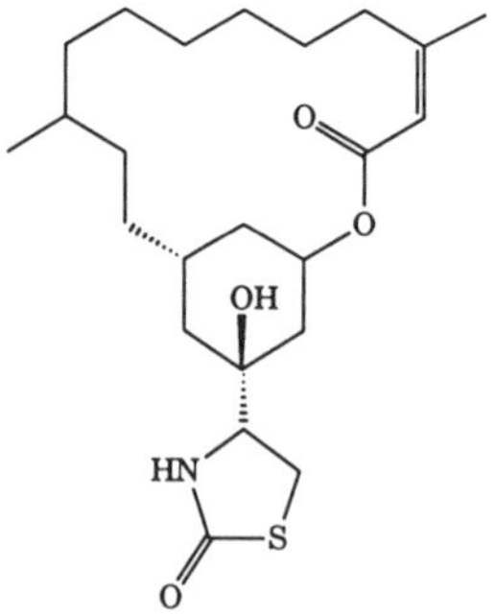

(**42**)

potent than cytochalasin, and affected different components of the actin-based cytoskeleton (Spector *et al.*, 1989). They currently represent the only alternative to the cytochalasins in pharmacological studies of both actin polymerization *in vitro* and actin organization and function in living cells.

Latrunculin A was also used to determine which fertilization processes are microfilament mediated in different species (G. Schatten *et al.*, 1986; H. Schatten and G. Schatten, 1986). The results suggested that the acrosome reaction of mouse sperm does not require microfilament activity, whereas that of sea urchins does. Latrunculin A was also found to inhibit macrophage phagocytosis without interfering with cell viability (de Oliveira and Mantovani, 1988), strengthening the case for participation of microfilaments in the mechanism of phagocytosis. Latrunculin A is expected to continue to make important contributions to the study of microfilament-mediated processes.

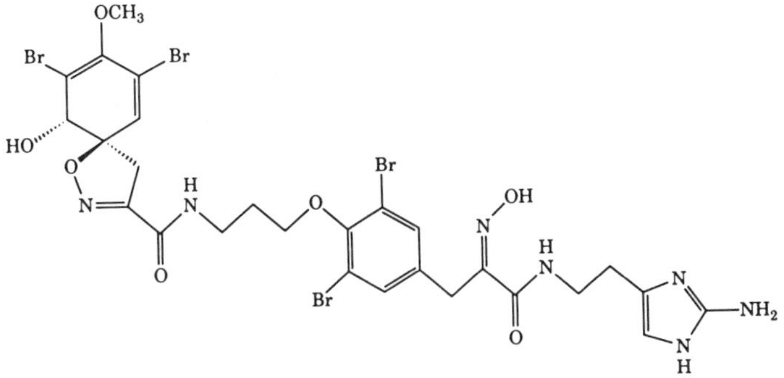

(43)

Purealin (**43**), isolated from the sponge *Psammaplysilla purpurea* (Nakamura *et al.*, 1985), was shown to affect various myosin ATPases. Purealin enhanced the stability of thick filaments of dephosphorylated gizzard myosin against the disassembling action of ATP, suggesting it acted on the ATP binding site of myosin (Takito *et al.*, 1986), possibly by directly affecting the myosin heads (Nakamura *et al.*, 1987). Furthermore, purealin increased the actin-activated ATPase activity of myosin (Takito *et al.*, 1987a). These results indicated that purealin inhibited myosin phosphorylation by acting as a calmodulin antagonist, inhibiting formation of the calmodulin–myosin light-chain kinase complex (Takito *et al.*, 1987b). Purealin has provided the research tool required to investigate the structure and conformation of various forms of myosin (Takito *et al.*, 1986).

Fucoidin, an L-fucose-rich sulfated heteropolysaccharide produced by brown algae of genus *Laminaria* (kelp), exhibited antithrombin (Church *et al.*, 1989), anticoagulant, fibrinolytic, oncoinhibitory (Maruyama *et al.*, 1987), and antitumor activities (Chida and Yamamoto, 1987), and was used as a probe of lymphocyte membranes (Brandley *et al.*, 1987). Perhaps the most interesting and useful attribute of fucoidin was its inhibition of sperm–egg binding. Fucoidin was useful in elucidating the molecular basis of mammalian sperm–egg recognition, as it inhibits guinea pig sperm–egg binding by interacting with the inner acrosomal membrane and equatorial segment domains of guinea pig spermatozoa (Huang and Yanagimachi, 1984). These results implied that the mechanisms of sperm–egg adhesion and acrosomal reaction were distinct. Further experiments with fucoidin in mice suggested that an L-fucose component of the sperm surface was involved in sperm–egg recognition (Boldt *et al.*, 1989). Thus, fucoidin provided important insight concerning the mechanism of sperm–egg recognition. Fucoidin was also found to be a potent inhibitor of tight binding of spermatozoa to the human zona pellucida in the human hemizona assay, an assay with demonstrated predictive value for human *in vitro* fertilization (Oehninger *et al.*, 1990).

8. THE FUTURE

It is easy to speculate about the future but much more difficult to be insightful. As the previous sections aptly showed, the vast resource of marine natural products played an important role in the explosive growth of biomedical science during the past two decades. This was particularly evident in the development and formulation of ion channel and tumor promotion models, but certainly not limited to these areas. Investigation of antitumor compounds also progressed to the extent that one compound, didemnin B, is now in clinical trials, and a second compound, bryostatin 1, will enter clinical trials shortly. In spite of these contributions it is clear that this resource has yet to be fully utilized.

In its infancy, the marine natural products field was dominated by chemical studies which generated literally thousands of novel structures, few of which were studied by marine pharmacologists. More recently the emphasis has shifted to an interdisciplinary approach, as chemists come to appreciate that chemical characterization of new compounds, no matter how exotic, is only a beginning, and pharmacologists realize that these metabolites can provide key insights into complex cellular events. Marine natural products will continue to play an important role as molecular probes in studying these types of biochemical events and unraveling their roles in cell regulation. The establishment of interactive collaborations between chemists and pharmacologists will be essential in the future to ensure that this resource meets its full potential.

REFERENCES

Abraham, S., Beatch, G. N., MacLeod, B. A., and Walker, M. J., 1989, Antiarrhythmic properties of tetrodotoxin against occlusion-induced arrhythmias in the rat: A novel approach to the study of the antiarrhythmic effect of ventricular sodium channel blockade, *J. Pharmacol. Exp. Ther.* **251:** 1166–1173.

Abramson, S. N., Culver, P., Kline, T., Li, Y., Guest, P., Gutman, L., and Taylor, P., 1988, Lophotoxin and related coral toxins covalently label the α-subunit of the nicotinic acetylcholine receptor, *J. Biol. Chem.* **263:**18568–18573.

Abramson, S. N., Li, Y., Culver, P., and Taylor, P., 1989a, An analog of lophotoxin reacts covalently with Tyr[190] in the α-subunit of the nicotinic acetylcholine receptor, *J. Biol. Chem.* **264:**12666–12672.

Abramson, S. N., Radic, Z., Manker, D., Faulkner, D. J., and Taylor, P., 1989b, Onchidal: A naturally occurring irreversible inhibitor of acetylcholinesterase with a novel mechanism of action, *Mol. Pharmacol.* **36:**349–354.

Albizati, K. F., Holman, T., Faulkner, D. J., Glaser, K. B., and Jacobs, R. S., 1987, Luffariellolide, an anti-inflammatory sesterterpene from the marine sponge *Luffariella* sp., *Experientia* **43**:949–950.

Albizati, K. F., Martin, V. A., Agharahimi, M. R., and Stolze, D. A., 1990, Synthesis of marine natural products, in: *Bioorganic Marine Chemistry*, Vol. 5 (P. J. Scheuer, ed.), Springer-Verlag, Berlin.

Baba, M., Snoeck, R., Pauwels, R., and de Clerq, E., 1988, Sulfated polysaccharides are potent and selective inhibitors of various enveloped viruses, including herpes simplex virus, cytomegalovirus, vesicular stomatitis virus, and human immunodeficiency virus, *Antimicrob. Agents Chemother.* **32**(11):1742–1745.

Baden, D. G., 1989, Brevetoxins: Unique polyether dinoflagellate toxins, *FASEB J.* **3**:1807–1817.

Bai, R., Pettit, G. R., and Hamel, E., 1990a, Dolastatin 10, a powerful cytostatic peptide derived from a marine animal. Inhibition of tubulin polymerization mediated through the vinca alkaloid domain, *Biochem. Pharmacol.* **39**:1941–1949.

Bai, R., Pettit, G. R., and Hamel, E., 1990b, Binding of dolastatin 10 to tubulin at a distinct site for peptide antimitotic agents near the exchangeable nucleotide and vinca alkaloid sites, *J. Biol. Chem.* **265**:17141–17149.

Baird-Lambert, J., Marwood, J. F., Davies, L. P., and Taylor, K. M., 1980, 1-Methylisoguanosine: An orally active marine natural product with skeletal muscle and cardiovascular effects, *Life Sci.* **26**: 1069–1077.

Ballesta, J. J., Palmero, M., Hidalgo, M. J., Gutierrez, L. M., Reig, J. A., Viniegra, S., and Garcia, A. G., 1989, Separate binding and functional sites for ω-conotoxin and nitrendipine suggest two types of calcium channels in bovine chromaffin cells, *J. Neurochem.* **53**:1050–1056.

Barzaghi, G., Sarace, H. M., and Mong, S., 1989, Platelet-activating factor-induced phosphoinositide metabolism in differentiated U-937 cells in culture, *J. Pharmacol. Exp. Ther.* **248**:559–566.

Benjamin, D., Hartmann, D. P., Bazar, L. S., and Jacobsen, R. J., 1986, Burkitt's cells can be triggered by teleocidin to secrete interferon-gamma, *Am. J. Hematol.* **22**:169–177.

Benjamin, D., Hooker, S., and Miller, J., 1990, Differential effects of teleocidin on TNF-alpha receptor regulation in human B cell lines: Relationship to coexpression of IL-2 and IL-1 receptors and to lymphokine secretion, *Cell. Immunol.* **125**:480–497.

Bergmann, W., and Burke, D. C., 1955, Contributions to the study of marine products. XXXIX. The nucleosides of sponges. III. Spongothymidine and spongouridine, *J. Org. Chem.* **20**:1501–1507.

Bergmann, W., and Feeney, R. J., 1951, Contributions to the study of marine products. XXXII. The nucleosides of sponges. I, *J. Org. Chem.* **16**:981–987.

Berkow, R. L., and Kraft, A. S., 1985, Bryostatin, a non-phorbol macrocyclic lactone, activates intact human polymorphonuclear leukocytes and binds to the phorbol ester receptor, *Biochem. Biophys. Res. Commun.* **131**:1109–1116.

Bialojan, C., and Takai, A., 1988, Inhibitory effect of a marine-sponge toxin, okadaic acid, on protein phosphatases, *Biochem. J.* **256**:283–290.

Bialojan, C., Rüegg, J. C., and Takai, A., 1988, Effects of okadaic acid on isometric tension and myosin phosphorylation of chemically skinned guinea-pig tenia coli, *J. Physiol.* **398**:81–95.

Billiar, R. B., Kalash, J., Romita, V., Tsuji, K., and Kosuge, T., 1988, Neosurugatoxin: CNS acetylcholine receptors and leutinizing hormone secretion in ovariectomized rats, *Brain Res. Bull.* **20**:315–322.

Biskupiak, J. E., and Ireland, C. M., 1984, Revised absolute configuration of dysidenin and isodysidenin, *Tetrahedron Lett.* **25**:2935–2936.

Boldt, J., Howe, A. M., Parkerson, J. B., Gunter, L. E., and Kuehn, E., 1989, Carbohydrate involvement in sperm–egg fusion in mice, *Biol. Reprod.* **40**(4):887–896.

Bontemps, F., and van den Berghe, G., 1989, Mechanism of adenosine triphosphate catabolism induced by deoxyadenosine and by nucleoside analogs in adenosine deaminase-inhibited human erythrocytes, *Cancer Res.* **49**:4983–4989.

Bourke, J. E., Bunn, S. J., Marley, P. D., and Livett, B. G., 1988, The effects of neosurugatoxin on evoked catecholamine secretion from bovine adrenal chromaffin cells, *Br. J. Pharmacol.* **93**:275–280.

Bouvier, M., 1990, Cross-talk between secondary messengers, *Ann. N. Y. Acad. Sci.* **594:**120–129.

Brandley, B. K., Ross, T. S., and Schnaar, R. L., 1987, Multiple carbohydrate receptors on lymphocytes revealed by adhesion to immobilized polysaccharides, *J. Cell. Biol.* **105**(2):991–997.

Bruno, I., Minale, L., and Riccio, R., 1990, Starfish saponins, Part 43, *J. Nat. Prod.* **53:**366–374.

Bryant, R. W., Granzow, C. A., Siegel, M. I., Egan, R. W., and Billah, M. M., 1990, Phorbol esters increase synthesis of decay-accelerating factor, a phosphatidylinositol-anchored surface protein, in human endothelial cells, *J. Immunol.* **144:**593–598.

Burres, N. S., and Clement, J. J., 1989, Antitumor activity and mechanism of action of the novel marine natural products mycalamide-A and -B and onnamide, *Cancer Res.* **49:**2935–2940.

Burres, N. S., Sazesh, S., Gunawardana, G. P., and Clement, J. J., 1989, Antitumor activity and nucleic acid binding properties of dercitin, a new acridine alkaloid isolated from a marine *Dercitus* species sponge, *Cancer Res.* **49:**5267–5274.

Cai, M., and Jordan, P. C., 1990, How does vestibule surface charge affect ion conduction and toxin binding in a sodium channel? *Biophys. J.* **57:**883–891.

Canonico, P. G., Pannier, W. L., Huggins, J. W., and Rinehart, K. L., 1982, Inhibition of RNA viruses *in vitro* and in rift valley fever-infected mice by didemnins A and B, *Antimicrob. Agents Chemother.* **22:**696–697.

Carmichael, W. W., Mahmood, N. A., and Hyde, E. G., 1990, Natural toxins from cyanobacteria (blue-green algae), in: *Marine Toxins* (S. Hall and G. Strichartz, eds.), American Chemical Society, Washington, D.C., pp. 87–106.

Carté, B., Mong, S., Poehland, B., Sarau, H., Westley, J. W., and Faulkner, D. J., 1989, 15-Acetylthioxy-furodysinin lactone, a potent LTB$_4$ receptor partial agonist from a marine sponge of the genus *Dysidea*, *Tetrahedron Lett.* **30:**2725–2726.

Catterall, W. A., 1980, Neurotoxins that act on voltage-sensitive sodium channels in excitable membranes, *Annu. Rev. Pharmacol. Toxicol.* **20:**15–43.

Charles, C., Braekman, J. C., Daloze, D., Tursch, B., and Karlson, R., 1978, Isodysidenin, a further hexachlorinated metabolite from the sponge *Dysidea herbacea*, *Tetrahedron Lett.* **17:**1519–1520.

Charles, C., Braekman, J. C., Dalose, D., and Tursch, B., 1980, The relative and absolute configuration of dysidenin, *Tetrahedron* **36:**2133–2135.

Chida, K., and Yamamoto, I., 1987, Antitumor activity of a crude fucoidan fraction prepared from the roots of kelp (*Laminaria* species), *Kitasato Arch. Exp. Med.* **60**(1–2):33–39.

Chun, H. G., Davies, B., Hoth, D., Suffness, M., Plowman, J., Flora, K., Grieshaber, C., and Leyland-Jones, B., 1986, Didemnin B: The first marine compound entering clinical trials as an antineoplastic agent, *Invest. New Drugs* **4:**279–284.

Church, F. C., Meade, J. B., Treanor, R. E., and Whinna, H. C., 1989, Antithrombin activity of fucoidin. The interaction of fucoidin with heparin cofactor II, antithrombin III, and thrombin, *J. Biol. Chem.* **264**(6):3618–3623.

Cohen, P., 1989, The structure and regulation of protein phosphatases, *Annu. Rev. Biochem.* **58:** 453–508.

Cohen, P., Klumpp, S., and Schelling, D. L., 1989a, An improved procedure for identifying and quantitating protein phosphatases in mammalian tissues, *FEBS Lett.* **250:**596–600.

Cohen, P., Schelling, D. L., and Stark, M. J. R., 1989b, Remarkable similarities between yeast and mammalian protein phosphatases, *FEBS Lett.* **250:**601–606.

Cohen, P., Holmes, C. F. B., and Tsukitani, Y., 1990, Okadaic acid: A new probe for the study of cellular regulation, *Trends Biochem. Sci.* **15:**98–102.

Cohen, S. S., 1977, The mechanisms of lethal action of arabinosyl cytosine (araC) and arabinosyl adenine (araA), *Cancer* **40:**509–518.

Crampton, S. L., Adams, E. G., Kuentzel, S. L., Li, L. H., Badiner, G., and Bhuyan, B. K., 1984, Biochemical and cellular effects of didemnins A and B, *Cancer Res.* **44:**1796–1801.

Cruz, L. J., Kupryszewski, G., LeCheminant, G. W., Gray, W. R., Olivera, B. M., and Rivier, J., 1989, μ-Conotoxin GIIIA, a peptide ligand for muscle sodium channels: Chemical synthesis, radio-labelling and receptor characterization, *Biochemistry* **28**:3437–3442.

Culver, P., and Jacobs, R. S., 1981, Lophotoxin: A neuromuscular acting toxin from the sea whip (*Lophogorgia rigida*), *Toxicon* **19**:825–830.

Culver, P., Fenical W., and Taylor, P., 1984, Lophotoxin irreversibly inactivates the nicotinic acetylcholine receptor by preferential association at one of the two primary agonist sites, *J. Biol. Chem.* **259**:3763–3770.

Culver, P., Bursch, M., Potenza, C., Wasserman, L., Fenical, W., and Taylor, P., 1985, Structure–activity relationships for the irreversible blockade of nicotinic receptor agonist sites by lophotoxin and congeneric diterpene lactones, *Mol. Pharmacol.* **28**:436–444.

Das, D. K., and Steinberg, H., 1988, Adenosine transport in the lung, *J. Appl. Physiol.* **65**:297–305.

Davies, L. P., Jamieson, D. D., Baird-Lambert, J. A., and Kaslauskas, R., 1984, Halogenated pyrrolopyrimidine analogues of adenosine from marine organisms: Pharmacological activities and potent inhibition of adenosine kinase, *Biochem. Pharmacol.* **33**:347–355.

Dawicki, D. D., Agarwal, K. C., and Parks, Jr., R. E., 1988, Adenosine metabolism in human whole blood, *Biochem. Pharmacol.* **37**:621–626.

De Andrea, M. L., de Camargo, B., and Melaragno, R., 1990, A new treatment protocol for childhood non-Hodgkin's lymphoma: Preliminary evaluation, *J. Clin. Oncol.* **8**:666–671.

Degnan, B. M., Hawkins, C. J., Lavin, M. F., McCaffrey, E. J., Parry, D. L., and Watters, D. J., 1989, Novel cytotoxic compounds from the ascidian *Lissoclinum bistratum*, *J. Med. Chem.* **32**:1355–1359.

De Oliveira, C. A., and Mantovani, B., 1988, Latrunculin A is a potent inhibitor of phagocytosis by macrophages, *Tetrahedron Lett.* **43**:1825–1830.

De Silva, E. D., and Scheuer, P. J., 1980, Manoalide, an antibiotic sesterterpenoid from the marine sponge *Luffariella variabilis* (Polejaeff), *Tetrahedron Lett.* **21**:1611–1614.

De Vries, G. W., Amdahl, L., Mobasser, A., Wenzel, M., and Wheeler, L. A., 1988, Preferential inhibition of 5-lipoxygenase activity by manoalide, *Biochem. Pharmacol.* **37**:2899–2905.

Dorr, A., Schwartz, R., Kuhn, J., Bayne, J., and Von Hoff, D. D., 1986, Phase I clinical trial of didemnin B (Abstract), *Proc. Am. Assoc. Cancer Res.* **5**:39.

Dorr, F. A., Kuhn, J. G., Phillips, J., and Von Hoff, D. D., 1988, Phase I clinical and pharmacokinetic investigation of didemnin B, a cyclic depsipeptide, *Eur. J. Cancer Clin. Oncol.* **24**:1699–1706.

Dowding, A. J., and Hall, Z. W., 1987, Monoclonal antibodies specific for each of the two toxin-binding sites of *Torpedo* acetylcholine receptor, *Biochemistry* **26**:6372–6381.

Dutar, P., Rascol, O., Lamour, Y., 1989, ω-Conotoxin GVIA blocks synaptic transmission in the CA1 field of the hippocampus, *Eur. J. Pharmacol.* **174**:261–266.

Edwards, R. A., Lutz, P. L., and Baden, D. G., 1989, Relationship between energy expenditure and ion channel density in the turtle and rat brain, *Am. J. Physiol.* **257**:R1354–1358.

Evans, J. S., Musser, E. A., Mengel, G. D., Forsblad, K. R., and Hunter, J. H., 1961, Antitumor activity of 1-β-D-arabinofuranosylcytosine hydrochloride, *Proc. Soc. Exp. Biol. Med.* **106**:350–353.

Evans, J. S., Musser, E. A., Bostwick, L., and Mengel, G. D., 1964, The effect of 1-β-D-arabinofuranosylcytosine hydrochloride on murine neoplasms, *Cancer Res.* **24**:1285–1293.

Famiani, V., 1987, *In vivo* effect of didemnin B on two tumors of the rat, *Oncology* **44**:42–46.

Farago, A., and Nishizuka, Y., 1990, Protein kinase C in transmembrane signalling, *FEBS Lett.* **268**:350–354.

Fattorusso, S., Minale, L., and Sodano, G., 1970, Aeroplysinin-I, a new bromo-compound from *Aplysina aerophoba*, *J. Chem. Soc. Perkin I* **12**:751–752.

Faulkner, D. J., 1991, Marine natural products, *Nat. Prod. Rep.* **8**:97–147.

Fautin, D. G. (ed.), 1988, *Biomedical Importance of Marine Organisms* (Memoirs of the California Academy of Sciences Number 13), California Academy of Sciences, San Francisco.

Félix, M., Cohen, P., and Karsenti, E., 1990, Cdc2 H1 kinase is negatively regulated by a type 2A phosphatase in the *Xenopus* early embryonic cell cycle: Evidence from the effects of okadaic acid, *EMBO J.* **9**:675–683.

Fenical, W., Okuda, R. K., Banduraga, M. M., Culver, P., and Jacobs, R. S., 1981, Lophotoxin: A novel neuromuscular toxin from pacific sea whips of the genus *Lophogorgia*, *Science* **212**:1512–1514.

Fredholm, B. B., 1990, Differential sensitivity to blockade by 4-aminopyridine of presynaptic receptors regulating [^{3}H]acetylcholine release from rat hippocampus, *J. Neurochem.* **54**:1386–1390.

Fujiki, H., and Sugimura, T., 1987, New classes of tumor promoters: Teleocidin, aplysiatoxin, and palytoxin, *Adv. Cancer Res.* **59**:223–264.

Fujiki, H., Suganuma, M., Nakayasu, M., Hakii, H., Horiuchi, T., Takayama, S., and Sugimura, T., 1986, Palytoxin is a non-12-*O*-tetradecanoylphorbol-13-acetate type tumor promoter in two-stage mouse skin carcinogenesis, *Carcinogenesis* **7**:707–710.

Fujiki, H., Suganuma, M., Suguri, H., Yoshizawa, S., Takagi, K., Uda, N., Wakamatsu, K., Yamada, K., Murata, M., Yasumoto, T., and Sugimura, T., 1988, Diarrhetic shellfish toxin, dino-physistoxin-1, is a potent tumor promoter on mouse skin, *Jpn. J. Cancer Res. (Gann)* **79**:1089–1093.

Fujiki, H., Suganuma, M., Suguri, H., Yoshizawa, S., Takagi, K., and Kobayashi, M., 1989a, Sarcophytols A and B inhibit tumor promotion by teleocidin in two-stage carcinogenesis in mouse skin, *J. Cancer Res. Clin. Oncol.* **115**:25–28.

Fujiki, H., Suganuma, M., Suguri, H., Yoshizawa, S., Takagi, K., Sassa, T., Uda, N., Wakamatsu, K., Yamada, K., Yasumoto, T., Kato, Y., Fusetani, N., Hashimoto, K., and Sugimura, T., 1989b, New tumor promoters from marine sources: The okadaic acid class, in: *Mycotoxins and Phycotoxins* (S. Natori, K. Hashimoto, and Y. Ueno, eds.), Elsevier, Amsterdam, pp. 453–460.

Fulmor, W., Van Lear, G. E., Morton, G. O., and Mills, R. D., 1970, Isolation and absolute configuration of the aeroplysinin 1 enantiomorphic pair from *Ianthella ardis*, *Tetrahedron Lett.* **52**:4551–4552.

Gavériaux, C., and Loor, F., Proliferation of interleukin 2 dependent cytotoxic T cell line cells, *Int. Arch. Allergy Appl. Immunol.* **88**:294–296.

Geller, R. B., Saral, R., Karp, J. E., Santos, G. W., and Burke, P. J., 1990, Cure of acute myelocytic leukemia in adults: A reality, *Leukemia* **4**:313–315.

Glaser, K. B., and Jacobs, R. S., 1986, Molecular pharmacology of manoalide, *Biochem. Pharmacol.* **35**:449–453.

Glaser, K. B., Vedvivk, T. S., and Jacobs, R. S., 1988, Inactivation of phospholipase A$_2$ by manoalide. Localization of the manoalide binding site on bee venom phospholipase A$_2$, *Biochem. Pharmacol.* **37**:3639–3646.

Glaser, K. B., de Carvalho, M. S., Jacobs, R. S., Kernan, M. R., and Faulkner, D. J., 1989, Manoalide: Structure activity studies and definition of the pharmacophore for phospholipase A$_2$ inactivation, *Mol. Pharmacol.* **36**:782–788.

Goris, J., Hermann, J., Hendrix, P., Ozon, R., and Merlevede, W., 1989, Okadaic acid, a specific protein phosphatase inhibitor, induces maturation and MPF formation in *Xenopus laevis* oocytes, *FEBS Lett.* **245**:91–94.

Gouiffès, D., Juge, M., Grimaud, N., Welin, L., Sauviat, M. P., Barbin, Y., Laurent, D., Roussakis, C., Henichart, J. P., and Verbist, J. F., 1988a, Bistramide A, a new toxin from the urochordata *Lissoclinum bistratum* sluiter: Isolation and preliminary characterization, *Toxicon* **26**:1129–1136.

Gouiffès, D., Moreau, S., Helbecque, N., Bernier, J. L., Hénichart, J. P., Barbin, Y., Laurent, D., and Verbist, J. F., 1988b, Proton nuclear magnetic study of bistramide A, a new cytotoxic drug isolated from *Lissoclinum bistratum* sluiter, *Tetrahedron* **44**:451–459.

Gschwendt, M., Kittstein, W., and Marks, F., 1987, Didemnin B inhibits biological effects of tumor promoting phorbol esters on mouse skin, as well as phosphorylation of a 100 kD protein in mouse epidermis cytosol, *Cancer Lett.* **34**:187–191.

Gunawardana, G. P., Kohmoto, S., Gunasekera, S. P., McConnell, O. J., and Koehn, F. E., 1988, Dercitin, a new biologically active acridine alkaloid from a deep water marine sponge, *Dercitus* sp., *J. Am. Chem. Soc.* **110**:4856–4858.

Gusovsky, F., and Daly, J. W., 1990, Maitotoxin: A unique pharmacological tool for research on calcium-dependent mechanisms, *Biochem. Pharmacol.* **39**:1633–1639.

Gustafson, K. R., Cardellina, J. H., Fuller, R. W., Weislow, O. S., Kiser, R. F., Snader, K. M., Patterson, G. M., and Boyd, M. R., 1989, AIDS-antiviral sulfolipids from cyanobacteria (blue-green algae), *J. Natl. Cancer Inst.* **81**(16):1254–1258.

Hall, S., Strichartz, G., Moczydlowski, E., Ravindran, A., and Reichardt, P. B., 1990, The saxitoxins: Sources, chemistry and pharmacology, in: *Marine Toxins* (S. Hall and G. Strichartz, eds.), American Chemical Society, Washington, D.C., pp. 29–65.

Hannun, Y. A., and Bell, R. A., 1989, Regulation of protein kinase C by sphingosine and lysophingolipids, *Clin. Chem. Acta* **185**:333–345.

Hatanaka, Y., Yoshida, E., Nakayama, H., and Kanaoka, Y., 1990, Synthesis of μ-conotoxin GIIIA: A chemical probe for sodium channels, *Chem. Pharm. Bull. (Tokyo)* **38**:236–238.

Hayashi, E., and Yamada, S., 1975, Pharmacological studies on surugatoxin, the toxin principle from Japanese ivory mollusc (*Babylonia japonica*), *Br. J. Pharmacol.* **53**:207–251.

Hayashi, E., Isogui, M., Kagawa, Y., Takayanagi, N., and Tamada, S., 1984, Neosurugatoxin, a specific antagonist of nicotinic acetylcholine receptors, *J. Neurochem.* **42**:1491–1494.

Haystead, T. A. J., Sim, A. T. R., Carling, D., Honnor, R. C., Tsukitani, Y., Cohen, P., and Hardie, D. G., 1989, Effects of the tumor promoter okadaic acid on intracellular protein phosphorylation and metabolism, *Nature* **337**:78–81.

Hennings, H., Blumberg, P. M., Pettit, G. R., Herald, C. L., Shores, R., and Yuspa, S. H., 1987, Bryostatin 1, an activator of protein kinase C, inhibits tumor promotion by phorbol esters in SENCAR mouse skin, *Carcinogenesis* **8**:1343–1346.

Herdon, H., and Nahorski, S. R., 1989, Investigations of the roles of dihydropyridine- and ω-conotoxin-sensitive calcium channels in mediating depolarization-evoked endogenous dopamine release from striatal slices, *Naunyn-Schmiedebergs Arch. Pharmacol.* **340**:36–40.

Hescheler, J., Mieskes, G., Rüegg, J. C., Takai, A., and Trautwein, W., 1988, Effects of a protein phosphatase inhibitor, okadaic acid, on membrane currents of isolated guinea-pig cardiac myocytes, *Pflügers Arch.* **412**:248–252.

Hirata, Y., and Uemura, D., 1986, Halichondrins—Antitumor polyether macrolides from a marine sponge, *Pure Appl. Chem.* **58**:701–710.

Hong, S. J., and Chang, C. C., 1989, Use of geographutoxin II (μ-conotoxin) for the study of neuromuscular transmission in mice, *Br. J. Pharmacol.* **97**:934–940.

Huang, T. T. F., and Yanagimachi, R., 1984, Fucoidin inhibits attachment of guinea pig spermatazoa to the zona pellucida through binding to the inner acrosomal membrane and equatorial domains, *Exp. Cell. Res.* **153**:363–373.

Hudson, J. B., Saboune, H., Abramowski, Z., Towers, G. H., and Rinehart, K. L., 1988, The photoactive antimicrobial properties of eudistomins from the Caribbean tunicate *Eudistoma olivaceum*, *Photochem. Photobiol.* **47**(3):377–381.

Ireland, C., and Faulkner, D. J., 1978, The defensive secretion of the opithobranch mollusc *Onchidella binneyi*, *Bioorg. Chem.* **7**:125–130.

Ireland, C. M., Roll, D. M., Molinski, T. F., McKee, T. C., Zabriskie, T. M., and Swersey, J. C., 1988, Uniqueness of the marine chemical environment: Categories of marine natural products from invertebrates, in: *Biomedical Importance of Marine Organisms* (Memoirs of the California Academy of Sciences Number 13; D. G. Fautin, ed.), California Academy of Sciences, San Francisco, pp. 41–57.

Ireland, C. M., Molinski, T. F., Roll, D. M., Zabriskie, T. M., McKee, T. C., Swersey, J. C., and Foster, M. P., 1989, Natural product peptides from marine organisms, in: *Bioorganic Marine Chemistry*, Vol. 3 (P. J. Scheuer, ed.), Springer-Verlag, Berlin, pp. 1–46.

Irie, K., Hagiwara, N., Tokuda, H., and Koshimizu, K., 1987, Structure–activity studies of the indole alkaloid tumor promoter teleocidins, *Carcinogenesis* **8:**547–552.

Ishihara, H., Martin, B. L., Brautigan, D. L., Karaki, H., Ozaki, H., Kato, Y., Fusetani, N., Watabe, S., Hashimoto, K., Uemura, D., and Hartshorne, D. J., 1989, Calyculin A and okadaic acid: Inhibitors of protein phosphatase activity, *Biochem. Biophys. Res. Commun.* **159:**871–877.

Issinger, O., Martin, T., Richter, W. W., Olson, M., and Fujiki, H., 1988, Hyperphosphorylation of N-60, a protein structurally and immunologically related to nucleolin after tumor-promoter treatment, *EMBO J.* **7:**1621–1626.

Itai, A., Kato, Y., Tomioka, N., Iitaka, Y., Endo, Y., Hasegawa, M., Shudo, K., Fujiki, H., and Sakai, S., 1988, A receptor model for tumor promoters: Rational superposition of teleocidins and phorbol esters, *Proc. Natl. Acad. Sci. USA* **85:**3688–3692.

Jacobs, R. S., Culver, P., Langdon, R., O'Brien, T., and White, S., 1985, Some pharmacological observations on marine natural products, *Tetrahedron* **41:**981–984.

Jacobson, P. B., Marshall, L. A., Sunf, A., and Jacobs, R. S., 1990, Inactivation of human sinovial fluid phospholipase A$_2$ by the marine natural product manoalide, *Biochem. Pharmacol.* **39:**1557–1564.

Jaken, S., 1990, Protein kinase C and tumor promoters, *Curr. Opin. Cell. Biol.* **2:**192–197.

Jan, C.-R., Titeler, M., and Schneider, A. S., 1990, Identification of ω-conotoxin binding sites on adrenal medullary membranes: Possibility of multiple calcium channels in chromaffin cells, *J. Neurochem.* **54:**355–358.

Jeffrey, A. M., and Liskamp, R. M. L., 1986, Computer assisted molecular modeling of tumor promoters: Rationale for the activity of phorbol esters, teleocidin B, and aplysiatoxin, *Proc. Natl. Acad. Sci. USA* **83:**241–245.

Jiang, T. L., Liu, R. H., and Salmon, S. E., 1983, Antitumor activity of didemnin B in the human tumor stem cell assay, *Cancer Chemother. Pharmacol.* **11:**1–4.

Kao, C. Y., 1966, Tetrodotoxin, saxitoxin, and their significance in the study of excitation phenomena, *Pharmacol. Rev.* **18:**997–1049.

Kao, C. Y., 1986, Structure–activity relations of tetrodotoxin, saxitoxin and analogues, in: *Tetrodotoxin, Saxitoxin and the Molecular Biology of the Sodium Channel* (Annals of the New York Academy of Science Volume 479; C. Y. Kao and S. R. Levinson, eds.), New York Academy of Science, New York, pp. 52–67.

Karschin, A., and Lipton, S. A., 1989, Calcium channels in solitary retinal ganglion cells from postnatal rat, *J. Physiol.* **418:**379–396.

Kashman, Y., Groweiss, A., and Shmueli, U., 1980, Latrunculin: A new 2-thiazolidinone macrolide from the marine sponge *Latrunculia magnifica*, *Tetrahedron Lett.* **21:**3629–3633.

Kaslauskas, R., Murphy, P. T., Wells, R. J., Baird-Lambert, J. A., and Jamieson, D. D., 1983, Halogenated pyrrolo [2,3-*d*] pyrimidine nucleosides from marine organisms, *Aust. J. Chem.* **36:**165–170.

Kather, H., 1990, Pathways of purine metabolism in human adipocytes, *J. Biol. Chem.* **265:**96–102.

Katao, Y., Fusetani, N., Matsunaga, S., and Hashimoto, K., 1968, Calyculin A, a novel antitumor metabolite from the marine sponge *Discodermia calyx*, *J. Am. Chem. Soc.* **108:**2780–2781.

Keith, R. A., Mangano, T. J., and Salama, A. I., 1989, Inhibition of *N*-methyl-D-aspartate- and kainic acid-induced neurotransmitter release by ω-conotoxin GVIA, *Br. J. Pharmacol.* **98:**767–772.

Kipreos, E. T., and Wang, J. Y. T., 1990, Differential phosphorylation of c-Abl in cell cycle determined by *cdc*2 kinase and phosphatase activity, *Science* **248:**217–220.

Kobayashi, M., Nakagawa, T., and Mitsuhashi, H., 1979, Marine terpenes and terpenoids. I. Structures of four cembrane-type diterpenes: sarcophytol-A, sarcophytol-A acetate, sarcophytol-

B, and sarcophytonin-A, from the soft coral, *Sarcophyton glaucum*, *Chem. Pharm. Bull. (Tokyo)* **27**:2382–2387.

Kobayashi, M., Kobayashi, J., and Ohizumi, Y., 1989, Cone shell toxins and the mechanisms of their pharmacological action, in: *Bioorganic Marine Chemistry*, Vol. 3 (P. J. Scheuer, ed.), Springer-Verlag, Berlin, pp. 71–84.

Kodama, I., Kondo, N., and Shibata, S., 1986, Electrochemical effects of okadaic acid isolated from black sponge in guinea-pig ventricular muscles, *J. Physiol.* **378**:359–373.

Kosuge, T., Tsuji, K., and Hirai, K., 1981, Isolation and structure determination of a new marine toxin, neosurugatoxin, from the Japanese ivory shell, *Babylonia japonica*, *Tetrahedron Lett.* **22**:3417–3420.

Kreuter, M. H., Bernd, A., Holzmann, H., Müller-Klieser, W., Maidhof, A., Weibmann, N., Kljajic, Z., Batel, R., Schröder, H. C., and Müller, W. E. G., 1989, Cytostatic activity of aeroplysinin-1 against lymphoma and epithelioma cells, *Z. Naturforsch.* **44c**:680–688.

Kume, H., Takai, A., Tokuno, H., and Tomita, T., 1989, Regulation of Ca^{2+}-dependent K^+-channel activity in tracheal myocytes by phosphorylation, *Nature* **341**:152–154.

Laporte, D. C., Wierman, B. M., *et al.*, 1980, Calcium-induced exposure of a hydrophobic surface on calmodulin, *Biochemistry* **19**:3814–3819.

Levine, L., and Fujiki, H., 1985, Stimulation of arachidonic metabolism by different types of tumor promoters, *Carcinogenesis* **6**:1631–1634.

Levine, L., Xiao, D., and Fujiki, H., 1986, A combination of palytoxin with 1-oleoyl-2-acetyl-glycerol (OAG) or insulin or interleukin-1 synergistically stimulates arachidonic acid metabolism, but combination of 12-*O*-tetradecanoylphorbol-13-acetate (TPA)-type tumor promoters with OAG does not, *Carcinogenesis* **7**:99–103.

Li, L. H., Timmins, L. G., Wallace, T. L., Krueger, W. C., Prairie, M. D., and Im, W. B., 1984, Mechanism of action of didemnin B, a depsipeptide from the sea, *Cancer Lett.* **23**:279–288.

Lin, Y.-Y., Risk, M., Ray, S. M., van Engen, D., Clardy, J. C., Golik, J., James, J. C., and Nakanishi, K., 1981, Isolation and structure of brevetoxin-B from the "red tide" dinoflagellate *Ptychodiscus brevis (Gymnodinium breve)*, *J. Am. Chem. Soc.* **103**:6773–6775.

Lister, M. D., Glaser, K. B., Ulevitch, R. J., and Dennis, E. A., 1989, Inhibition studies on the membrane-associated phospholipase A_2 *in vitro* and prostaglandin E_2 production *in vivo* of the macrophage-like P388D$_1$ cell, *J. Biol. Chem.* **264**:8520–8528.

Look, S. A., Fenical, W., Jacobs, R. S., and Clardy, J., 1986, The pseudopterosins: Anti-inflammatory and analgesic natural products from the sea whip *Pseudopterogorgia elisabethae*, *Proc. Natl. Acad. Sci. USA* **83**:6238–6240.

Loya, S., and Hizi, A., 1990, The inhibition of human immunodeficiency virus type 1 reverse transcriptase by avarol and avarone derivatives, *FEBS* **269**:131–134.

Luedke, E. S., 1990, The identification and characterization of the pseudopterosins: Anti-inflammatory agents isolated from the gorgonian coral *Pseudopterogorgia elisabethae*, Ph.D. Thesis, University of California, Santa Barbara, California.

Luetje, C. W., Wada, K., Rogers, S., Abramson, S. N., Tsuji, K., Heinemann, S., and Patrick, J., 1990, Neurotoxins distinguish between different neuronal nicotinic acetylcholine receptor subunit combinations, *J. Neurochem.* **55**:632–639.

MacKintosh, C., and Cohen, P., 1989, Identification of high levels of type 1 and type 2A protein phosphatases in higher plants, *Biochem. J.* **262**:335–339.

Maeda, N., Wada, K., Yuzaki, M., and Mikoshiba, K., 1989, Autoradiographic visualisation of a calcium channel antagonist, $[^{125}I]\omega$-conotoxin GVIA, binding site in the brains of normal and cerebellar mutant mice (*pcd* and *weaver*), *Brain Res.* **489**:21–30.

Maldonado, E., Lavergne, J. A., and Kraiselburd, E., 1982, Didemnin A inhibits the *in vitro* replication of dengue virus types 1, 2, and 3, *Puerto Rico Health Sci. J.* **1**:22–25.

Mangano, T. J., Patel, J., Salama, A. I., and Keith, R. A., 1990, Glycine-evoked neurotransmitter release from rat hippocampal brain slices: Evidence for the involvement of glutaminergic transmission, *J. Pharmacol. Exp. Ther.* **252:**574–580.

Marshall, I. G., and Harvey, A. L., 1990, Selective neuromuscular blocking properties of α-conotoxin *in vivo*, *Toxicon* **28:**231–234.

Maruyama, H., Nakajima, J., and Yamamoto, I., 1987, A study on the anticoagulant and fibrinolytic activities of a crude fucoidan from the edible brown seaweed *Laminaria religiosa*, with special reference to its inhibitory effect on the growth of sarcoma-180 ascites cells subcutaneously implanted in mice, *Kitasato Arch. Exp. Med.* **60**(3):105–121.

Mayer, A. M., and Jacobs, R. S., 1988, Manoalide: An antiinflammatory and analgesic marine natural product, in: *Biomedical Importance of Marine Organisms* (Memoirs of the California Academy of Sciences Number 13; D. G. Fautin, ed.), California Academy of Sciences, San Francisco, pp. 133–142.

Mayer, A. M. S., Glaser, K. B., and Jacobs, R. S., 1988, Regulation of eicosanoid biosynthesis *in vitro* and *in vivo* by the marine natural product manoalide: A potent inactivator of venom phospholipases, *J. Pharmacol. Exp. Ther.* **244:**871–878.

Minigo, H., Nemet, D., Planinc-Peraica, A., Bogdanic, V., Labar, B., Jaksic B., and Hauptmann, E., 1990, High dose ara-C as induction therapy of acute myeloid leukaemia, *Folia Haematol.* **117:** 135–140.

Mong, S., Votta, B., Sarau, H. M., Foley, J. J., Schmidt, D., Carte, B. K., Poehland, B., and Westley, J., 1990, 15-Acetylthioxy-furodysinin lactone, isolated from a marine sponge *Dysidea* sp., is a potent agonist to human leukotriene B_4 receptor, *Prostaglandins* **39:**89–97.

Montgomery, D. W., and Zukoski, C. F., 1985, Didemnin B: A new immunosuppressive cyclic peptide with potent activity *in vitro* and *in vivo*, *Transplantation* **40:**49–56.

Montgomery, D. W., Celniker, A., and Zukoski, C. F., 1987, Didemnin B: A new immunosuppressive cyclic peptide that stimulates murine antibody responses *in vitro* and *in vivo*, *Transplantation Proc.* **19:**1295–1296.

Moore, R. E., and Bartolini, G., 1981, Structure of palytoxin, *J. Am. Chem. Soc.* **103:**2491–2494.

Nakadate, T., 1989, The mechanism of skin tumor promotion caused by phorbol esters: Possible involvement of arachidonic acid cascade/lipoxygenase, protein kinase C and calcium/calmodulin systems, *Jpn. J. Pharmacol.* **49:**1–9.

Nakamura, H., Wu, H., Kobayashi, J., Nakamura, Y., Ohizumi, Y., and Hirata, Y., 1985, Purealin, a novel enzyme activator from the Okinawan marine sponge *Psammaplysilla purea*, *Tetrahedron Lett.* **26**(37):4517–4520.

Nakamura, Y., Kobayashi, M., Nakamura, H., Wu, H., Kobayashi, J., and Ohizumi, Y., 1987, Purealin, a novel activator of skeletal muscle actomyosin ATPase and myosin EDTA-ATPase that enhanced the superprecipitation of actomyosin, *Eur. J. Biochem.* **167:**1–6.

Nakamura, H., Yoshito, K., Pajares, M. A., and Rando, R. R., 1989, Structural basis of protein kinase C activation by tumor promoters, *Proc. Natl. Acad. Sci. USA* **86:**9672–9676.

Narahashi, T., 1974, Chemical tools in the study of excitable membranes, *Physiol. Rev.* **54:**813–889.

Narahashi, T., 1988, Mechanism of tetrodotoxin and saxitoxin action, in: *Handbook of Natural Toxins*, Vol. 3: *Marine Toxins and Venoms* (A. T. Tu, ed.), Marcel Dekker, New York, pp. 185–210.

Narisawa, T., Takahashi, M., Niwa, M., Fukaura, Y., and Fujiki, H., 1989, Inhibition of methylnitrosourea-induced large bowel cancer development in rats by sarcophytol A, a product from a marine soft coral *Sarcophyton glaucum*, *Cancer Res.* **49:**3287–3289.

Nishizuka, Y., 1988, The molecular heterogeneity of protein kinase C and its implications for cellular regulation, *Nature* **334:**661–665.

Nishizuka, Y., 1989, The Albert Lasker Medical Awards. The family of protein kinase C for signal transduction, *J. Am. Med. Assoc.* **262:**1826–1832.

Obara, K., Takai, A., Rüegg, J. C., and de Lanerolle, P., 1989, Okadaic acid, a phosphatase inhibitor, produces a Ca^{2+} and calmodulin-independent contraction of smooth muscle, *Pflügers Arch.* **414:** 134–138.

Oehninger, S., Acosta, A., and Hodgen, G. D., 1990, Antagonistic and agonistic properties of saccharide moieties in the hemizona assay, *Fertil. Steril.* **53**(1):143–149.

Ohizumi, Y., and Kobayashi, M., 1990, Ca-dependent excitatory effects of maitotoxin on smooth and cardiac muscle, in: *Marine Toxins* (S. Hall and G. Strichartz, eds.), American Chemical Society, Washington, D.C., pp. 133–143.

Olivera, B. M., Gray, W. R., Zeikus, R., McIntosh, J. M., Varga, J., Rivier, J., deSantos, V., and Cruz, L. J., 1985, Peptide neurotoxins from fish-hunting cone snails, *Science* **230:**1338–1343.

Olivera, B. M., Gray, W. R., and Cruz, L. J., 1988, Marine snail venoms, in: *Handbook of Natural Toxins,* Vol. 3: *Marine Toxins and Venoms* (A. T. Tu, ed.), Marcel Dekker, New York, pp. 327–354.

Olivera, B. M., Hillyard, D. R., Rivier, J., Woodward, S., Gray, W. R., Corpuz, G., and Cruz, L. J., 1990a, Conotoxins: Targeted peptide ligands from snail venoms, in: *Marine Toxins* (S. Hall and G. Strichartz, eds.), American Chemical Society, Washington, D.C., pp. 256–278.

Olivera, B. M., Rivier, J., Clark, C., Ramilo, C. A., Corpuz, G. P., Abogadie, F. C., Mena, E. E., Woodward, S. R., Hillyard, D. R., and Cruz, L. J., 1990b, Diversity of *Conus* neurotoxins, *Science* **249:**257–263.

Palczewski, K., McDowell, J. H., Jakes, S., Ingebritsen, T. S., and Hargrave, P. A., 1989, Regulation of rhodopsin dephosphorylation by arrestin, *J. Biol. Chem.* **264:**15770–15773.

Pardi, A., Galdes, A., Florance, J., and Maniconte, D., 1989, Solution structures of α-conotoxin G1 determined by two-dimensional NMR spectroscopy, *Biochemistry* **28:**5494–5501.

Perry, N. B., Blunt, J. W., Munro, M. H. G., and Pannell, L. K., 1988, Mycalamide A, an antiviral compound from a New Zealand sponge of the genus *Mycale, J. Am. Chem. Soc.* **110:**4850–4851.

Pettit, G. R., Herald, C. L., Doubek, D. L., and Herald, D. L., 1982, Isolation and structure of bryostatin 1, *J. Am. Chem. Soc.* **104:**6846–6848.

Pettit, G. R., Kamano, Y., Herald, C. L., Tuinman, A. A., Boettner, F. E., Kizu, H., Schmidt, J. M., Baczynskyj, L., Tomer, K. B., and Bontems, R. J., 1987, The isolation and structure of a remarkable marine animal antineoplastic constituent: Dolastatin 10, *J. Am. Chem. Soc.* **109:** 6883–6885.

Pettit, G. R., Singh, S. B., Hogan, F., Lloyd-Williams, P., Herald, D. L., Burkett, D. D., and Clewlow, P. J., 1989, The absolute configuration and synthesis of natural (−)-dolastatin 10, *J. Am. Chem. Soc.* **111:**5463–5465.

Pinchuk, L. N., and Pinchuk, G. V., 1990, Monoclonal antibody to brain cytoplasmic tetrodotoxin-sensitive protein detects an epitope associated with lymphocytes and involved in lymphocyte activation (Med Line Abstract), *J. Neuroimmunol.* **27:**71–78.

Plummer, M. R., Logothetis, D. E., and Hess, P., 1989, Elementary properties and pharmacological sensitivities of calcium channels in mammalian peripheral neurons, *Neuron* **2:**1453–1463.

Rapier, C., Lunt, G. G., and Wonnacott, S., 1990, Nicotinic modulation of [3H]-dopamine release from striatal synaptosomes: Pharmacological characterization, *J. Neurochem.* **54:**937–945.

Redpath, N. T., and Proud, C. G., 1989, The tumor promoter okadaic acid inhibits reticulocyte-lysate protein synthesis by increasing the net phosphorylation of elongation factor 2, *Biochem. J.* **262:** 69–75.

Reinhart, K. L., Holt, T. G., Fregeau, N. L., Stroh, J. G., Keifer, P. A., Sun, F., Li, L. H., and Martin, D. G., 1990, Ecteinascidins 729, 743, 745, 759A, 759B, and 770: Potent antitumor agents from the Caribbean tunicate *Ecteinascidia turbinata, J. Org. Chem.* **55:**4512–4515.

Rinehart, K. L., Gloer, J. B., Cook, J. C., Mizsak, S. A., and Scahill, T. A., 1981a, Structures of the didemnins, antiviral and cytotoxic depsipeptides from a Caribbean tunicate, *J. Am. Chem. Soc.* **103:**1857–1859.

Rinehart, K. L., Gloer, J. B., Hughes, R. G., Renis, H. E., McGovren, J. P., Swynenberg, E. B., Stringfellow, D. A., Kuentzel, S. L., and Li, L. H., 1981b, Didemnins: Antiviral and antitumor depsipeptides from a Caribbean tunicate, *Science* **212**:933–935.

Rinehart, K. L., Shaw, P. D., Shield, L. S., Gloer, J. B., Harbour, G. C., Koker, M. E. S., Samain, D., Schwartz, R. E., Tymiak, A. A., Weller, D. L., Carter, G. T., Munro, M. H. G., Hughes, R. G., Renis, H. E., Swyneberg, E. B., Stringfellow, D. A., Vavra, J. J., Coats, J. H., Zurenko, G. E., Kuentzel, S. L., Li, L. H., Bakus, G. J., Brusca, R. C., Craft, L. L., Young, D. N., and Connor, J. L., 1981c, Marine natural products as sources of antiviral, antimicrobial, and antineoplastic agents, *Pure Appl. Chem.* **53**:795–817.

Rinehart, K. L., Gloer, J. B., Wilson, G. R., Hughes, R. G., Li, L. H., Renis, H. E., and McGovren, J. P., 1983, Antiviral and antitumor compounds from tunicates, *Fed. Proc.* **42**:87–90.

Rossof, A. H., Johnson, P. A., Kimmell, B. D., Graham, J. E., and Roseman, D. L., 1983, *In vitro* phase II study of didemin B in human cancer (Abstract), *Proc. Am. Assoc. Cancer Res.* **24**:315.

Sakai, R., Ikeda, I., Kitani, H., Fujiki, H., Takaku, F., Rapp, U., Sugimura, T., and Nagao, M., 1989, Flat reversion by okadaic acid of *raf* and *ret*-II transformants, *Proc. Natl. Acad. Sci. USA* **86**:9946–9950.

Santostasi, G., Kutty, R. K., Bartorelli, A. L., Yasumoto, T., and Krishna, G., 1990, Maitotoxin-induced myocardial cell injury: Calcium accumulation followed by ATP depletion precedes cell death, *Toxicol. Appl. Pharmacol.* **102**:164–173.

Sassa, T., Richter, W. W., Uda, N., Suganuma, M., Suguri, H., Yoshizawa, S., Horita, M., and Fujiki, H., 1989, Apparent "activation" of protein kinases by okadaic acid class tumor promoters, *Biochem. Biophys. Res. Commun.* **159**:939–944.

Sawyer, P. J., Gentile, J. H., and Sasuer, Jr., J. J., 1968, Demonstration of a toxin from *Aphanizomenon flos-aque* (L.) Ralfs, *Can. J. Microbiol.* **14**:1199–1204.

Schatten, G., Schatten, H., Spector, H., Cline, C., Paweletz, N., Simerly, C., and Petzelt, C., 1986, Latrunculin inhibits the microfilament-mediated processes during fertilization, cleavage and early development in sea urchins and mice, *Exp. Cell Res.* **166**:191–208.

Schatten, H., and Schatten, G., 1986, Motility and centrosomal organization during sea urchin and mouse fertilization, *Cell Motil. Cytoskel.* **6**:163–175.

Schmitz, F. J., Prasad, R. S., Gopichand, Y., Hossain, M. B., and van der Helm, D., 1981, Acanthifolicin, a new episulfide-containing polyether carboxylic acid from extracts of the marine sponge *Pandaros acanthifolium*, *J. Am. Chem. Soc.* **103**:2467–2469.

Schneider, H. R., Mieskes, G., and Issinger, O., 1989, Specific dephosphorylation by phosphatases 1 and 2A of a nuclear protein structurally and immunologically related to nucleolin, *Eur. J. Biochem.* **180**:449–455.

Shibata, S., Ishida, Y., Kitano, H., Ohizumi, Y., Habon, J., Tsukitani, Y., and Kikuchi, H., 1982, Contractile effects of okadaic acid, a novel ionophore-like substance from black sponge, on isolated smooth muscles under the condition of Ca deficiency, *J. Pharmacol. Exp. Ther.* **223**:135–143.

Shimizu, Y., Chou, H.-N., Bando, H., and Duyne, G., and Clardy, J. C., 1986a, Structure of brevetoxin A (GB-1 toxin), the most potent toxin in the Florida Red Tide organism *Gymnodinium breve* (*Ptychodiscus brevis*), *J. Am. Chem. Soc.* **108**:514–515.

Shimizu, Y., Bando, H., Chou, H.-N., van Duyne, G., and Clardy, J. C., 1986b, Absolute configuration of brevetoxins, *J. Chem. Soc. Chem. Commun.* **1986**:1656–1658.

Sorenson, E. M., Culver, P., and Chiappinelli, V. A., 1987, Lophotoxin: Selective blockade of nicotinic transmission in autonomic ganglia by a coral neurotoxin, *Neuroscience* **20**:875–884.

Spector, I., and Shochet, N. R., 1983, Latrunculins: Novel marine toxins that disrupt microfilament organization in cultured cells, *Science* **219**:493–495.

Spector, I., Shochet, N. R., Blasberger, D., and Kashman, Y., 1989, Latrunculins—Novel marine macrolides that disrupt microfilament organization and affect cell growth: I. Comparison with cytochalasin D, *Cell Motil. Cytoskel.* **13**:127–144.

Stewart, J. A., Tong, W. P., Hartshorn, J. N., and McCormack, J. J., 1986, Phase I evaluation of didemnin B (NSC 325319) (Abstract), *Proc. Am. Soc. Clin. Oncol.* **5**:33.

Strichartz, G., and Castle, N., 1990, Pharmacology of marine toxins: Effects on membrane channels, in: *Marine Toxins* (S. Hall and G. Strichartz, eds.), American Chemical Society, Washington, D.C., pp. 2–20.

Strichartz, G., Rando, T., Hall, S., Gitscher, J., Hall, L., Magnani, B., and Hansen Bay, C., 1986, On the mechanism by which saxitoxin binds to and blocks sodium channels, in: *Tetrodotoxin, Saxitoxin and the Molecular Biology of the Sodium Channel* (Annals of the New York Academy of Science Volume 479; C. Y. Kao and S. R. Levinson, eds.), New York Academy of Science, New York, pp. 96–112.

Stroud, R. M., McCarthy, M. R., and Shuster, M., 1990, Nicotinic acetylcholine receptor superfamily of ligand-gated ion channels, *Biochemistry* **29**:11009–11023.

Stryer, L. (1981). *Biochemistry*, 2ed., Freeman, New York.

Suganuma, M., Fujiki, H., Suguri, H., Yoshizawa, S., Hirota, M., Nakayasu, M., Ojika, M., Wakamatsu, K., Yamada, K., and Sugimura, T., 1988, Okadaic acid: An additional non-phorbal-12-tetradecanoate-13-acetate-type tumor promoter, *Proc. Natl. Acad. Sci. USA* **85**:1768–1771.

Suganuma, M., Suttajit, M., Suguri, H., Ojika, M., Yamada, K., and Fujiki, H., 1989, Specific binding of okadaic acid, a new tumor promoter in mouse skin, *FEBS Lett.* **250**:615–618.

Suganuma, M., Fujiki, H., Furuya-Suguri, H., Yoshizawa, S., Yasumoto, S., Kato, Y., Fusetani, N., and Sugimura, T., 1990, Calyculin A, an inhibitor of protein phosphatases, a potent tumor promoter on CD-1 mouse skin, *Cancer Res.* **50**:3521–3525.

Sugawara, I., Itoh, W., Kimura, S., Mori, S., and Shimada, K., 1989, Further characterization of sulfated homopolysaccharides as anti-HIV agents, *Experientia* **45**(10):996–998.

Tachibana, K., Scheuer, P. J., Tsukitani, Y., Kikuchi, H., Van Engen, D., Clardy, J., Gopichand, Y., and Schmitz, F. J., 1981, Okadaic acid, a cytotoxic polyether from two marine sponges of the genus *Halichondria*, *J. Am. Chem. Soc.* **103**:2469–2471.

Takai, A., Bialojan, C., Troschka, M., and Rüegg, J. C., 1987, Smooth muscle myosin phosphatase inhibition and force enhancement by black sponge toxin, *FEBS Lett.* **217**:81–84.

Takai, A., Troschka, M., Mieskes, G., and Somlyo, A. V., 1989, Protein phosphatase composition in the smooth muscle of guinea-pig ileum studied with okadaic acid and inhibitor 2, *Biochem. J.* **262**: 617–623.

Takemura, M., Kishino, J., Yamatodani, A., and Wada, H., 1989a, Inhibition of histamine release from rat hypothalamic slices by ω-conotoxin CVIA, but not by nilvadipine, a dihydropyridine derivative, *Brain Res.* **496**:351–356.

Takemura, M., Kiyama, H., Fukui, H., Tohyama, M., and Wada, H., 1989b, Distribution of the ω-conotoxin receptor in rat brain. An autoradiographic mapping, *Neuroscience* **32**:405–416.

Takito, J., Nakamura, H., Kobayasi, J., Ohizumi, Y., Ebisawa, K., and Nonomura, Y., 1986, Purealin, a novel stabilizer of smooth muscle myosin filaments that modulates ATPase activity of dephosphorylated myosin, *J. Biol. Chem.* **261**(29):13861–13865.

Takito, J., Nakamura, H., Kobayasi, J., and Ohizumi, Y., 1987a, Enhancement of the actin-activated ATPase activity of myosin from canine cardiac ventricle by purealin, *Biochim. Biophys. Acta* **912**:404–407.

Takito, J., Ohizumi, Y., Kobayasi, J., Ebisawa, K., and Nonomura, Y., 1987b, The mechanism of inhibition of light-chain phosphorylation by purealin in chicken gizzard myosin, *Eur. J. Pharmacol.* **142**:189–195.

Tamplin, M. L., 1990, A bacterial source of tetrodotoxins and saxitoxins, in: *Marine Toxins* (S. Hall and G. Strichartz, eds.), American Chemical Society, Washington, D.C., pp. 78–86.

Taylor, P., Culver, P., Abramson, S., Wasserman, L., Kline, T., and Fenical, W., 1988, Use of selective toxins to examine acetylcholine receptor structure, in: *Biomedical Importance of Marine Organ-*

isms (Memoirs of the California Academy of Sciences Number 13; D. G. Fautin, ed.), California Academy of Sciences, San Francisco, pp. 109–114.

Thomson, C., and Wilkie, J., 1989, The conformations and electrostatic potential maps of phorbol esters, teleocidins and ingenols, *Carcinogenesis* **10**:531–540.

Trainer, V. L., Edwards, R. A., Szmant, A. M., Stuart, A. M., Mende, T. J., and Baden, D. G., 1990, Brevetoxins: Unique activators of voltage-sensitive sodium channels, in *Marine Toxins* (S. Hall and G. Strichartz, eds.), American Chemical Society, Washington, D.C., pp. 166–175.

Uemura, D., Ueda, K., and Hirata, Y., 1981, Further studies of palytoxin. II. Structure of palytoxin, *Tetrahedron Lett.* **22**:2781–2784.

Ulevitch, R. J., Watanabe, Y., Sano, M., Lister, M. D., Deems, R. A., and Dennis, E. A., 1988, Solubilization, purification, and characterization of membrane-bound phospholipase A_2 from the $P388D_1$ macrophage-like cell line, *J. Biol. Chem.* **263**:3079–3085.

Van Duyne, G. D., 1990, X-ray crystallographic studies of marine toxins, in: *Marine Toxins* (S. Hall and G. Strichartz, eds.), American Chemical Society, Washington, D. C., pp. 144–165.

Van Sande, J., Deneubourg, F., Beauivens, R., Breakman, J. C., Daloze, D., and Dumont, J. E., 1990, Inhibition of iodide transport in thyroid cells by dysidenin, a marine toxin, and some of its analogs, *Mol. Pharmacol.* **37**:583–589.

Wada, A., Uezono, Y., Arita, M., Tsuji, K., Yanagihara, N., Kobayashi, H., and Izumi, F., 1989, High affinity and selectivity of neosurugatoxin for the inhibition of ^{22}Na influx via nicotinic receptor-ion channel in cultured bovine adrenal medullary cells: Comparative study with histrionicotoxin, *Neuroscience* **33**:333–339.

Wattenberg, E. V., Fujiki, H,. and Rosner, M. R., 1987, Heterologous regulation of the epidermal growth factor receptor by palytoxin, a non-12-*O*-tetradecanoylphorbol-13-acetate-type tumor promoter, *Cancer Res.* **47**:4618–4622.

Wattenberg, E. V., Byron, K. L., Villereal, M. L., Fujiki, H., and Rosner, M. R., 1989a, Sodium as a mediator of non-phorbol tumor promoter action. Down-modulation of the epidermal growth factor receptor by palytoxin, *J. Biol. Chem.* **264**:14668–14673.

Wattenberg, E. V., McNeil, P. L., Fujiki, H., and Rosner, M. R., 1989b, Palytoxin down-modulates the epidermal growth factor receptor through a sodium-dependent pathway, *J. Biol. Chem.* **264**:213–219.

Watters, D., Marshall, K., Hamilton, S., Michael, J., McArthur, M., Seymour, G., Hawkins, C., Gardiner, R., and Lavin, M., 1990, The bistratenes: New cytotoxic marine macrolides which induce some properties indicative of differentiation in HL-60 cells, *Biochem. Pharmacol.* **39**:1609–1614.

Weinberg, J. M., Davis, J. A., Lawton, A., and Abarzua, M., 1988, Modulation of cell nucleotide levels in isolated kidney tubules, *Am. J. Physiol.* **254**:F311–F322.

Wender, P. A., Koehler, K. F., Sharkey, N. A., Dell'Aquila, M.L., and Blumberg, P. M., 1986, Analysis of the phorbol ester pharmacophore on protein kinase C as a guide to the rational design of new classes of analogs, *Proc. Natl. Acad. Sci. USA* **83**:4214–4218.

Wessler, I., Dooley, D. J., Osswald, H., and Schlemmer, F., 1990, Differential blockade by nifedipine and ω-conotoxin GVIA of α_1- and β_1-adrenoceptor-controlled calcium channel motor nerve terminals of the rat, *Neurosci. Lett.* **108**:173–178.

Wheeler, L. A., Sachs, G., Goodrum, D., Amdahl, L., Horowitz, N., and de Vries, W., 1988, Importance of marine natural products in the study of inflammation and calcium channels, in: *Biomedical Importance of Marine Organisms* (Memoirs of the California Academy of Sciences Number 13; D. G. Fautin, ed.), California Academy of Sciences, San Francisco, pp. 125–132.

Williams, M., Abreu, M., Jarvis, M. F., and Noronha-Blob, L., 1987, Characterization of adenosine receptors in the PC12 pheochromocytoma cell line using radioligand binding: Evidence for A-2 selectivity, *J. Neurochem.* **48**:498–502.

Wollner, D. A., Scheinman, R., and Catterall, W. A., 1988, Sodium channel expression and assembly during development of retinal ganglion cells, *Neuron* **1**:727–739.

Wright, A. W., Forleo, D, A., Gunawardana, G. P., Gunasekera, S. P., Koehn, F. E., and McConnell, O. J., 1990, Antitumor tetrahydroisoquinoline alkaloids from the colonial ascidian *Ectinascidia turbinata*, *J. Org. Chem.* **55**:4508–4512.

Wu, C. H., and Narahashi, T., 1988, Mechanism of action of novel marine neurotoxins on ion channels, *Annu. Rev. Pharmacol. Toxicol.* **28**:141–161.

Yoshikami, D., Bagabaldo, Z., and Olivera, B. M., 1989, The inhibitory effects of ω-conotoxins on Ca channel and synapses, *Ann. N. Y. Acad. Sci.* **560**:230–248.

Zabriskie, T. M., Mayne, C. L., and Ireland, C. M., 1988, Patellazole C: A novel cytotoxic macrolide from *Lissoclinum patella*, *J. Amer. Chem. Soc.* **110**:7919–7920.

Zafaralla, G. C., Ramilo, C., Gray, W. R., Karlstrom, R., Olivera, B. M., and Cruz, L. J., 1988, Phylogenetic specificity of cholinergic ligands: α-Conotoxin S1, *Biochemistry* **27**:7102–7105.

2

Isolation, Structural, and Mode-of-Action Studies on Bioactive Marine Natural Products

Yoko Naya, Kazuo Tachibana, and Koji Nakanishi

1. INTRODUCTION

Despite the current interest in bioactive marine compounds, our knowledge is limited because of the short history of this area of research. Moreover, the difficulties associated with the collection and isolation of marine samples in comparison to terrestrial samples have hampered progress. However, there are great possibilities for exciting discoveries, both academic and practical.

In the past, the study of many marine natural products was initiated because they were easy to isolate, and the compounds, subsequent to structure determination, were submitted to pharmacological assays, a format which is still followed. Marine natural products have also been submitted to general screening assays, as exemplified by the search for anticancer compounds, anti-AIDS compounds, antibiotics, and enzyme inhibitors. However, compounds having a direct bearing on the maintenance of the lives of marine organisms cannot be isolated by such

Yoko Naya • Suntory Institute for Bioorganic Research, Shimamoto, Mishima-gun, Osaka, Japan. *Kazuo Tachibana* • Suntory Institute for Bioorganic Research, Shimamoto, Mishima-gun, Osaka, Japan. Current address: Department of Chemistry, University of Tokyo, Hongo, Bunkyo, Japan. *Koji Nakanishi* • Suntory Institute for Bioorganic Research, Shimamoto, Mishima-gun, Osaka, Japan, and Department of Chemistry, Columbia University, New York, New York 10027.

Marine Biotechnology, Volume 1: Pharmaceutical and Bioactive Natural Products, edited by David H. Attaway and Oskar R. Zaborsky. Plenum Press, New York, 1993.

general screenings. Nevertheless, although the development of marine natural products has been more recent than that of products from terrestrial sources, historically many marine compounds have also been studied to clarify the cause of some conspicuous property, e.g., Tyrian purple for its color, tetrodotoxin for its toxicity, luciferin for its bioluminescence, and kainic acid for the anthelmintic property of the alga. This trend of monitoring the isolation by more specific assays related to the biology of an organism or the physiology of a factor, e.g., hormones, allelochemicals, or ecochemicals, is becoming increasingly popular. The isolation of such compounds is more challenging, but once accomplished, it offers us a means to understand the function or mechanisms of life on a molecular structural basis rather than being dependent on data on mixtures of unknown compounds. However, the isolation and characterization of a physiologically active principle is but the beginning of a far more challenging and important problem, namely, clarification of its mode of action on a molecular structural basis. Since physiological activity depends on the interaction between the bioactive compound and its receptor, usually proteins, nucleic acids, or saccharides, such mode-of-action studies invariably require a multidisciplinary approach involving organic chemists, biologists, neurobiologists, biochemists, biophysicists, physicists, etc.

Isolation of bioactive compounds in a pure state is usually the mandatory first hurdle for any natural products study. Unfortunately, the importance of this step is frequently underestimated and treated too casually, without realization that *success or failure of an entire project* may well be dictated by this first purification step. Even when a highly labile compound is isolated by ingenious manipulations, or a hormone/pheromone is isolated in minuscule amounts from huge quantities of starting material after years of frustrating biological assays, the isolation scheme is presented in only a few slides at symposia; however, because the particular scheme is usually applicable only to that specific compound, and because a challenging isolation depends to a great extent on the expertise of the scientist, the significance of isolation/purification tends to be overlooked by the general audience.

Isolation from marine sources is inherently more difficult than from terrestrial compounds: (i) If the isolation is from seawater, removal of salt presents a formidable problem, particularly if the amount of the bioactive factor is minuscule. (ii) Bioactive compounds transmitted through air, such as insect pheromones, can be collected by passing air through a column packed with absorbents, but this is not the case for seawater. (iii) Many marine compounds are water-soluble, which makes them more difficult to handle than lipophilic compounds. (iv) When the bioassay is based on behavior and/or life maintenance of marine organisms, our knowledge is very limited compared to studies of insects or terrestrial animals; furthermore, in ecology-based assays, it is difficult to mimic a marine environment.

Bioactive marine natural products are still largely unexplored and their activities are manifested in a huge variety. In the following we present accounts of three recent examples of isolation and structural studies and subsequent efforts to gain more knowledge of biological roles or modes of action in order to contribute to a better understanding of life.

2. CRUSTACEAN MOLT-INHIBITING REGULATORS

2.1. General

Crustaceans, like insects, molt periodically in order to grow. Hormones may function in the homeostatic regulation of physiological processes and/or may serve as signals for the initiation or regulation of developmental events. Chemical signals must reach their target tissues at the appropriate time and with the appropriate intensity. These requirements are met by a precise regulation of the effective hormone titer itself. This is also true in molting, metamorphosis, and reproduction of arthropods.

During the first decade of the twentieth century, biologists demonstrated that removal of eyestalks (ES) (Zeleny, 1905) from crustaceans promotes molting and leads to precocious ecdysis (Skinner, 1985), and that reimplantation of ES (Megusar, 1912) reverses this effect, thus suggesting the presence of a "molt-inhibiting hormone" (MIH) (Brown and Cunningham, 1939; Webster and Keller, 1989). In crustaceans, the postembryonic developmental molting is believed to be controlled (Kleinholz, 1985; Skinner, 1985) by at least two major types of hormones, the ecdysteroids or molting hormones (MH) (Lachaise, 1989), and MIH, the latter being released from the X-organ/sinus gland complexes in the eyestalks (ES) (Passano, 1960). A pair of Y-organs located in the anterior region of the cephalothorax (Passano, 1960) is the primary, if not the sole, physiological source of ecdysone (**1**; see Fig. 6, p. 58), a precursor of 20-hydroxyecdysone (**2**; Fig. 6) (Lachaise and Feyereisen, 1976; Jegla *et al.*, 1983), the circulating major molting hormone (Chang *et al.*, 1976; Skinner, 1985) in various target tissues peripheral to the Y-organ. A similarity in the physiological role between Y-organs of certain crustaceans and prothoracic glands of insects has been noted (Gabe, 1953). A progressive increase in the titer of 20-hydroxyecdysone in hemolymph initiates the events generically referred to as molting. More recent studies indicate that juvenile hormone-III (**3**; Fig. 6) may also be involved in developmental events in crustaceans (Borst *et al.*, 1987; Laufer *et al.*, 1987).

The injection of ES-extract (ES-X) delays the induced molting of eyestalk-ablated crustaceans. With advances in Y-organ culture and the development of ecdysteroid radioimmunoassay in the mid-1970s (Chang and O'Connor, 1977), it has been demonstrated that MIH extracted from ES decreases the basal rate of ecdysone secretion, although the rates of ecdysone synthesis or secretion by the Y-organ vary. The actual molting hormone titer is the result of biosynthesis and catabolism, and therefore the critical period of the molting is controlled not only by MIH but by multiple factors (Kleinholz, 1985; Skinner, 1985; Borst *et al.*, 1987; Laufer *et al.*, 1987), even though the negative regulation of ecdysteroidogenesis seems to have a great impact on the hormone titer in crustaceans. Although our knowledge on the subject is still limited, analogy to Insecta (S. L. Smith, 1985) suggests that ecdysteroidogenesis in Crustacea involves a putative cerebral peptide or Y-organotropic hormone and hemolymph proteins (a sterol carrier protein) (Watson and Spaziani, 1985; Spaziani, 1988). In addition, the ecdysteroidogenesis is probably mediated by the actions of protein kinase C (Mattson and Spaziani, 1987) and cytochrome P450 biocatalysts. In Insecta, cytochrome P450s are known to be involved in hydroxylations at C-2, C-22, and C-25 in the biosynthesis of ecdysone from cholesterol (**4**; Fig. 6) and also in the C-20 hydroxylation of ecdysone (Kappler *et al.*, 1988). Each enzyme involved in the chain of events leading to molt must perform a well-defined precise role. The ecdysone titers in crustaceans are dependent on enzymatic activities due to season, locality, sex, maturity, etc., the interrelations of which are not clear; it is this unreliability of assay crustaceans that has been a major obstacle in the progress in MIH characterization.

Despite many efforts, the chemical nature of MIH from ES has until recently remained obscure. Considerable evidence indicates (Webster and Keller, 1989) that MIH-active heat-stable peptides, not entirely species-specific and with molecular weights ranging from 4000 to 8000 Da, are released by secretory neurons in crustacean ESs. Using molt cycle intervals and assays of circulating ecdysteroid titers to monitor eluted HPLC fractions, Chang *et al.* (1990) have recently obtained a "doublet" reverse-phase HPLC peak from the sinus glands of the lobster *Homarus americanus* (Hoa); this has been characterized as a peptide with 71 residues (mol. wt. 8483) with molt-inhibiting and hyperglycemic (CHH) activities. This Hoa-MIH was shown to be different from Hoa-CHH. The presence of two different compounds with closely related amino acid sequences is consistent with the presence in the shore crab, *Carcinus maenas*, of both Cam-MIH and Cam-CHH (Webster and Keller, 1986), and the presence in the Mexican crayfish, *Procambarus bouvieri*, of Prb-MIH and Prb-CHH (Huberman and Aguilar, 1989). Although the molecular mechanisms of the inhibition are unknown, it is suggested that the molting regulation could be associated with hyperglycemia an also with 5-hydroxytryptamine in stress-mediated inhibition (Mattson and Spaziani, 1986). Further work is necessary to understand the control of molting, which, like the proposal of Newcom *et al.* (1985), cannot be based on a "one-hormone-target-response" concept. In the following we describe the low-molecular-weight compound(s) that exert molt-inhibiting activity *in vitro* (Naya *et al.*, 1988, 1989; Ohnishi *et al.*, 1991; Sonobe *et al.*, 1991) as well as *in vivo* (Naya *et al.*, 1989). The active compound *in vivo* was found to be identical to "an indole alkylamine" reported by Soyez and Kleinholz (1977), the factor inhibiting molt in *Orchestia*, isolated from the ES of *Pandalus*.

2.2. Materials and EBI Assay

We first attempted to distinguish direct effects on Y-organs, i.e., ecdysone biosynthesis inhibition (EBI) by eyestalk extracts (ES-X), from peripheral effects (Freeman, 1980) of MIH activity, i.e., morphologically defined endpoints (ecdysis). The EBI was assayed by the method of conventional homogenized Y-organ culture (Naya *et al.*, 1988), in which ecdysone was determined by use of HPLC in the presence or absence (control) of ES-X. This protocol for ecdysteroidogenesis and/or bioassay, illustrated in Fig. 1, was a modification of the Y-organ culture established by Chang and O'Connor (1977).

After immobilization of crustaceans by chilling on ice, their ESs were excised and frozen with dry-ice and lyophilized for storage as the MIH pool. The destalked crabs were killed by acute freezing with liquid nitrogen and then stored below $-80°C$ until used as the Y-organ source. Although ecdysteroidogenesis potency undergoes slow deterioration during storage, about 90% of the animals thus treated were found to be usable for about 3 months as the source of Y-organs. In the case of crayfish, *Procambarus clarkii*, after excision of the bilateral ESs (day 0) and a simultaneous forced autotomy of the chelea, the activated Y-organs (Naya *et al.*, 1989; Naya and Sonobe, 1990; Sonobe *et al.*, 1991) were dissected out at the appropriate time (days 4–6, see below) and submitted to EBI assay.

As it is known that the level of ecdysone secretion varies widely between

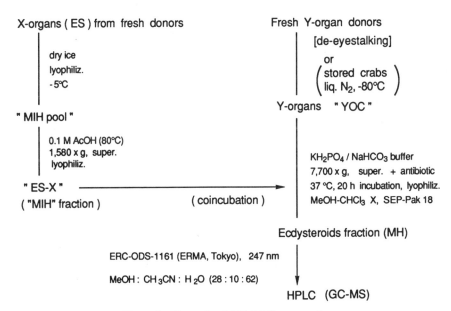

Figure 1. Conventional EBI (MIH) assay system.

species and also in the same species between the individual stages of the molting cycle (Kleinholz, 1985; Skinner, 1985), this procedure using homogenates of Y-organs with their adherent tissues (Y-organ complex: YOC) rather than individual Y-organs was effective for leveling the variations among animals; this allowed a reduction in the number of control experiments. The inhibitory activity manifested itself even when the ES-X and YOC were taken from different families of crabs or crayfish. The EBI(s) thus appears to be species-nonspecific between ES and YOC donors over a wide range; the cross-reactivities were found over examined crabs (Naya *et al.*, 1988), *Callinectes sapidus*, *Portunus trituberculatus*, *Charybdis japonica*, *Geothelphusa dehaai*, and the crayfish (Naya *et al.*, 1989) *Procambarus clarkii*, collected off different coasts of the United States and Japan. Based on immunological and spectroscopic methods (Sonobe *et al.*, 1991), the Y-organs of the crayfish, *P. clarkii*, have since been anatomically and morphologically characterized (Naya and Sonobe, 1990) and confirmed as the ecdysial gland (see Fig. 4a below) having the capacity of ecdysteroidogenesis *in vitro*. This freshwater crayfish was easy to rear in the laboratory and led us to a clearer understanding in the molting regulation in crustaceans.

2.3. Isolation of EBI

Following the bioassay, 700 μg of EBI was isolated from 16 g of ES-X of *C. sapidus* (Fig. 2), and determined to be a metabolite of L-tryptophan (**5**; Fig. 6), i.e., 3-hydroxy-L-kynurenine (**6**; Fig. 6) (L-3OHK), by UV, circular dichroic spectrum (CD), EI-MS, NMR, electrophoresis, and amino acid analysis: EI-MS m/z 322 (as diacetyl methyl ester); UV (in 0.05 M HCl) 224 nm (ϵ 18400), 266 nm

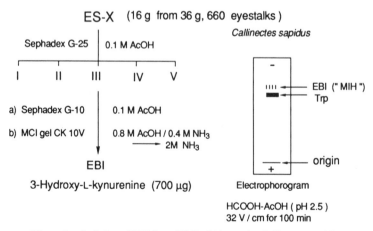

Figure 2. Isolation of EBI from ES-X of blue crab, *Callinectes sapidus*.

(ϵ 7600), 370 nm (ϵ 4000); CD (in 0.05 M HCl) 321 nm $\Delta\epsilon$ +1.0), 273 nm ($\Delta\epsilon$ +0.5), 370 ($\Delta\epsilon$ −0.2). The compound **6** (Fig. 6) gave an electrophorogram closely resembling that reported as an "indole alkylamine" (Soyez and Kleinholz, 1977); judging from the reported chemical properties, the latter is also likely to be L-3OHK (**6**; Fig. 6).

The assay of authentic L-3OHK showed the EBI activity to be comparable to that of isolated natural compound. On the other hand, racemic DL-3OHK [a gift from S. Senoh (Kotake *et al.*, 1951)] was less potent. These experiments suggested that a metabolite of L-3OHK might be responsible for the inhibition. Indeed, a crude enzyme preparation obtained from the YOC indicated this assumption to be indeed the case. An active search for xanthurenic acid (**7**; Fig. 6) (XA), a metabolite of 3-OHK, then led to its detection in ES-X. Namely, ES-X was dissolved in aqueous ammonium acetate (0.17 M, pH 5, a solvent system found to be efficient for XA extraction) and centrifuged (1580 × g, 15 min). The supernatant was placed on an Asahipak GS 320 column (Asahi Chem. Ind.), and L-3OHK was eluted with 0.17 M ammonium acetate and XA with 10% acetonitrile/0.17 M ammonium acetate.

XA inhibited ecdysteroidogenesis *in vitro* more potently than did L-3OHK. The ED_{50} of crude ES-X (*C. sapidus*) was about 1 mg. In this amount of ES-X, equivalent to 1/25 of ES, 136 ng of 3-OHK and 78 ng of XA were detected. Even when assuming a 100% conversion of 3-OHK into XA, this combined amount did not account for the full potency of ES-X; the detailed dose–response study is still under investigation. The difficulty in securing definite ED_{50} values is due to the seasonally variable inhibitory effect of the ecdysteroidogenetic potency of Y-organs (Naya *et al.*, 1989; Sonobe *et al.*, 1991).

2.4. Enzymatic Transformation of 3-OHK into XA

The body fluid (Naya *et al.*, 1988, 1989) extracted from crustaceans with 0.17 M KH_2PO_4/$NaHCO_3$ buffer (pH 7.1) was treated with a large excess of casein (to inhibit proteolysis) and further with saturated aqueous ammonium sulfate. After centrifugation, the residue was dialyzed for 24 hr against water. The nondialyzed fraction was passed through Sephadex G-200 (Pharmacia) with the buffer and used as a crude enzyme preparation. This crude enzyme preparation was incubated with L-3OHK in the buffer (0.17 M KH_2PO_4/$NaHCO_3$, pH 7.1) at 37°C for 3 hr. Upon termination of the incubation, the mixture was lyophilized and dissolved in 0.1 M ammonium hydroxide, then dialyzed overnight against 0.05 M ammonium hydroxide. The dialyzable solution was lyophilized, and the quantity of XA in the residue was determined by HPLC under the conditions described above. Transformation of L-3OHK into XA was accomplished in 61% yield. On the other hand, the transformation of D-3OHK (obtained by a chiral column separation from the DL compound) and DL-3OHK resulted in 1 and 9% yields, respectively; the transformation is thus L-enantio-specific and is inhibited by the D-enantiomer. The crude enzyme preparations obtained from their ES and YOC were also capable of transforming 3-OHK into XA.

2.5. Physiological Change Induced by Eyestalk Ablation

An equal volume of hemolymph was drawn from the abdomen of each of 15 crayfish, *P. clarkii* (bilaterally destalked and forced to autotomize at day 0), at 24 hr intervals beginning on day 1. The hemolymph collected on each day was combined and used for the crude enzyme preparation (see above). Aliquots of the supernatant containing transaminase were incubated at 37°C for 1 hr with gentle stirring in the presence or absence (control) of exogenous L-3OHK. Upon termination of the incubation, each aliquot was worked up according to the procedure outlined above for XA analysis on HPLC. The capacity of hemolymph to convert L-3OHK into XA peaked after a period of several days after de-eyestalking, depending upon the size of the crayfish. The enzyme that converts L-3OHK into XA can be regarded as representing the apparent 3-OHKase.

Although it is not known whether exogenous L-3OHK is metabolized along the same lines as the endogenous substrate, the peak in the apparent 3-OHKase activity was followed by a peak in the titer of 20-hydroxyecdysone. A decrease in the enzymatic activity or in the XA titer presumably signals the timing for a sharp increase in the titer of ecdysone, and subsequently of 20-hydroxyecdysone, the active molting hormone in the hemolymph. The transient increase in the apparent 3-OHKase activity induced by eyestalk ablation was not accompanied by a change in the total amount of the endogenous aminotransferase substrates. After removal of proteinaceous compounds in the hemolymph, the total concentration of aminotransferase substrates was determined by ninhydrin colorimetry and amino acid analysis. The apparent 3-OHKase in the crude enzyme preparation was somewhat inhibited by a general aminotransferase substrate, L-glutamic acid, although inhibition by other amino acids was not investigated.

Interestingly, 5-hydroxytryptophan (10^{-7} M) provided excitatory input to MIH-containing (producing) neurosecretory cells (Mattson and Spaziani, 1986). This is possibly caused by a feedback regulation among the tryptophan metabolites. An injection of 20-hydroxyecdysone given to the crayfish *Procambarus simulans* (Lowe et al., 1968) 24 hr after eyestalk removal did not induce molting as effectively as an injection given 48 or 72 hr after operation. Twenty-four hours after eyestalk removal, the Y-organs of crabs *Cancer antennarius* (Spaziani, 1988) displayed a 30-fold higher uptake of ^{14}C-cholesterol than did the Y-organs of unablated premolts. These reports, together with our results, indicate that de-eyestalking not only induces acute and long-term abnormalities in several physiological parameters, but also directly affects the biochemical processes associated with late premolt stages.

2.6. In Vivo Effects of L-3OHK and XA on Crayfish Molting

Eyestalk ablation and forced autotomy of chelae induced precocious ecdysis in *P. clarkii* (Naya *et al.*, 1989) which received daily injections of crustacean saline (100 μl/animal, control); eyed animals that received the same treatment did not molt for 22 days. Each of ten animals in another group received daily injections of L-3OHK solution in saline (30 ng/100 μl) at 24 hr intervals after the operation; this delayed the onset of the first molt by 2 days, and lengthened the interval between the first and second molts by 3 days (Fig. 3). On the other hand, daily injection of XA solution in sale (10 and 30 ng/100 μl) induced no significant effects regarding the molt-inhibiting properties before the animals died. These results indicated that the circulating L-3OHK was accumulated in the Y-organs through a putative L-3OHK receptor, and then transformed into XA to inhibit ecdysteroidogenesis. Indeed, the presence of the chemoreception system for L-3OHK in the Y-organ (*in vitro*) was shown by the specific incorporation of [3H]-L-3OHK as followed by radioautography (unpublished data; see Fig. 4b).

2.7. Hormonal Characteristics of EBI; Titer Changes of MH and EBI

For each collection date, six crabs of *C. japonica* (Naya *et al.*, 1989) were randomly selected, and the titers of L-3OHK, XA, ecdysone, and 20-hydroxy-ecdysone were determined by HPLC analysis. The quantities of each compound in

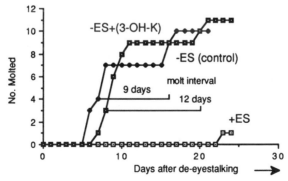

Figure 3. Effect of L-3OHK injection on crayfish, *Procambarus clarkii*. −ES: de-eyestalked crayfish; +ES: eyed crayfish.

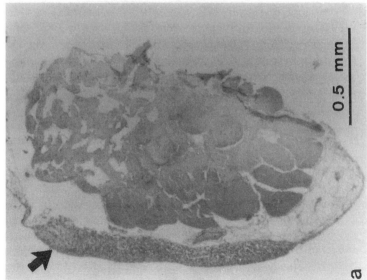

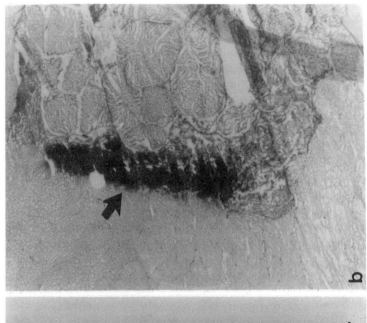

Figure 4. (a) Y-organ (vertical section) of crayfish, *P. clarkii* (photograph taken by H. Sonobe); (b) specific incorporation of [³H]-L-3OHK in the Y-organ observed by radioautography (photograph by Institute of Whole Body Metabolism).

the ES, YOC, and hemolymph were recorded for 12 months. Staggered correlations between the XA and ecdysteroid titer were found in the YOC (Fig. 5a) and hemolymph (Fig. 5b). When the YOC was incubated in November/December, the period of high XA titer (Fig. 5a), the ecdysone titer increased 100-fold as the inhibitory effect of XA was overridden. In February, the period following the high titer of XA, the ecdysone level rose to ca. 30ng/YOC. Incubation of the YOC in this period reduced the ecdysone level to 3 ng/YOC. A high concentration of 3-OHK prior to incubation (ca. 1000 ng/YOC and 150 ng/ml hemolymph) presumably resulted in the genesis of XA, the agent responsible for the inhibition of ecdysteroidogenesis. Thus, incubation accelerated the metabolism of ecdysone, and XA generated from L-3OHK by incubation suppressed ecdysteroidogenesis. These experimental data also suggested the presence of a putative L-3OHK receptor in the Y-organ (see above). The apparent 3-OHKase activity, estimated as the ratio [XA]/([L-3OHK] + [XA]), was an indicator of seasonal physiological change in the animals; *in vivo* aminotransferase capacity varied significantly

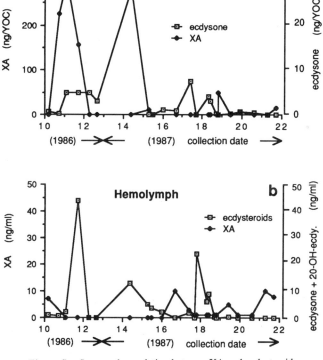

Figure 5. Staggered correlation between XA and ecdysteroids.

during the year. The apparent 3-OHKase capacity was not always parallel among the ES, YOC, and hemolymph, as a result of localization of the enzymatic capacity.

It is interesting to note that the annual profiles of the hemolymph lipoproteins (high-density type) in the crab *Cancer antennarius* (Spaziani, 1988) show concentration variation, the peak in November being on the average fourfold higher than the lows of early summer. Cholesterol is bound to the hemolymph lipoproteins (Spaziani, 1988) to be carried and taken up by Y-organ cells. The results of this study give a clear rationalization for the dramatic increase in ecdysone titer upon YOC incubation (see above); the accelerated activity of the multienzyme complex in the YOC in November was synchronous with the increase in the concentration of carried substrate. The concentration of lipoprotein-bound cholesterol may be the rate-determining factor in ecdysteroidogenesis.

2.8. Inhibition Mechanism

Our approach to understanding the inhibitory mechanism incorporates the working hypothesis (Naya *et al.*, 1988, 1989) that the ecdysteroidogenic cascade is mediated by the biocatalyst cytochrome P450s (Kappler *et al.*, 1988). Experiments (Ohnishi *et al.*, 1991) have been performed to prove this hypothesis; the ecdysteroidogenesis *in vitro* of the excised prothoracic glands (PGs) of silk worms, *Bombyx mori*, were shown to be inhibited by high XA concentrations in a reversible manner, indicating ligand exchange and/or hydrophobic interactions with cytochrome P450s (Testa and Jenner, 1981). The inhibitory effects were increased in the presence of the detergent Tween 80, but were independent of changes in the intracellular calcium ion levels caused by addition of the calcium ionophore A23187. At lower XA concentrations, ecdysteroidogenesis was stimulated, but in contrast, the effects of L-3OHK were not clear-cut, possibly reflecting the activity of the 3-OHKase.

Incubation of a model enzyme, cytochrome $P450_{PB}$ (induced by phenobarbital in rat), with XA resulted in a difference UV spectrum exhibiting a shift from 417 to 448 nm (Ohnishi *et al.*, 1991). An increase in absorbance at 448 nm, the Soret peak (γ) arising from the formation of the cytochrome $P450_{PB}$ ferrous-XA forms, was observed in a dose-dependent manner. Although changes in the α-band (569 nm) and β band (535 nm) were obscure, the peak at 448 nm indicated that the hydroxyl group most likely constituted the sixth ligand of cytochrome $P450_{PB}$. An analogue, kynurenic acid, which bears no hydroxyl group at C-8, showed no evidence of binding to cytochrome $P450_{PB}$. No spectroscopic evidence of interaction between L-3OHK and the biomolecules was found either. It was thus suggested that the 8-OH of XA is involved in ligand exchange at the iron porphyrin of cytochrome $P450_{PB}$. These results are in agreement with those from ESR (Naya *et al.*, 1988), which indicated a typical substrate binding between cytochrome c (as a model compound) and XA; $Fe^{3+}-O^-$, $g = 2.30, 2.12, 1.88, 77°K$.

2.9. Conclusion

XA inhibited ecdysone biosynthesis, but did not interfere in the C-20 hydroxylation of ecdysone; compounds involved in crustacean molting are illustrated in Fig. 6. Experiments with insect tissues *in vitro* (T. Fujita, Kyoto University, personal communication) and ecdysone 20-monooxygenase from insect sources (S. L. Smith, Bowling Green University, personal communication) demonstrated that neither L-3OHK nor XA inhibited the 20-hydroxylation of ecdysone, suggesting diversity among the oxygenases involved in ecdysteroidogenesis. When eyestalk-ablated lobsters received two injections of L-3OHK or XA (4 μg/animal), no significant effect was observed (E. S. Chang, Bodega Marine Laboratory, University of California, personal communication). In this study, the "injection effect" was defined as the log of the preinjection 20-hydroxyecdysone concentration divided by the log of the postinjection value, where the 20-hydroxyecdysone concentration before and after injection was determined by radioimmunoassay. On the other hand, we observed that daily injection of L-3OHK into eyestalk-ablated crayfish led to a delay of 2–3 days in the significant increase of the 20-hydroxyecdysone titer and subsequent shedding of the exoskeleton. The average weight of gastrolith pairs from ten animals on the day of ecdysis was increased by L-3OHK injection (470 ± 130 mg) as compared to saline injection (338 ± 114 mg), this increase being a result of the lengthening of the proecdysial period. The injection effect, which was related to the quick metabolic turnover of L-3OHK and L-3OHKase capacity, varied as a function of injection timing and animal size.

In conclusion, our results indicate that XA functions as an inhibitory reagent in ecdysteroidogenesis (*in vitro*) in both excised molting glands from the silkworm (10^{-2} M) and the crayfish ($<10^{-6}$ M). This suggests a similar molecular mechanism for the inhibition in the silkworm and crayfish, but through different modes of transportation into their respective molting glands. In crustaceans, endogenous L-3OHK was accumulated in the Y-organ through the chemoreception system (a countertransportation) and transformed into XA to suppress the biocatalysts. In contrast, in insects, XA was transported into PG in a passive manner due to the absence of a putative receptor, thus suggesting no *in vivo* function in the insect. Expression of the biochemical function is therefore dependent on the membrane structures of the molting glands. The roles of neuropeptides (Webster and Keller, 1986; Huberman and Aguilar, 1989; Chang *et al.*, 1990) having MIH effects and that of JH (juvenile hormone) (Borst *et al.*, 1987; Laufer *et al.*, 1987) in crustaceans remain to be elucidated.

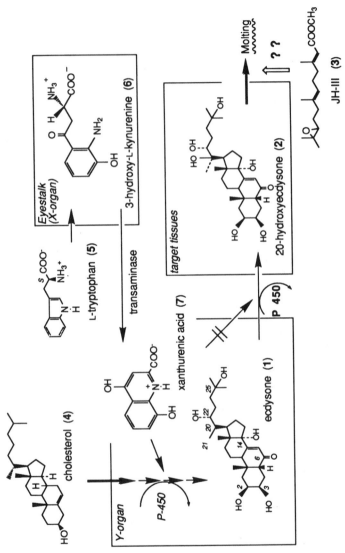

Figure 6. Compounds involved in crustacean molting.

3. SHARK REPELLENTS IN THE DEFENSE SECRETION OF PARDACHIRID SOLES

It is well documented that the shark detects its prey from hundreds of meters away with its olfactory and acoustic senses. This keen sensory system is perhaps essential for this carnivore, which sustains itself by locating large living objects in seawater through which the light reaches only a limited distance. In addition to these and the visual sense, which is also highly evolved (Gruber, 1977), elasmobranch fishes possess ahead of their nostrils an electroreceptive complex called Lorenzini's ampullae, which is used in the location of living prey at close distances (Kalmijin, 1966). This electroreceptive sense occasionally leads them to erroneous attacks on oceanic cables, giving another motivation for shark repellent research. In contrast to such an elaborate contrivance for prey detection, little is known on how they sense danger. It may even be that no such thing exists, because they have no predator except for humans, who were included in their ecosystem only in the very recent phase of their evolution. Nevertheless, there has been much effort on finding ways to evade shark attacks by evoking aversive behavior in them.

The first organized effort to develop a chemical shark repellant was made by the U.S. Navy during World War II, when the morale of their servicemen was a serious problem in the naval battle in the western Pacific. Their imagined fear of a potential shark attack upon being thrown into the sea may have far exceeded their fear of Japanese troops (Zahuranec and Baldridge, 1983). The effort resulted in the invention of Shark Chaser, a packet containing copper acetate as the "active" ingredient and nigrosine dye to generate a smoke screen. Postwar reevaluation, however, determined that copper acetate is not an effective shark repellent, and the U.S. Navy later terminated the delivery of Shark Chaser.

A small righteye sole, *Pardachirus marmoratus*, has been regarded among the inhabitants of Israel as a shark repellent fish, as indicated by its local name, Moses sole, after Moses, who led the Jewish people safely across the Red Sea from Egypt back to Israel in the Old Testament. The fish was introduced to the rest of the world by Clark (1974) along with experimental evidence. She stated in her subsequent account (Clark and George, 1979) that soles of this genus are characterized by their sacciform secretory organs along the base of dorsal and anal fin spines (apparent left and right sides, respectively) at the expense of a volume of swimming muscle. Upon external disturbance, they emit a detergentlike ichthyotoxic secretion. Flatfishes in general are known indeed as favorite preys of many sharks and rays, which can detect them with electroreception despite their otherwise successful camouflage under a layer of sand. It is thus not unlikely that the chemical defense of pardachirid soles was evolved particularly against sharks and rays.

Chemical investigation on the shark repellent secretion of Moses sole was undertaken by researchers in Jerusalem following Clark's experiment, resulting in the isolation and partial characterization of an ichthyotoxic "protein" therefrom, named pardaxin (Primor *et al.*, 1978). These results led to an international symposium on the theme of "shark repellents from the sea," under the auspices of the American Association for the Advancement of Science, at Toronto in 1981 (Zahuranec, 1983). The main conceptual thrust was that in searching for shark repellents one should look for clues from the behavior of denizens of the marine habitat which out of necessity may have acquired something to avert sharks.

The Moses sole, *P. marmoratus*, is replaced around India by another species, *P. pavoninus*, the latter reaching the southern coast of Japan. Chemical investigation of the corresponding secretion from this species collected on the Ryukyu Archipelago revealed the existence of a series of lipophilic ichthyotoxins in addition to pardaxin (Fig. 7). They were found to be monoglycosidic cholestanoids, constituting more than 80% in weight of the total lipidic contents in the

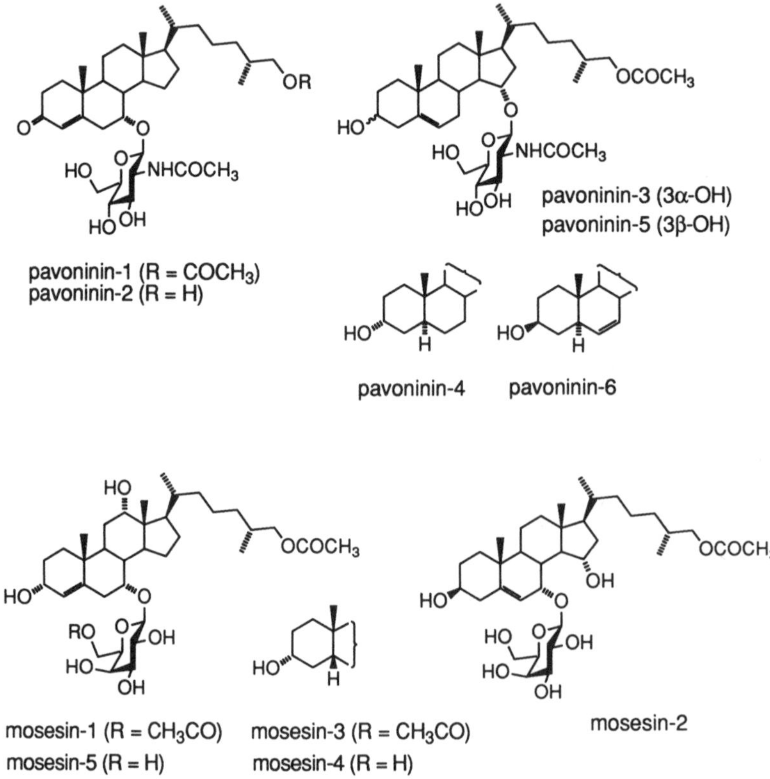

Figure 7. Pavoninins and mosesins.

secretion, and named pavoninins (Tachibana *et al.*, 1985). Ichthyotoxicity among respective pavoninins did not differ much, all being lethal to Japanese killifish at 5–10 ppm. In the natural abundance mixture, they also elicited positive aversive action from lemon sharks when dosed as an aqueous ethanol solution in the oral cavity, though at much higher concentrations.

The discovery of pavoninins led to chemical reinvestigation of the Moses sole secretion, resulting in the isolation of a similar but different series of mono-glycosidic cholestanoids, named mosesins (Tachibana and Gruber, 1988). Al-though *N*-acetylglucosamine in pavoninins is replaced by galactose or its 6-acetate in mosesins, they correspond in contents as well as in ichthyotoxicity. Mosesins are more shark repellent by a small margin. A notable feature in the molecular profile of the steroid glycosides is the amphiphilic arrangement of atoms flanking the steroidal plane (Fig. 8). This amphiphilic dipole is somewhat tilted around the traverse axis of the steroid skeleton in the 15α-glycoside series of pavoninins.

It has been difficult to determine the relative contributions to shark repellency or ichthyotoxicity of these secretions between pavoninins or mosesins and par-daxin, because the ichthyotoxicity of both lyophilized crude secretions and the acetone precipitates prepared therefrom is liable to quick diminution in aqueous media even during the bioassay. The ichthyotoxicity of the acetone precipitates, however, was stabilized by gel filtration through Sephadex G-150 at the expense of a partial loss of pardaxin to a known extent, indicating that proteolytic enzyme(s) coexist in the secretion. Purification of pardaxin from *P. pavoninus* proceeded from this ichthyotoxic gel filtrate, leading to the isolation of three ichthyotoxic compo-nents, named pardaxins P-1 to P-3 in decreasing order in contents, as the chemical entity of pardaxin in the Pacific sole. Conventional amino acid analysis and sequential analysis with Edman degradation on each component, together with an enzymolytic analysis of carboxyl termini, determined their complete sequences as

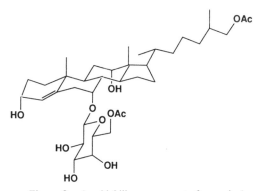

Figure 8. Amphiphilic arrangement of mosesin-1.

		5	10
Pardaxin	P-1	Gly-Phe-Phe- Ala -Leu- Ile -Pro-Lys- Ile - Ile -Ser-	
Pardaxin	P-2		
Pardaxin	P-3	-Phe-	
Pardaxin	M-1		
Lazarovici *et al.*			

		15	20
Pardaxin	P-1	-Ser- Pro -Leu-Phe- Lys -Thr-Leu-Leu- Ser -Ala- Val -	
Pardaxin	P-2	- Ile -	
Pardaxin	P-3		
Pardaxin	M-1		
Lazarovici *et al.*			

		25	30
Pardaxin	P-1	-Gly- Ser - Ala -Leu- Ser -Ser-Ser- Gly -Glu -Gln-Glu	
Pardaxin	P-2		-Gly -
Pardaxin	P-3		
Pardaxin	M-1		-Asp-
Lazarovici *et al.*			-Gly -

Figure 9.   Amino acid sequences of pardaxins.

in Fig. 9 (Thompson *et al.*, 1986). Pardaxins P-1 and P-2 have subsequently been chemically synthesized by a solid-phase method. Chromatographic and spectroscopic comparisons between natural and synthetic samples confirmed that the peptide sequences are conjugated with no other chemical component (Thompson *et al.*, 1987). Following the same isolation scheme, the major pardaxin of the Moses sole secretion was also isolated, tentatively named pardaxin M-1, and sequenced to be different from pardaxin P-1, the major *P. pavoninus* pardaxin, by a single amino acid residue (Fig. 9) (Thompson *et al.*, 1988). Lazarovici *et al.* (1990) reported another sequence, differing from ours at the same residue (Fig. 9), for what they call pardaxin, from their independent investigation. We did not recognize ichthyotoxicity in the corresponding fractions of higher pI from the Moses sole secretion that we obtained, unlike the case for pardaxin P-2 in the *P. pavoninus* secretion.

The ichthyotoxicity of each of the purified pardaxins is comparable to that of the pavoninins and mosesins when compared in equal molar concentrations. The shark repellency does not seem to differ much either, based on an assay result for the stabilized crude sample and the pardaxin content thereof, though direct comparison on a single bioassay series has yet to be made. Although we do not know the original quantity of pardaxin in the crude secretions, it is estimated roughly to be similar to that of the pavoninins or mosesins. Therefore, both factors are inferred to participate with similar importance in the shark repellency of the secretions.

Pardaxins largely take a random conformation in distilled water, as indicated by the CD spectra, but they become more ordered in the increased presence of either organic solvents or inorganic salts, and also with an increase in their own concentration. This is maximized in the presence of detergent micelle or phospholipid vesicles, the CD spectra giving an estimated 40% α-helix and 20% β-sheet (Thompson *et al.*, 1987). Conventional calculations with statistical parameters (Chou and Fasman, 1978) predict the helix to be located at the middle of the sequence. This helix would then be faultlessly made amphiphilic along its equatorial direction (Fig. 10) (Thompson *et al.*, 1987), explaining the strong surfactant nature of the molecule as well as of the secretion, as in the case of mosesins shown above (see Fig. 8). The location as well as rigidity of this helix were determined by the observation of adequate interresidual NOE cross-peaks in two-dimensional ¹H-NMR measurements on pardaxin P-2 in aqueous trifluoro-ethanol, where the CD spectrum indicated the corresponding conformation (Thompson *et al.*, 1987), together with molecular dynamics calculations using distance constraints based on the NOE data incorporated in the XPLOR program (Zagorski *et al.*, 1991). This computation also indicated that the amino-terminal segment takes the form of a somewhat irregular and mobile helix, separated by a bend of ca. 80° around proline-13. Lazarovici *et al.* (1990) reported a similar architecture of pardaxin derived from energy minimization using the Amphi and Delphi programs and the subsequent use of computer graphics guided by inter-

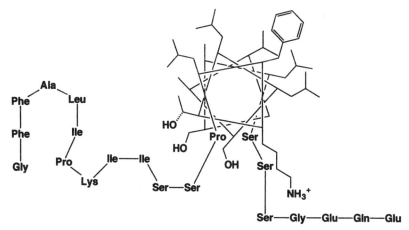

Figure 10. Axial projection of the α-helix estimated in pardaxin P-1. The other pardaxins are similarly estimated.

molecular ridges-into-grooves packing at the hydrophobic surface. The oligomeric behavior of pardaxins in buffer solutions had in fact been exhibited through gel filtration and ultracentrifugation (Primor *et al.*, 1978; Thompson *et al.*, 1986), which led to the erroneous categorization of pardaxin as a protein. The apparently contradictory formation of the helix with stronger ionic strength, as indicated by the CD changes, is also explained by the oligomeric formation, segregating its large hydrophobic surface from water molecules as do organic additives. A similar behavior, as well as the corresponding amphiphilic structure, are known for the peptidic bee venom melittin in aqueous solutions (Brown *et al.*, 1980) and in a crystal prepared therefrom (Terwillinger and Eisenberg, 1982).

Various forms of bioactivity have been reported on pardaxin or the crude secretion of Moses sole in addition to shark repellency and ichthyotoxicity, such as hemolysis, smooth muscle contraction, and Na^+/K^+-ATPase inhibition [summarized with original references in Lazarovici *et al.* (1986)]. All, including shark repellency, may well be explained by a disturbance of the lipid bilayer structure of protoplasmic membranes and/or consequent depolarization of transmembranal electric potential. Pardaxins indeed release entrapped carboxyfluorescein from artificial liposomes at an effective concentration of 10^{-8} M (Zlotkin and Barenholz, 1983; Tachibana *et al.*, 1989). Further, on the basis that electrical conductance (Moran *et al.*, 1984) and permeability to inorganic cations (Lazarovici *et al.*, 1990) across a planar phospholipid bilayer are increased by lower concentrations of pardaxin, though on a longer time scale, it has been speculated that several molecules of pardaxin form an ion channel across the lipid bilayer at concentrations down to 10^{-10} M.

Stabilized gel filtrates from *P. pavoninus* secretion contained des-(Gly^1–Ala^4)-pardaxins P-1 and P-2, that is, lacking four amino-terminal residues, in varied amounts over different batches as a possible series of autolyzed products from respective pardaxins (Thompson *et al.*, 1986). Interestingly, they were found to be devoid of ichthyotoxicity despite their retention of strong surfactancy as well as intact existence of the amphiphilic α-helix as determined by the CD spectra. In the more sensitive dye leakage assay using artificial phospholipid liposomes, however, the des peptides exhibited some activity, though markedly decreased, the effective concentration being 30 times higher than that of pardaxins (Tachibana *et al.*, 1989). Other des pardaxins P-1, where each of the segments Pro^7–Ser^{12}, Leu^{13}–Gly^{23}, and Ser^{24}–Glu^{33} were deleted, were chemically synthesized and their disruptive activity to the liposome was measured for each. While deletion of Pro^7–Ser^{12} or the helical segment Leu^{13}–Gly^{23} diminished the activity by four times or, not surprisingly, nullified it, deletion of the hydrophilic carboxyl terminus affected it little. This deletion instead caused a marked decrease in aqueous solubility of the peptide. From the structure–activity relationships thus obtained, it has been tentatively concluded that the middle helical segment is required for the permeabilizing action, while the hydrophobic amino terminus assists by binding the molecule to the membrane. The hydrophilic carboxyl terminus is insignificant once the molecule binds to the membrane surface, but it perhaps facilitates transportation of the oligomer through seawater to provide the shark repellency of the secretion. Since these results left the role of the segment Pro[7]–Ser[12] somewhat vague, other analogues were synthesized in the same manner with various modifications in the sequence of this region. The activity was found to be fairly sensitive to modification, rather than mere deletion, of this segment (Barrow *et al.*, 1992). It seems that the geochemical arrangement between the amino terminus and the helix is of some importance.

The steroidal repellents pavoninins and mosesins permeabilize the phospholipid liposome only at concentrations higher by two orders of magnitude than those of pardaxins (unpublished results), in contrast to the near equipotency between the two chemical classes with regard to ichthyotoxicity and hemolysis (Tachibana *et al.*, 1985; Thompson *et al.*, 1986). This can be examined by the speculation, yet to be experimentally tested, that the steroid glycosides perturb biological membranes by interacting with another component thereof in addition to phospholipid, most likely cholesterol in the case of animal cells.

Putting together all the results above, one can propose the following hypothetical scheme for the cooperative action responsible for the shark repellency of the defensive secretion. The steroidal components reach the target membrane as a comicelle with pardaxins or similar peptidic carrier, and associate with cholesterol molecules there to make the bilayer structure brittle to the wedging intervention of the helical segment of pardaxin, which is kept proximate to the membrane surface in appropriate orientation by the anchoring action of the hydrophobic amino terminus. Once transmembranal depolarization takes place at the suitable cells connected to the sensory nervous system of the shark, it may well arouse an unpleasant or painful sensation. If pardaxins form channels at the less exposed protoplasmic membrane, this may intensify learning of the individual shark with physiological discomfort as a delayed effect. It has not been determined yet where the shark senses the repellents. Primor (1985) experimentally demonstrated that the target of pardaxin is located between the pharynx and the gills, by devising a barrier in the oral cavity of a dogfish shark.

Based on the strong surfactant property of the Moses sole secretion, representative commercial detergents were tested for their shark repellent activity with sodium dodecyl sulfate (SDS) as the most potent among those tested (Gruber, 1981; Zlotkin and Gruber, 1984), marginally surpassing the Moses sole secretion and the respective active components therein and being far more economical. It still requires a concentration of some hundred parts per million in the shark's oral cavity to exhibit repellency, and thus is of little practical utility in controlled constant release. The Moses sole, however, manages to deliver the needed concentration by patiently waiting for the last moment when it is already in the wide open jaws of an attacking shark (Clark, 1974).

4. TUNICHROMES, THE BLOOD PIGMENTS OF TUNICATES (SEA SQUIRTS)

4.1. General

The tunicates (Ascidiacea or ascidians), or sea squirts, are common filter-feeding animals inhabiting all oceans, at all depths. They selectively accumulate metals such as vanadium (in species such as *Ascidia nigra*) or iron (*Molgula manhattensis*) in their blood cells, occasionally up to concentrations of 1 M. However, the nature of this metal concentration was an enigma for 80 years, because of the difficulty of isolating the pigments supposedly responsible for the accumulation, the "tunichromes" (TC), because of their sensitivity to oxidation. The use of strictly anaerobic conditions from the initial collection from blood through the end of the extraction process coupled with the use of centrifugal partition chromatography (CPC), etc., finally led to the characterization of several tunichromes, An-1, -2, -3 and Mm-1, -2 (Bruening *et al.*, 1985; Oltz *et al.*, 1988).

It now appears that a principal activity of tunichrome is that of a metal-ion complexing/reducing agent, but the physiological role of vanadium is unknown. Free tunichrome and vanadium have been found to coexist in at least one subpopulation of vanadium-containing cells called the morula, although the metal, and possibly the ligand, is also present in signet ring cells (Oltz *et al.*, 1989). On a dry-weight basis, tunichrome can constitute 20–50% of the blood in *Ascidia nigra* and other species (Oltz *et al.*, 1988). The difficulties inherent to clarifying the physiological roles of tunichromes and vanadium are outlined below and under-score the importance of an interdisciplinary approach; synthetic and natural products chemistry have been integrated with inorganic, physiological, and ecological studies to better comprehend the prevalence of O_2-sensitive tunichrome and V(III) in the living organism. A variety of analytically pure synthetic and natural tunichromes are now available, thereby permitting several key investigations to proceed.

4.2. Isolation and Structure of Tunichromes

Attempts to isolate unprotected tunichromes were impeded by their tendencies to decompose via oxidation, hydrolysis, polymerization, or exposure to mild alkaline conditions. A further difficulty was the tendency of tunichromes to adhere tenaciously to most solid-phase packing materials, including silica, alumina, cellulose, reversed-phase, and lipophilic matrices (Bruening *et al.*, 1985, 1986). A 5-year effort ultimately yielded a new class of polyphenolic blood pigments, the tunichromes, in their free forms. Originally, several unusual chromatographic methods such as anhydrous, anaerobic extraction chromatography on freeze-dried blood cells (thereby avoiding contact with O_2, vanadium, and water) and centrifugal partition chromatography as well as reversed-phase HPLC had to be employed (Bruening *et al.*, 1986). Spectroscopic and chemical studies on free tunichromes and their peracetates showed that they consisted of a family of homologous hydroxy-Dopa peptides (Bruening *et al.*, 1985; Oltz *et al.*, 1988). The initial class of compounds, An-1, -2, and -3 isolated from the vanadium-sequestering *Ascidia nigra*, differed in their degree of ring hydroxylation, while subsequent isolations of Mm-1 and -2 from the iron-accumulating *Molgula manhattensis* led to simpler structures containing only two polyphenolic rings (Fig. 11) (Oltz *et al.*, 1988).

The direct isolation of tunichrome gave very poor overall yields of <1% because of the final HPLC purification step (Bruening *et al.*, 1986). A fortuitous spinoff from the total syntheses of An-1 (Horenstein and Nakanishi, 1989) and Mm-1 and -2 (Kim *et al.*, 1990) was the discovery of an efficient and simple means to obtain pure natural tunichrome from tunicate blood pellets on a preparative scale (Kim *et al.*, 1991). The application of the synthetic protection/deprotection scheme for the hydroxyl and amino substituents of tunichrome, using *tert*-butyldimethylsilyl (TBDMS) and *tert*-butyloxycarbonyl (BOC) protecting groups, respectively, proved remarkably effective. Previously the isolation of limited quantities of pure natural or synthetic tunichrome was a formidable task. Now the relatively labile and chromatographically intractable tunichromes can be isolated efficiently by derivatizing the blood pellets directly, and separating the derivatized tunichromes simply by flash chromatography and HPLC. The clean separation of the TBDMS and BOC derivatives of An-1, -2, and -3, or Mm-1 and -2, is now routine, even by preparative TLC; both protecting groups are then deprotected by methods used in the syntheses (Horenstein and Nakanishi, 1989). This protection/

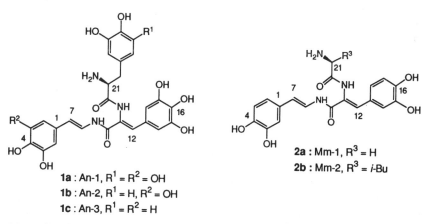

Figure 11. (1) Tunichome An-1 and related congeners isolated from *Ascidia nigra*, a vanadium-accumulator. (2) Tunichrome Mm-1 and -2 isolated from *Molgula manhattensis*, an iron-accumulator.

deprotection method should be applicable to other unstable compounds containing hydroxyl and/or amino groups.

4.3. Synthesis of Tunichromes

The total syntheses of free An-1 (Horenstein and Nakanishi, 1989) and Mm-1 and -2 (Kim *et al.*, 1990) have been completed (Scheme 1). As in An-1 synthesis (Horenstein and Nakanishi, 1989), the two olefinic functions were introduced by selenoxide elimination and Horner–Emmons Wittig condensation. The key start-ing material, phosphonoglycinate (**2**, Scheme 1), was prepared from the known methyl carbobenzoxyphosphonoglycinate (**1**) (Schmidt *et al.*, 1988) in three steps. The addition of TMS cyanide to TBDMS catecualdehyde (**3**) afforded TMS-cyanohydrin, which upon LAH reduction, followed by protection of the amino group with di-*tert*-butyldicarbonate, treatment with *o*-nitrophenylselenocyanate and tri-*n*-butylphosphine, and deprotection of the BOC group gave TBDMS-protected selenodopamine (**4**). Phosphonoglycine (**2b**) was then coupled to **4**, the product was treated with H_2O_2, and the BOC group was deprotected with TFA to yield the *E*-enamide (**5**), a common intermediate for the synthesis of Mm-1 and Mm-2 and other analogs.

Coupling of amine **5** (Scheme 1) with BOC-glycine provided phosphonate **6** in 85% yield. The Horner–Emmons Wittig reaction (Schmidt *et al.*, 1988) of **6** with aldehyde **3** afforded protected Mm-1 (see **7a**) as a 5/1 mixture of E and Z isomers, which upon irradiation (300-nm Rayonet lamps) gave a 1/3 mixture of the desired Z isomer. The BOC and TBDMS protecting groups were removed using dry 20% trifluoroacetic acid/CH_2Cl_2 and 48% aq. HF/pyridine, respectively (under Ar in degassed solvents). The separation of deprotected Mm-1 (**7a**) from the crude mixture was achieved by employing the fractional precipitation technique (Horenstein and Nakanishi, 1989) in which precipitation of the highly polar tunichrome was induced by addition of CH_2Cl_2 and hexane. Three repetitions of MeOH dissolution and precipitation with CH_2Cl_2 afforded bright yellow tuni-chrome Mm-1. For the synthesis of Mm-2 (**7b**), BOC-glycine was replaced by (*s*)-(+)-BOC-leucine. A comparison of the CD of peracetylated synthetic and natural-Mm-2 determined the absolute stereochemistry at C-21 to be *S*.

4.4. Assays for Metal Ions and Tunichrome

All tunicates accumulate iron(II) to some degree and within a relatively narrow concentration range (<30-fold), whereas accumulated vanadium levels vary greatly (approximately 10^4- fold). Species that assimilate vanadium(V) store the element as vanadium(III) and vanadium (IV), and have blood cells with fivefold to over tenfold more vanadium than iron (hence are "vanadium accumula-

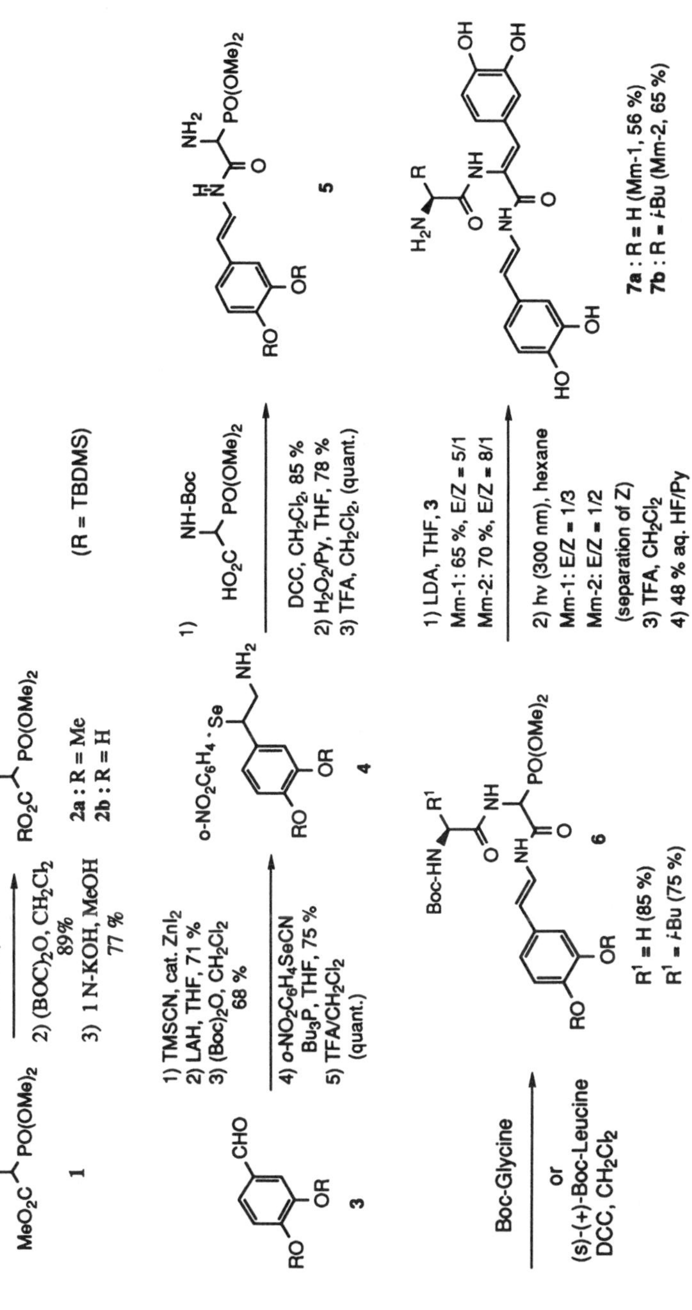

NH-CBZ
MeO2C PO(OMe)2
1
1) H2, 5%-Pd/C, MeOH, 95%
2) (BOC)2O, CH2Cl2 89%
3) 1 N-KOH, MeOH 77 %
NH-Boc
RO2C PO(OMe)2
2a : R = Me
2b : R = H
(R = TBDMS)
CHO
RO OR
3
1) TMSCN, cat. ZnI2
2) LAH, THF, 71 %
3) (Boc)2O, CH2Cl2 68 %
4) o-NO2C6H4SeCN Bu3P, THF, 75 %
5) TFA/CH2Cl2 (quant.)
o-NO2C6H4 · Se
NH2
4
1) NH-Boc HO2C PO(OMe)2
DCC, CH2Cl2, 85 %
2) H2O2/Py, THF, 78 %
3) TFA, CH2Cl2, (quant.)
NH2
PO(OMe)2
5
Boc-Glycine
or
(s)-(+)-Boc-Leucine
DCC, CH2Cl2
Boc-HN
R1
6
R1 = H (85 %)
R1 = i-Bu (75 %)
1) LDA, THF, 3
Mm-1: 65 %, E/Z = 5/1
Mm-2: 70 %, E/Z = 8/1
2) hν (300 nm), hexane
Mm-1: E/Z = 1/3
Mm-2: E/Z = 1/2
(separation of Z)
3) TFA, CH2Cl2
4) 48 % aq. HF/Py
H2N
R
OH
7a : R = H (Mm-1, 56 %)
7b : R = i-Bu (Mm-2, 65 %)

Scheme 1

tors"). Vanadium is stored in such cells at levels as high as 0.5–1.1 M, which is 10 million times its concentration in the ocean.

Measurements in *A. nigra* blood using a superconducting quantum interference device (SQUID susceptometer) indicate that ca. 90% of the vanadium is present as V(III) and 10% as V(IV) (Lee *et al.*, 1988). The distributions of *free* tunichrome and total vanadium in *A. nigra* and *A. ceratodes* blood cells have been determined by sorting the cell by fluorescence-activated cell sorting (FACS; flow cytometry) and performing subsequent microanalyses on the sorted blood cells (Oltz *et al.*, 1989). The total vanadium was analyzed by atomic absorption, while the tunichromes were analyzed by peracetylation of blood cells and characterization as their acetates; note that this peracetylation method is only valid for free tunichromes since metal-complexed pigments cannot be acetylated. Unexpectedly, the yellow to yellow-green morula cells (containing several spherical vacuoles) held nearly all of the free tunichrome and up to 30% of the vanadium detected, whereas the green-gray signet ring cells (characterized by one large vacuole) yielded only a trace of free tunichrome and the bulk of the vanadium (Oltz *et al.*, 1989). The widely accepted inference that signet sing cells give rise to morula cells is difficult to reconcile with our data. If true for either *A. nigra* or *A. ceratodes*, it would require that signet ring cells excrete virtually all vanadium (sequestered at no small expense) and proceed to accumulate a potent ligand, tunichrome.

4.5. Intracellular Vanadium Environment

Even though tunichrome is a formidable contender for the endogenous metal ion complexing/reducing agent (Oltz *et al.*, 1988; M. J. Smith, 1989), no direct evidence linking this or any other compound (Frank *et al.*, 1987) to metal ion assimilation has been secured, and understandably so; the unambiguous identification of a particular (labile) metal complex among a dynamic collection of homologous metal complexes (in fragile, developing cells) is no small feat.

Now that a diverse array of natural (Oltz *et al.*, 1988) and synthetic (Horentstein and Nakanishi, 1989; Kim *et al.*, 1990) tunichromes (affiliated with both vanadium- and iron-accumulators) and analogues (Kime-Hunt *et al.*, 1988; Bulls *et al.*, 1990) are available, investigations directed toward the coordination chemistry of these ligands may proceed. However, the design of such reconstitution studies *in vitro* is challenging, especially if the aim is to simulate the complex chemistry of tunicate blood cells accurately.

Another view of vanadium complexation suggests that the metal is coordinated as the aqua ion $[V(III)SO_4(H_2O)_{4-5}]^+$ at a pH < 3, based on extensive ^{1}H-NMR (Carlson, 1975), electron paramagnetic resonance (EPR) (Frank *et al.*, 1986), and XAS (Lee *et al.*, 1988) analyses of *A. ceratodes* blood cells. If intracellular vanadium is present as an aqua complex such as $[V(III)SO_4(H_2O)_{4-5}]^+$, there should be a strongly acidic environment, since above pH 3, in the absence of other stabilizing agents, V(III) exists as insoluble black V_2O_3. Since the cytosolic pH of virtually all living cells is maintained near neutrality, the pH of the milieu around vanadium has thus been the subject of much controversy. A medium stabilizing simple hydrated, lower-valence vanadium can be rationalized by postulating the existence of an intracellular "pH < 3 compartment." Such a compartment might arise under three distinct and mutually compatible conditions: (1) a subcellular vacuole, (2) a hydrophobic region like the interior of a membrane, and (3) a damaged or senescent cell itself. The pH of unsorted and sorted *A. ceratodes* blood cells has been reexamined recently, using equilibration experiments with radiolabeled 5,5-dimethyloxazolidine-2,4-dione and methylamine (Lee *et al.*, 1990). The average pH of resuspended blood cells was 6.98 ± 0.15, whereas that for the cytoplasm and vacuoles of morula subpopulations was 7.1 ± 0.2 and 5.0 ± 0.1, respectively. The absence of similar pH determinations on the vanadium-laden signet ring cells, owing to their ease of lysis and resulting low recovery, admits the possible existence of a "pH < 3 compartment" therein. Nevertheless, a considerable fraction of the assimilated intracellular V(III)/V(IV) still appears to be stabilized by chelation at a neutral pH (at least in morula cells).

4.6. Mechanistic Considerations of Vanadium Assimilation

An intracellular excess of free tunichrome over vanadium in morula cells may thus prove to be the mechanistic key (M. J. Smith, 1989) to assimilating of vanadium and maturation of vanadocytes (cells assimilating vanadium), whereby vanadium uptake might occur at the expense of tunichrome oxidation and give rise to signet ring cells possessing abundant vanadium but little free tunichrome. One should thus bear in mind that the cellular milieu around free tunichrome must be nonoxidative to a significant degree. The crux of vanadium assimilation in tunicates thus consists of the series of chemical reactions that drive vanadium up a concentration gradient and reduce it to V(III). Three key factors are the energy requirements, the ligands involved, and the attendant pH. How might V(V) be driven into a cell and stored as V(III)? A mechanistic variation of the chemiosmotic relation gave rise to a so-called "vanadium trapping model," made up of two components (Macara *et al.*, 1979), a membrane transport system and tunichrome, a refined version of which has been proposed by M. J. Smith (1989). A subsequent finding that neither inhibitors of glycolysis nor oxidative phosphorylation blocks vanadate uptake indicates that vanadium assimilation in tunicates is not an active transport process, requiring a direct energy input, e.g., ATP hydrolysis. Instead, the "passive sequestration" or "facilitated diffusion" of vanadate against an *apparent* 10 million-fold concentration gradient can be achieved in two ways (Dingley *et al.*, 1981): (1) vanadate uptake (influx) is coupled to the virtual efflux of ions with equivalent charge (e.g., phosphate/OH$^-$ antiport); (2) intracellular vanadate is removed from solution by being bound or converted to some other chemical species. Both alternatives may obtain, since vanadate utilizes the "phosphate transporter" and undergoes a two-step reduction to V(III). Should the vanadium complexes so incorporated become "trapped" as nonpermeable ions, insoluble polymers, or membrane-bound, this would further shift the trans-membrane vanadium distribution in favor of vanadate uptake.

4.7. Biological Roles of Tunichrome and Vanadium

The biosynthesis of a resilient covering or tunic for tunicates has long been postulated to be the rationale for the vanadium-accumulating blood cells (vanado-cytes); parallel activities presumably exist within iron-accumulating species. The mode of action of vanadocytes includes migration to and lysis within tissue lining the tunic. Tunic biogenesis may thus be analogous to the hardening or maturation (sclerotization) of insect cuticle, which occurs via oxidation of catecholamine cross-linking agents (Sugumaran, 1987; Schaffer *et al.*, 1987).

As the most primitive chordates in existence, tunicates may have evolved at a time when conditions on earth were less oxidizing (more anaerobic). The primor-dial appearance of O_2 in the atmosphere created a double jeopardy for many forms of life: (1) a diminished availability of essential metal ions like vanadium and iron, due to their precipitation as hydroxide polymers; and (2) exposure to the hazardous oxidizing agent O_2. Three organismal responses to this dilemma appear to have been to (1) elaborate high-affinity metal-ion transport systems (Neilands, 1984), (2) maintain alternative respiratory options (Hochachka and Guppy, 1987, p. 10), and (3) protect cells from overexposure to O_2 (Hochachka and Guppy, 1987, p. 10). The characteristics of tunicate blood are its vivid yellow-green color, its high reactivity, and the anomalous content of tunichrome and vanadium. Clarification of the unusual chemical ecology of tunicates may thus derive from continued efforts to determine the chemical mode of action of its blood. The isolation, characterization, and synthesis of a predominant constituent, tunichrome, has been instrumental. Now that adequate quantities of analytically pure synthetic and natural tunichromes are available, several key investigations may proceed.

ACKNOWLEDGMENT. Partial funding was provided by National Institutes of Health grant AI 10187 (K.N.)

REFERENCES

Barrow, C. J., Nakaniski, K., and Tachibana, K., 1992, Structural considerations for the pore forming properties of pardaxin, *Biochim. Biophys. Acta* (Submitted).
Borst, D. W., Laufer, H., Landau, M., Chang, E. S., Hertz, W. A., Baker, F. C., and Schooley, D. A.,

1987, Methyl farnesoate and its role in crustacean reproduction and development, *Insect Biochem.* **37:**1123–1127.

Brown, F. A., Jr., and Cunningham, O., 1939, Influence of the sinus gland of crustaceans on normal viability and ecdysis, *Biol. Bull.* **77:**104–114.

Brown, L. R., Lauterwein, J., and Wüthrich, K., 1980, High resolution 1H-NMR studies of self-aggregation of melittin in aqueous solution, *Biochim. Biophys. Acta* **622:**231–244.

Bruening, R. C., Oltz, E. M., Furukawa, J., Nakanishi, K., and Kustin, K., 1985, Isolation and structure of tunichrome B-1, a reducing blood pigment from the tunicate *Ascidia nigra*, *J. Am. Chem. Soc.* **107:**5298.

Bruening, R. C., Oltz, E. M., Furukawa, J., Nakanishi, K., and Kustin, K., 1986, Isolation of tunichrome B-1, a reducing blood pigment of the sea squirt, *Ascidia nigra*, *J. Nat. Prod.* **49:**193–204.

Bulls, A. R., Pippin, C. G., Hahn, F. E., and Raymond, K. N., 1990, *J. Am. Chem. Soc.* **112:**2627.

Carlson, R. M. K., 1975, *Proc. Natl. Acad. Sci. USA* **72:**2217.

Chang, E. S., and O'Connor, J. D., 1977, Secretion of a-ecdysone by crab Y-organs *in vitro*, *Proc. Natl. Acad. Sci. USA* **74:**615–618.

Chang, E. S., Sage, B. A., and O'Connor, J. D., 1976, The qualitative and quantitative determinations of ecdysones in tissues of the crab, *Pachygrapsus crassipes*, following molt induction, *Gen. Comp. Endocrinol.* **30:**21–33.

Chang, E. S., Prestwich, G. D., and Bruce, M. J., 1990, Amino acid sequence of a peptide with both molt-inhibiting and hyperglycemic activities in the lobster, *Homarus americanus*, *Biochem. Biophys. Res. Commun.* **171:**818–826.

Chou, P. Y., and Fasman, G. D., 1978, Empirical predictions of protein conformation, *Annu. Rev. Biochem.* **47:**251–276.

Clark, E., 1974, The Red Sea's shark proof fish, *Natl. Geogr. Mag.* **146:**719–727.

Clark, E., and George, A., 1979, Toxic soles, *Pardachirus marmoratus* from the Red Sea and *P. pavoninus* from Japan, with notes on other species, *Environ. Biol. Fishes* **4:**103–123.

Dingley, A. L., Kustin, K., Macara, I. G., and McLeod, G. C., 1981, Accumulation of vandium by tunicate blood cells occurs via a specific anion transport system, *Biochim. Biophys. Acta* **649:**493–502.

Frank, P., Carlson, R. M. K., and Hodgson, K. O., 1986, Vandyl ion EPR as a noninvasive probe of pH in intact vanadocytes from *Ascidia ceratodes*, *Inorg. Chem.* **25:**470–478.

Frank, P., Hedman, B., Carlson, R. M. K., Tyson, T. A., Roe, A. L., and Hodgson, K. O., 1987, A large reservoir of sulfate and sulfonate resides within plasma cells from *Ascidia ceratodes*, revealed by absorption near-edge structure spectroscopy, *Biochemistry* **26:**4975–4979.

Freeman, J. A., 1980, Hormonal control of chitinolytic activity in the integument of *Balanus amphitrite*, *in vitro*, *Comp. Biochem. Physiol.* **65A:**13–17.

Gabe, M., 1953, Sur l'existence, chez quelques crustacés malacostracés, d'un organe comparable à la glande de la mue des insectes, *C. R. Acad. Sci. Paris* **237:**1111–1113.

Gruber, S. H., 1977, The visual system of sharks, *Am. Zool.* **17:**453–469.

Gruber, S. H., 1981, Shark repellents: Perspectives for the future, *Oceanus* **24(4):**72–76.

Hochachka, P. W., and Guppy, M., 1987, *Metabolic Arrest and the Control of Biological Time*, Harvard University Press, Cambridge, Massachusetts.

Horenstein, B. A., and Nakanishi, K., 1989, Synthesis of unprotected ($\pm$)-tunichrome An-1, a tunicate blood pigment, *J. Am. Chem. Soc.* **111:**6242–6246.

Huberman, A., and Aguilar, M. B., 1989, A neuropeptide with molt-inhibiting hormone activity from the sinus gland of the Mexican crayfish *Procambarus bouvieri* (Ortmann), *Comp. Biochem. Physiol.* **93B:**299–305.

Jegla, T. C., Ruland, C., Kegel, G., and Keller, R., 1983, The role of the Y-organ and cephalic gland in ecdysteroid production and the control of molting in the crayfish *Orconectes limosus*, *J. Comp. Physiol.* **152B:**91–95.

Kalmijin, A. J., 1966, Electro-perception in sharks and rays, *Nature* **212**:1232–1233.

Kappler, C., Kabbouh, M., Hetru, C., Durst, F., and Hoffmann, J. A., 1988, Characterization of three hydroxylases involved in the final steps of biosynthesis of the steroid hormone ecdysone in *Locusta migratoria* insecta orthoptera, *J. Steroid Biochem.* **31**:891–898.

Kim, D., Li, Y., and Nakanishi, K., 1991, Isolation of unstable tunichromes from tunicate blood via protection–deprotection, *J. Chem. Soc. Chem. Commun.* **1991**:9–10.

Kim, D., Li, Y., Horenstein, B. A., and Nakanishi, K., 1990, Synthesis of tunichromes Mm-1 and Mm-2, blood pigments of the iron-assimilating tunicate, *Molgula manhattensis*, *Tetrahedron Lett.* **31**:7119–7122.

Kime-Hunt, E., Spartalian, K.,and Carrano, C. J., 1988, Models for vanadium-tunichrome interactions, *J. Chem. Soc. Chem. Commun.* **1988**:1217–1218.

Kleinholz, L. H., 1985, Biochemistry of crustacean hormones, in: *The Biology of Crustacea*, Vol. 9 (D. E. Bliss and L. H. Mantel, eds.), Academic Press, New York, pp. 463–522.

Kotake, M., Sakan, T., and Senoh, S., 1951, Studies on amino acids IV. The synthesis of 3-hydroxy-kynurenie, *J. Am. Chem. Soc.* **73**:1832–1834.

Lachaise, F., 1989, Diversity of molting hormones in Crustacea, in: *Ecdysone, From Chemistry to Mode of Action* (J. Koolman, ed.), Thieme Medical, New York, pp. 313–318.

Lachaise, F., and Feyereisen, R., 1976, Métabolisme de l'ecdysone par divers organes de *Carcinus maenas* L. incubés *in vitro*, C. R. Acad. Sci. Paris **283**:1445–1448.

Laufer, H., Landau, M., Homola, E., and Borst, D. W., 1987, Methyl farnesoate: Its site of synthesis and regulation of secretion in a juvenile crustacean, *Insect Biochem.* **17**:1129–1131.

Lazarovici, P., Primor, N., and Loew, L. M., 1986, Purification and pore-forming activity of two hydrophobic polypeptides from the secretion of the Red Sea Moses sole (*Pardachirus marmoratus*), *J. Biol. Chem.* **261**:16704–16713.

Lazarovici, P., Primor, N., Gennaro, J., Fox, J., Shai, Y., Lelkes, P. I., Caratsch, C. G., Raghunathan, G., Guy, H. R., Shih, Y. L., and Edwards, C., 1990, Origin, chemistry, and mechanisms of action of a repellent, presynaptic excitatory, ionophore polypeptide, in: *Marine Toxins: Origin, Structure, and Molecular Pharmacology* (S. Hall and G. Strichartz, eds.), American Chemical Society, Washington, D.C., pp. 347–364.

Lee, S., Kustin, K., Robinson, W. E., Frankel, R. B., and Spartalian, K., 1988, Magnetic properties of tunicate blood cells. I. *Ascidia nigra*, *J. Inorg. Biochem.* **33**:183–192.

Lee, S., Nakanishi, K., and Kustin, K., 1990, The intracellular pH of tunicate blood cells: *Ascidia ceratodes* whole blood, morula cells, vacuoles and cytoplasm. *Biochem. Biophys. Acta* **1033**:311–317.

Lowe, M. E., Horn, D. H. S., and Galbraith, M. N., 1968, The role of crustecdysone in the molting crayfish, *Experientia* **24**:518–519.

Macara, I. G., McLeod, G. C., and Kustin, K., 1979, Isolation, properties and structural studies on a compound from tunicate blood cells that may be involved in vanadium accumulation, *Biochem. J.* **181**:457–465.

Mattson, M. P., and Spaziani, E., 1986, Regulation of the stress-responsive X-organ–Y-organ axis by 5-hydroxy-tryptamine in the crab. *Cancer antennarius, Gen. Comp. Endocrinol.* **62**:419–427.

Mattson, M. P., and Spaziani, E., 1987, Demonstration of protein kinase C activity in crustacean Y-organs and partial definition of its role in regulation of ecdysteroidogenesis, *Mol. Cell. Endocrinol.* **49**:159–172.

Megusar, F., 1912, Experimente über den Farbwechsel der Crustaceen, *Arch. Entwickl.-Mech.* **33**:462.

Moran, A., Korchak, Z., Moran, N., and Primor, N., 1984, Surfactant and channel-forming activities of the Moses sole toxin, in: *Toxins, Drugs, and Pollutants in Marine Animals* (L. Bolis, J. Zadunaisky, and R. Gilles, eds.), Springer-Verlag, Berlin, pp. 13–25.

Naya, Y., and Sonobe, H., 1990, Molt-regulating factors: Comparative study in crustaceans and insects, *Heredity* **44**:62–67 [in Japanese].

Naya, Y., Kishida, K., Sugiyama, M., Murata, M., Miki, W., Ohnishi, M., and Nakanishi, K., 1988, Endogenous inhibitor of ecdysone synthesis in crabs, *Experientia* **44**:50–52.

Naya, Y., Ohnishi, M., Ikeda, M., Miki, W., and Nakanishi, K., 1989, What is molt-inhibiting hormone? The role of an ecdysteroidogenesis inhibitor in the crustacean molting cycle, *Proc. Natl. Acad. Sci. USA* **86**:6826–6829.

Neilands, J. B., 1984, Siderophores of bacteria and fungi, *Microbiol. Sci.* **1**:9–14.

Newcom, R. W., Stuenkel, E. L., and Cooke, I. M., 1985, Characterization, biosynthesis and release of neuropeptides from the X-organ-sinus gland system of the crab *Cardisoma carnifex*, *Am. Zool.* **25**:157–171.

Ohnishi, M., Nakanishi, K., and Naya, Y., 1991, Down-regulation of ecdysteroidogenesis: Physiological dynamics of prothoracic and Y-glands, *CHIMICAoggi* **9**:53–56.

Oltz, E. M., Bruening, R. C., Smith, M. J., Kustin, K., and Nakanishi, K., 1988, The tunichromes. A class of reducing blood pigments from sea squirts: Isolation, structures, and vanadium chemistry, *J. Am. Chem. Soc.* **110**:6162–6172.

Oltz, E. M., Pollack, S., Delohery, T., Smith, M. J., Ojika, M., Lee, S., Kustin, K., and Nakanishi, K., 1989, Distribution of tunichrome and vanadium in sea squirt blood cells sorted by flow cytometry, *Experientia* **45**:186–190.

Passano, L. M., 1960, Molting and its control, in: *The Physiology of Crustacea*, Vol. 1 (T. H. Waterman, ed.), Academic Press, New York, pp. 473–536.

Primor, N., 1985, Pharyngeal cavity and the gills are the target organ for the repellent action of pardaxin in shark, *Experientia* **41**:693–695.

Primor, N., Parness, J., and Zlotkin, E., 1978, Pardaxin, the toxic factor from the skin secretion of the flatfish *Pardachirus marmoratus* (Soleidae), in: *Toxins: Animal, Plant and Microbial* (P. Rosenberg, ed.), Pergamon Press, Oxford, pp. 539–547.

Schaffer, J., Kramer, K. J., Garbow, J. R., Jacob, G. S., Stejskal, E. O., Hopkins, T. L., and Speirs, R. D., 1987, Aromatic cross-links in insect cuticle: Detection by solid-state ^{13}C and ^{15}N NMR, *Science* **235**:1200–1204.

Schmidt, U., Lieberknect, A., and Wild, J., 1988, Didehydroamino acid (DDAA) and didehydropeptides (DDP), *Synthesis* **1988**:159–172.

Skinner, D. M., 1985, Molting and regulation, in: *The Biology of Crustacea*, Vol. 9 (D. E. Bliss and L. H. Mantel, eds.), Academic Press, New York, pp. 43–146.

Smith, M. J., 1989, Vanadium biochemistry: The unknown role of vanadium-containing cells in ascidians (sea squirts), *Experientia* **45**:452–457.

Smith, S. L., 1985, Regulation of ecdysteroid titer: Synthesis, in: *Comprehensive Insect Physiology, Biochemistry and Pharmacology* (G. A. Kerkut and L. I. Gilbert, eds.), Pergamon Press, Oxford, Vol. 7, pp. 295–341.

Sonobe, H., Kanba, M., Ohta, K., Ikeda, M., and Naya, Y., 1991, *In vitro* secretion of ecdysteroids by Y-organs of the crayfish, *Procambarus clarkii*, *Experientia* **47**:948–952.

Soyez, D., and Kleinholz, L. H., 1977, Molt-inhibiting factor from the crustacean eyestalk, *Gen. Comp. Endocrinol.* **31**:233–242.

Spaziani, E., 1988, Serum high-density lipoprotein in the crab, *Cancer antennarius* Stimpson: II. Annual Cycles, *J. Exp. Zool.* **246**:315–318.

Sugumaran, M., 1987, Quinone methide sclerotization: A revised mechanism for b-sclerotization of insect cuticle, *Bioorg. Chem.* **15**:194–211.

Tachibana, K., and Gruber, S. H., 1988, Shark repellent lipophilic constituents in the defense secretion of the Moses sole (*Pardachirus marmoratus*), *Toxicon* **26**:839–853.

Tachibana, K., Sakaitani, M., and Nakanishi, K., 1985, Pavoninins, shark-repelling and ichthyotoxic steroid *N*-acetylglucosaminides from the sole *Pardachirus pavoninus* (Soleidae), *Tetrahedron* **41**:1027–1037.

Tachibana, K., Xu, W.-H., Barrow, C. J., and Imoto, S., 1989, Action of pardaxins and their des analogues on artificial liposomes, in: *Peptide Chemistry 1988* (M. Ueki, ed.), Protein Research Foundation, Osaka, pp. 279–282.

Terwillinger, T. C., and Eisenberg, D., 1982, The structure of melittin: Interpretation of the structure, *J. Biol. Chem.* **257**:6016–6022.

Testa, B., and Jenner, P., 1981, Inhibitors of cytochrome P-450s and their mechanism of action, in: *Drug Metabolism Reviews* (F. J. Di Carlo, ed.), Marcel Dekker, New York, Vol. 12(1), pp. 1–117.

Thompson, S. A., Tachibana, K., Nakanishi, K., and Kubota, I., 1986, Melittin-like peptides from the shark-repelling defense secretion of the sole *Pardachirus pavoninus*, *Science* **233**:341–343.

Thompson, S. A., Minakata, H., Xu, W.-H., Tachibana, K., Nakanishi, K., and Kubota, I., 1987, Melittin-like amphiphilic peptides in the defense secretion of the sole *Pardachirus pavoninus*: Isolation, structures, and bioactivity, in: *Peptide Chemistry 1986* (T. Miyazawa, ed.), Protein Research Foundation, Osaka, pp. 181–184.

Thompson, S. A., Tachibana, K., Kubota, I., and Zlotkin, E., 1988, Amino acid sequence of pardaxin in the defense secretion of the Red Sea Moses sole, in: *Peptide Chemistry 1987* (T. Shiba and S. Sakakibara, eds.), Protein Research Foundation, Osaka, pp. 127–130.

Watson, R. D., and Spaziani, E., 1985, Effects of eyestalk removal on cholesterol uptake and ecdysone secretion by crab *Cancer antennarius* Y-organs *in vitro*, *Gen. Comp. Endocrinol.* **57**:360–370.

Webster, S. G., and Keller, R., 1986, Purification, characterization and amino acid composition of the putative moult-inhibiting hormone (MIH) of *Carcinus maenas* (Crustacea, Decapoda), *J. Comp. Physiol.* **156B**:617–624.

Webster, S. G., and Keller, R., 1989, Molt inhibiting hormone, in: *Ecdysone, From Chemistry to Mode of Action* (J. Koolman, ed.), Thieme Medical, New York, pp. 211–216.

Zagorsky, M. G., Norman, D. G., Barrow, C. J., Iwashita, T., Tachibana, K., and Patel, D. J., 1991, Solution structure of pardaxin P-2, *Biochemistry* **30**:8009–8017.

Zahuranec, B. J. (ed.), 1983, *Shark Repellents from the Sea: New Perspectives*, Westview Press, Boulder, Colorado.

Zahuranec, B. J., and Baldridge, Jr., H. D., 1983, Shark research and the United States Navy, in: *Shark Repellents from the Sea: New Perspectives* (B. J. Zahuranec, ed.), Westview Press, Boulder, Colorado, pp. 1–10.

Zeleny, C., 1905, Compensatory regulation, *J. Exp. Zool.* **2**:1–102.

Zlotkin, E., and Barenholz, Y., 1983, On the membranal action of pardaxin, in: *Shark Repellents from the Sea: New Perspectives* (B. J. Zahuranec, ed.), Westview Press, Boulder, Colorado, pp. 157–171.

Zlotkin, E., and Gruber, S. H., 1984, Synthetic surfactants: A new approach to the development of shark repellents, *Arch. Toxicol.* **56**:55–58.

3

Pharmacological Studies of Novel Marine Metabolites

*Robert S. Jacobs, Mary A. Bober, Isabel Pinto,
Allen B. Williams, Peer B. Jacobson,
and Marianne S. de Carvalho*

1. INTRODUCTORY REMARKS

Fundamental to the complex process associated with drug discovery is the establishment of a carefully defined collaborative relationship between academic bioorganic chemists and cellular or molecular pharmacologists. We mention this as a new emerging direction in research that should be particularly emphasized in an academic setting. The collaborative goal of the academic chemists and pharmacologists should be to contribute in some specific way to the advancement of each other's science. Educational goals are not served well through routine programs for isolating and screening natural products. Screening programs of this kind are well developed in U.S. and foreign drug companies. In fact, new molecular receptor assays that allow a wide variety and large number of screens to be executed rapidly are becoming available. Academic pharmacologists have the responsibility to develop knowledge fundamental to advancing biomedical science by using bioactive substances as probes to better understand normal and patholog-

Robert S. Jacobs, Mary A. Bober, Isabel Pinto, Allen B. Williams, Peer B. Jacobson, and Marianne S. de Carvalho • Marine Science Institute, and Department of Biological Sciences, University of California, Santa Barbara, California 93106.

Marine Biotechnology, Volume 1: Pharmaceutical and Bioactive Natural Products, edited by David H. Attaway and Oskar R. Zaborsky. Plenum Press, New York, 1993.

ical biological processes, to define in rigorous fashion the properties of these probes, and to determine their biochemical limitations and how they can be used properly and safely.

Results of academic research serve as a stimulus and a starting point for new research projects in the pharmaceutical and chemical industries. The applications of progressive collaborative efforts are not always known. However, it is gratifying to visit a drug company and find that one's research has spawned new ideas or renewed thinking about almost intractable problems in medical research. Academic pharmacologists also play a most important role in developing skilled students who will continue to advance pharmacological science.

The following sections summarize the recent research findings of our academic group, which specializes in the study of novel bioactive substances from marine sources. We have not included our earlier work with manoalide (Glaser and Jacobs, 1986; de Freitas *et al.*, 1984), lophotoxin (Culver and Jacobs, 1981; Fenical *et al.*, 1981), stypoldione (O'Brien *et al.*, 1986), psuedopterosin (Ettouati and Jacobs, 1987; Look *et al.*, 1986), and other important marine natural products, but instead focus this chapter on our most recent investigations. All of our research was conducted in close collaboration with bioorganic chemists, particularly John Faulkner, William Fenical, Francis Schmitz, and Valerie Paul, some of whose efforts are detailed elsewhere in this book. This group, particularly Faulkner and Fenical, and more recently Schmitz and Paul, have unselfishly committed their resources and intellectual expertise in bioorganic chemistry necessary to assure advances on issues of mutual interest.

2. PHARMACOLOGICAL MODELS OF CELLULAR AND MOLECULAR PROCESSES

One of the most exciting areas in marine pharmacology is the investigation of chemical regulators of cellular and molecular processes in marine organisms. Marine models have long been employed to work out details of mammalian processes. For example, the squid axon has been used in many fundamental studies in neurophysiology (Fishman and Lipicky, 1991; Allen, 1990). The electric organ of the skate is used as a rich source of acetylcholine receptors and enzymes (Fox *et al.*, 1990). The understanding of nervous system function has been eloquently advanced through studies of the sea hare *Aplysia* (Busselberg *et al.*, 1991; Calignano *et al.*, 1991).

Understanding the unique differences between marine organisms and their terrestrial counterparts at the biochemical and molecular level can be useful in understanding comparable processes in humans and their relationship to disease. Investigation of marine cellular and molecular processes is also fundamental to understanding the biosynthesis of natural products. Within this context and to better understand the process of inflammation and chemical defense, we have in our group undertaken studies of eicosanoid biosynthesis in tunicates and in coralline red algae. These phylogenetically remote classes of organism have both been shown to metabolize arachidonic acid and to produce prostaglandin and leukotriene-like compounds. The goal of our studies is to determine the biochemical regulators of eicosanoid synthesis in these organisms as clues to new classes of anti-inflammatory drugs, and if possible determine the role eicosanoids play in these organisms.

This chapter will discuss our diverse yet cohesive approach to the study of eicosanoid biosynthesis and marine natural products. Our investigations consist of pharmacological models of cellular and molecular processes and site and mechanism of action of marine natural products. The former category includes studies on eicosanoid biosynthesis in the marine tunicate and coralline red algae. The site and mechanism of action of marine natural products target specifically the biological activities of fucoside and scalaradial as well as a cellular cancer model using the multiple-drug-resistance CH^R-C5 cells for screening the potential anticarcinogenic activity of marine natural products.

2.1. Eicosanoid Biosynthesis in Tunicates

The metabolites from arachidonic acid metabolism are ubiquitous compounds that require attention as the multifarious scope of their function becomes elucidated. The enzymes of arachidonic acid metabolism are currently considered to include cyclooxygenases, lipoxygenases, and more recently a cytochrome P450 enzyme (Schwartzman *et al.*, 1987). Cyclooxygenases and lipoxygenases metabolize arachidonic acid to form the inflammatory products of prostaglandins and leukotrienes, respectively. Interestingly, the P450 enzymes are now linked to the formation of epoxyeicosatrienoic acids (EETs) *and* hydroxyeicosatetraenoic acid (HETE) metabolites from arachidonic acid (Schlondorff *et al.*, 1987). The epoxy compounds are capable of increasing local blood pressure in the eye and kidney, while 20-HETE, a metabolite from a P450 enzyme located in the medullary thick ascending Loop of Henle, clearly has an inhibitory effect on the local influx of cellular K^+ (Escalante *et al.*, 1991). The complexity of the arachidonic acid cascade increases as the scope of its metabolism begins to include other enzymes, such as the cytochrome P450 enzymes, which are present at multiple sites. In addition, the P450 pathway for arachidonic acid metabolism yields a heretofore unstudied epoxide metabolite. Current studies also are suggesting a "new" role for arachidonic metabolites, including local ion and pump regulation. Metabolic pathways of arachidonic acid metabolism are important in mammalian systems for homeostasis, inflammatory responses, and osmoregulation. However, imbalances in eicosanoid metabolism can occur, leading to pathological conditions such as can be found in the cornea, the kidney, and tumorous growths. These "imbalances" can result in numerous abnormalities, including local inflammation and high blood pressure and other chronic problems.

We wanted to study a primitive vertebrate system that we hoped would provide a simpler and more direct approach to understanding eicosanoid processes than studies on complex mammalian systems. Marine tunicates are readily available and are the most primitive class of vertebrate. A tunicate offers an interesting model for studying eicosanoid processes and evolutionary changes because of its relatively simplistic structure compared to other members of the family Urochordata. Its life cycle includes both an invertebrate and a vertebrate stage; thus, one can envision the tunicate as an evolutionary bridge between the invertebrate and vertebrate. Its first larval stage is mobile and has a rudimentary notochord. The notochord is eventually absorbed and the hermaphroditic tunicate settles on a surface to live most of its life in colonies or individually as an invertebrate. The colonial tunicates are further divided between those that have a shared circulatory system and those that live communally but maintain separate circulatory systems. One might say the former tunicates are aspects of a homogeneous mass, while the latter tunicates are mutually exclusive entities of the community. Are eicosanoids present in primitive vertebrates? What enzymatic pathways are responsible for their formation? Most importantly, what purpose(s) do these compounds serve?

The survival of the primitive marine vertebrate may be intimately related to the presence of the eicosanoid compounds. Considering that survival of a marine organism is associated with osmoregulation as well as reproduction, the possibility of a link between these functions and arachidonic acid metabolism deserves attention. Consider recent data showing that the isolated mammalian renal glomeruli, glomerular epithelial cells, cortical tubules, and toad bladder are known to be affected by metabolites of the lipoxygenase and/or P450 pathways (Schlondorff *et al.*, 1987). In addition, the corneal microsomes contain a Na^+/K^+ ATPase which can be inhibited by a cytochrome P450 arachidonate metabolite. These data indicate that osmoregulation, not only at the level of the kidney, may be linked to eicosanoid production. The tunicate may prove to be a useful organism for defining the early pathways of eicosanoid formation and regulation as well as their physiological roles.

A preliminary comparative study is described for two types of colonial tunicates, *Clavelina huntsmani*, the light bulb tunicate (LBT), which has an unshared circulatory system, and an unidentified species (UCT) with a shared circulatory system. The study encompassed the determination of UV-active compounds from an organic extract, measurements of phospholipase A_2 (PLA$_2$), prostaglandin E_2 (PGE$_2$), leukotriene B_4 (LTB$_4$), and leukotriene C_4 (LTC$_4$), and the screening of cellular fractions for cross-reactivity with mammalian P450 antibodies.

2.1.1. PLA₂ Activity

PLA$_2$ activity is present in both species of tunicates and in all three samples—10K supernatant, cytosolic, and microsomal. The UCT has greater endogenous PLA$_2$ activity than the LBT. For both the UCT and the LBT, the majority of activity is found in microsomes; however, a significant amount of PLA$_2$ activity is distributed in the cytosol for both species (see Table I). The assay also indicates that the summation of the PLA$_2$ activity in the cytosolic and microsomal fractions accounts for approximately 100% of the activity found in the 10K homogenate for both species of tunicates. The PLA$_2$ assay results show that the LBT and UCT both have an endogenous phospholipase enzyme. This enzyme could yield the product arachidonic acid (AA), which would then subsequently be the required substrate in the arachidonic acid cascade for a lipoxygenase, P450, and/or cyclooxygenase enzymes.

Next, results from three radioimmunoassays are presented: LTB$_4$, LTC$_4$ products of AA metabolism via a lipoxygenase enzyme, and PGE$_2$, a product of AA metabolism via a cyclooxygenase enzyme.

2.1.2. LTB₄

The assay is designed as a 2×3 Latin square, two species of tunicates and three homogenates. The endogenous levels of leukotriene B$_4$ in the LBT are significantly higher (Fig. 1a) than those of the UCT (Fig. 1b). When the amount of leukotriene is adjusted for homogenate protein concentration [in both the arachidonic acid (AA)-stimulated and nonstimulated samples], the LBT consistently shows a more highly enzymatically active preparation. Thus, the LBT produces more endogenous LTB$_4$ than the UCT. A comparison of cytosolic and microsomal fractions indicates that for both the LBT and UCT the cytosolic fraction is more responsive to arachidonic acid stimulation, suggesting that the cytosolic lipoxygenase enzyme is more active or in higher concentrations than in the microsomal preparations.

Table I. Phospholipase Activity (pmole/hr)

Tunicate[a]	$10,000 \times g$	Cytosol	Microsomes
UCT	406	225	150
LBT	117	51.4	58.9

[a]UCT, unidentified colonial tunicate; LBT, light bulb tunicate, *Clavelina huntsmani*.

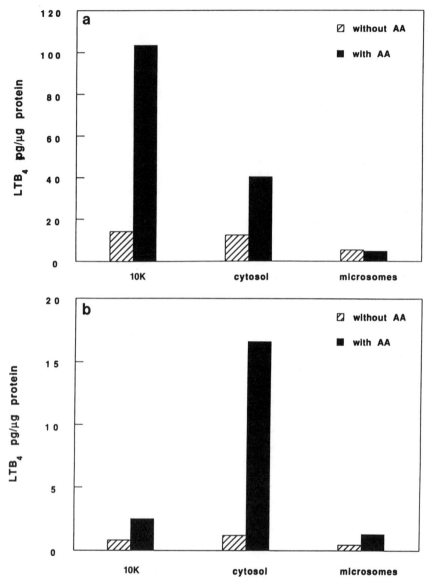

Figure 1. (a) Leukotriene B_4 (pg/μg protein) measured in the 10,000 × *g* supernatant (10K), cytosolic, and microsomal fractions of (a) the tunicate *Clavelina huntsmani* and (b) our unidentified colonial tunicate (UCT), using radioimmunoassay. A control (untreated) and an AA-treated sample (0.34 mM) were assayed. One sample was incubated for 15 min with AA. All samples were extracted with ethyl acetate. The extract was dried under nitrogen, reconstituted in assay buffer, and then 100-μl aliquots (in duplicate) were assayed for LTB_4.

2.1.3. LTC_4

LTC_4 is also increased with the addition of AA in the 10,000 $\times$ g and cytosolic homogenates of the LBT and UCT. In these studies, the increase occurs primarily in the cytosolic fraction in the LBT (Fig. 2a) and in the 10,000 $\times$ g fraction in the UCT species (Fig. 2b). LTC_4 can be decreased with the addition of the selective P450 inhibitor SKF 525A.

2.1.4. PGE_2

PGE_2 also was detected in *Clavelina huntsmani*, with the highest levels found in the cytosolic fraction (Fig. 3a). The addition of indomethacin, a cyclooxygenase inhibitor, did not decrease the levels of PGE_2 when AA was added to the homogenate; however, we have sampled *Clavelina huntsmani* from other locations (data not shown) and have found indomethacin to be a potent inhibitor of PGE_2 synthesis. Conversely, the UCT had significantly less endogenous PGE_2 in the 10K, cytosolic, and microsomal preparations (Fig. 3b).

2.1.5. Additional Observations

Initial studies on the LBT provide evidence for the presence of both dienes and trienes, which can be further enhanced with the addition of arachidonic acid and eicosapentaenoic acid (EPA), both substrates of lipoxygenases. Conjugated dienes and trienes would be present as metabolites of arachidonic acid metabolites via a lipoxygenase enzyme. The UV absorbance spectra of the organic extracts of the LBT were measured. The diene and triene peaks absorb at 270 and 232 nm, respectively. These peaks can be significantly diminished by heating the samples at 100°C for 20 min. This response is indicative of enzymatic activity. Furthermore, thin-layer chromatography indicates that the LBT contains a compound migrating in the region of an 11- or 12-HETE. This can be seen clearly when developed with iodine. When 14C-AA is used, a faint band can be observed in the region of the 11- or 12-HETE. This is of particular interest since 12-HETE is a metabolite that has been associated with a P450 enzyme in the cornea (Schwartzman et al., 1987). Slight cross-reactivity with an anti-cytochrome P450IIB was found in two preparations of the LBT (10K and cytosol). However, this finding is not surprising, since cytochrome P450 enzymes isolated from many aquatic species are 3-methylchloranthracene (3-MC)- rather than phenobarbital-inducible systems. Studies performed by Stegeman and Lech (1991) as well as Melancon et al. (1986) have shown the cross-reactivity of antibodies directed against 3-MC cytochrome P450 enzymes among aquatic species is significantly higher than would be achieved with antibodies directed against phenobarbital-inducible cytochrome P450 enzymes. These data suggest that tunicate preparations should be

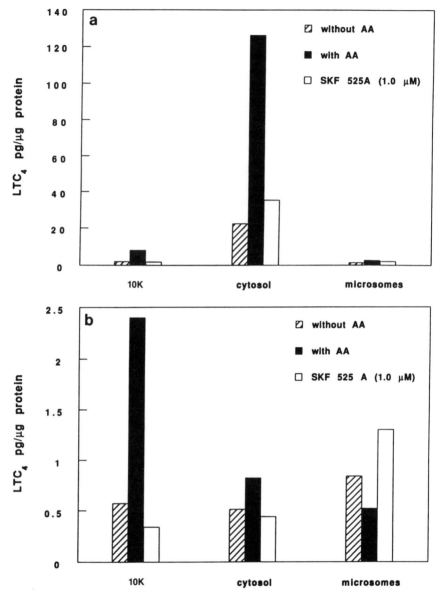

Figure 2. Leukotriene C_4 (pg/µg protein) measured in the 10,000 × *g* supernatant, cytosolic, and microsomal fractions of (a) the tunicate *Clavelina huntsmani* and (b) our unidentified colonial tunicate (UCT), using a radioimmunoassay. A control (untreated), AA-treated (0.34 mM) and inhibitor, SKF 525A, pretreated sample were assayed. One sample was incubated for 15 min with AA, and another was pretreated for 30 min with inhibitor prior to the addition of AA. All samples were extracted with ethyl acetate. The extract was dried under nitrogen, reconstituted in assay buffer, and then 100-µl aliquots (in duplicate) were assayed for LTC_4.

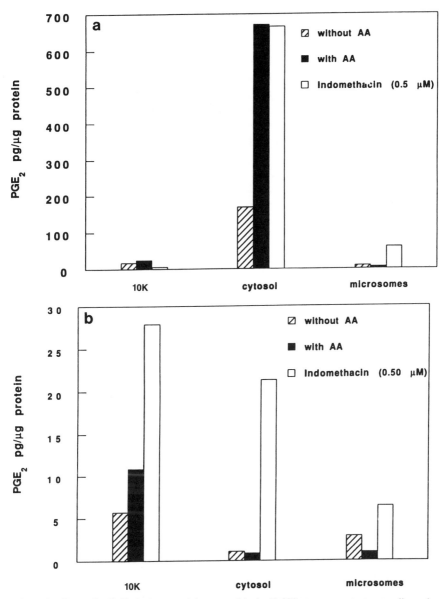

Figure 3. Prostaglandin E_2 (pg/μg protein) measured in the 10,000 $\times$ g supernatant, cytosolic, and microsomal fractions of (a) the tunicate *Clavelina huntsmani* and (b) our unidentified colonial tunicate (UCT), using a radioimmunoassay. A control (untreated), AA-treated (0.34 mM), and inhibitor, indomethacin, pretreated sample were assayed. One sample was incubated for 15 min with AA, and another was pretreated for 30 min with inhibitor prior to the addition of AA. All samples were extracted with ethyl acetate. The extract was dried under nitrogen, reconstituted in assay buffer, and then 100-μl aliquots (in duplicate) were assayed for PGE_2.

screened with numerous other antibodies and especially with the 3-MC cytochrome P450-directed antibodies.

The role of eicosanoids in reproduction has been well documented. While the emphasis of this early research targeted the products of the cyclooxygenase pathway (i.e., prostaglandins) and reproduction, recently metabolites of the lipoxygenase and P450 pathways are receiving more attention. We believe that the data collected herein offer strong evidence that the marine tunicate may serve as an excellent nonmammalian model to study eicosanoid synthesis, because we have found that tunicates exhibit pathways for releasing and metabolizing AA that appear similar to those of mammals. Studies on marine products of eicosanoid biosynthesis include, however, not only research on the primitive marine vertebrate, but also surveys of eicosanoid biosynthesis in tropical and temperate seaweeds. Of particular interest are the biochemical similarities that may have been conserved at the cellular level between these marine plants and the primitive marine vertebrate.

2.2. Eicosanoid Biosynthesis in Coralline Red Algae

In the course of systematic surveys of the natural products chemistry of tropical and temperature seaweeds, Moghaddam *et al.* (1990) found that many macroalgae, especially rhodophytes, have the enzymatic capacity to metabolize polyunsaturated fatty acids in ways analogous to the lipoxygenase pathway in mammalian systems. They also found that the concentrations of these compounds were higher in algal tissues than in animals and that the algal compounds were often structurally more complex.

Until recently, little was known of the role of eicosanoid metabolism in algae, but, because these compounds are produced in some algae in relatively large amounts (0.5–5.0% of crude lipid extract) (Moghaddam *et al.*, 1990), it is logical to speculate that this metabolism is important to physiological processes in the algae.

In recent studies of eicosanoid metabolism in *Bossiella orbigniana*, we have shown the conversion of added archidonic acid (AA) to a previously unknown eicosapentaenoic acid product, bosseopentaenoic acid (BPA) (Burgess *et al.*, 1991). In order to investigate the cellular and physiological roles of BPA in this alga, we investigated the bioavailability of AA when added to intact and homogenized plants. In addition, we investigated oxygen consumption during BPA biosynthesis and the effect of photostimulation.

Increased oxygen uptake was observed upon the addition of AA to both plant homogenates and isolated thallus segments. This increased uptake was coincident with an increased formation of BPA. No significant differences were found between incubations in light and dark conditions. Experiments using a new

5-lipoxygenase inhibitor, L-651,896 (Chan *et al.*, 1988), resulted in marked increases in biosynthesis or decreased metabolism of BPA.

With these results, conditions are now set to begin to investigate the physiological and cellular roles of BPA in this alga. We recently noted that a lipoxygenase product from arachidonic acid was produced in addition to BPA and that a cytochrome P450 enzyme was present. Such evidence raises the possibility that, in addition to some fundamental physiological process being involved in eicosanoid biosynthesis, the P450 enzyme(s) may be involved in chemical defense (i.e., oxidative metabolism of environmental toxins). We chose to study *Bossiella* because coralline red algae in this genus are among the most primitive of marine organisms. Their calcified fossils have contributed to the infrastructure of coral reefs for millions of years. Furthermore, these organisms seem to have evolved little. Thus, through this study, we are in a position to study arachidonic acid metabolism in one of the earliest phylogenetic forms.

The investigations on the coralline red algae have indicated the presence of a novel, previously unreported metabolite from AA metabolism. While the endogenous function of the marine natural products are of primary interest, an equally critical question is whether these compounds possess selective activities that may afford them a biomedical potential. A means of linking research investigations (similar to those previously described) to relevant biomedical studies can be realized with screening studies using a cellular model that is representative of a specific critical dysfunction.

3. SITE AND MECHANISM OF ACTION OF MARINE NATURAL PRODUCTS

3.1. Fuscoside: A Novel Inhibitor of Leukotriene Biosynthesis Isolated from the Caribbean Soft Coral Eunicea fusca

Conventional strategies for managing inflammatory diseases, such as arthritis, emphysema, myocardial ischemia, and reperfusion injury, and chronic bowel disease have relied almost exclusively upon the use of steroidal and nonsteroidal anti-inflammatory agents such as cortisone, dexamethasone, aspirin, and indomethacin, respectively. These agents specifically inhibit the formation of bioactive prostaglandins and thromboxanes which have been implicated in a wide number of inflammatory diseases. Studies on the mechanism of action of the steroidal agents have been inconclusive, but generally include either drug-induced biosynthesis of lipocortin (Ghiara *et al.*, 1984), which is postulated to be an endogenous inhibitor of phospholipase A_2, or indirect inhibition of prostaglandin synthetase (cyclooxygenase) biosynthesis (Raz *et al.*, 1989). The nonsteroidal drugs inhibit prostaglandin production by directly inactivating the enzyme cyclooxygenase. While these drugs have proven effective against a number of inflammatory diseases, their significant untoward effects have stimulated a search for more potent and specific anti-inflammatory agents with fewer side effects. New areas of inflammatory research have included inhibitors of phospholipase A_2, the rate-dependent enzyme in arachidonic acid release, direct inhibitors of 5-lipoxygenase or its translocation, specific eicosanoid receptor antagonists, inhibitors of cytokine release, and inhibitors of proteases found within several types of immunocompetent cell types such as polymorphonuclear leukocytes (PMNs) and eosinophils. As the pathophysiology of inflammatory diseases is better understood, it is expected that new drugs can be developed which target these and other sites with increased specificity and potency and thereby alleviate a number of the problems faced with conventional anti-inflammatory therapies.

One of the newer areas of research is the search for and development of drugs which inhibit leukotriene biosynthesis. The leukotrienes are bioactive metabolites of arachidonic acid which have been implicated in a wide variety of inflammatory processes. Leukotriene B_4 (LTB_4) is a potent chemotactic agent for PMNs and has been found in elevated concentrations in skin lesions of psoriatic patients where large numbers of PMNs are concomitantly found (Bauer, 1986). LTs and elevated concentrations of PMNs have also been identified at regions of myocardial infarction following reperfusion (Werns and Lucchessi, 1987) and at sites of severe ulcerative colitis (Hermanowicz *et al.*, 1985). Leukotrienes have additionally been shown to induce bronchoconstriction, mucous hypersecretion, vasoconstriction, and microvascular edema (Lewis and Austen, 1984).

The first step in the biosynthesis of leukotrienes is the peroxidation of arachidonic acid by 5-lipoxygenase to form 5-hydroxyeicosatetraenoic acid (5-HPETE), which is subsequently dehydrated by the same enzyme to the unstable epoxide leukotriene A_4 (LTA_4) LTA_4 is subsequently metabolized, depending upon the enzymes available, to $(5S),(12R)$-dihydroxyeicosatetraenonic acid (LTB_4), as in neutrophils, or it may be converted to the spasmogenic peptidoleukotriene LTC_4 by the addition of glutathione, as evidenced in macrophages. Significant information has recently been obtained regarding the regulation of 5-lipoxygenase. 5-LO is an unstable, 80-kDa enzyme which requires multiple stimulatory factors for activation, including calcium, ATP, phosphatidylcholine, and another, as yet unidentified, cellular component shown to stimulate activity (Rouzer *et al.*, 1986). It has also been demonstrated that following cellular stimulation, changes in intracellular calcium levels promote the translocation of 5-LO from the cytosol to the membrane, where it is presumed to form an activated complex with a regulatory protein called FLAP (5-lipoxygenase activating protein) (Dixon *et al.*, 1990). Following the formation of leukotrienes, the membrane-associated complex undergoes irreversible inactivation. During the last few years, the pharmaceutical industry has focused increasing attention on the development of specific lipoxygenase inhibitors and agents which modulate leukotriene production. While several compounds have shown potency *in vivo* as inhibitors of leukotriene biosynthesis, future developments in this area will greatly depend upon the clinical demonstration of efficacy in diseases such as arthritis, emphysema, and inflammatory bowel disease.

One of the more recent and exciting discoveries in our marine natural products program has been the characterization of an inhibitor of leukotriene biosynthesis called fuscoside, an arabinose-containing diterpenoid isolated from the Caribbean soft coral, *Eunicea fusca*. Fuscoside (FSD) effectively inhibits phorbol myristate acetate (PMA)-induced edema (1 μg PMA/ear) in mouse ears at a level comparable with indomethacin (INDO) (IC_{50}: 82 μg FSD/ear vs. 75 μg INDO/ear) over a 3.3-hr exposure period. FSD has also been shown to be significantly more effective than INDO at a single dose over 24 hr in the PMA model (80% inhibition at 200 μg FSD/ear vs. 30% inhibition with 200 μg INDO/ear). FSD doses below 500 μg/ear were not effective against arachidonic acid (AA)-induced inflammation in mouse ears, indicating FSD may act at the level of PLA_2. However, FSD (up to 500 μM) does not inhibit bee venom or human synovial fluid PLA_2. Histological preparations and quantification of the neutrophil-specific marker myeloperoxidase in PMA-treated ears demonstrate that FSD inhibits neutrophil infiltration into edematous and inflamed regions. FSD selectively inhibits LTC_4 synthesis and release (IC_{50} = 8 μM) from calcium ionophore (A23187) or opsonized zymosan-stimulated mouse peritoneal macrophages (30 min preincubation), yet has negligible effects on PGE_2 synthesis and release (IC_{50} > 50 μM). In adherent or suspended human polymorphonuclear leukocytes stimulated with A23187, FSD inhibits LTB_4 synthesis and release in a dose-dependent manner (IC_{50} = 10 μM) under standard conditions with no apparent effect on degranulation as measured by myeloperoxidase or lactoferrin release. Using 500 μg protein from the 10K supernatant of homogenized human neutrophils as the source of active 5-lipoxygenase, FSD inhibits AA conversion to LTB_4 (IC_{50} = 40 μM). In purified ram seminal vesicle cyclooxygenase preparations, FSD (500 μM) failed to inhibit PGE_2 synthesis from AA. Preliminary kinetic studies indicate that FSD is an irreversible inhibitor of leukotriene synthesis. Unlike the reversible lipoxygenase inhibitor L-651,896, the inhibitory effects of FSD can be diluted by increasing the cell concentration in the assay but can be restored by increasing the concentration of FSD proportionate to the number of cells/protein. Additionally, the inhibitory effects of FSD cannot be reversed after 10 min incubation, even after repeated washings with buffer or 0.1% albumin. Quantification by RP-HPLC of cellular uptake/binding of FSD further demonstrates a time-dependent correlation among drug loss from supernatant, drug uptake/binding by cells, and inhibition of leukotriene synthesis. Mass spectral analysis of HPLC elutants also indicates that fuscoside is not metabolized by cultured neutrophils or macrophages. Other studies utilizing noninflammatory cell types such as the GH3 pituitary tumor cell line have examined the effects of FSD on calcium mobilization following stimulation by thyrotropin-releasing hormone (TRH) or potassium chloride (KCl). FSD had minimal effects on the fura-2 signal elicited by TRH, but completely blocked the KCl-generated signal at submicromolar concentrations. These studies indicate that FSD affects calcium influx through voltage-dependent calcium channels in these cells, and may provide a clue to this drug's mechanism of inhibiting the calcium-dependent generation of leukotrienes in inflammatory cells. While further studies on the exact mechanism of action of FSD are in progress, this marine natural product has demonstrated efficacy in several models of inflammation and may prove to be clinically useful against certain disease states, such as inflammatory skin diseases or inflammatory bowel disease. Furthermore, FSD represents a new chemical class of leukotriene biosynthesis inhibitors and is the first group of marine natural products to have a demonstrated ability to modulate the lipoxygenase pathway of arachidonic acid metabolism in mammalian cells.

3.2. Mechanism of Action of Scalaradial

Before the beginning of modern pharmacology, natural products from terrestrial plants were used as therapeutic agents against inflammatory disorders. For centuries in England, extracts from the bark of the willow tree were known to be effective in the abatement of fever. Salicylic acid was identified in the 1800s as one of the compounds responsible for the antipyretic effects of this extract (Goodman Gilman *et al.*, 1985) and a more potent derivative of salicylic acid, acetylsalicylic acid (aspirin), has been used since the early 1900s as an anti-inflammatory agent in humans. It was not until the 1970s that the target for aspirin, cyclooxygenase, was identified by Vane (1971). He used aspirin as a probe and linked its anti-inflammatory action to inhibition of prostaglandin production.

Marine invertebrates and marine plants have provided a new source of pharmacological probes for the study of the mechanisms mediating inflammatory disease processes. New structural classes of anti-inflammatory compounds have been discovered by our research group from pharmacological evaluation of pure marine natural products isolated from marine sponges (Glaser *et al.*, 1989; Kernan *et al.*, 1989; Look *et al.*, 1986). The identification of the biochemical targets for these anti-inflammatory natural products will provide insight into the mechanisms of inflammation and will provide lead compounds to use for the rational design of better therapeutic anti-inflammatory agents.

The scalaradials, marine natural products extracted from the sponge of the genus *Cacospongia*, are an extremely important class of anti-inflammatory compounds. These compounds were first isolated in the 1970s by Cimino *et al.* (1974). We originally observed the anti-inflammatory properties of this group of marine natural products in an *in vivo* assay against PMA-induced inflammation. In *in vivo* screening the parent compound, scalaradial (SLD) was a potent inhibitor with an IC_{50} of 51 µg/ear. SLD was also active when given systemically, which suggested it might provide useful information for the design of therapeutic agents. In contrast to aspirin and indomethacin, SLD appeared to act by a different mechanism, as it did inhibit arachidonic acid (AA)-induced inflammation, as would be expected if it acted to inhibit cyclooxygenase. These observations led us to hypothesize that SLD was acting to inhibit phospholipase A_2 (PLA_2), the step prior to cyclooxygenase. We had previously identified a number of PLA_2 inhibitors from marine invertebrates. Manoalide (MLD) is a PLA_2 inhibitor from the sponge *Luffariella variabilis* which we studied extensively (Glaser *et al.*, 1986).

We found that SLD was a potent inhibitor of a purified extracellular form of PLA_2 from bee venom (IC_{50} of 0.7 µM). SLD differed structurally from MLD, providing an additional structure to use to study the chemical moieties needed to design effective PLA_2 inhibitors for therapeutic use.

SLD kinetic properties also differed from MLD. Kinetic studies showed that SLD formed an apparent noncovalent complex with PLA_2 before binding to the enzyme in an irreversible manner (de Carvalho and Jacobs, 1991). These data suggested that SLD had an apparent affinity for the enzyme and suggested that it might bind to the active site. Further studies showed that inactivation could be slowed if substrate was present during the inactivation period. The above data suggested that SLD might be a useful active-site probe for PLA_2 in inflammatory cells.

Additional studies were performed in the *in vitro* phorbol myristate acetate (PMA)-stimulated murine peritoneal macrophage model. PMA stimulates the release of AA through activation of PLA_2 in macrophages (Wijkander and Sundler, 1989). The released AA is further metabolized into prostaglandin E_2 (PGE_2) (Bonney *et al.*, 1980). SLD inhibited the release of PGE_2 from PMA-stimulated macrophages, $IC_{50} = 0.5$ µM, with 84% inhibition occurring at 1 µM. This inhibition of PGE_2 production appeared to be due to a decrease in precursor AA, as SLD inhibited AA release with 86% inhibition at 1 µM. Since SLD inhibited the release of AA from PMA-stimulated macrophages, this suggested that it inhibited macrophage PLA_2. In order to determine whether SLD directly inactivated the macrophage PLA_2, PLA_2 activity was directly assayed in macrophage homogenates. A calcium-dependent PLA_2 activity was characterized in macrophage homogenates which was increased after stimulation of macrophages of PMA. SLD directly inhibited this PLA_2 activity by 86% at 20 µM when it was added to homogenates obtained from PMA-stimulated cells, suggesting that SLD might be a useful probe for PLA_2 in inflammatory cells.

PLA_2 is one of the enzymes which regulates the release of arachidonic acid, the precursor to the proinflammatory prostaglandins and leukotrienes. This enzyme is an attractive target for therapeutic agents. The study of marine natural products like SLD will provide new insight into the role of PLA_2 in inflammatory diseases.

3.3. Multidrug Resistance in CH^R-C5 Cells

In chemotherapy, one major impediment to successful treatment is the inherent ability of cancer cells to evade or overcome the cytotoxic nature of anticancer drugs (Beck, 1987; Bellamy *et al.*, 1988; Bruno *et al.*, 1990). Although there may be diverse methods of circumventing cytotoxicity, it has recently been suggested that drug resistance may be mediated, at least in part, by a phenomenon known as multidrug resistance or MDR (Beck, 1987). MDR is the process whereby cytotoxins are actively extruded from the interior of resistant cells. A membrane-bound glycoprotein (p-glycoprotein or p-170) acts to pump a wide variety of structurally and functionally unrelated cytotoxins from the cell's interior. These compounds include, among others, vinca alkaloids, anthracyclines, colchicine, and epipodophyllotoxins (Biedler and Meyers, 1981; Gottesman, 1988). In tissue culture, cells exposed to one of more of these agents not only become resistant to the initial selecting agent, but also display cross-resistance to other compounds the cells have yet to encounter (Pastan *et al.*, 1987). Gel electrophoresis and immunocytochemical analyses show elevated levels of p-glycoprotein in drug-resistant cells compared to parental sensitive cells (Chan *et al.*, 1988). Furthermore, the degree of resistance is proportional to the amount of p-170 expressed in these cells.

It is interesting to note that normal human tissues contain detectable levels of *mdr1* RNA. Some tissues express *mdr1* RNA at levels that are found in drug-resistant tumor samples, which suggests a normal functioning of the *mdr1* gene product. In particular, the colon, liver, kidney, and adrenal gland contain elevated levels of both *mdr1* product and p-glycoprotein. The efflux pump present in many mammalian sources bears structural homology to transmembrane proteins in certain bacteria and some insects. If we are able to demonstrate the presence of p-glycoprotein in marine organisms, it may help in further elucidating the role of this efflux pump in normal tissue. Like cytochrome P450, the p-170 protein may serve an important role in chemical defense.

That the MDR phenotype can be modulated by increased efflux of cytotoxic agents can be demonstrated by activation of protein kinase C (PKC). Some data suggest p-glycoprotein is more efficient in extruding cytotoxins when the pump is hyperphosphorylated, which can apparently be modulated by PKC (Center, 1988). Several reports also implicate phosphatidyl inositol metabolism in p-170 efflux activity, although this phenomenon is not well understood (Staats *et al.*, 1990). We are studying these stimuli as well as examining novel marine natural products for their ability to reverse drug resistance and restore sensitivity in select cell lines.

Researchers have shown that the MDR phenotype can be overcome using certain agents in combination with a known cytotoxin such as adriamycin (Ford and Hait, 1990). Some of the best reversing agents include calcium channel blockers and calmodulin inhibitors such as verapamil and trifluoperazine. Verapamil is probably the best studied of these drugs and has been put to use in clinical trials as part of a chemotherapy regimen (Ozols *et al.*, 1987). Although verapamil is active *in vivo*, the levels of this drug needed to maintain resistance also induce cardiac failure (Ozols *et al.*, 1987). It was previously thought that calcium inhibitors worked through a direct action on intracellular Ca^{2+} levels. Two reports have since indicated that this is not the mechanism of action of these drugs, but the true mechanism remains unclear (Huet and Robert, 1988; Kessel and Wilberding, 1984). Subsequently, there has been a drive by investigators to identify other drugs which have little or no intrinsic activity of their own other than to antagonize p-170.

We have started research in this field by investigating a limited number of marine natural products utilizing an MTT proliferation assay to quantify the ability of these compounds to modulate the efflux activity of p-glycoprotein. We have chosen the CH^R-C5 cell line for this study because of the high level of resistance imparted to these cells via p-glycoprotein. Incubation of CH^R-C5 cells in 1% EtOH, 1 µg/ml verapamil, or in 1 µg/ml colchicine had no apparent effect on cell growth (Fig. 4). These conditions were used as control groups. Coincubation

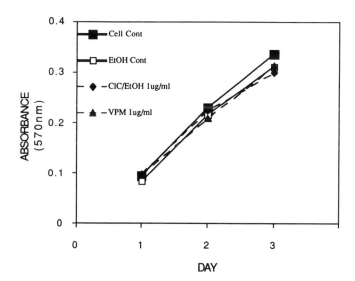

Figure 4.   Incubation of CH^R-C5 cells in 1% EtOH, 1 µg/ml verapamil, or 1 µg/ml colchicine showed no apparent effect on cell growth.

of 10 μg/ml verapamil and 1 μg/ml colchicine resulted in reversal of the MDR phenotype (not shown). Partial reversal was seen when 1 μg/ml verapamil was incubated with 1 μg/ml colchicine, in agreement with reports which indicate that this is approximately the IC_{50} of colchicine in CH^R-C5 cells. Presumably, at this concentration there is sufficient intracellular verapamil to compete with colchicine at the binding site of p-170. This competition results in a decreased efflux of colchicine, thus allowing more to bind to its target, thereby inhibiting cell division.

The marine natural products tested in this report are characterized as (1) active, (2) partially active, (3) inactive, or (4) cytotoxic as a result of their ability to inhibit proliferation with or without colchicine. Inactive compounds are those which did not appreciably inhibit growth with or without coincubation of 1 μg/ml colchicine. Examples of these compounds include **1** and **2**. Natural products classified as active are characterized by their ability to reverse resistance similar to

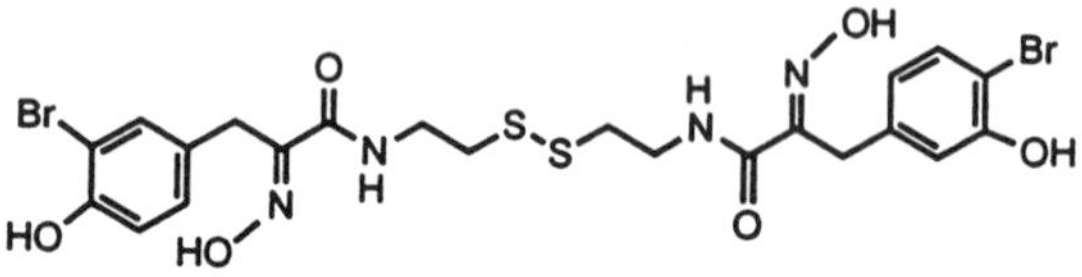

(1)

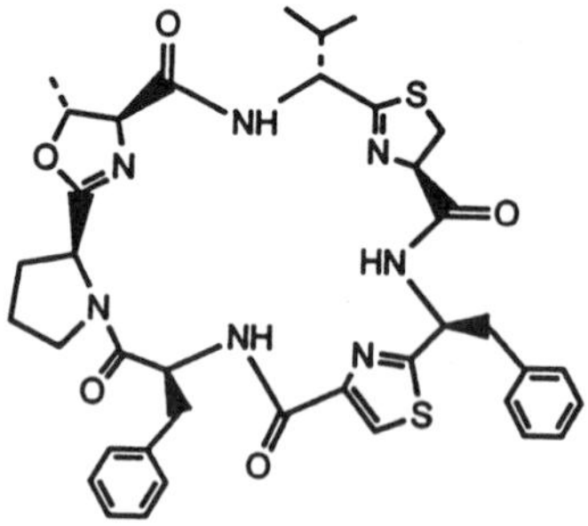

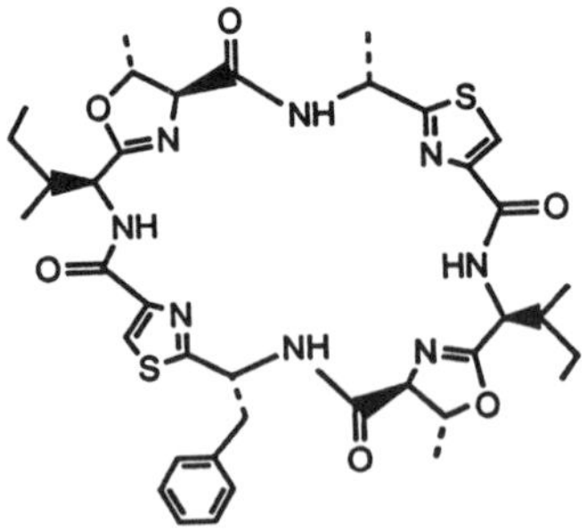

(2)   (3)

verapamil in combination with colchicine but show no activity when given alone. Only one of the compounds tested (**3**) showed a level of reversal activity similar to verapamil (Fig. 5). The cytotoxic compounds WF 240 and WF 241 (related to didemnin B) inhibited cell growth even in the absence of colchicine (Fig. 6). Several compounds tested demonstrated partial activity (not shown) in modulating p-170-mediated efflux.

The initial findings of this report suggest that products isolated from marine organisms may serve as a useful source of pharmacological tools for the further elucidation of multidrug resistance in cultured cells. Although a wide variety of chemical structures may be active in restoring sensitivity to resistant cell lines, we

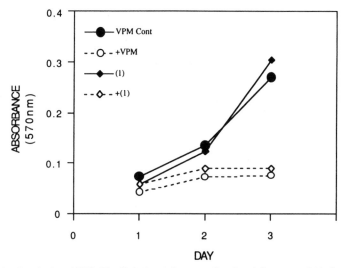

Figure 5. Incubation of CH[R]-C5 cells in 1 μg/ml verapamil or 1 μg/ml compound **1** had no effect on cell growth. However, coincubation of these compounds with colchicine demonstrated drug-resistance-reversal activity.

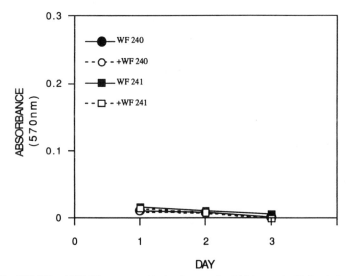

Figure 6. WF 240 and WF 241 were tested in the absence of colchicine, and with 1 μg/ml colchicine (+WF 240 and +WF 241). The cytotoxic compounds WF 240 and WF 241 (related to didemnin B) inhibited cell growth even in the absence of colchicine.

have had greater initial success with some of the cyclic peptides. With the exception of cyclosporin A, this class of natural products has not previously been reported to act selectively on MDR cells. One report even shows cyclosporin A to be significantly more active than verapamil in reversing MDR. The octapeptide we tested was previously isolated from the ascidian *Lissoclinum patella*, whereas cyclosporin A is an undecapeptide isolated from a fungus and may not alter drug accumulation in resistant cells, but may act at the membrane level or via calmodulin perturbation. According to Degnan *et al.* (1989a, b), some peptides similar to our own octapeptides have been shown to be cytotoxic in their assay. However, one of their compounds corresponding to the structure of **3** apparently is not toxic at 1 μg/ml. These peptides and other natural products demonstrating activity in our model will eventually be tested in the parental sensitive line, Aux B1, and in other resistant lines to determine the extent of p-170 efflux inhibition.

The occurrence of a structure homologous to p-glycoprotein has been demonstrated in some bacteria and insects as well as a wide range of mammalian species, but there have been no citations of p-glycoprotein expression in marine organisms. To protect themselves from predation, many marine species have developed chemical defense mechanisms. Some of these products (e.g., didemnins) are extremely toxic in various bioassays and may serve to defend against grazers and predators. However, not all organisms show deleterious effects of exposure to these compounds. Some actually share a mutually beneficial existence with cytotoxin-producing organisms, while others may sequester similar compounds in their tissues without ill effects or even utilize the products for their own defense. Using the cytoplasmic-binding antibody C219 in Western blotting, we will attempt to detect p-glycoprotein in sponges, corals, algae, tunicates, and other marine species of interest. A molecule structurally or functionally related to p-170 that is evolutionarily conserved may indicate a natural role for this protein in preserving species.

4. SUMMARY

Our research has taken two paths for the study of marine natural products and metabolites. One is the use of marine organisms as pharmacological models of cellular and molecular processes, while the second approach targets the site and mechanism of action of marine natural products. These two directions have aided us in forming a more cohesive and complete picture with regard to the eicosanoid biosynthesis and cellular activity of marine organisms. The data indicate that both primitive marine vertebrates and coralline red algae can metabolize AA to yield a variety of eicosanoid metabolites. These metabolites can in some cases cross-react with mammalian prostaglandin and leukotriene antibodies. Furthermore, marine natural products may function to selectively inhibit pathways involved in eicosanoid metabolism as shown with the marine compounds fucoside and scalaradial. The vast possibilities of marine products in the biomedical field are especially highlighted by the screening studies performed using marine natural products which tested their ability to modulate the activity of cellular p-170 in CH^R-C5 cells.

It is clear from our studies that marine resources will continue to play a pivotal role in the discovery of new, chemically diverse structures with unique pharmacological profiles. Furthermore, as the ecological and physiological significance of these compounds in marine systems is further characterized, a better understanding of mammalian biochemistry and physiology will be achieved.

REFERENCES

Allen, T. J., 1990, The effects of manganese and changes in internal calcium on Na$^+$–Ca$^+$ exchange fluxes in the intact squid giant axon, *Biochim. Biophys. Acta* **1030**:101–110.

Bauer, F. W., van der Kerkhof, P. C., and Massen-de Grood, R. M., 1986, Epidermal hyperproliferation following the induction of microabscesses by leukotriene B$_4$, *Br. J. Dermatol.* **114**:409–412.

Beck, W. T., 1987, The cell biology of multiple drug resistance, *Biochem. Pharmacol.* **36**(18):2879–2887.

Bellamy, W. T., Dalton, W. S., Kailey, J. M., Gleason, M. C., McCloskey, T. M., Dorr, R. T., and Alberts, D. S., 1988, Verapamil reversal of doxorubicin resistance in multidrug-resistant human myeloma cells and association with drug accumulation and DNA damage, *Cancer Res.* **48**(22): 6365–6370.

Biedler, J. L., and Meyers, M. B., 1989, Multidrug resistance (vinca alkaloids, actinomycin D, and anthracycline antibiotics), *Drug Resist. Mamm. Cell* **2**:57–58.

Bonney, R. J., Wightman, P. D., Dahlgren, M. E., Davies, P., Kuehl, F. A., and Humes, J., 1980, Effect of RNA and protein synthesis inhibitors on the release of inflammatory mediators by macrophages responding to phorbol myristate acetate, *Biochim. Biophys. Acta* **633**:410–421.

Burgess, J. R., de la Rosa, R. I., Jacobs, R. S., and Butler, A., 1991, A new eicosapentaenoic acid formed from arachidonic acid in the coralline red algae *Bossiella orbigniana*, *Lipids* **26**:162–165.

Busselberg, D., Evans, M. L., Rahmann, H., and Carpenter, D., 1991, Lead and zinc block a voltage-activated calcium channel of *Aplysia* neuron, *J. Neurophysiol.* **65**:786–795.

Bruno, N. A., Carver, L. A., and Slate, D. L., 1990, Isolation and characterization of doxorubicin-resistant Lewis lung carcinoma variants, *Cancer Commun.* **2**(4):151–158.

Calignano, A., Piomelli, D., Sacktor, T. C., and Swartz, J. H., 1991, A phospholipase A$_2$ stimulating protein regulated by protein kinase C in *Aplysia* neurons, *Brain Res. Mol. Brain Res.* **9**(4):347–351.

Center, M. S., 1988, Mechanisms regulating cell resistance to adriamycin, *Biochem. Pharmacol.* **34**: 1471–1476.

Chan, H. S., Bradley, G., Thorner, P., Haddad, G., Gallie, B. L., and Ling, V., 1988, Methods in laboratory investigation: A sensitive method for immunocytochemical detection of P-glycoprotein in multidrug-resistant human ovarian-carcinoma cell-lines, *Lab. Invest.* **59**(6):870–875.

Cimino, G., De Stefano, S., and Minale, L., 1974, Scalaradial, a third sesterterpene with the tetracarbocyclic skeleton of scalarin, from the sponge *Cacospongia mollior*, *Experientia* **30**:846–847.

Culver, P., and Jacobs, R. S., 1981, Lophotoxin: A neuromuscular toxin from the sea whip (*Lophogorgia rigida*), *Toxicon* **6**:825–830.

De Carvahlo, M. S., and Jacobs, R. S., 1991, Two step inactivation of bee venom PLA2 by scalardial, *J. Biochem. Pharmacol.* **42**(B):1621–1626.

De Freitas, J. C., Blankemeier, L. A., and Jacobs, R. S., 1984, *In vitro* inactivation of the neurotoxic action of β-bungarotoxin by the marine natural product, manoalide, *Experientia* **40**(7):864–865.

Degnan, B. M., Hawkins, C. J., Lavin, M. F., McCaffrey, E. J., Parry, C. L., Vandenbrenk, A. L., and Watters, D. J., 1989a, New cyclic-peptides with cyto-toxic activity from the ascidian *Lissoclinum-Patella*, *J. Med. Chem.* **32**(6):1349–1354.

Degnan, B. M., Hawkins, C. J., Lavin, M. F., McCaffrey, E. J., Parry, C. L., and Watters, D. J., 1989b, Novel cyto-toxic compounds from the ascidian *Lissoclinum-Bistratum*, *J. Med. Chem.* **32**(6): 1354–1359.

Dixon, R. A., Diehl, R. E., Opas, E. E., Rands, E., Vickers, P. J., Evans, J. F., Gillard, J. W., and Miller, D. K., 1990, Requirement of a 5-lipoxygenase-activating protein for leukotriene synthesis, *Nature* **343**:282–284.

Escalante, B., Erlij, D., Falck, J. R., and McGiff, J. C., 1991, Effect of cytochrome P450 arachidonate metabolites on ion transport in rabbit kidney loop of Henle, *Science* **251**:799–802.

Ettouati, W. S., and Jacobs, R. S., 1987, Effect of pseudopterosin A on cell division, cell cycle progression, DNA, and protein synthesis in cultured sea urchin embryos, *Mol. Pharmacol.* **31**: 500–505.

Fenical, W., Okuda, R. K., Bandurraga, M. M., Culver, P., and Jacobs, R. S., 1981, Lophotoxin: A novel neuromuscular toxin from Pacific sea whips of the genus *Lophogorgia*, *Science* **212**:1512.

Fishman, H. M., and Lipicky, R. J., 1991, Determination of K^+-channel relaxation times in squid axon membranes by Hodgkin–Huxley and direct linear analysis, *Biophys. Chem.* **39**:177–190.

Ford, J. M., and Hait, W. N., 1990, Pharmacology of drugs that alter multidrug resistance in cancer, *Pharmacol. Rev.* **42**:155–190.

Fox, G. W., Kriebel, M. E., and Pappas, G. D., 1990, Morphological, physiological and biochemical observation on skate electric organ, *Anat. Embryol.* **18**:305–315.

Ghiara, P., Meli, T., Parente, L., and Persico, P., 1984, Distinct inhibition of membrane bound and lysosomal phospholipase A_2 by glucocorticoid-induced proteins, *Biochem. Pharmacol.* **33**:1445–1450.

Glaser, K. B., and Jacobs, R. S., 1986, Molecular pharmacology of manoalide; inactivation of bee venom phospholipase A_2, *Biochem. Pharmacol.* **35**:449–453.

Glaser, K. B., de Carvalho, M. S., Jacobs, R. S., Kernan, M. R., and Faulkner, D. J., 1989, Manoalide: Structure–activity studies and definition of the pharmacophore for phospholipase A_2 inactivation, *Mol. Pharmacol.* **36**:782–788.

Goodman Gilman, A., Goodman, S. S., Rall, W. T., and Murad, F., 1985, *The Pharmacological Basis of Therapeutics*, 7th ed., Macmillan, New York.

Gottesman, M. M., 1988, Multidrug resistance during chemical carcinogenesis: A mechanism revealed, *J. Natl. Cancer Inst.* **80**(17):1352–1353.

Hermanowicz, A., Gibson, P. R., and Jewell, D. P., 1985, The role of phagocytes in inflammatory bowel disease, *Clin. Sci.* **69**:2241–2249.

Huet, S., and Robert, J., 1988, The reversal of doxorubicin resistance by verapamil is not due to an effect on calcium channels, *Int. J. Cancer* **41**(2):283–286.

Kernan, M. R., Faulkner, D. J., Parkanyi, L., Clardy, J., de Carvalho, M. S., and Jacobs, R. S., 1989, Luffolide, a novel anti-inflammatory terpene from the sponge *Luffariella* sp., *Experientia* **45**: 388–390.

Kessel, D., and Wilberding, C., 1984, Mode of action of calcium antagonists which alter anthracycline resistance, *Biochem. Pharmacol.* **33**(7):1157–1160.

Lewis, R. A., and Austen, K. F., 1984, The biologically active leukotrienes: Biosynthesis, metabolism, receptors, functions, and pharmacology, *J. Clin. Invest.* **73**:889–897.

Look, S. A., Fenical, W., Jacobs, R. S., and Clardy, J., 1986, The pseudopterosins: Anti-inflammatory and analgesic natural products from the sea whip *Pseudopterogorgia elisabethae*, *Proc. Natl. Acad. Sci. USA* **83**:6238–6240.

Melancon, M. J., Yeo, S. E., and Lech, J., 1986, Induction of hepatic microsomal monooxygenase activity in fish by exposure to river water, *Environmental Toxicology and Chemistry* **6**:127–135.

Moghaddam, M. F., Gerwick, W. H., and Ballantine, D. L., 1990, Discovery of the mammalian insulin release modulator, hepoxilin B_3, from the tropical red algae *Platysiphonia miniata* and *Cottoniella filamentosa*, *J. Biol. Chem.* **265**:6126–6130.

O'Brien, E. T., Asai, D. J., Groweiss, A., Lipshutz, B. H., Fenical, W., Jacobs, R. S., and Wilson, L., 1986, Mechanism of action of the marine natural product stypoldione: Evidence for reaction with sulfhydryl groups, *J. Med. Chem.* **29**:1829–1851.

Ozols, R. F., Cunnion, R. E., Klecker, R. W., Hamilton, T. C., Ostchega, Y., Parrillo, J. E., and Young, R. C., 1987, Verapamil and adriamycin in the treatment of drug-resistant ovarian-cancer patients, *J. Clin. Oncol.* **5**(4):641–647.

Pastan, I., Gottesman, M., Kahn, C. R., Flier, J., and Eder, P., 1987, Multiple-drug resistance in human cancer, *N. Engl. J. Med.* **316**(22):1388–1393.

Piomelli, D., 1991, Metabolism of arachidonic acid in the nervous system of marine mollusk, *Aplysia californic*, *Am. J. Physiol.* **260**(5Pt2):R844–848.

Raz, A., Wyche, A., and Needleman, P., 1989, Temporal and pharmacological division of fibroblast cyclooxygenase expression into transcriptional and translational phases, *Proc. Natl. Acad. Sci. USA* **86**:1657–1661.

Rouzer, C. A., Matsumoto, T., and Samuelsson, B., 1986, Single protein from human leukocytes possesses 5-lipoxygenase and leukotriene A_4 synthase activities, *Proc. Natl. Acad. Sci. USA* **83**:857–861.

Schlondorff, D., Petty, E., Oates, J. A., Jacoby, M., and Levine, S. D., 1987, Epoxygenase metabolites of arachidonic acid inhibit vasopressin response in toad bladder, *Am. J. Physiol.* **253**:F464–F470.

Schwartzman, M. L., Balazy, M., Masferrer, J., Abraham, N. G., McGiff, J. C., and Murphy, R. C., 1987, 12 (R)-hydroxyicosatetraenoic acid: A cytochrome-P450-dependent arachidonate metabolite that inhibits Na^+,K^+-ATPase in the cornea, *Proc. Natl. Acad. Sci. USA* **84**:8125–8129.

Staats, J., Marquardt, D., and Center, M. S., 1990, Characterization of a membrane-associated protein kinase of multidrug-resistant HL60 cells which phosphorylates p-glycoprotein, *J. Biol. Chem.* **265**:4084–4090.

Stegeman, J. J., and Lech, J., 1991, P-450 monooxygenase systems in aquatic species: Carcinogen metabolism and biomarkers for carcinogen and pollutant exposure, *Environmental Health Perspectives* **90**:101–190.

Vane, J. R., 1971, Inhibition of prostaglandin synthesis as a mechanism for aspirin-like drugs, *Nature New Biol.* **231**(25):232–235.

Werns, S. W., and Lucchessi, B. R., 1987, Inflammation and myocardial infarction, *Br. Med. Bull.* **43**:460–471.

Wijkander, J., and Sundler, R. A., 1989, A phospholipase A_2 hydrolyzing arachidonyl-phospholipids in mouse peritoneal macrophages, *FEBS Lett.* **244**:51–56.

4

Eicosanoids and Related Compounds from Marine Algae

William H. Gerwick and Matthew W. Bernart

1. INTRODUCTION

One of the more surprising groups of natural products to find wide distribution in the marine environment are the eicosanoids and related fatty acids. In mammalian systems, this assemblage of diverse structures is of seminal importance to the maintenance of normal physiology. Furthermore, enhanced or aberrant production of metabolites in this structural class underlies a number of diseases related to inflammation. In the late 1960s and throughout the 1970s, occasional discoveries were made of metabolites generally describable as "eicosanoid-like" from diverse marine life. However, in the 1980s there has been an enormous increase in the number of eicosanoid-like metabolites discovered from these creatures, particularly in the red algae (Rhodophyta) and corals. Despite the widespread occurrence of eicosanoids among marine life forms and their central importance to mammalian physiology and biochemistry, very little is known about what role these compounds play in the ecology or physiology of the producing organisms. Further, in only a few cases have investigators sought to probe the biosynthetic origins of these fat-derived substances, a feature of their mammalian occurrence which has been of extreme interest to mechanistic chemists and central impor-

William H. Gerwick and Matthew W. Bernart • College of Pharmacy, Oregon State University, Corvallis, Oregon 97331.

Marine Biotechnology, Volume 1: Pharmaceutical and Bioactive Natural Products, edited by David H. Attaway and Oskar R. Zaborsky. Plenum Press, New York, 1993.

tance in medicinal considerations. Finally, to the extent that marine-derived eicosanoid-like substances have been evaluated for useful pharmacological properties, they are, as is expected, a potently active class.

With this growth in the number of eicosanoid-like substances reported from marine life and subsequent in-depth biosynthetic and pharmacological followup studies, it is timely to review this never previously reviewed subject. However, given the large number of reports in recent years, it is necessary to restrict the scope in this present chapter to discoveries with marine algae. Within certain confines concerning the definition of "eicosanoid-like" we have sought to review this subject comprehensively for the marine plants. The term eicosanoid strictly defined refers to 20-carbon fatty acids with at least one site of oxidation in addition to the carboxyl group. However, it is often found that enzymes capable of metabolizing 20-carbon polyunsaturated fatty acids can also metabolize 18- or 22-carbon analogs. Indeed, a careful chemical analysis of "eicosanoid"-containing species often reveals the co-occurrence of these similarly metabolized acids of different chain length. Principally, this is a nomenclature problem, one which has been recognized by mammalian, higher plant, and marine chemists for some time, and several suggestions are under current review for a more inclusive term (Smith and Willis, 1987; Smith *et al.*, 1990).

Specifically excluded from comprehensive treatment in this review are reports on the compositions of fatty acid, glycolipid, phospholipid, and other complex lipid species in various marine plants. Further, we have chosen to exclude from consideration the examples of polyacetylenic fatty acids from red algae (Paul and Fenical, 1980), which likely owe their origin to substantially different metabolic pathways. Similarly, the C15 "enyne" chemistry from *Laurencia* (Erickson, 1983) and the phloroglucinol-containing (Gerwick and Fenical, 1982) and resorcinol-containing (Gregson *et al.*, 1977) fatty acids from brown algae have been excluded because of their substantially different biogenesis. Many of these excluded classes of related compounds have been previously covered in comprehensive reviews (Faulkner, 1990).

2. EICOSANOIDS AND RELATED COMPOUNDS FROM MARINE ALGAE

2.1. Historical

It has long been recognized that marine plants are abundant producers of polyunsaturated fatty acids, and are particularly well known for their elevated levels of 20- and 22-carbon ω-3 fatty acids (most commonly, eicosapentaenoic and docosahexenoic acids) (Stefanov *et al.*, 1988). These two acids are generally believed to be the components of fish and fish oils which are responsible for the health benefits of elevated levels of these in the diet (Lands, 1986). This is believed to occur as a result of the acceptance of these arachidonic acid analogs by enzymes which normally convert arachidonic acid to prostaglandins, thromboxanes, leuko-trienes, and related compounds. The resultant substances have both subtly differ-ent structures and biological properties. Up until the early 1980s, there were only a few reports (*Lyngbya majuscula*, *Gracilaria* spp., and *Laurencia hybrida*) that marine plants also metabolized these fatty acids via various oxidative pathways to substances analogous to the eicosanoids. However, research in the last few years has shown that a great many different marine plants are capable of these transfor-mations. While most of these reports are for algae of the Rhodophyta, there have been a sprinkling of reports for algae of the Chlorophyta, Phaeophyta, and Cyanobacteria as well. This may suggest that all algae are capable to one degree or another of this type of metabolism. This treatment of the eicosanoids and related fatty acids of marine plants is phylogenetically arranged.

2.2. Cyanobacteria

The cyanobacteria, or blue-green algae, have only been sporadically reported to contain eicosanoid-like natural products. However, based on the structural novelty of these (**1–7**), this group would most certainly benefit from a more systematic evaluation for this class of chemistry. As a general trend, metabolism of shorter-chain fatty acids appears to be the rule in the cyanobacteria, which is in keeping with their predominant complement of C16 and C18 nonoxygenated fatty acids (Moore, 1981). Only hydroxy fatty acids and no carbocyclic or oxycyclic eicosanoid-like products are known from these prokaryotes.

2.2.1. Lyngbya majuscula

The first report of an oxygenated fatty acid metabolite from a cyanobacterium came from work with *Lyngbya majuscula* in 1978 (Cardellina *et al.*, 1978). It was found that the C14 methoxy-containing fatty acid **1** was present as a major

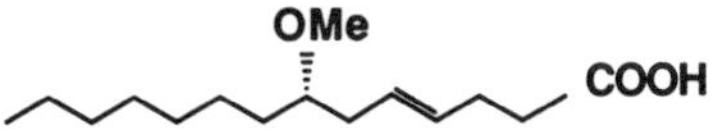

(1)

metabolite in many collections of this blue-green from the Pacific (Moore, 1981) as well as from the Caribbean (Gerwick *et al.*, 1987). The structure was deduced from spectroscopic analysis and the 7*S* absolute stereochemistry by ozonolysis to both the C10 methoxy acid and C10 hydroxy acid (following demethylation of **1** with BF$_3$ and ethanethiol) followed by comparison of optical rotations with model compounds. Subsequently, this C14 methoxy acid has been found in amide linkage

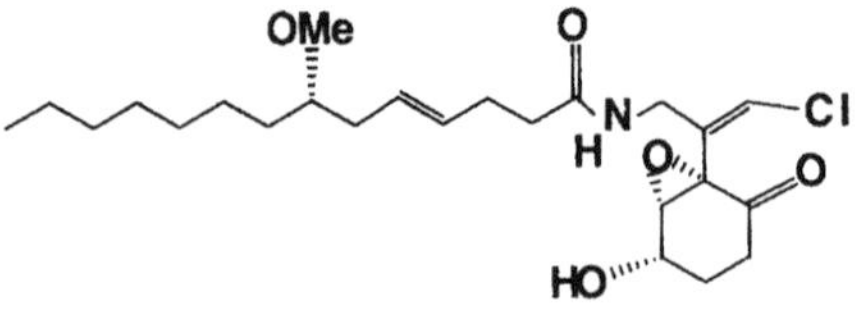

(2)

with a variety of interesting N-containing units which presumably derive from amino acids (i.e., **2**) (Ainslie *et al.*, 1985).

Continued work with *L. majuscula* by Cardellina and Moore (1980) led to the isolation of another hydroxy fatty acid, malyngic acid (**3**), which contained three hydroxy groups and two olefins on a C18 chain. The gross structure of **3** was

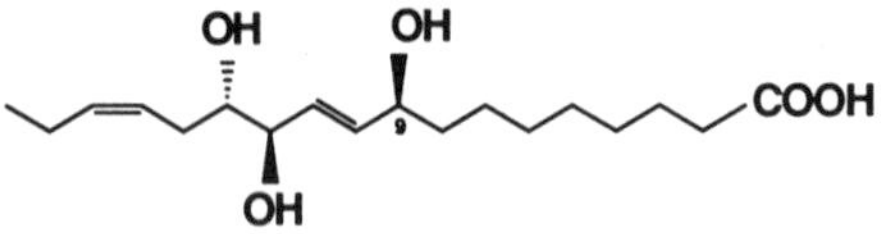

(3)

deduced based on the spectroscopic properties of the natural product and two key derivatives, a methyl ester triacetate and C12–C13 acetonide. Geometry of the two olefins were established from IR absorptions and coupling-constant analysis, and the locations of these followed from interpretation of ^{1}H NMR and mass spectrometry (MS) data. Stereochemistry was determined by optical rotary and spectroscopic measurement of several derivatives in comparison with standards. The structure of malyngic acid (**3**) closely parallels that of several trihydroxy C18 acids that have subsequently been isolated from a variety of sources, including higher plants (Herz and Kulanthaivel, 1985; Kato *et al.*, 1986) and fungi (Hamberg *et al.*, 1986), and recently synthesized (Gurjar and Reddy, 1990). Most likely, malyngic acid derives from the action of a 13-lipoxygenase on α-linolenic acid followed by intramolecular rearrangement to the 12*S*,13*S*-epoxy-9*S*-hydroxy-10(E),15(Z)-octadecadienoic acid. Nucleophilic opening of this epoxide at C12 with inversion would provide the observed malyngic acid structure (**3**) (Fig. 1).

2.2.2. *Schizothrix calcicola/Oscillatoria nigroviridis*

A structurally interesting C13 metabolite (**4**) was reported from a 1:1 mix of two cyanobacteria, *Schizothrix calcicola* and *Oscillatoria nigroviridis*, collected from Enewetak Atoll in the Pacific (Mynderse and Moore, 1978). In addition to

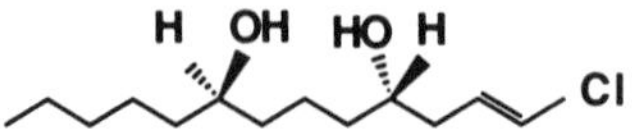

(4)

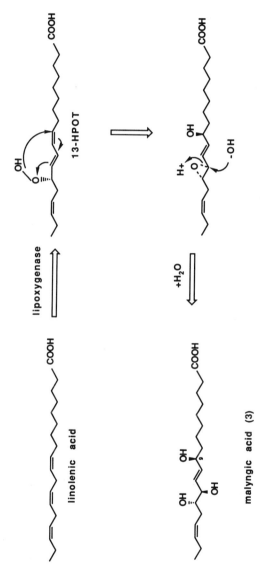

Figure 1. Potential biogenesis of malyngic acid (3) by *Lyngbya majuscula.*

two hydroxy groups positioned at C6 and C8, this metabolite possesses a very interesting C1 vinyl chloride of unprecedented structure. The stereochemistries of the two alcohols were deduced as 6*R*,9*R* by comparison of the optical rotation of catalytically hydrogenated **4** with that of model compounds, while the *trans* geometry of the C1 olefin was determined by ^{1}H NMR coupling constants. While the algal mixture was highly toxic to mice, metabolite **4** was not, but rather showed antimicrobial activity to *Mycobacterium smegmatis*.

2.2.3. Enzyme Studies with Cyanobacteria

Recently, a pair of reports appeared which describe lipoxygenase (Beneytout *et al.*, 1989) and hydroperoxide lyase (Andrianarison *et al.*, 1989) enzyme systems from *Oscillatoria* sp. The lipoxygenase was purified from the homogenate of log-phase harvested cultures of *Oscillatoria* by centrifugation, NH_4SO_4 precipitation, DEAE-Tris-acryl column chromatography, and Sephadex G-150 column chromatography (81-fold purification). An uptake of O_2 was recorded upon incubation with linoleic acid and this could be inhibited by either heat denaturation or treatment with the lipoxygenase inhibitor esculetin. Carbon monoxide, a cytochrome P450 inhibitor, was not inhibitory. The molecular weight of the *Oscillatoria* lipoxygenase was estimated at 124,000 with maximal activity at pH 8.8. This prokaryotic lipoxygenase metabolizes exogenously supplied linoleic acid to an approximately equal mixture of 13*S*-hydroperoxy-9(Z),11(E)-octadecadienoic acid (**5**) and 9*S*-hydroperoxy-10(E),12(Z)-octadecadienoic acid (**6**) as determined by chiral-phase high-performance liquid chromatography (HPLC).

               (**5**)   (**6**)

Subsequently, a hydroperoxide metabolizing enzyme (Andrianarison *et al.*, 1989) was isolated from later fractions in the final Sephadex G-150 column chromatography of the above lipoxygenase from *Oscillatoria*. Purification of this activity was 45-fold relative to the crude homogenate, and the molecular weight was estimated at 56,000. The product of incubation of 13-hydroperoxylinoleic acid (pH optimum 6.4) with this partially purified enzyme preparation was analyzed by UV and by gas chromatography–mass spectrometry (GC-MS) of the methyl ester, trimethylsilyl (TMS) ether, following $NaBH_4$ reduction. These data were consistent with a 13-oxo-9,11-dienoic acid (**7**) structure, indicating a hydroperoxide lyase-type reaction. Gas chromatography of the volatile compounds produced during incubation with 13-hydroperoxyoctadecadienoic acid (13-HPOD) led to the identification of pentanol as the C14–C18 fragment, suggesting a different mechanism for C–C scission than in *Chlorella* (Vick and Zimmerman,

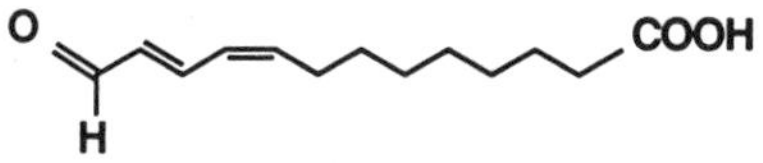

(7)

1989) (see below). It is proposed that this enzyme catalyzes a potentially intra-molecular rearrangement of the 13-hydroperoxide to yield two oxidized fragments, **7** and pentanol.

2.3. Rhodophyta

2.3.1. Cryptonemiales

2.3.1a. Constantinea simplex. Over several years our research group based in Oregon had been collecting the mushroom-shaped alga *Constantinea simplex* as part of an evaluation of these plants for new biomedicinals (W. H. Gerwick, M. W. Bernart, M. F. Moghaddam, D. Nagle, M. L. Solem, P. J. Proteau, Z. D. Jiang, and M. Wise, work in progress). Although we had some interest in this organism as a result of reports of antiviral activity associated with the aqueous solubles (Ehresmann *et al.*, 1977), it was not until we chanced to make a spring collection of this plant that we discovered it to be an abundant producer of interesting secondary metabolites (Nagle and Gerwick, 1990). Following stabilization by methylation and acetylation, vacuum silica gel chromatography followed by normal-phase HPLC led to the efficient recovery of a number of new and known eicosanoids. The known compounds, 12-S-hydroxyeicosatetraenoic acid, (12-S-HETE) (**8**), 12-S-hydroxy-eicosapentaenoic acid (12-S-HEPE) (**9**), and 12-oxododecatrienoic acid (**10**), were all consistent with the metabolism of polyunsaturated fatty acids by a 12-lipoxygenase.

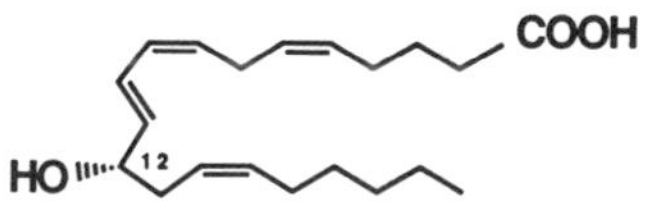

(8)

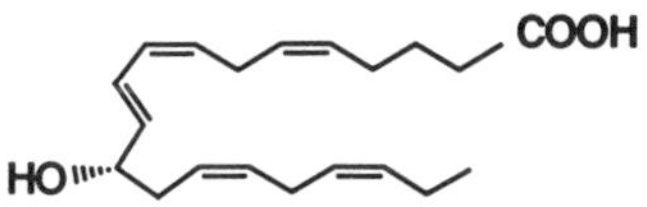

(9)

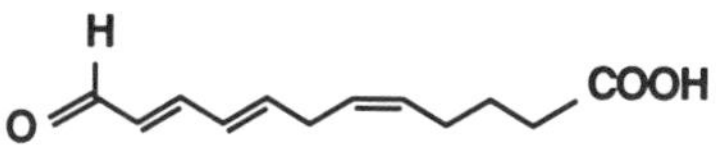

(10)

The two new compounds (**11, 12**) from *C. simplex* contained many of the protons characteristic of 12-HETE; however, they additionally contained a new hydroxyl, an ester, and a cyclopropyl ring. Extensive use of two-dimensional (2D) NMR techniques, which was supported by MS information, led to the sequencing of these functional groups in the linear fatty acid spin system, and identified

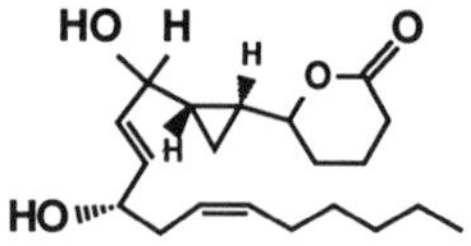

(11) 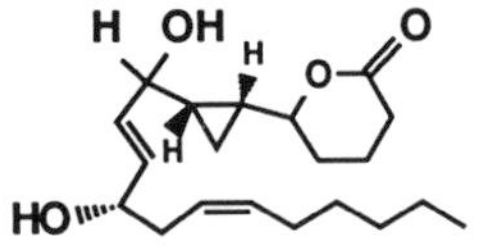   (12)

hydroxy group equivalents at C5, C9, and C12 with a cyclopropyl bridging C6 and C8. The two new compounds, termed constanolactones A (**11**) and B (**12**), were shown to be C9 epimers by a careful comparison of their ^{1}H and ^{13}C NMR spectral data. The co-occurrence of several known 12-lipoxygenase metabolites in this alga, coupled to our previous findings of many epoxy alcohols of the hepoxilin type from red algae, led us to propose a 12-lipoxygenase-initiated route to these novel cyclopropyl-lactones which involves an intermediate epoxy cation (Fig. 2). This is distinct from the 9-lipoxygenase pathway proposed for similar metabolites from coral (Baertschi *et al.*, 1989). Interestingly, there have been a number of closely related metabolites recently isolated from coral (Baertschi *et al.*, 1989), sponge (Niwa *et al.*, 1989), and molluscan sources (Ojika *et al.*, 1990).

Recently, we have been able to form acetone powder preparations of this alga which are quite efficient in their metabolism of exogenously supplied arachidonic acid to a mixture of hydroxyeicosanoids and constanolactones A (**11**) and B (**12**) (Fig. 2) (Nagle and Gerwick, work in progress). Biosynthetic efforts employing this preparation are currently examining this proposed 12-lipoxygenase metabolic pathway. In biological testing, a crude mixture of constanolactones has been found to possess only a weak ability to inhibit human polymorphonuclear leukocyte (PMNL) phospholipase A_2 (PLA_2) (IC_{50} ca. 30 μg/ml).

2.3.1b. Farlowia mollis. In screening efforts, the crude lipid extract of *Farlowia mollis* collected from the Oregon coast was found to possess anti-microbial activity to *Staphylococcus aureus* and *Escherichia coli* (Solem *et al.*, 1989; Gerwick, Bernart, Moghaddam, Nagle, Solem, Proteau, Jiang, and Wise, work in progress). Subsequently, we found this alga to have an abundance of three new dihydroxyeicosanoid-like substances (**13–15**), isolated as diacetate-methyl

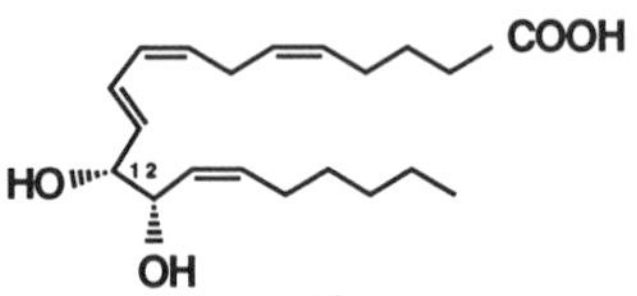

(13) 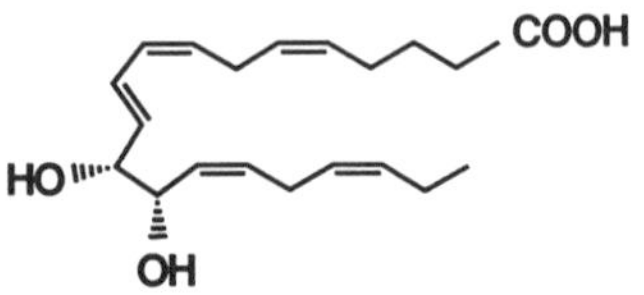   (14)

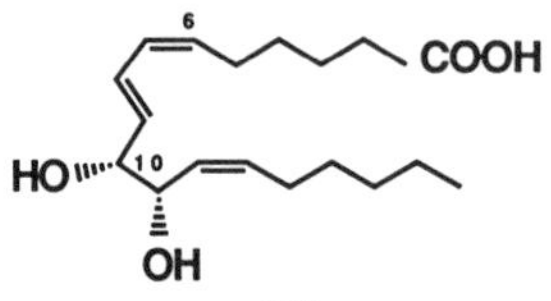

(15)

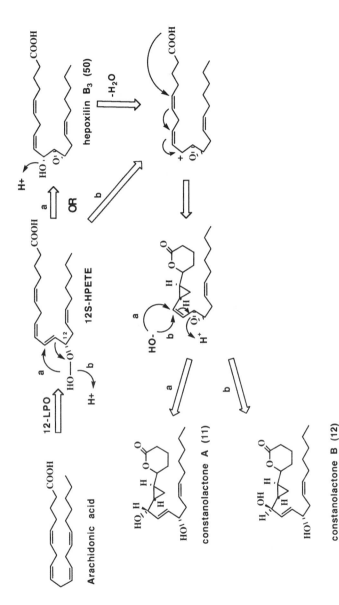

Figure 2. Proposed 12-lipoxygenase-initiated metabolism of arachidonic acid in *Constantinea simplex* to produce constanolactones A (**11**) and B (**12**) and related eicosanoids.

esters (Solem *et al.*, 1989). Their constitutive structures were deduced from extensive NMR and MS studies. Geometry of the olefinic bonds and relative stereochemistry of the vicinal diol were deduced from coupling constants and ^{13}C NMR shifts. Use of ^1H NMR coupling constants to deduce the C12–C13 relative stereochemistry as *threo* assumed that the two acyl chains would determine the favored rotomer by assuming a maximal distance from one another. This assumption was later shown to be incorrect by synthesis of both the *threo* and *erthyro* stereoisomers of metabolite **13** (Lumin and Falck, 1990), and independently confirmed with the natural product by forming the corresponding acetonide and observation of nOe between both H12 and H13 and only one of the two methyl groups of the acetonide (Jiang and Gerwick, 1990). An absolute stereochemistry of *R* at C12 was correctly given by formation of the 12,13-bis (*p*-nitrobenzoate) derivative of metabolite **13** and observation of a positive split Cotton effect in its circular dichroism (CD) spectrum. This was independently determined by the unequivocal total synthesis of **13** in which 2-deoxy-D-galactose was used as starting material (Lumin and Falck, 1990).

A mixture of metabolites **13** and **14** was shown to possess several interesting properties typical of hydroxyeicosanoids, including a weak inhibition of 5-lipoxygenase activity in A23187-stimulated human PMNLs and a weak inhibition of isolated dog kidney Na^+/K^+ ATPase. Further, in additional testing with human-derived PMNLs, this mixture was shown to be a weak primary stimulator of superoxide anion production while also inhibiting its release by formylmethionyl-leucylphenylalanine (fMLP) stimulation. Recent preliminary testing has suggested that metabolite **14** is also able to effect progesterone levels following ovarian arterial administration in sheep with concomitant early reentry to estrous (F. Stormshak, K. E. Orwig, S. Leers-Sucheta, M. F. Moghaddam, Z. D. Jiang, and W. H. Gerwick, work in progress).

Recently, biosynthetic experiments have been carried out with enzyme extracts from another red alga, *Gracilariopsis laemaneiformis*, which produces two of these same dihydroxyeicosanoids, 12*R*,13*S*-diHETE (**13**) and 12*R*,13*S*-diHEPE (**14**) (Jiang and Gerwick, 1990) (see discussion below). Presumably, the same pathways (12-lipoxygenation followed by intramolecular rearrangement of the intermediate hydroperoxide; see Fig. 4) are utilized by both algae.

2.3.1c. Lithothamnion corallioides/L. calcareum.

A recent report on two species of the calcareous alga *Lithothamnion* describes the structures of several new and interesting eicosanoids (Guerriero *et al.*, 1990). These algae were collected from the Brittany coast and stored in EtOH for a prolonged period. Evaporation of the EtOH and extraction of the aqueous residue with petroleum ether gave an extract that was chromatographed by reverse-phase-flash, Amberlyst-A21, and high-performance liquid chromatography to yield three new and four known compounds. All of these compounds were isolated as ethyl esters, which

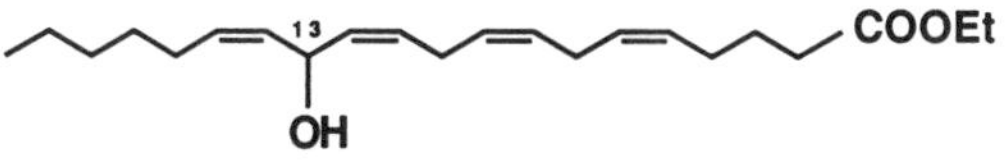

(16)

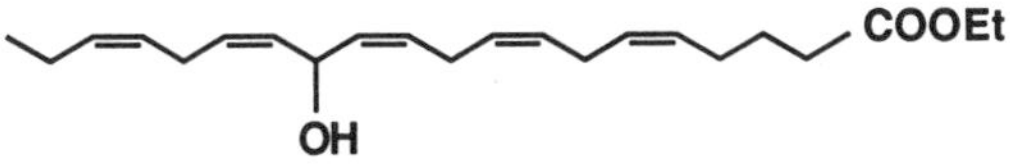

(17)

are presumably artifacts of the storage and extraction process. The structures of the new metabolites are based on spectroscopic analysis, largely 2D NMR. The two major compounds, **16** (0.02%) and **17** (0.01%), are olefin homologs of one another which derive from arachidonic acid (AA) and eicosapentaenoic acid (EPA), respectively, and contain an unusual 13-hydroxy functionality which is bis-allylic to two isolated Z-olefins. This is an unusual functionality within an eicosanoid structure, as the lack of a conjugated diene suggests a mechanism other than normal lipoxygenation for introduction of the hydroxyl group. A third metabolite (**18**) was distinctly different in that it possessed a hydroxy group

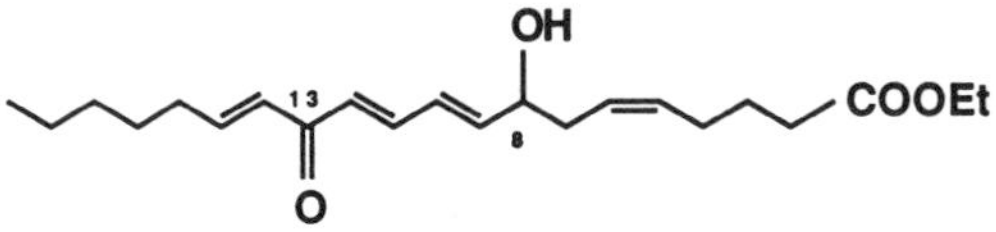

(18)

at C8 and a keto functionality at C13. Two E-olefins span the intervening carbon atoms, and a C14–C15 olefin also possesses the E geometry. This is of particular interest since an identical relationship of functional groups was earlier found in another red algal metabolite, ptilodene (**44**) from *Ptilota filicina* (Lopez and Gerwick, 1988). The stereochemistry of the hydroxy groups in the new compounds (**16–18**) was not determined. The four known compounds, all isolated as ethyl esters [11*R*-hydroxy- 5(Z),8(Z),12(E),14(Z)-eicosatetraenoic acid (**19**), 15*S*-hydroxy-5(Z),8(Z),11(Z),13(E)-eicosatetraenoic acid (**20**), 5*S*-hydroxy-6(E),8(Z), 11(Z),14(Z)-eicosatetraenoic acid (**21**), and 12*S*-hydroxy-5(Z),8(Z),10(E),14(Z)-

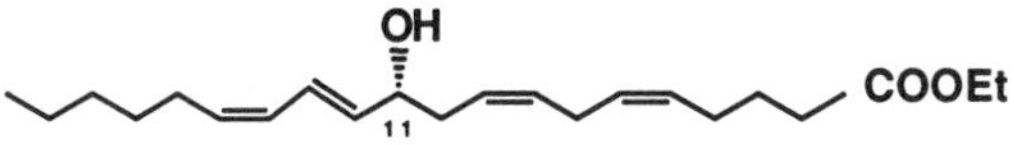

(19)

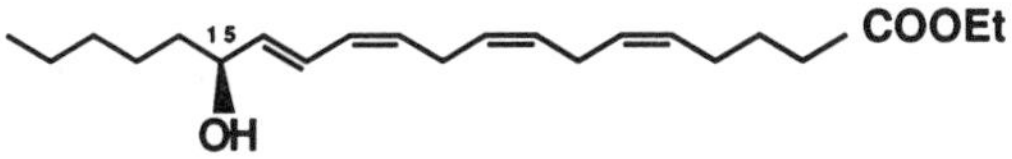

(20)

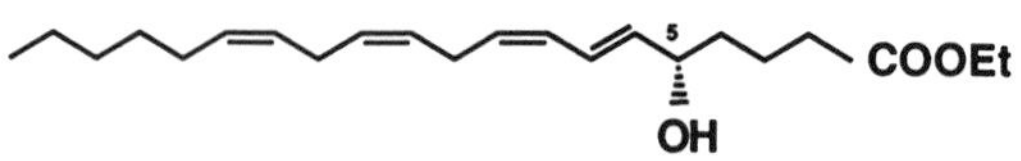

(21)

eicosatetraenoic acid (**8**)] were identified by comparison with synthetic reference compounds and analysis of their spectroscopic features.

2.3.1d. Bossiella orbigniana. Another coralline red alga, *Bossiella orbigniana* has recently been studied for the ability of its cell-free buffer extract to metabolize arachidonic acid (Burgess *et al.*, 1991). The supernatant fraction from low-speed centrifugation was efficient in its metabolism of exogenously supplied arachidonic acid, as measured by ultraviolet absorption, O_2 utilization, and HPLC, to a novel metabolite (bosseopentaenoic acid, **22**). The product of this incubation

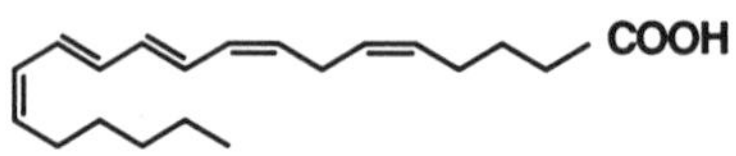

(22)

(**22**) was also identified, albeit in lower yield, by HPLC of the organic extract from freshly homogenized algal tissue, thus demonstrating its natural occurrence. Further, heat pretreatment completely inactivated the AA-metabolizing capacity of *B. orbigniana* buffer extracts. Large-scale incubations gave 5.2 mg of the new AA metabolite (**22**), which was subsequently characterized by fast atom bombardment (FAB) MS, UV, ^1H and ^{13}C NMR, and ^1H–^1H correlated spectroscopy (COSY) as an eicosapentaenoic acid with a conjugated tetraene. The C8–C15 location of the tetraene followed from connectivities established from ^1H–^1H COSY, while geometries of the olefins were deduced from coupling constants and ^{13}C NMR shifts. It is proposed that bosseopentaenoic acid (**22**) arises via dehydration of a 12-HETE (**8**)-type intermediate, itself produced either by 12-lipoxygenase or cytochrome P450 metabolism (Fig. 3).

2.3.2. Gracilariales

2.3.2a. Gracilaria lichenoides. One of the more interesting groups of red algae relative to their eicosanoid content are those in the order Gracilariales. Several species of the type genus, *Gracilaria*, have been studied from various locations around the world, most notably from Australia and Japan.

The first studied of these (Gregson *et al.*, 1979), *G. lichenoides* from Eastern Australia, displayed a potent antihypertensive activity in the hypertensive rat model. Purification of the antihypertensive agent from the aqueous extract was achieved by use of this assay at each step of the purification. However, final purification of these seaweed-derived materials required derivatization to the

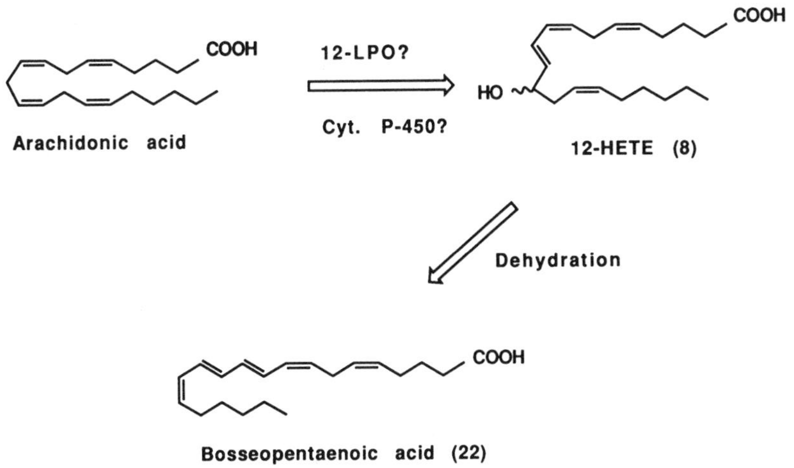

Bosseopentaenoic acid (22)

Figure 3. Proposed biogenesis of bosseopentaenoic acid (**22**) by *Bossiella orbigniana*.

corresponding methyl esters and preparative thin-layer chromatography (TLC). The structures of prostaglandin E_2 (PGE$_2$) (**23**) and PGF$_{2\alpha}$ (**24**) were established by comparison of various spectroscopic properties of their methyl esters and TMS ethers with the authentic substances. Absolute stereochemistry was deduced by comparison of optical rotations obtained for the seaweed eicosanoids with authentic materials. From bioassay data, the concentrations of PGE$_2$ and PGF$_{2\alpha}$ in *G. lichenoides* were estimated as 0.05–0.07% and 0.07–0.10% of the dry weight, respectively. The authors speculate that the occurrence of these substances in the algae may be correlated with the small numbers of epiphytic plants on *G. lichenoides*.

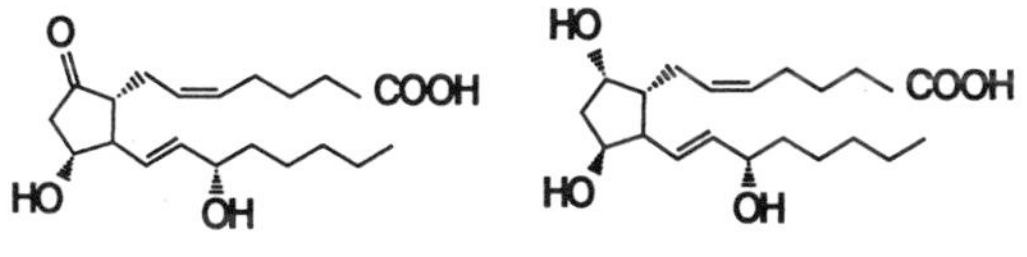

In what appears to have been an exceptional insight, the authors speculated in this early paper that these seaweed prostaglandins may in fact arise from the action of lipoxygenase metabolism on arachidonic acid. This was in contrast to the then established cyclooxygenase route to these compounds in mammalian systems. While the biosyntheses of these has never been pursued, it seems likely to be true, given the preponderance of obviously lipoxygenase-derived metabolites which have been subsequently isolated from other red algae and the few biosynthetic experiments performed in this area to date.

2.3.2b. Gracilaria verrucosa. Prostaglandin E_2 was subsequently isolated from a Japanese species of Gracilaria, *Gracilaria verrucosa*, which has been the source of "Ogonori" poisoning among humans in Japan (Fusetani and Hashimoto, 1984). This syndrome, which apparently arises following a water-soaking treatment of the algae, is typified by diarrhea and nausea and in extreme cases can be fatal. A mouse bioassay was used to follow the chromatographic isolation of two diarrhea-producing substances from the aqueous extract. The more potent of these was shown by spectroscopic means as well as chemical degradation to be identical with PGE_2 (**23**); however, the absolute stereochemistry was not investigated. The less potent compound was similarly shown to be identical with PGA_2 (**25**), though

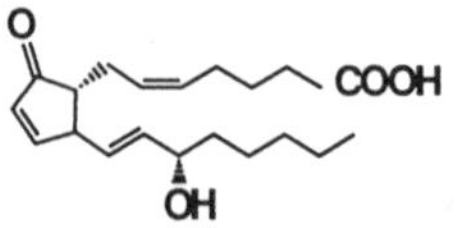

(25)

this was suggested to be at least partially of artifactual origin arising from acid-catalyzed dehydration of PGE_2. The summed concentrations of these prostanoids in the algae are estimated at 0.01% of its fresh weight, which is roughly comparable with that found for *G. lichenoides* from Australia. In the same year as this first report on *G. verrucosa*, a patent appeared which details the commercial isolation of 48 mg of PGE_2 from 500 g of the minced algae (ca. 0.01%) (Thermo Company, 1984). Orally administered authentic PGE_2 was shown to produce the same toxicity symptoms in the mouse assay as found for the algae-derived prostanoids, and these were fully consistent with the intoxication observed in humans. Because it appears that soaking the algae in fresh water is required to produce these prostanoids, it is speculated that these algae have endogenous inhibitors of prostanoid production which require removal before prostaglandins can be detected. Subsequently, an examination of the lipids of *G. verrucosa* (Kinoshita *et al.*, 1986) showed it to have an extraordinary high content of arachidonic acid, principally in the triglyceride [triarchidonin (**26**) = 20–44%] and phosphatidylcholine [diarachidonoylphosphotidylcholine (**27**) = 56–64%] lipid classes. Arachidonic acid is also a constituent of glycolipid species recently

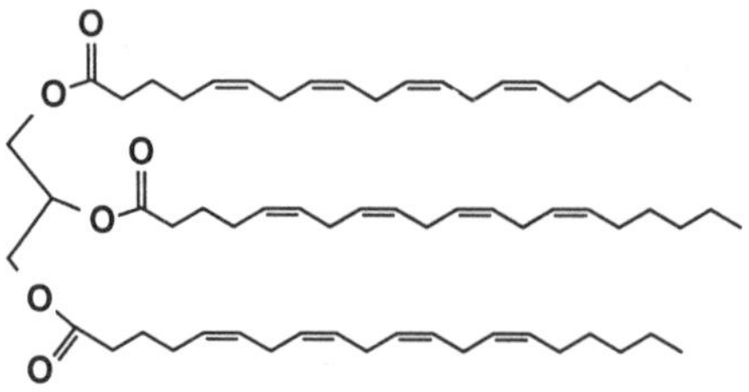

(26)

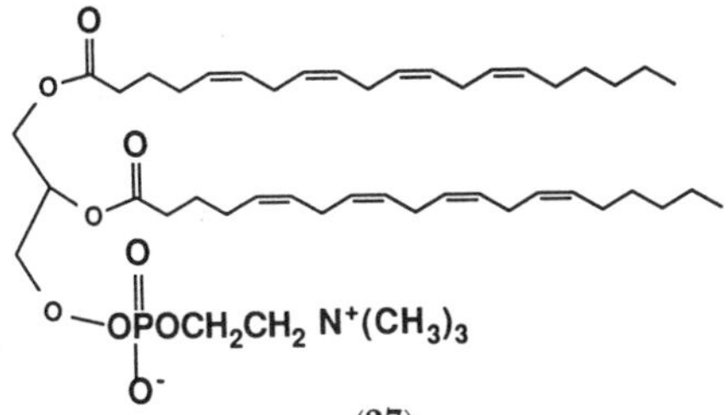

(27)

reported from this seaweed collected in Korea (**28, 29**) (Son, 1990). If control of prostanoid production in seaweeds parallels that in mammals, then the water-soluble inhibitors in this seaweed may have as their site of action the release of arachidonic acid from these arachidonic acid-rich lipid classes.

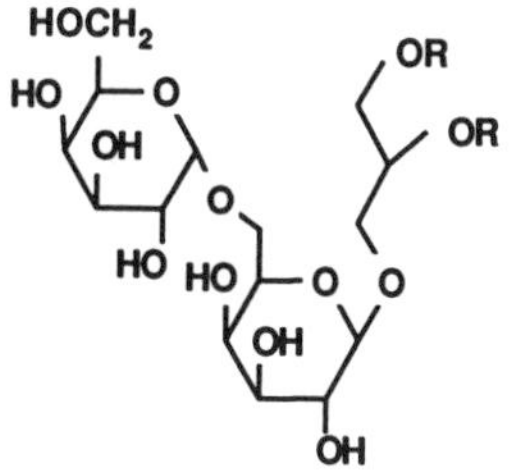

(28) R = b:c:d = 5:1:4

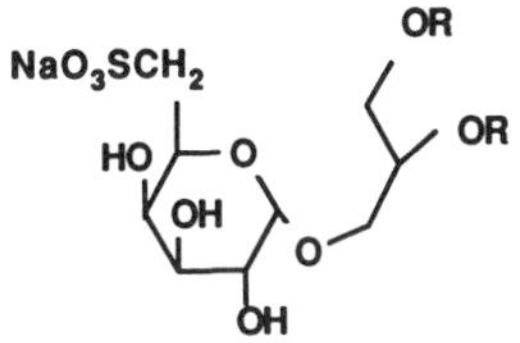

(a) $-CO(CH_2)_{12}CH_3$
(b) $-CO(CH_2)_{14}CH_3$
(c) $-CO(CH_2)_7CH=CH(CH_2)_7CH_3$
(d) $-CO(CH_2)_2(CH_2CH=CH)_4(CH_2)_4CH_3$

(29) R = a:b:c: = 4:15:1

2.3.2c. Gracilariopsis lemaneiformis. While the above work with aqueous extracts from *Gracilaria* spp. yielded carbocyclic eicosanoids, work with the lipid extract from the related alga *G. lemaneiformis* from Oregon has yielded a large number of acyclic eicosanoids and related compounds (Jiang and Gerwick, 1990, 1991). The predominant theme in these is oxidation at C12 in a 20-carbon precursor or C13 of an 18-carbon precursor by a presumed 12-lipoxygenase. What is especially unique about the eicosanoid-like metabolites of *G. lemaneiformis* is the co-occurrence of a number of different eicosanoids and related fatty acids in both free form and as components of two related classes of complex plant lipids, monogalactosyldiglycerides and digalactosyldiglycerides. The ready availability of this algae from the Oregon coast combined with its abundance of eicosanoid-like natural products have made it a prime organism for initial biosynthetic experiments.

Initially, this alga was investigated only as a result of TLC observations, which showed it to be rich in unusual lipid natural products. Observation that some of these were highly polar and did not chromatograph well in conventional solvent systems led us to derivatize the crude extract as peracetyl-methyl esters. From this derivatized mixture a series of complex galactolipids was first isolated (**35**, **36**) (Jiang and Gerwick, 1990). More recently (Jiang and Gerwick, 1991), a complex assortment of free fatty acid metabolites has also been described, including 12*S*-HETE (**8**) and 12*S*-HEPE (**9**), metabolites we have found in several other red algae, and 12*R*,13*S*-diHETE (**13**) and 12*R*,13*S*-diHEPE (**14**), which we had previously isolated from *Farlowia mollis* (Solem *et al.*, 1989). Additionally (Jiang and Gerwick, 1991), we found in this alga a new but related diol containing an additional position of oxidation, a C18 ketone in conjugation with a C14–C17 diene (**30**). A series of biosynthetically related C18 oxidized fatty acids were also isolated, including the new compound 10-hydroxyoctadectetraeonoic (**31**) acid and the known metabolites 13*S*-hydroxyoctadecadienoic acid (HODE) (**32**), 13-ketoocta-dienoic acid (KODE) (**33**), and 13*S*-hydroxyoctadecatrienoic acid (HOTE) (**34**). As described below, the 18-carbon precursor fatty acids (linoleic and linolenic acids)

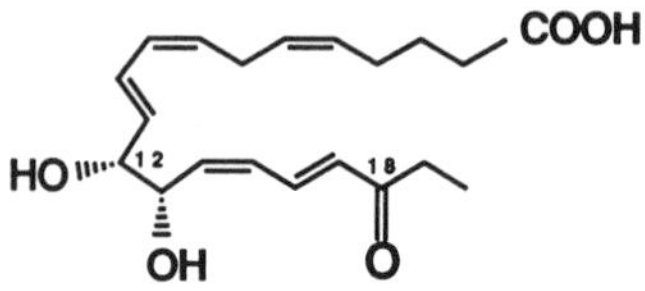

(**30**)

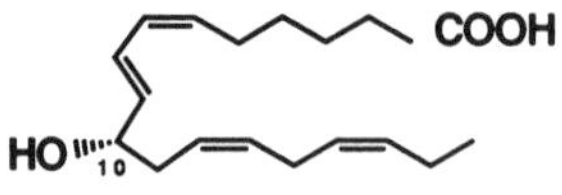

(**31**) 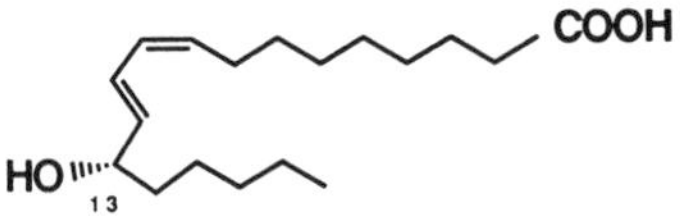   (**32**)

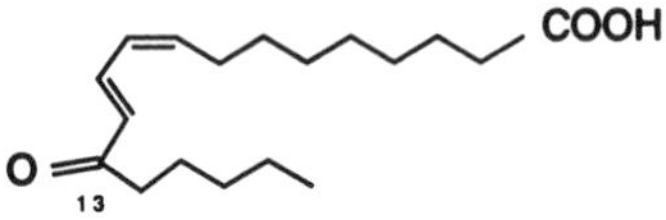

(**33**) 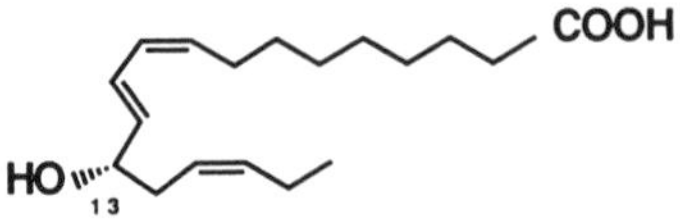   (**34**)

are poor but acceptable substrates for the lipoxygenase system present in *G. lemaneiformis*, and yield these C13-oxidized acids (W. H. Gerwick and M. Hamberg, work in progress). Further, the chain-cleaved metabolite 12-hydroxy-dodecatrienoic acid was recovered, which likely results from the action of hydroperoxide lyase on a 12-hydroperoxy-containing (i.e., 12*S*-HPETE) eicosa-noid. These various hydroxy- and ketone-containing fatty acids ranged in concentration from 0.01 to 1% of the crude lipid extract.

    Additional insight into the biochemistry and potential functional roles of these eicosanoid-like substances was gained by our isolation of two complex lipid species which contained eicosanoids or eicosanoid fragments as components (Jiang and Gerwick, 1990). Again, these unusually oxidized metabolites were worked with as the peracetyl derivatives, and purified by a combination of silica gel chromatography and HPLC and collectively accounted for approximately 2% of the lipids. The structures of three of these have been determined by extensive use of 2D NMR methodology in combination with some chemical degradations. The first, metabolite **35**, an example of a monogalactosyldiglyceride, contains two

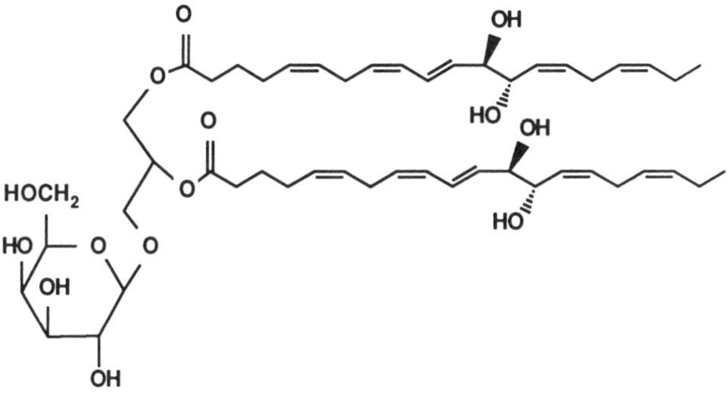

(35)

20-carbon polyunsaturated acyl chains at *sn*-1 and *sn*-2, defining this metabolite to be of cytoplasmic origin. An additional feature of this molecule which makes it unique is that both acyl chains derive from eicosapentaenoic acid and have hydroxyl functionalities as C12 and C13. Therefore, the eicosanoid 12*R*,13*S*-diHEPE (**14**) occurs in both free and bound forms in *G. lemaneiformis*. Galactose at *sn*-3 was shown to be *D*, while the stereochemistry of the glycerol backbone was not determined. By a combination of NMR and chemical manipulations, a second glycerol-based metabolite (**36**) was shown to contain a digalactosyl unit at *sn*-3, a palmitate unit at *sn*-2, and a 12*R**,13*S*-diHEPE unit at *sn*-1 (Jiang and Gerwick, 1990). The third defined galactolipid metabolite (**37**) also contained a digalactosyl unit at *sn*-3 and a palmitate at *sn*-2, but the oxidized species at *sn*-1 showed some

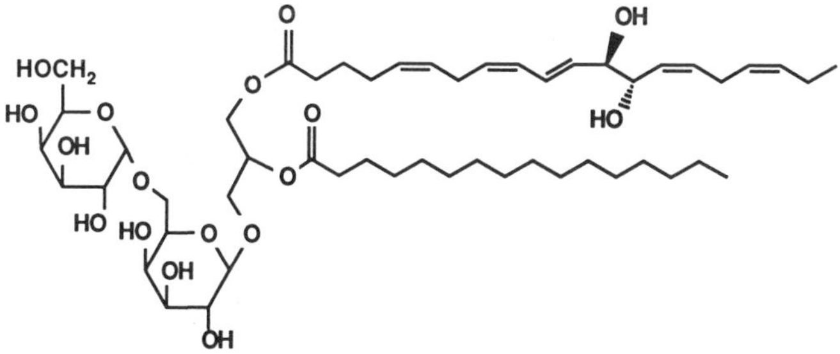

(36)

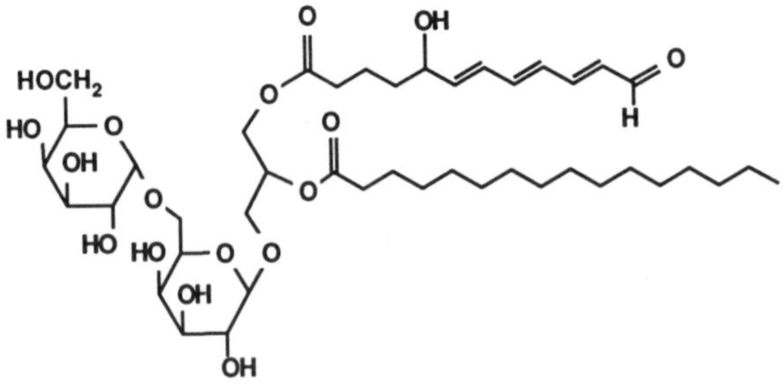

(37)

interesting modifications (Jiang and Gerwick, work in progress). Easily defined by
^{1}H–^{1}H COSY and confirmed by UV as well as following cleavage and isolation,
this oxidized acyl chain was fragmented at C12, yielding an aldehyde. Furthermore, this was in conjugation with C6–C11 *t,t,t* triene, while at C5 there was an
alcohol functionality. The presence of two points of oxidation, C5 and C12,
separated by a conjugated triene, is highly reminiscent of the leukotriene B$_4$-type
structure. However, that metabolite 37 is cleaved between what was C12 and C13
in the acyl precursor suggests the action of a hydroperoxide lyase on a C12
hydroperoxide. This later feature is inconsistent with most mammalian biosynthetic pathways leading to the leukotrienes and suggests that this algal metabolite
(37) may be produced by a double dioxygenation sequence followed by isomerization of the C8 olefin.

It is well established that galactolipids are specifically localized to chloroplastic membranes in higher plants (Van Hummel, 1975). We speculate that in *G.
lemaneiformis* these galactolipids act as sources of substrate arachidonic acid for
the biosynthesis of these complex oxidized lipids. Subsequently, and perhaps in
response to specific metabolic stimulus, these eicosanoids are released from these
glycerol lipids. Further, we speculate that the biosynthesis of these marine plant
eicosanoids occurs in chloroplasts and may be coupled to the production of
molecular oxygen by photosynthesis. Experimentation probing these intriguing
points has yet to be performed.

Recent experimentation aimed at probing the lipoxygenase capacity of red
algae and the biosynthetic origin of the unusual C13 oxidation in 12*R*,13*S*-diHETE
has been undertaken by our group in Oregon in concert with M. Hamberg at the
Karolinska Institute in Sweden. Initially, we were able to develop an enzymatically active cell-free preparation from *G. lemaneiformis* which could
metabolize exogenously supplied arachidonic acid to a variety of oxidized
eicosanoids or related substances, including 12*S*-HETE (8), 12*R*,13*S*-diHETE, and
12-oxododecatrienoic acid, the first time that such a system has been developed for
a marine plant (Moghaddam and Gerwick, 1990; Moghaddam and Gerwick, work
in progress). That the biosynthetic system was cell free was additionally advan-

tageous in that this circumvented the problem of getting exogenously supplied substrates into the cells, a common problem encountered in biosynthetic studies with marine macrophytes (Marchant and Fowke, 1977; Barrow, 1983; Garson, 1989).

Using this system, we have now probed in considerable detail the mechanism by which the unusual position of oxidation (C13) is introduced (Gerwick and Hamberg, work in progress, Fig. 4). Working principally with the C18-derived substrate, 13S-hydroperoxy octadecadienoic acid, which is a good substrate for the diol-producing enzyme to form 13R,14S-dihydroxy-9(Z),11(E)-octadecadienoic acid, we have shown that the oxygen at C14 derives from exclusive intramolecular rearrangement of the distal oxygen of this C13 hydroperoxide functionality. Furthermore, this is introduced stereospecifically to produce a 14S stereochemistry without reference to the stereochemistry of the C13 hydroperoxide (i.e., the 13S-hydroperoxide gives the *erthyro* 13,14 diol, while the 13R-hydroperoxide yields the *threo* 13,14 diol). Since the diol-forming enzyme has an absolute requirement for a hydroperoxide precursor (i.e., 13S-HOD is not a substrate), the lipoxygenase reaction occurs prior to the unusual C14 oxidation. Further, as 9S-hydroperoxy octadecadienoic acid was efficiently metabolized to the corresponding 8,9 diol, the diol-forming enzyme does not distinguish between carboxyl and methyl termini in the substrate.

2.3.3. Gigartinales

2.3.3a. Agardhiella tenera, Anatheca montagnei, Euchema cottonii, E. spinosum, and Meristotheca senegalensis. A recent and thorough study (Miralles *et al.*, 1990) of the fatty acid composition of five red algae from the coast of Senegal (*Agardhiella tenera, Anatheca montagnei, Euchema cottonii, E. spinosum*, and *Meristotheca senegalensis*) revealed the presence of several unique compounds in these algae of potential chemotaxonomic significance (family Solieriaceae). Based on a GC-MS comparison with an authentic sample, a known but highly unusual cyclopentyl-containing fatty acid (**38**) was described from the

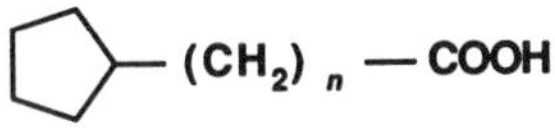

$$\text{cyclopentyl}-(\text{CH}_2)_n-\text{COOH}$$

(**38**) n = 10
(**39**) n = 12

crude mixture of fatty acid methyl esters (FAMEs). A new and homologous cyclopentyl-containing fatty acid (**39**) was also described from this mixture, as were two unique ω-5-containing fatty acids [18:1 ω-5 (**40**), 16:1 ω-5, (**41**)]. These latter acids are proposed as precursors of the cyclic species by direct cyclization.

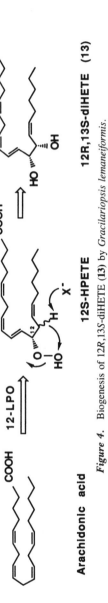

Figure 4. Biogenesis of 12*R*,13*S*-diHETE (**13**) by *Gracilariopsis lemaneiformis*.

(**40**) n = 3
(**41**) n = 4

2.3.4. Ceramiales

2.3.4a. Ceramiaceae

Ptilota filicina In 1984, TLCs of the crude and antimicrobial lipid extract of the Oregon intertidal alga *Ptilota filicina* showed it to contain a number of unique-appearing natural products. Initially, two new eicosapentaenoic acids (**42**, **43**) were isolated as their methyl ester derivatives (Lopez and Gerwick, 1987). These were isolated as UV-active but optically inactive substances which gave richly detailed NMR spectra that were thoroughly explored by 2D methods. We speculate that the Z,E,E isomer (**42**) is the first formed and perhaps true natural product of the seaweed. Subsequently, and perhaps only artifactually, the E,E,E triene (**43**)

(**42**) 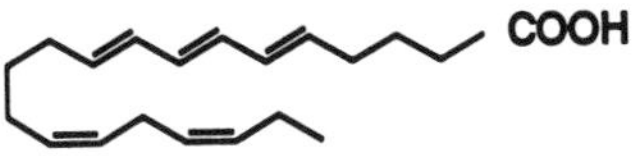   (**43**)

is produced from isomerization of **42**. By analogy with the biosynthetic pathways established for conjugated trienes of related structure from higher plants (Crombie and Holloway, 1985), we speculate that a radical mechanism is responsible for the stepwise movement of the homoconjugated double bonds in an EPA precursor to their C5 to C10 positioning in these molecules. An approximate 1:1 mixture of these double-bond isomers was moderately effective at inhibiting the action of dog kidney-derived Na^+/K^+ ATPase.

Subsequently, we isolated another eicosanoid from *P. filicina* which contains two unique points of oxidation in the fatty acid chain and apparently also derives from EPA (Lopez and Gerwick, 1988). The isolation of this later substance, named ptilodene (**44**), was complicated by its instability to silica gel chromatography and derivatization with CH_2N_2. However, a small yield of the methyl ester monoacetate was finally obtained and purified by HPLC. Again, 2D NMR techniques were employed as the principal structure elucidation tool, and defined ptilodene as a 20-

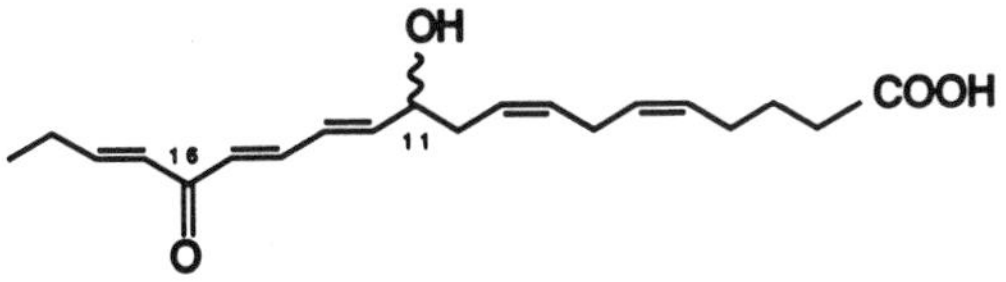

(**44**)

carbon EPA-derived molecule with a hydroxy functionality at C11 and a ketone at C16. Furthermore, the intervening carbon atoms between these oxygenated sites were present as a *trans, trans* diene. The ω-3 olefin was also shown by coupling constant analysis to be of *trans* geometry, a further unusual modification in this algal eicosanoid. While entirely speculative at this point, it seems reasonable that ptilodene is produced by the sequential oxidations by an 11-lipoxygenase and a 16-oxidase, in analogy with the biogenesis of 12,13-diHETE (**13**) by another Oregon seaweed, *Gracilariopsis lemaneiformis*. Adjustment of the oxidation state at these two sites and isomerization of the 14,15- and 17,18-olefins yields ptilodene (**44**) in a reasonable number and series of steps.

2.3.4b. Rhodomelaceae

Laurencia hybrida A back-to-back pair of reports in 1981 described several interesting eicosanoid-like structures from *Laurencia hybrida* collected on the south coast of Great Britain. The first of these (Higgs, 1981) described one new C12 eicosanoid fragment (**10**) and one new acyclic eicosanoid (**45**), while the latter (Higgs and Mulheirn, 1981) reported on the structure elucidation of two novel eicosapentaenoic acid-derived, cyclopentyl-containing lactones (**46, 47**).

The crude chloroform-soluble materials from a MeOH extract of the fresh seaweed was found to be antimicrobial by the paper disc-agar plate method to several pathogens, including *Staphylococcus aureus*, *Bacillus subtilis*, *Escherichia coli*, and *Cladosporium cucumerinum*. By bioautography and bioassay-guided fractionation, two *C. cucumerinum*-active substances were isolated in approximately 1% yield each and found active in pure form at 20 μg per disc. Final purification of each of these was by HPLC of the corresponding semisynthetic methyl esters (Higgs, 1981).

The simpler of the two was easily characterized by MS, IR, UV, and ^{1}H NMR with extensive homonuclear decouplings as 12-oxo-5(Z),8(E),11(E)-dodecatrienoic acid (**10**). Overlap of signals in the olefin region precluded a first-order assessment of the olefin geometry; however, on the basis of computer simulation, the 5(Z),8(E),10(E) configuration was assigned [unfortunately, the 8(Z) stereoisomer was drawn in error (Higgs, 1981)]. Short-chain aldehydes such as this logically arise from the action of hydroperoxide lyase on a C12 hydroperoxide, itself arising by action of a 12-lipoxygenase on arachidonic acid.

The second metabolite in this first report (Higgs, 1981) on the antimicrobial substances of *L. hybrida* was of more surprising structure. The methyl ester derivative was characterized by IR, UV, MS, optical rotation, and extensive ^{1}H NMR spectral data. A 20-carbon parent structure with five double bonds and an allylic alcohol were easily deduced. However, the placements of these functional groups in the chain were more difficult, especially in the absence of any useful MS fragmentations with this derivative. Hence, ^{1}H NMR decoupling data were used exclusively for these positional assignments and yielded a surprising structure

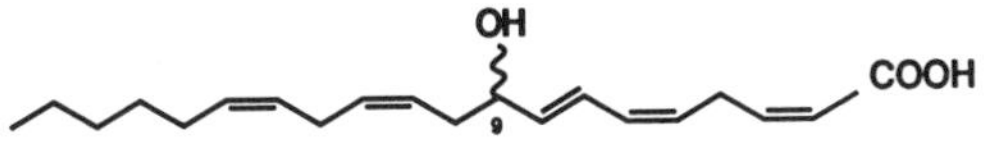

(45)

with an olefin at C2–C3 and a 9-hydroxy group (**45**). It was noted in this report that several of the assigned protons possessed unexpected chemical shifts, in particular, those of the olefin at C2–C3 (Higgs, 1981). Subsequently, we isolated 12*S*-hydroxyeicosa-5(Z),8(Z),10(E),14(Z),17(Z)-pentaenoic acid (12*S*-HEPE, **9**) from a related red algae, *Murrayella periclados* (see below), and based on comparisons between our material and the original data set for this 9-HEPE (**45**), we reassigned the structure of this *L. hydrida* metabolite to 12-HEPE (**9**) as well (Bernart and Gerwick, 1988). Unfortunately, both positive and negative rotations were given for this *L. hybrida*-derived 12-HEPE such that its absolute stereochemistry cannot be established.

A second paper on *L. hybrida* described the structures of two novel cyclopentyl lactones which were neither antimicrobial nor possessed UV chromophores (Higgs and Mulheirn, 1981). These were isolated by HPLC in approximately 0.5% yield for the major compound, hybridalactone (**46**), and 0.01% for the minor and

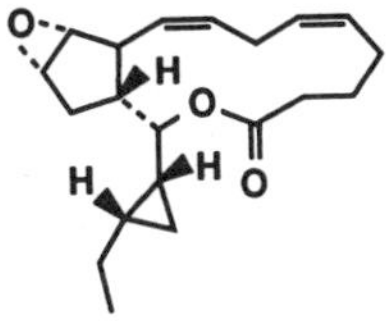

(46) 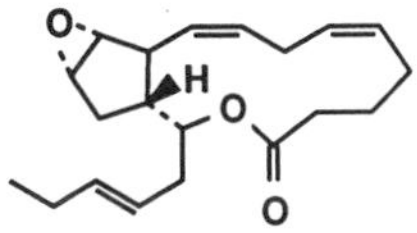   (47)

unnamed homolog (**47**). Again, a combination of spectroscopic data, IR, optical rotation, MS, ^{13}C NMR, and ^{1}H NMR was used to formulate the basic structure of hybridalactone as a 20-carbon chain containing two isolated olefins and lactone, epoxide, cyclopentyl, and cyclopropyl rings. Location of these groups within the 20-carbon chain was again accomplished by careful ^{1}H NMR decoupling experiments and relative stereochemistry suggested from a combination of coupling constant data for the natural product and a derivative formed from methanolysis (C12) of the epoxide. The minor substance (**47**) was incompletely characterized owing to its small yield from the alga; however, based on a strong analogy to ^{1}H NMR data obtained for hybridalactone (**46**), it was deduced to have the C16–C18 cyclopropyl group replaced with a simple C17–C18 *trans*-olefin.

These fascinating structures have several features of obvious structural similarity to the prostaglandins; however, the positions of rings in the chain and nature of the oxidized atoms suggest they have a substantially different biosynthetic origin. These intriguing features prompted Corey *et al.* (1984) a couple of years later to propose a unique biosynthetic origin for hybridalactone (**46**) which was based on initiation by a 12-lipoxygenase. In this proposal (Fig. 5), it is envisioned that cationic oxirane formation is followed by carbocyclization to a cyclopentyl allylic cation. This undergoes a complex cyclopropyl–cyclobutyl–cyclopropyl rearrangement to give a C15 cation which is quenched by lactonization, yielding hybridalactone (**46**). In combination with conformational analysis calculations for the relative configuration at C14 and C15, this biogenetic proposal predicted, beginning with a normal 12S-hydroperoxide, a 10S,11R,12S,14R,15S,16R,17S absolute stereochemistry for hybridalactone (**46**). Subsequently, this hypothesis was employed as a guide in a stereospecific total synthesis of 10S,11R,12S,14R, 15S,16R,17S-hybridalactone (Corey and De, 1984). Synthetic and natural hybridalactone samples showed identical ^{1}H NMR, IR, MS, TLC mobilities, and optical rotations. This is an excellent example of how biogenetic reasoning and conformational analysis can be powerful tools to aid in structure elucidations and synthetic efforts. These biogenetic and synthetic studies were independently corroborated by parallel x-ray crystallographic investigations on a heavy-atom derivative (11-hydroxy, 12-bromo) of reisolated hybridalactone (**46**) (Corey *et al.*, 1984). Apparently, in *L. hybrida* the initial 12-hydroperoxide formed by action of 12-lipoxygenase on an EPA precursor can be further metabolized by one of three potential manifolds: (1) 12-hydroperoxide lyase to form 12-oxo-5(Z),8(E),10(E)-dodecatrienoic acid (**10**), (2) 12-hydroperoxidase to form 12-HEPE (**9**), or (3) a series of cationic oxirane and carbocyclizations to give hybridalactone (**46**) (Fig. 5).

Laurencia spectabilis Owing in part to these exciting findings in *L. hybrida*, we decided to investigate a related Pacific species, *Laurencia spectabilis* (Bernart and Gerwick, work in progress). Collections of this alga from the central Oregon coastline were extracted in standard fashion and a mixture of hydroxy-acids esters was obtained by preparative TLC. Using the analytical system previously developed for quantitative assessment of 12-HETE (**8**) and 12-HEPE (**9**) (see discussion with *Murrayella periclados*), these two hydroxyeicosanoids were found in *L. spectabilis* as approximately 0.075% and 0.045% of the lipid extract, respectively. Hence, the 12-lipoxygenase pathway seems a consistent feature of fatty acid metabolism in both flattened (*L. spectabilis*) and terete (*L. hybrida*) forms of the genus *Laurencia*.

Murrayella periclados The Rhodomelacean alga *Murrayella periclados* grows in the intertidal zone on mangrove roots in Puerto Rico and many other tropical locations. Collections of this alga by our laboratory in 1984 showed several unique-appearing substances by TLC; however, by 2D TLC these were found to be unstable to silica gel chromatography and hence were derivatized to peracetyl and methyl esters (Bernart and Gerwick, 1988). Following our recognition that this alga is rich in various acarbocyclic eicosanoids, we have isolated these same substances in both underivatized form as well as simple methyl esters. The first eicosanoid obtained from this alga, protected as a peracetyl-methyl ester, was the

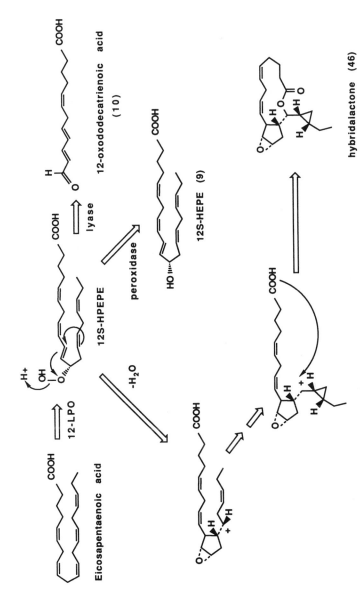

Figure 5. Proposed biogenesis of 12S-HEPE (**9**), 12-oxo-5(Z),8(E),11(E)-dodecatrienoic acid (**10**), and hybridalactone (**46**) by 12-lipoxygenase-initiated metabolism of eicosapentaenoic acid in *Laurencia hybrida*.

well-known EPA-derived mammalian autacoid 12S-HEPE (**9**) (Bernart and Gerwick, 1988). In mammals, this autacoid has several physiological properties, including an ability to stimulate chemotaxis and superoxide anion production by neutrophils and inhibit platelet aggregation. The structure was relatively easily deduced from a variety of spectroscopic data, including IR, UV, optical rotation, MS, and ^{1}H and ^{13}C NMR. However, the single most powerful tool in this structure elucidation was use of ^{1}H–^{1}H COSY. The linear and continuous nature of the spin system present in these lipoxygenase-derived metabolites provides an ideal system for analysis by this technique. Our isolation of this eicosanoid from *M. periclados* was the first ever from a plant source. Furthermore, it was obtained from the alga in relatively high yield (ca. 1–2%) such that we had several orders of magnitude more (ca. 30 mg) for our spectroscopic analyses than is normally available in work on eicosanoids. This allowed us to record spectroscopic features of this well-known molecule that were previously unapproachable (i.e., ^{13}C NMR). Further, biological testing made possible with this new source of 12S-HEPE has led to the discovery of new neutrophil (human)-degranulatory and Na^+/K^+ (dog kidney) and H^+/K^+ (hog gastric mucosa) ATPase-inhibitory properties of this eicosanoid (Bernart and Gerwick, 1988). Additionally, we have recently found 12S-HEPE to be a weak inhibitor of human phospholipase A_2 ($IC_{50} = 7.2$ μg/ml).

As discussed with *Laurencia hybrida*, our isolation of 12S-HEPE (**9**) from *M. periclados* and recording of its spectroscopic features led to our reassignment of the structure of a reported 9-HEPE (**45**) compound from the former seaweed (Higgs, 1981). We found the *L. hybrida* compound had identical spectroscopic features to our *M. periclados* 12-HEPE (**9**), a point which was of additional interest in that it provided further support to a proposal by Corey on a 12-lipoxygenase-initiated biogenesis of hybridalactone (**46**) (Corey *et al.*, 1984).

In continuing work with *M. periclados* we have isolated a number of other eicosanoids which logically arise from a 12-lipoxygenase manifold (Fig. 6) (Gerwick *et al.*, 1990; Bernart and Gerwick, work in progress). This included the closely related arachidonic acid-derived substance 12S-HETE (**8**), also a well-known and potent mediator of various physiological processes in humans. Again using algae-derived materials, we have recently detected a weak inhibitory activity by 12S-HETE of both human 5-lipoxygenase ($IC_{50} > 20$ μg/ml) and human phospholipase A_2 ($IC_{50} = 10$ μg/ml). Its structure elucidation from this, as well as other red algae (see discussion for *Platysiphonia miniata*, *Cottoniella filamentosa*, and *Gracilariopsis lemaneiformis*), similarly made major use of ^{1}H–^{1}H COSY, MS, and UV data.

Having isolated and characterized these two eicosanoids, 12-HETE (**8**) and 12-HEPE (**9**), from *M. periclados*, we became interested as to whether they were present in all phases of the life history of the plant (male, female, tetrasporophyte) or whether their concentrations varied in some systematic fashion. Preliminary findings with these three separately cultured phases of the plant followed by

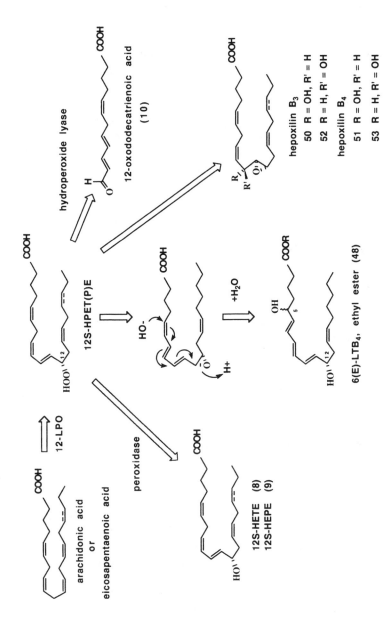

Figure 6. Eicosanoid metabolism in the red alga *Murrayella periclados.*

quantitative determination of 12-HETE and 12-HEPE by HPLC has revealed that all produce these substances in appreciable quantities (M. W. Bernart, W. H. Gerwick, D. L. Ballantine, and V. Cruz, work in progress). While there are variations in the amounts produced, the pattern and significance of this to the possible role these compounds play in the physiology of the alga are still uncertain.

Of greater structural chemical excitement and interest in our work with *M. periclados* has been the isolation of two other classes of 12-lipoxygenase-derived eicosanoid, leukotrienes and hepoxilins (Gerwick *et al.*, 1990; Bernart and Gerwick, work in progress). That this primitive marine plant produces a leukotriene is a striking example of the parallels in eicosanoid metabolism between terrestrial and marine creatures. The seaweed-derived substance, 6-*trans*-leukotriene B_4 (LTB_4), ethyl ester (**48**), was defined as a mixture of diastereomers,

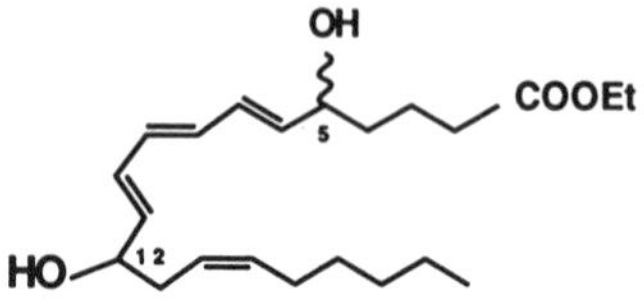

(48)

probably at C5, on the basis of a full set of spectroscopic features. It has previously been shown that 6-*trans*-LTB_4 is formed by nonenzymatic hydrolysis of either, and most commonly, a 5,6-epoxy leukotriene (5,6-LTA_4) (Samuelsson and Funk, 1989), or less commonly, an 11,12-epoxy leukotriene (11,12-LTA_4) (Westlund *et al.*, 1988). Nonenzymatic hydrolysis results in the formation of a mixture of diastereomers in either case. Given the preponderance of 12-lipoxygenase-derived substances in this *M. periclados*, it seems likely that this seaweed-derived diastereomeric mixture of 6-*trans*-LTB_4 comes from the nonenzymatic hydrolysis of an 11,12-LTA_4 substance (**49**) (Fig. 6).

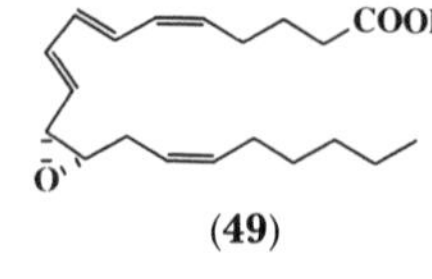

(49)

Most recently, we have isolated four closely related hepoxilins (**50–53**) from the crude methylated fractions of *M. periclados* (Bernart and Gerwick, work in progress). All four of these epoxy-alcohols logically derive from initial reaction of either arachidonic acid or EPA with 12-lipoxygenase. With mammalian-derived hepoxilins, which can elicit the release of insulin in the pancreas (Pace-Asciak *et al.*, 1986) and may serve other neuroendocrine roles (Pace–Asciak *et al.*, 1990), it has been shown that they are formed from the corresponding hydroperoxide by an intramolecular rearrangement (Pace-Asciak, 1984). Again, a complex interplay of normal spectroscopic features along with various 2D NMR experiments and

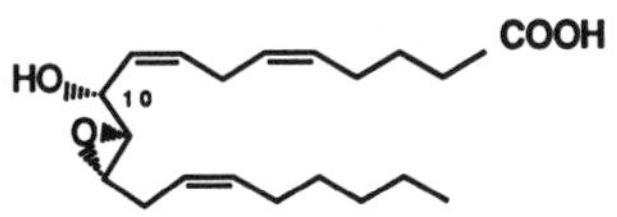

(50)

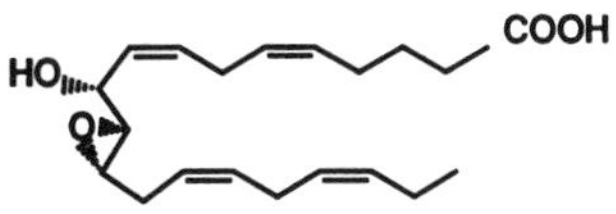

(51)

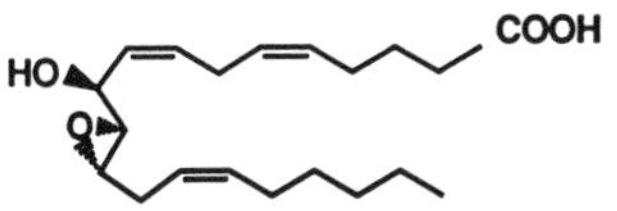

(52)

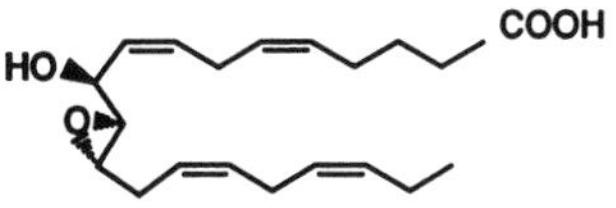

(53)

MS analyses were used to solve these structures. In these algae-derived hepox-ilins, the epoxide in all cases has a *trans* stereochemistry (by ^{1}H coupling constant) while the secondary alcohol at C10 is in both an *erythro* and *threo* disposition (approximately a 1:1 mixture, determined by chemical shift, coupling constant, and relative polarity on TLC) in both the ω-6 (arachidonic acid-derived) and ω-3 (eicosapentaenoic acid-derived) series. These findings are consistent with the product mixture obtained by the hemoglobin-catalyzed decomposition of fatty acid hydroperoxides to a mixture of epoxy-alcohols (Dix and Marnett, 1985), and raises the question as to whether the seaweed-derived epoxy-alcohols are of true natural occurrence (**50–53**). The absolute stereochemistry at C12, while currently unknown but under investigation, is expected to derive from the 12S-hydroperoxide (Fig. 6). The methyl ester of the *threo* isomer of hepoxilin B$_3$ (**50**) is weakly inhibitory to human phospholipase A$_2$ (IC$_{5)}$ = 12 μg/ml).

2.3.4c. Delesseriaceae

Platysiphonia miniata/Cottoniella filamentosa As a result of a survey of tropical algae from the environs of Puerto Rico for antimicrobial properties (Ballantine *et al.*, 1987), toxicity to brine shrimp, or thin-layer chromatographic interest, two red algae were targeted for further study, *Platysiphonia miniata* and *Cottoniella filamentosa*. These two red algae are occupants of a deep-water (15–30 m) algal community offshore from southwestern Puerto Rico and had not been previously investigated for secondary metabolites prior to this work. The crude lipid extracts of both showed antimicrobial activity to the Gram-positive bacte-rium *Staphylococcus aureus* as well as possessed numerous potentially interesting metabolites by TLC. However, based on TLC analysis, some of these potentially interesting metabolites were observed to have very poor chromatographic charac-teristics upon silica gel chromatography, and thus the crude extracts were deri-vatized with acetic anhydride to peracetates and with CH$_2$N$_2$ to methyl esters. These were subsequently chromatographed over silica gel in the vacuum mode and then by HPLC to yield the same two eicosanoids from both algae.

The first isolated (Moghaddam *et al.*, 1989), 12S-HETE (**8**) as a semisyn-thetic acetate, methyl ester, was obtained in 0.35% yield from *P. miniata* and ca.

0.2% from *C. filamentosa*. Its structure was easily deduced based on ^{1}H–^{1}H COSY in concert with MS and UV data, and absolute stereochemistry determined by comparing its optical rotation with published data. As discussed above, 12S-HETE is a physiologically important substance in mammalian systems, being involved in mediating the action of polymorphonuclear leukocytes in inflammatory reactions and inhibiting platelet aggregation. Our isolation of 12S-HETE from *P. miniata* was the first time that this important compound was isolated from a nonmammalian source. Continued study of these two algae led to the isolation of a second and more minor derivatized metabolite from both, 10-hydroxy-11,12-*trans*-epoxy-5(Z),8(Z),14(Z)-eicosatrienoic acid, also known as hepoxilin B_3 (**50** or **52**) (Moghaddam *et al.*, 1990). Its structure was determined again by a combination of spectroscopic techniques, principally ^{1}H–^{1}H COSY, ^{13}C NMR, and MS. However, the orientation of the epoxy-alcohol portion of the molecule in the chain was uncertain based on these data, mainly because of accidental degeneracy in the ^{1}H NMR spectrum of the C9 and C14 protons. This dilemma was resolved by use of ^{1}H–^{1}H relay transfer COSY, in which new correlations could be observed between the protons at C8 and C10, thus defining the position of the hydroxy group at C10. While the *trans* nature of the epoxide and all-*cis* configuration of the olefins could be deduced from coupling constants and ^{13}C NMR shifts, we were unable to determine the relative stereochemistry of the C10 alcohol in this isolate of hepoxilin B_3. Hence, in this review, the hepoxilin B_3 obtained from these two latter algae must be assigned structure **50** or **52**. In latter isolates of this substance (see above in discussion with *M. periclados*) (Bernart and Gerwick, work in progress), both stereoisomers at C10 were obtained and in substantially higher yield such that relative stereochemistry at C10 was assignable.

At the time of these isolations from *P. miniata* and *C. filamentosa*, hepoxilin B_3 had only been previously isolated from mammalian tissues (Pace-Asciak *et al.*, 1983) and the neurons of the sea hare *Aplysia* (Piomelli *et al.*, 1987). The full range of its biological functions remains uncertain, though it has been shown to possess insulin secretagogue activity in mammals (Pace-Asciak *et al.*, 1986) and play a possible second messenger role in *Aplysia* sensory cells (Piomelli *et al.*, 1987). As described above with the discussion for *Murrayella periclados*, this and related epoxy alcohols seem to be relatively common metabolites of red algae. In mammalian tissues they arise from the intramolecular rearrangement of 12-hydroperoxyeicosatetraenoic acid (12-HPETE), itself formed by action of 12-lipoxygenase on arachidonic acid (Pace-Asciak, 1984) (Fig. 6).

Polyneura latissima Recently, we have discovered yet another Delesserian red alga, *Polyneura latissima*, which produces a rich assortment of interesting eicosanoids (Jiang and Gerwick, work in progress). Initially, this alga was collected from just north of San Francisco, California, and later from Puget Sound, Washington, as part of our systematic evaluation of algae from the Pacific coast of the United States for new biomedicinals. It was targeted for further study because a number of potentially interesting compounds were observed in the TLC of the crude lipid extract. The crude extract from both the California and Washington collections was derivatized with CH_2N_2 to form methyl esters and then chromatographed in a conventional manner over silica gel followed by HPLC. From both, a new hydroxyeicosanoid was obtained which has only been rarely reported previously, 9-hydroxy-5(Z),7(E),11(Z),14(Z)-eicosatetraenoic acid (9-HETE; **54**) (Miyamoto *et al.*, 1987; Spector *et al.*, 1988; Pace-Asciak and Asotra, 1989). Its structure was deduced principally from a combination of ^1H NMR data, including extensive homonuclear ^1H NMR decouplings, and MS data. The electron impact mass spectrometry (EIMS) for the TMS ether, methyl ester gave a characteristic and major cleavage between C9 and C10, thus defining C9 as the hydroxyl-bearing site.

More polar fractions from chromatography of the CH_2N_2-derivatized crude extract from the California collection yielded two new and related eicosanoids which contained both hydroxy and epoxy oxygen functional groups. These functional groups were again located in the fatty acid chain of both molecules through the combination of MS fragmentation data and $^1H-^1H$ COSY information.

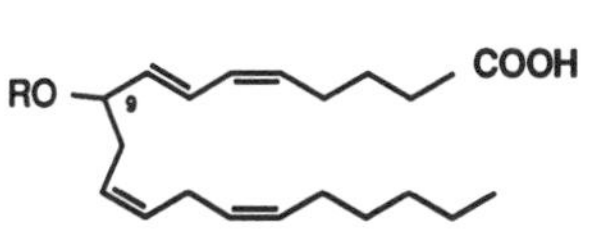

(54) R = H
(58) R = OH

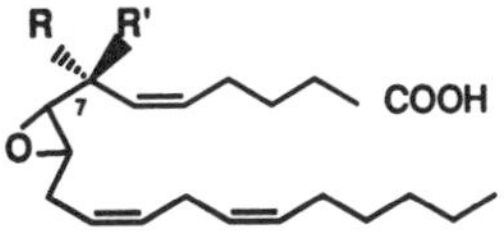

(55) R = OH, R' = H
(56) R = H, R' = OH

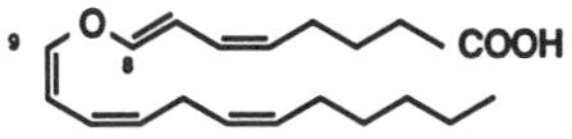

(57)

In these "hepoxilins," the epoxy group is located at C8 and C9 with an adjacent hydroxyl group at C7. In both of these, the epoxide is *trans* and all olefins are in the *cis* configuration. Hence, the two new *P. latissima* "hepoxilins" are C7 epimers, as expected based on previous findings with mammalian- (Corey *et al.*, 1983) and seaweed-derived hepoxilins (Bernart and Gerwick, work in progress). The co-occurrence of 9-HETE and these two hepoxilins (**55**, **56**) is again strong circum-stantial evidence for an intermediary 9-hydroperoxide, which is either reduced to form 9-HETE (**54**) or intramolecularly rearranged to form the two new hepoxilins **55** and **56** (Fig. 7).

However, the major compound from the Washington collection of *P. latissima* was another novel eicosanoid-like substance (**57**) which contained a highly unusual bis-enol ether functionality (Jiang and Gerwick, work in progress). This substance, isolated as the methyl ester derivative in approximately 2% yield from the lipid extract, possessed a beautifully dispersed ^1H NMR spectrum which was

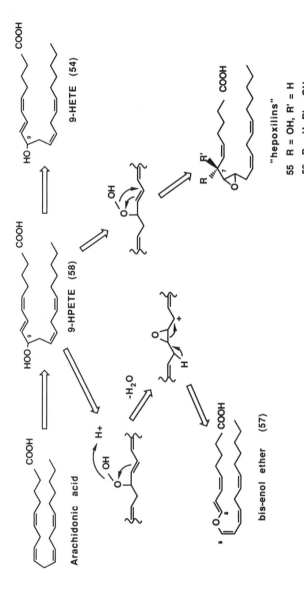

Figure 7. Eicosanoid metabolism in the red alga *Polyneura latissima*.

very effectively probed by ^{1}H–^{1}H COSY and ^{1}H-^{13}C heteronuclear correlated spectroscopy (HETCOR) experiments. Two distinct spin systems were apparent for protons on C2–C8 and C1′–C12′. These spin systems were linked by a bis-enol ether as shown by ^{1}H and ^{13}C NMR shifts and consideration of the molecular formula. Additionally, characteristic fragmentation of the C–O bond to either side of this bis-enol ether oxygen gave ions diagnostic for its position in the chain. Assignment of the stereochemistry of the double bonds was accomplished by coupling constant analysis and consideration of model compounds (Galliard *et al.*, 1973). While analogous C18 bis-enol ethers have been previously isolated from the potato (Galliard *et al.*, 1973), these are the first isolates of 20-carbon-derived bis-enol ethers. Furthermore, in the previous isolates from potato, both olefins bearing the oxygen atom are of *trans* geometry, while in the seaweed compound only the C7 olefin is *trans*. In potato, these have been shown to arise from a unique rearrangement of an intermediate epoxy cation which itself comes from a lipoxygenase-introduced hydroperoxide (Fahlstadius and Hamberg, 1990). Presuming an analogous mechanistic pathway for the *P. latissima* bis-enol ether, then its occurrence similarly indicates the presence in the seaweed of the 9-lipoxygenase product 9-hydroperoxyeicosatetraenoic acid (9-HPETE) (**58**) as an intermediate (Fig. 7). Our findings with *P. latissima* are unique at present in that it appears to have a well-developed 9-lipoxygenase pathway rather than the 12-lipoxygenase pathway characteristic of other red algae.

2.3.5. Rhodymeniales

2.3.5a. Rhodymenia pertusa. One of the most recent red algae that we have studied for its eicosanoid content is the foliaceous Washington alga *Rhodymenia pertusa* (Jiang and Gerwick, work in progress). Silica gel fractionation of the crude methylated lipid extract yielded several fractions with diol-containing natural products by NMR analysis. Subsequently, these were purified by HPLC and characterized, principally by ^{1}H–^{1}H COSY, MS, UV, and optical rotation analyses. The structures of the two major compounds were easily deduced when pure as 5,6-dihydroxy-7(E),9(E),11(Z),14(Z)-eicosatetraenoic acid (**59**) and 5,6-dihydroxy-7(E),9(E),11(Z),14(Z),17(Z)-eicosapentaenoic acid (**60**). Based on opti-

(59)   (60)

cal rotation [all four possible stereoisomers have been synthesized (Kugel *et al.*, 1989)], these seaweed diHETE and diHEPE natural products both have a 5*R*,6*S* stereochemistry. The structure of one minor compound has also been solved based on a similar set of data and is that of 5*R**,12(*R*,*S*)-dihydroxy-6(E),8(E),10(E),

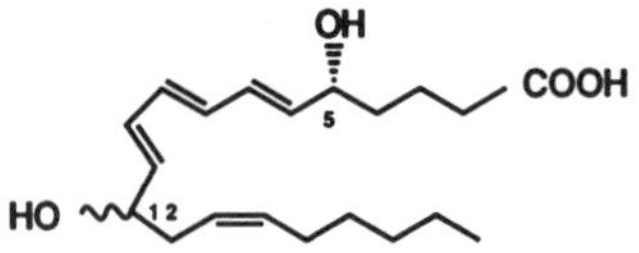

(61)

14(Z)-eicosatetraenoic acid (**61**; compare with the C5 epimerized mixture of dia-stereomers isolated from *Murrayella periclados*, **48**). Further, the co-occurrence of both 5,6-diHETE and 5,12-diHETE is strong circumstantial evidence for the occurrence of a 5,6-epoxytriene species (i.e., LTA_4; see Fig. 8). It has been shown in other systems that nonenzymatic hydrolysis of LTA_4 leads to the production of both 5,6-diHETE and 5,12-diHETE, the latter as a mixture of diastereomers resulting from formation of both epimers at C12 (Samuelsson *et al.*, 1987).

2.4. Chlorophyta

The natural products chemistry of the green algae is generally less well characterized (Paul and Fenical, 1987), and so it is perhaps not surprising that there are only a couple of very recent findings of eicosanoid-like substances from algae of this group. This should not be construed, however, to suggest that this class of chemistry is less abundant or less interesting in the Chlorophyta, as the following examples illustrate.

2.4.1. Enzyme Studies

Prior to our own work with two species of macrophytic marine chlorophytes, there were no reports of lipoxygenase, lipoxygenase products, or other eicosanoid-like substances from marine green algae. However, there have been a couple of reports of such activity in freshwater chlorophytes of unicellular form. The earliest (Zimmerman and Vick, 1973) explored the ability of cell homogenates of *Chlorella pyrenoidosa* cultures to metabolize exogenous linoleic and linolenic acids to oxidized products. While no such metabolism was detected when the cells were homogenized in the presence of oxygen, in the rigorous exclusion of oxygen a pronounced lipoxygenase-type activity was observed. The homogenate had a pH optimum of 7.4 and was effectively inhibited by either heat denaturation or 2×10^{-4} M nordihydroguaiaretic acid, a lipoxygenase-specific inhibitor. Two products from incubations with linoleic acid were characterized by MS following reduction and methylation as methyl 13-hydroxy stearate (80%) and methyl 9-hydroxy stearate (20%), indicating the presence of 13-HPOD (**5**) as a major product and 9-HPOD (**6**) as a minor product of this lipoxygenase. Oleic acid was not a sub-strate for the enzyme.

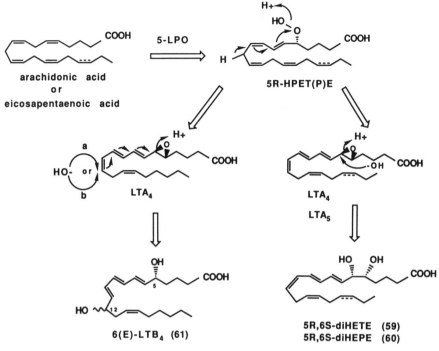

Figure 8. Proposed routes of formation of 5,6-diHETE (**59**) and 5,12-diHETE [6(E)-LTB$_4$; **61**] in *Rhodymenia pertusa*.

Recently (Vick and Zimmerman, 1989), it was shown by incubating [^{14}C]-13-hydroperoxylinolenic acid with a partially purified enzyme preparation from *C. pyrenoidosa* that this alga contains a unique hydroperoxide lyase activity, in this case yielding 13-oxo-9(E),11(E)-tridecadienoic acid (**7**) and pentene. Product **7** was identified by analysis of its spectroscopic features (UV, IR, MS, and ^{1}H NMR), while *cis*-2-pentene was identified by GC comparison with standards. This lyase activity was easily separated from the lipoxygenase activity on Sephadex G-200. Two other major products of this incubation experiment were the simple reduction product, 13-hydroxyoctadecatrienoic acid (**34**), and a novel substance, 12,13-*trans*-epoxy-9-oxo-10(E),15(Z)-octadecadienoic acid (**62**). Its structure was established from spectroscopic methods and by reduction and methylation of the

(**62**)

C9-oxo group to the trideuteromethoxy derivative which could be confidently positioned at C9 by MS analysis. Metabolite **62**, while homologous to known non-enzymatic products of 13-hydroperoxylinoleic acid, was shown by use of appropriate controls to be formed enzymatically by the algal enzyme. In these same studies, *C. pyrenoidosa* was found to contain several of the enzymes involved in the conversion of linolenic acid to jasmonic acid, a plant growth regulator. However, in the apparent absence of any endogenous production of jasmonic acid, the significance of these latter findings is uncertain.

The eicosanoid content of a unicellular protist, *Euglena gracilis*, was evaluated by serological and immunochromatographic methods (Levine *et al.*, 1984). This protist has many algae-like characteristics, including chloroplasts. While the methodology employed (HPLC, immunoassay, UV spectroscopy) did not provide a rigorous determination of structure, a good case is developed in this paper for an abundant endogenous production of prostanoid (PGE_2, $PGF_{2\alpha}$), hydroxyeicosanoid (12-HETE, **8**) and leukotriene (LTC_4, LTB_4)-like compounds in this protist. However, production of these substances was greater in dark-grown cells than in light-grown cells, suggesting that their production in this protist may be associated with its animal-like rather than plant-like metabolic capacity.

Another freshwater unicellular green algae, *Dunaliella acidophila*, which is a commercial source of lipid natural products including β-carotene, has been recently characterized for its hydroxy-fatty acid content (Pollio *et al.*, 1988). Two such hydroxy acids, 12*R*-hydroxy-9(Z),13(E),15(Z)-octadecatrienoic acid (**63**) and 9*S*-hydroxy-10(E),12(Z),15(Z)-octadecatrienoic acid (**64**), were characterized as

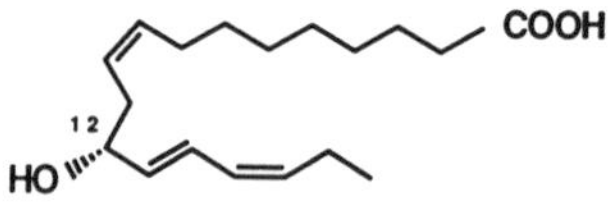

(63) 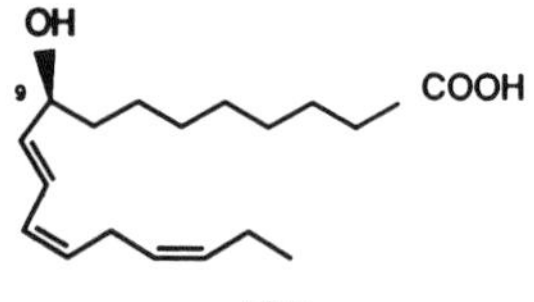   (64)

major metabolites (3.4 and 4.2%, respectively) from the acidified lipid extract of axenic cultures of *D. acidophila*. Their overall structures followed from an analysis of MS, ^{1}H NMR, ^{13}C NMR, and IR data (UV not reported), while absolute stereochemistry was indicated by a close correspondence of measured optical rotations with that of homologous model compounds.

An apparent α-oxidase activity has been characterized from freshly collected plants as well as bacteria and epiphyte-free thalli cultures of the marine green alga *Ulva pertusa* (Fujimura *et al.*, 1990). The resultant long-chain aldehydes [8(Z),11(Z)-heptadecadienal] formed by decarboxylation of these α-oxidized acids are responsible for much of the flavor and odorant properties of the essential oil obtained from this alga.

2.4.2. Acrosiphonia coalita

Our first work (M. W. Bernart, G. Whatley, and W. H. Gerwick, work in progress) with a marine chlorophyte was with a common inhabitant of the intertidal region along the Pacific West Coast of the United States, *Acrosiphonia coalita*. The crude lipid extract from this alga showed antimicrobial activity to several Gram-positive bacteria as well as numerous and relatively minor metabolites of potentially novel structure by TLC. Subsequently, a standard sequence of vacuum silica gel chromatography followed by HPLC yielded a novel and chemically unstable aldehyde-containing metabolite (**65**). The structure of this

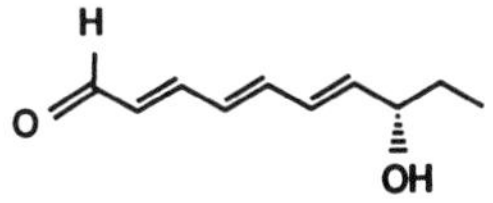

(**65**)

short-chain aldehyde (approximately 0.7%) was easily deduced by analysis of its UV, IR, MS, and ^{1}H NMR spectra, including ^{1}H–^{1}H COSY. Formation of the *p*-methoxybenzoate derivative followed by CD spectroscopy allowed assignment of an 8S stereochemistry based on observation of a characteristic positive split Cotton effect, thus defining this new compound as 8S-hydroxy-2(E),4(E),6(E)-decatrienal. Re-isolation efforts of this and related compounds led to the HPLC isolation of a closely related substance (**66**) that was shown to convert sponta-

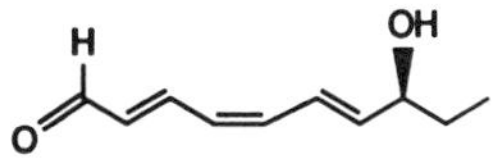

(**66**)

neously to the all-*trans* trienal (**65**) upon UV irradiation (1 hr, 254 nm). This related substance (**66**) was isolated in approximately 0.2% yield and characterized principally by ^{1}H–^{1}H COSY, ^{1}H NMR coupling constant analysis, and ^{13}C NMR as the 4(Z) isomer of trienal (**65**). However, we now believe that the major short-chain aldehyde-containing product of the alga is the 2(E),4(Z),6(E) isomer (**66**), which isomerizes to the 2(E),4(E),6(E) isomer upon extraction and purification. The occurrence of these short-chain aldehydes suggests a carbon-chain cleavage metabolism of oxidized polyunsaturated fatty acids in this seaweed [i.e., action of hydroperoxide lyase on a 9-hydroperoxy-10(E),12(Z),15(Z)-octadecatrienoic acid precursor]. However, additional insights into the oxidative chain cleavage capacities of this alga were gained by characterization of the metabolites described below.

A more polar chromatography fraction was subjected, following derivatization to methyl esters with CH_2N_2, to extensive HPLC separation to yield three novel eicosanoid-like substances (Bernart and Gerwick, work in progress). The simplest of these was easily characterized on the basis of MS, UV, and ¹H NMR as the methyl ester derivative of an 18-carbon fatty acid with three double bonds and a hydroxy and ketone functionalities. By ¹H-¹H COSY and UV measurements, the locations of these functionalities were easily placed in the chain and subsequently confirmed by MS fragmentation patterns, and described this new compound as 16-hydroxy-9-keto-10(E),12(Z),14(E)-octadecatrienoic acid (**67**). While isolation of

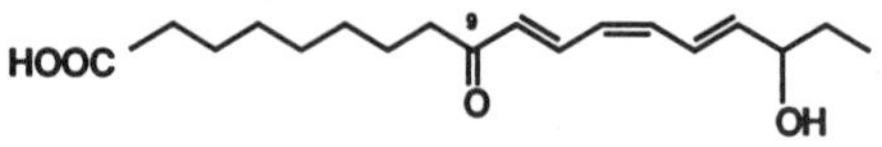

(67)

this substance was of considerable interest in that it provided insight into the formation of the C10 trienals **65** and **66**, it was of even greater excitement to isolate the two additional eicosanoid-like substances described below.

Immediately upon ¹H NMR spectroscopic analysis of purified methyl ester derivatives of metabolites **68** and **69**, it was clear that these substances were wholly unique in that they contained three distinct termini—a carbomethoxy ester

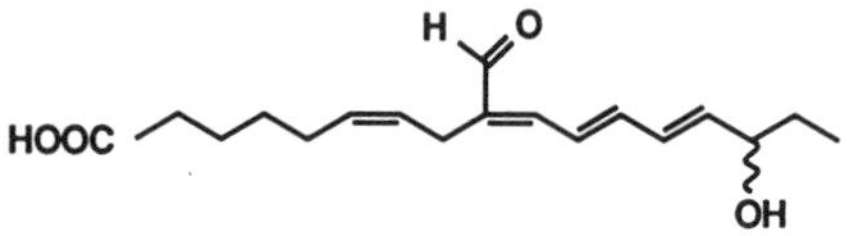

(68)

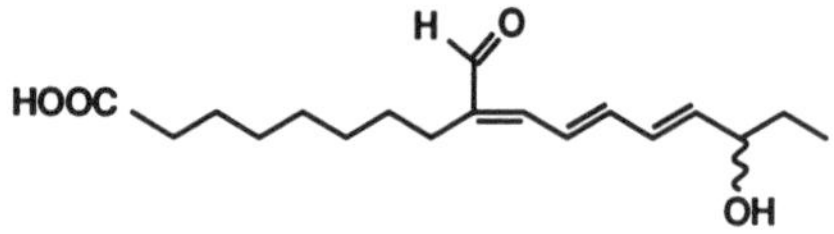

(69)

of partial synthetic origin, a triplet high-field methyl group, and a singlet aldehyde proton. By UV measurement, the characteristic 320-nm absorption of a trienal was observed as seen for metabolites **65–67**. Furthermore, by ¹³C NMR and ¹H–¹H COSY, it was readily apparent that the C1 carbomethoxy ester through to a bis-allylic methylene at C8 were present in compound **68**. Similarly, the methyl-terminus region through the C11 olefin proton (numbering for metabolite **67**) was also present. Missing from the spectrum relative to ketone-containing metabolite **67** was the C10 olefin proton, which was consistent with the change in coupling pattern of C11 from a doublet of doublets in **67** to a broadened doublet in **68**. Hence, the aldehyde substituent on this C17 chain in metabolite **68** was placed on vinyl carbon C9, and this placement confirmed by nOe studies. Mass spectral fragmentations of the corresponding TMS ether of this methoxy derivative strongly support the positioning of these substituents. As metabolite **68** showed no appreciable optical rotation at a number of wavelengths, the absolute stereo-chemistry of the C15 alcohol was evaluated by GC and GC-MS analysis of the fragment produced from O_3 cleavage of the (−)-menthoxycarbonyl derivative followed by methylation. By this unequivocal method, the C15 alcohol in metabolite **68** was found to be completely racemized. The structure of metabolite **69** followed similar lines of data interpretation and was the corresponding C6–C7 dihydro analog of metabolite **68**.

That these two metabolites (**68** and **69**) are ultimately derived from stearidonic acid and α-linolenic seems certain, particularly given the co-occurrence of metabolite **67**; however, the metabolic routes by which this occurs are very uncertain. Further, their isolation raises the further possibility that metabolites **65**–**67** as well are produced by unprecedented and unexpected pathways.

2.4.3. Cladophora columbiana

Recently, we have investigated another chlorophyte common to the Pacific coast, *Cladophora columbiana*, which showed Gram-positive antimicrobial activity (*B. subtilis*) in its lipid extract (Jiang and Gerwick, work in progress). Vacuum silica gel chromatography followed by HPLC led to the isolation of one identifiable product, that being the relatively simple short-chain unsaturated acid (**70**). While

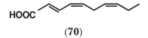

(70)

not of structural interest in itself, its occurrence illustrates that green algae on a more general level oxidatively metabolize polyunsaturated fatty acids, and do so by potentially unique metabolic routes to produce novel substances (see discussion above with *Acrosiphonia coalita*) It is interesting to note that metabolite **70** conceivably arises from chain scission of 9-hydroperoxylinolenic acid, the same intermediate logically involved in the formation of the above *A. coalita* metabolites.

2.5. Phaeophyta

Until quite recently, eicosanoid and homologous substances were virtually unknown among brown algae. However, in the last couple of years there has been a surge in findings of this class of chemistry in the Phaeophyta, and interestingly, mainly among those which have classically not yielded the more traditional classes of interesting natural product (i.e., terpenes). Further, in addition to the unique eicosanoid-like products described below, there is also a suggestion in the literature that C18 and C20 polyunsaturated fatty acids are the progenitors of the various C8 and C11 hydrocarbons which play such an important role as gamete attractants in the life history of many brown algae (Neumann and Boland, 1990). Polyunsaturated fatty acids have also been identified as the active antimicrobial agents present in the extracts of several brown algae, including *Alaria marginata*, *Desmarestia ligulata* var. *ligulata*, *Egregia menziesii*, *Eisenia arborea*, *Fucus distichus*, *Laminaria saccharina*, *Macrocystis integrefolia*, *Nereocystis luetkeana*, and *Pleurophycus garderi* (Rosell and Strivastava, 1987).

2.5.1. Chordariales

2.5.1a. Chordariaceae

Cladisiphon okamuranus During the course of cultivation efforts with the edible brown seaweed *Cladisiphon okamuranus*, it was discovered that seawater in which cut fragments of the algae were placed became discolored and was rendered unsuitable for the growth of any other seaweed for a prolonged period (Kakisawa *et al.*, 1988). Bioassay-guided fractionation of the MeOH extract from this alga lead to the isolation of a single substance responsible for this activity. This was easily characterized by IR and ¹H NMR of the natural product and MS of the semi-synthetic methyl ester as 6(Z),9(Z),12(Z),15(Z)-octadecatetraenoic acid (**71**). This

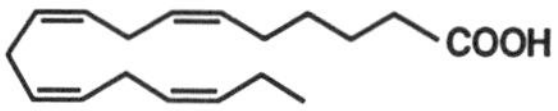

(**71**)

polyunsaturated fatty acid was subsequently evaluated for toxicity to 36 species of microalgae and the gametes of two species of macroalgae. Approximately 50% of the microalgae were found to be sensitive to the toxic effects of **71** at the 5 ppm level. The limited evaluation of macroalgal gametes showed a similar trend when the gametes were free from protective parent structures (i.e., released). In further testing, arachidonic acid and eicosapentaenoic acid were found to have similar levels of toxicity (2 ppm) to one of the more sensitive microalgae, *Heterosigma akashiwo*. It is proposed that this fatty acid is produced allelopathically and confers a survival advantage to *C. okamuranus* by inhibiting the growth of nearby competing seaweeds.

2.5.1b. Notheiaceae

Notheia anomala The first report (Warren *et al.*, 1980) on an unusual eicosanoid-like lipid from a brown alga was from the Australian alga *Notheia anomala*, which grows either epiphytically or parasitically on another brown alga, *Hormosira banksii*. Standard silica gel chromatography of the hexane extract gave crystalline **72** as a major metabolite (25%). Its linear structure was determined by spectroscopic means coupled with extensive chemical degradation, while the relative stereochemistry was determined by x-ray crystallography on the dihydro

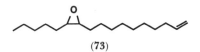

(72) R = R' = H

(79) R = CO(CH$_2$)$_{14}$CH$_4$, R' = H   (83) R = H, R' = CO(CH$_2$)$_{14}$CH$_3$

(80) R = CO(CH$_2$)$_{10}$CH$_3$, R' = H   (84) R = H, R' = CO(CH$_2$)$_{10}$CH$_3$

(81) R = CO(CH$_2$)$_{12}$CH$_4$, R' = H   (85) R = H, R' = CO(CH$_2$)$_{12}$CH$_3$

(82) R = CO(CH$_2$)$_{16}$CH$_3$, R' = H   (86) R = H, R' = CO(CH$_2$)$_{16}$CH$_3$

derivative. Absolute stereochemistry of the C7 alcohol was determined by the Horeau method performed on the C10-monoacetate derivative, yielding the complete structure of **72** as (6*S*,7*S*,9*R*,10*R*)-6,9-epoxynonadec-18-ene-7,10-diol. Presumably, and as supported by subsequent isolations of related materials (see below), the carbon chain of **72** arises from decarboxylation of an eicosanoic acid (C20) precursor.

A follow-up and comprehensive reinvestigation of the lipid natural products of *N. anomala* has recently appeared (Barrow and Capon, 1990). In addition to a reisolation of metabolite **72**, one new C17 lipid, 13 new C19 lipids, and one new and two known C21 lipids were isolated by silica gel chromatography followed by HPLC. Their isolation and description provides some interesting insights into the possible biosynthesis of the earlier isolated tetrahydrofuran metabolite (**72**) as well as of the metabolic capabilities of brown algae in general.

(73)

The new C17 lipid (**73**), isolated as 0.015% of the dry weight of the seaweed, was deduced to possess one epoxide functionality and a terminal olefin. Location of the epoxide followed from analysis of MS fragmentation data, and absolute stereochemistry suggested by comparison of molecular rotation data with that of model compounds.

The new C19 lipids (**74–87**) are a mixture of tetrahydrofuran derivatives, epoxides, vicinal diols, and hydrocarbons. Their structures are all based on analysis of spectroscopic data and in some cases formation of informative derivatives. Two new C19 hydrocarbon derivatives were isolated and defined, principally on the basis of ^{1}H and ^{13}C NMR data and MS data, as 6(Z),9(Z),18-nonadecatriene (**74**, 0.006%) and 3(Z),6(Z),9(Z),18-nonadecatetraene (**75**, 0.003%) (numbering as for **72**). These presumably arise through decarboxylation of eicosadienoic and

(74)   (75)

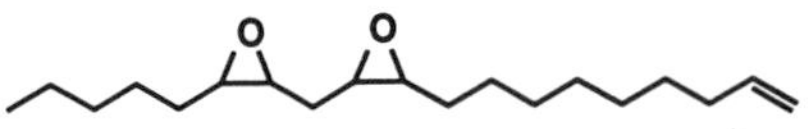

(76) 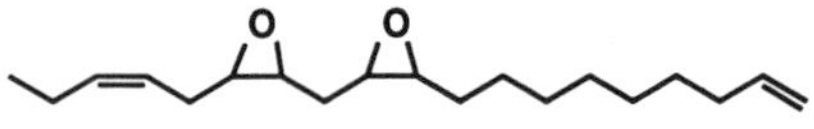   (77)

eicosatrienoic acids, respectively, and make attractive precursors for the other C19 lipids isolated from *Notheia* (Fig. 9). Two C19 diepoxides were isolated, **76** and **77** in 0.03 and 0.15% yield, respectively, and their structures determined spectroscopically and interrelated by catalytic hydrogenation. In this manner, compound **77** was defined as the 3,4-didehydro derivative of **76**. Furthermore, treatment of **76** with aqueous acetic acid in tetrahydrofuran gave a 50% yield of tetrahydrofuran derivative (**72**). By consideration of the probable mechanism of this transformation, the observed relative stereochemistry of the product (by ^{1}H NMR), and the precedence for the exo-carbon sites as the preferred sites for nucleophilic opening in bis-epoxides of this sort, a 6R,7S,9R,10S stereochemistry for metabolite **76** was tentatively assigned. A new C19 vicinal diol (**78**) was isolated in 0.01% yield and

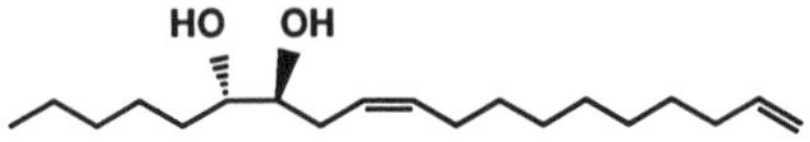

(78)

characterized by spectroscopic investigation of the natural product and several key derivatives. Its relative and absolute stereochemistry are proposed on mechanistic grounds, assuming an epoxide precursor, in concert with rotational data as compared with that of model compounds. Finally, eight new derivatives of the previously described tetrahydrofuran derivative **72** were characterized by spectroscopic means combined with partial degradations (**79–86**; see Fig. 9). These were isolated as two series of homologous and inseparable natural products—those with fatty acid esters attached at C7 and those with esters attached at C10. The position of attachment was characterized by ^{1}H and ^{13}C shifts as well as MS fragmentation data of the semisynthetic monoacetates. Hydrolysis of this acyl component and characterization by GC and GC-MS showed palmitate (**79, 83**) to predominate with minor amounts of laurate (**80, 84**), myristate (**81, 85**), and stearate (**82, 86**) in both the C7 and C10 series.

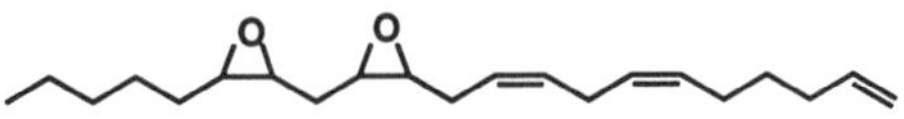

(87)

The structure of the new C21 *N. anomala* metabolite (**87**) was determined by ^{13}C NMR shift reasoning and MS of the per-hydro derivative to be another diepoxide derivative with olefins at C12, C15, and C20. The high degree of comparability of the ^{13}C NMR spectrum with that obtained for **76** combined with ^{13}C shifts indicative of two *cis* olefins provided the overall structure. A 6R,7S,9R,10S absolute stereochemistry is suggested based on the similarity of

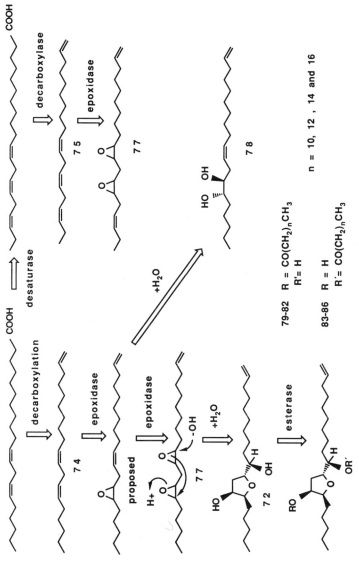

Figure 9. Eicosanoid-like metabolism in the brown alga *Notheia anomala*.

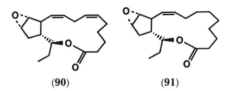

(88)

(89)

rotational data obtained for the per-hydro derivative of **87** as compared with the per-hydro derivatives obtained from **76** and **77**. The structures of the two known C21 hydrocarbon compounds (**88, 89**) were determined by comparison with literature data and consideration of their various spectroscopic features (MS, ^{1}H and ^{13}C NMR). These had been previously isolated and characterized from another brown algae, *Fucus vesiculosus*. A biogenetic pathway interrelating these various *Notheia* metabolites has been proposed (Fig. 9).

2.5.2. Laminariales

2.5.2a. Ecklonia stolonifera. Quite recently, a fascinating report appeared on a novel prostanoid-like natural product from the kelp *Ecklonia stolonifera* (Kurata *et al.*, 1989). An investigation of this alga was originally undertaken in order to discover which specific natural products were responsible for the observed abalone antifeedant properties of this species. Two compounds were isolated which had weak abalone antifeedant activity, and their structures were determined as ecklonialactones A (**90**) and B (**91**).

(90)   (91)

Ecklonialactones A and B were isolated as minor metabolites (0.65 and 0.33%, respectively) by a combination of column chromatography over alumina followed by HPLC. The structure of ecklonialactone A was determined by a detailed analysis of its IR, EIMS, ^{13}C NMR, and ^{1}H NMR spectra, in which the presence of two disubstituted olefins, a lactone, an epoxide, and a cyclopentyl ring were deduced. The ^{1}H–^{1}H COSY experiment provided the connectivity of this spin system and overall planar structure for **90**. This structure was confirmed by an x-ray crystallographic analysis which also gave the relative stereochemistry of ecklonialactone A. The structure of ecklonialactone B (**91**) was deduced as 6,7-dihydro ecklonialactone A from its very comparable data set. This was confirmed by comparing the relevant spectroscopic features for the per-hydro derivatives produced from both compounds by catalytic hydrogenation.

The relative stereochemistry in ecklonialactones A (**90**) and B (**91**) is the same as that found for hybridalactone (**46**), a remarkably similar structure found from *Laurenica hybrida* (see above), and suggests a similar biosynthetic origin. It is important to note the difference in chain length of the two classes of metabolites (18 carbons for **90** and **91**; 20 carbons for **46**) with consequent differences in the positions of double bond, oxidations, and cyclizations. However, by analogy to the pathway proposed and partially substantiated for the biosynthesis of hybridalactone, it seems likely that ecklonialactone A (**90**) derives from stearidonic acid (**71**), while ecklonialactone B (**91**) derives from linolenic acid. Presumably, their biosynthesis is initiated by a lipoxygenase reaction which introduces a hydroperoxide functionality at C13. This then undergoes a similar sequence of cyclizations to form the cyclopentanoid and epoxide rings, however, with a direct trapping of the C16 carbonium ion via lactone formation (Fig. 10).

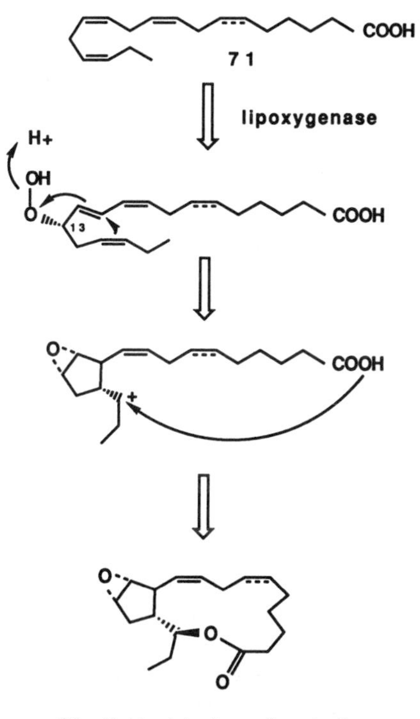

90 Ecklonialactone A, Δ 6

91 Ecklonialactone B, Δ 9

Figure 10. Proposed biosynthesis of the ecklonialactones A (**90**) and B (**91**) by *Ecklonia stolonifera*.

2.5.2b. Laminaria sinclairii. Over a period of several years, we have observed a variable antimicrobial activity associated with the Oregon coastal kelp *Laminaria sinclairii* (Gerwick, Bernart, Moghaddam, Nagle, Solem, Proteau, Jiang, and Wise, work in progress). While this is in part expected given the documented antimicrobial activities of polyphenolics which are found ubiquitously in these kelps, the lipid extract of *L. sinclairii* also showed several minor and UV-active natural products of potentially novel structure. Recently (Proteau and Gerwick, work in progress), one of these substances was isolated by our laboratory and shown by comparison of the relevant spectroscopic features to be similar to the bis-enol ether natural product (**57**; see Fig. 7) obtained from *Polyneura latissima* (Jiang and Gerwick, work in progress). Isolation of this lipoxygenase-derived substance from a kelp-type seaweed is encouraging in that it suggests that these larger brown algae may generally be a rich source of novel eicosanoid-like natural products.

3. CONCLUSION

3.1. Metabolic Themes

Several hypotheses can be presented on the trends in the eicosanoid metabolism of marine plants based on the findings summarized above. This is best supported in the Rhodophyta, which yielded the majority of reports on this class of chemistry in algae. Speculations on such trends for the other groups, based on the relatively few numbers of reports to date, are less certain.

The few eicosanoid-like natural products reported to date from the cyanobacteria suggests that these prokaryotes generally metabolize C18 and shorter fatty acids to simple hydroxy compounds or via rearrangement of hydroperoxy intermediates to triols analogous to mammalian trioxilins. Additionally, these prokaryotes often modify C1 of these C13–C18 acids by introduction of unusual substituents (rearranged amino acids, **2**; chlorine, **4**).

The prevailing trend in red algae is the 12-lipoxygenase-initiated metabolism of 20-carbon polyunsaturated fatty acids (arachidonic and eicosapentaenoic acids). Subsequent metabolism of the intermediate hydroperoxide is quite diverse: reduction to the simple 12-hydroxy eicosanoids, rearrangement to the hepoxilin-type epoxy alcohols, lyase-type scission reactions of the carbon chain, presumed formation of 11,12-LTA$_4$ species with subsequent 1,8 addition of water to LTB$_4$-like species, and more complex rearrangements of the chain to form carbocycles, including cyclopropyl and cyclopentyl species.

An alternate trend observed in fewer numbers of red algae is the metabolism of AA or EPA precursors by presumed 5-lipoxygenase, 9-lipoxygenase, and 11-

lipoxygenase enzymes. The action of a 5-lipoxygenase is indicated by the structures of metabolites isolated from *Gracilariopsis lemaneiformis* (**37**) and *Rhodymenia pertusa* (**59, 60**) which in the latter case is apparently dehydrated to a 5,6-LTA$_4$ species. The proposed 9-hydroperoxyeicosanoid intermediate in *Polyneura latissima* also has several potential metabolic fates, giving rise to hydroxyeicosanoids (**54**), hepoxilin-type epoxy alcohols (**55, 56**), and bis-enol ethers (**57**). The eicosanoids of *Ptilota filicina* (**42–44**), appear to be produced by a diversity of routes; however, ptilodene (**44**), which contains oxygen functionalities at two positions, likely arises from initial reaction by an 11-lipoxygenase.

The trend that emerges from consideration of the metabolites isolated to date from the Chlorophyta is that 18-carbon acids are the predominant substrate. These often become oxidized at C9, C13, and C16. In the microalgae, the C13 hydroperoxide is metabolized to simple alcohols (**34**), rearranged to epoxy alcohol-derived substances (**62**), or subject to hydroperoxide lyase scission reactions to produce C13 aldehydes (**7**). In the macroalgae studied to date, it appears that C9 and C16 oxidation predominates, with most undergoing some type of carbon-chain scission between C8 and C9 (**66, 68, 70**).

In the Phaeophyta, a diversity of metabolic pathways is reflected in the metabolite composition reported to date. By invoking reactions analogous to those proposed for hybriadalactone (**46**) (Corey *et al.*, 1984), it would appear that in the biosynthesis of the ecklonialactones (**90, 91**), *Ecklonia stolonifera* possesses a lipoxygenase capable of producing a 13-hydroperoxide in an 18:4 precursor fatty acid (**71**). This same fatty acid (**71**) is reported from the brown alga *Cladisiphon okamuranus*. However, in *Laminaria sinclairii*, it appears that a 9-lipoxygenase may be present as found in some red algae. Finally, a very different metabolic theme is seen in the genus *Notheia*, in which 20- and 22-carbon fatty acids are decarboxylated to C19 and C21 metabolites and oxidized by cytochrome P450-like processes to a series of epoxides (**76**) and epoxide-derived products (**72**).

3.2. Role of Eicosanoids in Marine Plants

There is very little information on what role eicosanoid-like compounds play in the physiology or ecology of the seaweeds, which is not unlike our current understanding of the role of similar substances in higher plants (possible hormonal and antifungal roles). In *Cladisiphon okamuranus*, the only alga which produces eicosanoid-like compounds and for which there is a putative role for this chemistry, it appears that these compounds may enhance survival of the alga by inhibiting the growth of other algal competitors (Kakisawa *et al.*, 1988). A reproductive role seems ruled out in *Murrayella periclados*, in which preliminary quantitative assessment shows that all life phases of the plant produce comparable levels of hydroxyeicosanoids (M. W. Bernart, W. H. Gerwick, and D. L. Ballantine, work in progress). Many of the eicosanoid-like substances isolated to date show some antibacterial or antifungal activity; however, it is not known if this represents their major ecological function. Another common property of many alga-derived eicosanoid-like substances is their ability to inhibit the function of animal-derived ion-pumping ATPases, and hence, it is possible they play a similar role in the physiology of algae (Bernart and Gerwick, 1988; Solem *et al.*, 1989). Clearly, more research is urgently needed in this exciting area.

3.3. Marine Eicosanoids and Related Compounds in Medicine and Physiological Research

Eicosanoids and related compounds from marine plants have not had a large impact on medicine or physiological research. As a new field, we are just now discovering that such metabolic pathways are in operation in these organisms. These pathways are in the process of being outlined through a description of the structures which they produce. However, many of these metabolites contain novel structural features which represent unique analogs of physiologically important mammalian eicosanoids. Hence it can be expected that as they are more widely evaluated in appropriate biological systems, some will be found as useful biochemical probes through their modulation of mammalian eicosanoid biosynthesis and action. In the limited testing to date, several of these alga-derived eicosanoid-like substances have the ability to inhibit the mammalian Na^+/K^+ and H^+/K^+ ATPase enzymes, modulate the release of superoxide anion by neutrophils, and inhibit the metabolism of arachidonic acid in human cell lines. Further, an exploration of their biological roles in the algae should give valuable insights into the range of physiological functions that eicosanoid-like substances can possess, some of which may have medical application. As has been the case in studies of the eicosanoid-like substances from higher plants (e.g., studies of lipoxygenase metabolism in the soybean provided the foundation for the discovery of similar metabolic pathways in humans), studies of this metabolism in the algae may ultimately be instructive in helping to identify other pathways of fatty acid metabolism in animals. In this regard, a detailed knowledge is needed of the biosynthetic pathways by which these algal eicosanoids and related substances are produced.

ACKNOWLEDGMENTS. We gratefully acknowledge the many helpful discussions with Mats Hamberg of the Karolinska Institute, Stockholm, Sweden, in the preparation of this manuscript. The Syntex Research Corporation, Palo Alto, California (Drs. Mike Ernest and David Waterbury) has provided many of the biological testing results reported herein, and we further thank the Karolinska Institute for help in the preparation of this manuscript.

REFERENCES

Ainslie, R. D., Barchi, Jr., J. J., Kuniyoshi, M., Moore, R. E., and Mynderse, J. S., 1985, Structure of malyngamide C, *J. Org. Chem.* **50**:2859–2862.

Andrianarison, R.-H., Beneytout, J.-L., and Tixier, M., 1989, An enzymatic conversion of lipoxygenase products by a hydroperoxide lyase in blue-green algae (*Oscillatoria* sp.), *Plant Physiol.* **91**:1280–1287.

Baertschi, S. W., Brash, A. R., and Harris, T. M., 1989, Formation of a cyclopropyl eicosanoid via an allene oxide in the coral *Plexaura homomalla*: Implications for the biosynthesis of 5,6-*trans*-prostaglandin A$_2$, *J. Am. Chem. Soc.* **111**:5003–5005.

Ballantine, D. L., Gerwick, W. H., Velez, S. M., Alexander, E., and Guevara, P., 1987, Antibiotic activity of lipid soluble extracts from Caribbean marine algae, *Hydrobiologia* **151/152**:463–469.

Barrow, K. D., 1983, Biosynthesis of marine natural products, in: *Marine Natural Products*, Vol. 5 (P. J. Scheuer, ed.), Academic Press, New York, pp. 51–94.

Barrow, K. D., and Capon, R. J., 1990, Epoxy lipids from the Australian epiphytic brown alga *Notheia anomala*, *Aust. J. Chem.* **43**:895–911.

Beneytout, J.-L., Andrianarison, R.-H., Rakotoarisoa, Z., and Tixier, M., 1989, Properties of a lipoxygenase in blue-green algae (*Oscillatoria* sp.), *Plant Physiol.* **91**:367–372.

Bernart, M. W., and Gerwick, W. H., 1988, Isolation of 12-(S)-HEPE from the red marine alga *Murrayella periclados* and revision of structure of an acyclic icosanoid from *Laurencia hybrida*. Implications to the biosynthesis of the marine prostanoid hybridalactone, *Tetrahedron Lett.* **29**: 2015–2018.

Burgess, J. R., de la Rosa, R. I., Jacobs, R. S., and Butler, A., 1991, A new eicosapentaenoic acid formed from arachidonic acid in the coralline red algae *Bossiella orbigniana*, *Lipids* **26**:162–165.

Cardellina II, J. H., and Moore, R. E., 1980, Malyngic acid, a new fatty acid from *Lyngbya majuscula*, *Tetrahedron* **36**:993–996.

Cardellina II, J. H., Dalietos, D., Marner, F.-J., Mynderse, J. S., and Moore, R. E., 1978, (−)-*trans*-7(S)-Methoxytetradec-4-enoic acid and related amides from the marine cyanophyte *Lyngbya majuscula*, *Phytochemistry* **17**:2091–2095.

Corey, E. J., and De, B., 1984, Total synthesis and stereochemistry of hybridalactone, *J. Am. Chem. Soc.* **106**:2735–2736.

Corey, E. J., Kang, J., Laguzza, B. C., and Jones, R. L., 1983, Total synthesis of 12-(S)-10-hydroxy-*trans*-11,12-epoxyeicosa-5,9,14-(Z)-trienoic acids, metabolites of arachidonic acid in mammalian blood platelets, *Tetrahedron Lett.* **24**:4913–4916.

Corey, E. J., De, B., Ponder, J. W., and Berg, J. M., 1984, The stereochemistry and biosynthesis of hybridalactone, an eicosanoid from *Laurencia hybrida*, *Tetrahedron Lett.* **25**:1015–1018.

Crombie, L., and Holloway, S. J., 1985, The biosynthesis of calendic acid, octadeca-(8E,10E,12Z)-trienoic acid, by developing marigold seeds: Origins of (E,E,Z) and (Z,E,Z) conjugated triene acids in higher plants, *J. Chem. Soc. Perkin Trans. I* **1985**:2425–2434.

Dix, T. A., and Marnett, L. J., 1985, Conversion of linoleic acid hydroperoxide to hydroxy, keto, epoxyhydroxy, and trihydroxy fatty acids by hematin, *J. Biol. Chem.* **260**:5351–5357.

Ehresmann, D. W., Deig, E. F., Hatch, M. T., Disalvo, L. H., and Vedros, N. A., 1977, Antiviral substances from California marine algae, *J. Phycol.* **13**:37–40.

Ehresmann, D. W., Deig, E. F., and Hatch, M. T., 1979, Anti-viral properties of algal polysaccharides and related compounds, in: *Marine Algae in Pharmaceutical Science* (H. A. Hoppe, T. Levring, and Y. Tanaka, eds.), W. de Gruyter, Berlin, pp. 293–363.

Erickson, K. L., 1983, Constituents of *Laurencia*, in: *Marine Natural Products*, Vol. 5 (P. J. Scheuer, ed.), Academic Press, New York, pp. 131–257.

Fahlstadius, P., and Hamberg, M., 1990, Stereospecific removal of the *pro-R* hydrogen at C-8 of (9S)-

hydroperoxyoctadecadienoic acid in the biosynthesis of colnelic acid, *J. Chem. Soc. Perkin Trans. I* **1990**:2027–2030.

Faulkner, D. J., 1990, Marine natural products, *Nat. Prod. Rep.* **7**:269–309.

Fujimura, T., Kawai, T., Shiga, M., Kajiwara, T., and Hatanaka, A., 1990, Long-chain aldehyde production in thalli culture of the marine green alga *Ulva pertusa*, *Phytochemistry* **29**:745–747.

Fusetani, N., and Hashimoto, K., 1984, Prostaglandin E2: A candidate for causative agent of "Ogonori" poisoning, *Bull. Jpn. Soc. Sci. Fish.* **50**:465–469.

Galliard, T., Phillips, D. R., and Frost, D. J., 1973, Novel divinyl ether fatty acids in extracts of *Solanum tuberosum*, *Chem. Phys. Lipids* **11**:173–180.

Garson, M. J., 1989, Biosynthesis of marine natural products, *Nat. Prod. Rep.* **6**:143–170.

Gerwick, W. H., and Fenical, W., 1982, Phenolic lipids from related marine algae of the order Dictyotales, *Phytochemistry* **21**:633–637.

Gerwick, W. H., Reyes, S., and Alvarado, B., 1987, Two malyngamides from the Caribbean cyanobacterium *Lyngbya majuscula*, *Phytochemistry* **26**:1701–1704.

Gerwick, W. H., Bernart, M. W., Moghaddam, M. F., Jiang, Z. D., Solem, M. L., and Nagle, D. G., 1990, Eicosanoids from the Rhodophyta: New metabolism in the algae, *Hydrobiologia* **204/205**:621–628.

Gregson, R. P., Kazlauskas, R., Murphy, P. T., and Wells, R. J., 1977, New metabolites from the brown alga *Cystophora torulosa*, *Aust. J. Chem.* **30**:2527–2532.

Gregson, R. P., Marwood, J. F., and Quinn, R. J., 1979, The occurrence of prostaglandins PGE$_2$ and PGF$_{2\alpha}$ in a plant—The red alga *Gracilaria lichenoides*, *Tetrahedron Lett.* **20**:4505–4506.

Guerriero, A., D'Ambrosio, M., and Pietra, F., 1990, Novel hydroxyicosatetraenoic and hydroxyicosapentaenoic acids and a 13-oxo analog. Isolation from a mixture of calcareous red algae *Lithothamnion corallioides* and *Lithothamnion calcareum* of Brittany waters, *Helv. Chim. Acta* **73**:2183–2189.

Gurjar, M. K., and Reddy, A. S., 1990, Stereoselective synthesis of (9S,12R,13S)-trihydroxyoctadeca (10E,15Z)-dienoic acid, *Tetrahedron Lett.* **31**:1783–1784.

Hamberg, M., Herman, C. A., and Herman, R. P., 1986, Novel transformations of 15-L-hydroperoxy-5,8,11,14-eicosatetraenoic acid, *Biochim. Biophys. Acta* **877**:447–457.

Herz, W., and Kulanthaivel, P., 1985, Trihydroxy-C18-acids and a labdane from *Rudbeckia fulgida*, *Phytochemistry* **24**:89–91.

Higgs, M. D., 1981, Antimicrobial components of the red alga *Laurencia hybrida* (Rhodophyta, Rhodomelaceae), *Tetrahedron* **37**:4255–4258.

Higgs, M. D., and Mulheirn, L. J., 1981, Hybridalactone, an unusual fatty acid metabolite from the red alga *Laurencia hybrida* (Rhodophyta, Rhodomelaceae), *Tetrahedron* **37**:4259–4262.

Jiang, Z. D., and Gerwick, W. H., 1990, Galactolipids from the temperate red marine alga *Gracilariopsis lemaneiformis*, *Phytochemistry* **29**:1433–1440.

Jiang, Z. D., and Gerwick, W. H., 1991, Eicosanoids and other hydroxylated fatty acids from the marine alga *Gracilariopsis lemaneiformis* (Rhodophyta), *Phytochemistry* **30**:1187–1190.

Kakisawa, H., Asari, F., Kusumi, T., Toma, T., Sakurai, T., Oohusa, T., Hara, Y., and Chihara, M., 1988, An allelopathic fatty acid from the brown alga *Cladosiphon okamuranus*, *Phytochemistry* **27**:731–735.

Kato, T., Yamaguchi, Y., Ohnuma, S., Uyehara, T., Namai, T., Kodama, M., and Shiobara, Y., 1986, Structure and synthesis of 11,12,13-trihydroxy-9Z,15Z-octadecadienoic acids from rice plant suffering from rice blast disease, *Chem. Lett.* **1986**:577–580.

Kinoshita, K., Takahashi, K., and Zama, K., 1986, Triarachidonin and diarachidonylphosphotidylcholine in "Ogonori" *Gracilaria verrucosa*, *Bull. Jpn. Soc. Sci. Fish.* **52**:757.

Kugel, C., Lellouche, J.-P., and Beaucourt, J.-P., 1989, Stereospecific total synthesis of (5,6)-diHETE isomers, *Tetrahedron Lett.* **30**:4947–4950.

Kurata, K., Taniguchi, K., Shiraishi, K., Hayama, N., Tanaka, I., and Suzuki, M., 1989,

Ecklonialactone-A and -B, two unusual metabolites from the brown alga *Ecklonia stolonifera* Okamura, *Chem. Lett.* **1989**:267–270.

Lands, W. E. M., 1986, *Fish and Human Health*, Academic Press, Orlando, Florida.

Levine, L., Sneiders, A., Kobayashi, T., and Schiff, J. A., 1984, Serologic and immunochromato-graphic detection of oxygenated polyenoic acids in *Euglena gracilis var. bacillaris*, *Biochem. Biophys. Res. Commun.* **120**:278–285.

Lopez, A., and Gerwick, W. H., 1987, Two new icosapentaenoic acids from the Oregon red seaweed *Ptilota filicina*, *Lipids* **22**:190–194.

Lopez, A., and Gerwick, W. H., 1988, Ptilodene, a novel icosanoid inhibitor of 5-lipoxygenase and Na$^+$/K$^+$ ATPase from the red marine alga *Ptilota filicina* J. Agardh, *Tetrahedron Lett.* **29**:1505–1506.

Lumin, S., and Falck, J. R., 1990, Synthesis and stereochemical revision of a bioactive dihydroxy-eicosanoid isolated from the red marine alga *Farlowia mollis*, *Tetrahedron Lett.* **31**:2971–2974.

Marchant, M. F., and Fowke, L. C., 1977, Preparation, culture and regeneration of protoplasts from filamentous green algae, *Can. J. Bot.* **55**:3080–3086.

Miralles, J., Aknin, M., Micouin, L., Gaydoou, E.-M., and Kornprobst, J.-M., 1990, Cyclopentyl and ω-5 monounsaturated fatty acids from red algae of the Solieraceae, *Phytochemistry* **29**:2161–2163.

Miyamoto, T., Lindgren, J. A., and Samuelsson, B., 1987, Isolation and identification of lipoxygenase products from the rat central nervous system, *Biochim. Biophys. Acta* **922**:372–378.

Moghaddam, M. F., and Gerwick, W. H., 1990, 12-Lipoxygenase activity in the red marine alga *Gracilariopsis lemaneiformis*, *Phytochemistry* **29**:2457–2459.

Moghaddam, M. F., Gerwick, W. H., and Ballantine, D. L., 1989, Discovery of 12-(S)hydroxy-5,8,10-14-icosatetraenoic acid [12-(S)-HETE] in the tropical red alga *Platysiphonia miniata*, *Prostaglandins* **37**:303–308.

Moghaddam, M. F., Gerwick, W. H., and Ballantine, D. L., 1990, Discovery of the mammalian insulin release modulator hepoxilin B$_3$ from the tropical red algae *Platysiphonia miniata* and *Cottoniella filamentosa*, *J. Biol. Chem.* **265**:6126–6130.

Moore, R. E., 1981, Constituents of blue-green algae, in: *Marine Natural Products*, Vol. 4 (P. J. Scheuer, ed.), Academic Press, New York, pp. 1–52.

Mynderse, J. S., and Moore, R. E., 1978, The isolation of (−)-E-1-chlorotridec-1-ene-6,8-diol from a marine cyanophyte, *Phytochemistry* **17**:1325–1326.

Nagle, D. G., and Gerwick, W. H., 1990, Isolation and structure of constanolactones A and B, new cyclopropyl hydroxy-eicosanoids from the temperate red alga *Constantinea simplex*, *Tetrahedron Lett.* **31**:2995–2998.

Neumann, C., and Boland, W., 1990, Stereochemical studies on algal pheromone biosynthesis: A model study with the flowering plant *Senecio isatideus* (Asteraceae), *Eur. J. Biochem.* **191**:453–459.

Niwa, H., Wakamatsu, K., and Yamada, K., 1989, Halicholactone and neohalicholactone, two novel fatty acid metabolites from the marine sponge *Halichondria okadai* Kodota, *Tetrahedron Lett.* **30**:4543–4546.

Ojika, M., Yoshida, Y., Nakayama, Y., and Yamada, K., 1990, Aplydilactone, a novel fatty acid metabolite from the marine mollusc *Aplysia kurodai*, *Tetrahedron Lett.* **31**:4907–4910.

Pace-Asciak, C. R., 1984, Arachidonic acid epoxides: Demonstration through [^{18}O]-oxygen studies of an intramolecular transfer to the terminal hydroxyl group of (12S)-hydroperoxyeicosa-5,8,10,14-tetraenoic acid to form hydroperoxides, *J. Biol. Chem.* **259**:8332–8337.

Pace-Asciak, C. R., and Asotra, S., 1989, Biosynthesis, catabolism and biological properties of HPETEs, hydroperoxide derivatives of arachidonic acid, *Free Radical Biol. Med.* **7**:409–433.

Pace-Asciak, C. R., Mizuno, K., and Yamamoto, S., 1983, Resolution by DEAE-cellulose chroma-tography of the enzymatic steps in the transformation of arachidonic acid into 8,11,12 and 10,11,12 trihydroxyeicosatetraenoic acid by the rat lung, *Prostaglandins* **25**:79–84.

Pace-Asciak, C. R., Martin, J. M., and Corey, E. J., 1986, Hepoxilins, potential endogenous mediators of insulin release, *Prog. Lipid Res.* **25**:625–628.

Pace-Asciak, C. R., Laneuville, O., Su, W.-G., Corey, E. J., Gurevich, N., Wu, P., and Carlen, P. L., 1990, A glutathione conjugate of hepoxilin A_3: Formation and action in the rat central nervous system, *Proc. Natl. Acad. Sci. USA* **87**:3037–3041.

Paul, V. J., and Fenical, W., 1980, Toxic acetylene-containing lipids from the red marine alga *Liagora farinosa* Lamouroux, *Tetrahedron Lett.* **21**:3327–3330.

Paul, V. J., and Fenical, W., 1987, Natural products chemistry and chemical defense in tropical marine algae of the phylum Chlorophyta, in: *Bioorganic Marine Chemistry*, Vol. 1 (P. J. Scheuer, ed.), Springer-Verlag, Berlin, pp. 1–29.

Piomelli, D., Feinmark, S. J., and Swartz, J. H., 1987, Metabolites of arachidonic acid in the nervous system of *Aplysia*: Possible mediators of synaptic modulation, *J. Neurosci.* **7**:3675–3686.

Pollio, A., Greca, M. D., Monaco, P., Pinto, G., and Previtera, L., 1988, Lipid composition of the acidophilic alga *Dunaliella acidophila* (Volvocales, Chlorophyta) I. Non-polar lipids, *Biochim. Biophys. Acta* **963**:53–60.

Rosell, K.-G., and Srivastava, L. M., 1987, Fatty acids as antimicrobial substances in brown algae, *Hydrobiologia* **151/152**:471–475.

Samuelsson, B., and Funk, C. D., 1989, Enzymes involved in the biosynthesis of leukotriene B_4, *J. Biol. Chem.* **264**:19469–19472.

Samuelsson, B., Dahlen, S.-E., Lindgren, J. A., Rouzer, C. A., and Serhan, C. N., 1987, Leukotrienes and lipoxins: Structures, biosynthesis, and biological effects, *Science* **237**:1171–1176.

Smith, W. L., and Willis, A. L., 1987, A suggested shorthand nomenclature for the eicosanoids, *Lipids* **22**:983–986.

Smith, W. L., Borgeat, P., Hamberg, M., Roberts II, L. J., Willis, A. L., Yamamoto, S., Ramwell, P. W., Rokach, J., Samuelsson, B., Corey, E. J., and Pace-Asciak, C. R., 1990, Nomenclature, *Meth. Enzymol.* **187**:1–9.

Solem, M. L., Jiang, Z. D., and Gerwick, W. H., 1989, Three new and bioactive icosanoids from the temperate red alga *Farlowia mollis*, *Lipids* **24**:256–260.

Son, B. W., 1990, Glycolipids from *Gracilaria verrucosa*, *Phytochemistry* **29**:307–309.

Spector, A. A., Gordon, J. A., and Moore, S. A., 1988, Hydroxyeicosatetraenoic acids (HETEs), *Prog. Lipid Res.* **27**:271–323.

Stefanov, K., Konaklieva, M., Brechany, E. Y., and Christie, W. W., 1988, Fatty acid composition of some algae from the Black Sea, *Phytochemistry* **27**:3495–3497.

Thermo Company, Ltd., 1984, Prostaglandin E_2, *Chem. Abs.* **101**:14827n.

Van Hummel, H. C., 1975, Chemistry and biosynthesis of plant galactolipids, *Fortschr. Chem. Org. Naturst.* **32**:267–293.

Vick, B. A., and Zimmerman, D. C., 1989, Metabolism of fatty acid hydroperoxides by *Chlorella pyrenoidosa*, *Plant. Physiol.* **90**:125–132.

Warren, R. G., Wells, R. J., and Blount, J. F., 1980, A novel lipid from the brown alga *Notheia anomala*, *Aust. J. Chem.* **33**:891–898.

Westlund, P., Edenius, C., and Lindgren, J. A., 1988, Evidence for a novel pathway of leukotriene formation in human platelets, *Biochim. Biophys. Acta* **962**:105–115.

Zimmerman, D. C., and Vick, B. A., 1973, Lipoxygenase in *Chlorella pyrenoidosa*, *Lipids* **8**:264–266.

5

Marine Proteins in Clinical Chemistry

Gurdial M. Sharma and Mukesh K. Sahni

1. INTRODUCTION

When we were invited to contribute to this volume a single-spaced typewritten article up to 25 pages long on clinical applications of marine proteins, we were delighted and accepted the assignment with great pleasure. Later, we became a little concerned about this venture, because compared to the literature on the secondary metabolites of marine organisms, there appeared to be a paucity of information on biomedical applications of marine proteins. Second, most of the literature that was available on this topic until 1990 had been reviewed by Kaul and Daftari (1986) and by Wu and Narahashi (1988). With the availability of these two reviews, it appeared that we might not have much new to offer on clinical applications of marine proteins. This concern seemed to be all the more real when a 1989 update on marine natural products and biomedicine cited only three references on biological properties of marine polypeptides (Scheuer, 1989). Two of these references were on the detection of endotoxins by *Limulus* amebocyte lysates and the third citation was on a cyclic depsipeptide which had potent antiviral and anticancer activities. Since then, structures and biological properties of a few more proteins, isolated from sponges, tunicates, hemocytes, and hemolymph of

Gurdial M. Sharma and Mukesh K. Sahni • Department of Chemistry, William Paterson College, Wayne, New Jersey 07470.

Marine Biotechnology, Volume 1: Pharmaceutical and Bioactive Natural Products, edited by David H. Attaway and Oskar R. Zaborsky. Plenum Press, New York, 1993.

horseshoe crabs, have been reported (Nakamura *et al.*, 1988; Topham *et al.*, 1988; Miyata *et al.*, 1989; Kawano *et al.*, 1990; Rinehart *et al.*, 1990a, b; Enghild *et al.*, 1990; Fusetani, 1990). Other than this, we have not come across any new developments on marine proteins which could warrant the writing of a 25-page article.

Luckily, while conducting a literature search on the topic of this article, it became apparent that there is a class of marine proteins whose biomedical implications had not been reviewed lately for the marine audience. Some of these proteins have been called "the cell recognition molecules" by their investigators (Cohen, 1984; Sharon and Lis, 1989). We, however, propose to delete the word "cell" and suggest the general designation "recognition molecules" for all those nonenzymatic natural peptides which noncovalently bind/agglutinate various types of vitamins, hormones, trace elements, carbohydrates, and other substances, including cells with high specificity and affinity. Major medical applications of these proteins are in the area of clinical diagnostics. Apart from their clinical significance, these proteins may serve as tools with which to study the chemical and physiological processes that take place in living systems. Such studies lead to a better understanding of key cellular events upon which depend the growth, differentiation, and reproduction of all organisms. A memoir on these proteins should therefore be a very appropriate contribution to this first volume of the series Advances in Marine Biotechnology. Before proceeding further, we would like to clarify that antibodies, although well qualified to be called recognition molecules, are not included in this article.

In organizing the contents of this chapter, we have taken the liberty of presenting first the work that we have carried out on marine proteins which recognize and bind vitamin B_{12} and its analogues with high specificity and avidity. These proteins have the potential of being useful as binders for the determination of vitamin B_{12} and unchelated calcium ions in ocean waters and in biological specimens by radioisotope dilution techniques (RID techniques or simply radioassays). We then pass on to endotoxin-recognizing peptides and lectins of marine origin. Endotoxin-recognizing peptides are useful in the clinical evaluation of endotoxemia. Lectins recognize carbohydrate specifically and therefore they play a key role in the control of various normal and pathological processes in living organisms. We will conclude this chapter by assessing the relative merits and benefits of research on marine versus terrestrial recognition proteins.

2. VITAMIN B_{12}-BINDING PROTEINS OF MARINE ORGANISMS

In this section, the antipernicious anemia factor, vitamin B_{12}, which is α-(5,6-dimethyl-benzimidazolyl)-cobamide cyanide, and its more than a dozen naturally occurring analogues will be collectively called cobalamins or corrinoids. These molecules occur in nature in small amounts and are produced exclusively by microorganisms. Among the cobalamins thus far discovered, the parent molecule vitamin B_{12} (or simply B_{12}) is the most potent micronutrient known. It is required for the growth of most organisms, including the marine phytoplankton that constitute the first trophic level of the marine food chain. Most prokaryotes and simple eukaryotes can synthesize B_{12} *de novo*. All higher animals and some microorganisms (e.g., marine phytoplankton) require that B_{12} be supplied in their diets or in their growth media. Current information on the chemistry, biochemistry nutrition, ecology, and medical aspects of corrinoids has been reviewed by Schneider and Stroinski (1987).

Uptake of exogenous cobalamins and their cellular accumulation and transport in humans, other terrestrial animals, and microorganisms are mediated by several types of glycopeptides called cobalamin-binding proteins. Among the cobalamin-binding proteins, the one called mammalian intrinsic factor binds B_{12} exclusively with an affinity constant 1×10^{10} M^{-1}. Most of the other corrinoid-binding proteins associate B_{12} and its analogues indiscriminately with binding constants ranging from $1 \times 10,^{11}$ to 3×10^{11} M^{-1}. Cobalamin-binding proteins occur at an extremely low concentration in living organisms and quite often a purification factor of 10^3–10^6 may be needed before homogeneous preparations of these molecules are obtained (Allen, 1973; Sennett and Rosenberg, 1981; Schneider and Stroinski, 1987).

The intrinsic factor, because it complexes with cyanocobalamin exclusively, is used as a B_{12}-specific binder in the radioisotope dilution techniques (RID techniques or simply radioassays) for the direct determination of the vitamin in the human blood plasma (Liu and Sullivan, 1972; Bain *et al.*, 1982; Schneider and Stroinski, 1987). Intrinsic factor-based B_{12} radioassays are rapid, accurate, highly specific, and economical to perform. Consequently, they have replaced the costly and time-consuming B_{12} bioassays, which, unlike radioassays, give false low values for the vitamin when inhibitors are present in the samples to be analyzed. Radioassays, however, are significantly less sensitive than the bioassays. The detection limit of the radioassay is around 20 pg B_{12}/ml, while that of the bioassays has been claimed to be around 0.1 pg B_{12}/ml (Liu and Sullivan, 1972; Carlucci, 1973). Accurate determination of B_{12} at levels <20 pg/ml by radioassays will require proteins that bind the vitamin specifically with affinity constant in excess of 10^{11} M^{-1}.

Some years ago we suggested that proteins having higher affinity and specificity for B_{12} should be present in marine organisms (Sharma *et al.*, 1985). The rationale behind this view was that ocean waters contain cobalamins at concentrations which, according to bioassays, range from undetectable levels to 12 pg/ml for B_{12} and 2–20 pg/ml for the analogues. Although these concentrations are extremely low, some marine organisms are able to satisfy their biological need for

B_{12} and/or its analogues by absorbing them directly from the water column. These organisms must be accomplishing this remarkable feat of sequestering B_{12} from a medium containing extremely low levels of the vitamin with the help of proteins that should bind B_{12} and its analogues differentially with affinity constants greater than that of the intrinsic factor. These proteins may be isolated and used in B_{12} radioassays to increase the low-end sensitivity of the standard curve in the range of clinical interest. Recently, several publications have stressed the need for improving the detection limit of radioassays to unequivocally diagnose B_{12} deficiency in geriatric patients (Hayes *et al.*, 1985).

New B_{12} radioassays having detection limits at least an order of magnitude lower than that of the current RID techniques may also be suitable for the rapid and accurate determination of the vitamin in ocean waters. Concentrations of B_{12} and its analogues in seawater vary both seasonally and geographically (Swift and Guillard, 1977; Bruno and Staker, 1978; Nishijima and Yoshiniko, 1979; Swift, 1980; Schneider and Stroinski, 1987). Variations in the levels of cobalamins should effect the growth and distribution of phytoplankton, because these primary producers require the micronutrient for their growth. During the 1960s and 1970s there was considerable interest in elucidating the role that cobalamins might be playing in the growth and distribution of phytoplankton in the marine environment, but only a limited amount of field work has been done on this project (Carlucci, 1973; Carlucci and Cuhel, 1977; Droop, 1982; Fiala and Oriol, 1984; Schneider and Stroinski, 1987). Interest in marine B_{12} work seems to have subsided due to the lack of analytical techniques by which a large number of environmental samples could be analyzed for the concentration of B_{12} rapidly, accurately, and economically. Radioisotope dilution techniques capable of measuring as little as 1 pg B_{12}/ml accurately will be ideally suited to the marine B_{12} work. For the development of these techniques, all one needs is a protein that should bind B_{12} selectively with an affinity constant in excess of 10^{11} M^{-1}. All other ingredients ($^{57}CoB_{12}$ and dextran-coated charcoal) needed for RID techniques are already commercially available (Becton Dickinson Co., 1985).

Arguments in favor of searching marine organisms for the presence of proteins capable of binding B_{12} selectively with affinity constants in excess of 10^{11} M^{-1} have already been presented. To test the idea, body fluids, tissue extracts, culture media, and/or cell extracts of the marine organisms listed in Table I were analyzed for the presence of cobalamin-binding proteins. Radiolabeled B_{12} ($^{57}CoB_{12}$) and dextran-coated charcoal assay techniques were used to achieve this goal (Sharma *et al.*, 1985). Although all biological specimens were found to contain B_{12}-binding proteins, we decided to work with the corrinoid-binding proteins of the horseshoe crab first for several reasons: (1) *Limulus* blood plasma and hemocytes were found to be relatively richer in cobalamin-binding proteins. Consequently, this organism may serve as a commercially preferred source of B_{12}-binding proteins that are used in radioassay kits for the determination of B_{12} in

Table I. Marine Organisms Tested for Cobalamin-Binding Proteins

Marine organism	Biological specimen	B_{12}-Binding activity	Specificity
Horseshoe crab	Blood plasma	+	S*
	LAL	+	S*
Blue crab	Blood plasma	−	−
Clams	Whole animal tissue	+	−
Barnacles	Whole animal tissue	+	N
Sea urchin	Whole animal tissue	+	N
Strombus gigas	Whole animal tissue	+	N
Gorgonia ventalina	Whole animal tissue	+	N
Sponges (3 spp.)	Whole animal tissue	+	N
Phytoplankton (2 spp.)	Growth media	+	S

Abbreviations: +, shows B_{12}-binding activity; −, no B_{12}-binding activity; S*, specificity with season; S, specific for B_{12}; N, nonspecific for B_{12}; LAL, *Limulus* amebocyte lysate.

biological fluids; (2) Since horseshoe crabs are "living fossils," a comparative study of the B_{12}-binding proteins of this animal and highly evolved terrestrial mammals may reveal the structural changes that might have occurred in these proteins during the course of evolution, and finally, (3) *L. polyphemus* may serve as an ideal model for understanding mechanisms which facilitate the uptake of B_{12} by cells. The uptake of B_{12} is mediated by cobalamin-binding proteins, but the mechanisms involved are not fully understood (Sennett and Rosenberg, 1981).

2.1. Vitamin B_{12}-Binding Proteins of the Horseshoe Crab

We have already published a short communication on the isolation of two types of B_{12}-binding proteins from *Limulus* blood plasma (Sharma *et al.*, 1985). A brief description of the published results is essential to the presentation of new data in the proper context.

For the isolation work, *Limulus* whole blood was obtained from crabs that were captured from New Jersey beaches during the breeding season, which lasts from May until July of each year. The hemolymph was collected in a citrate buffer (pH 4.7) to prevent clumping and lysing of amebocytes. The blood specimens thus collected were centrifuged to pellet amebocytes. The clear, pale blue supernatants, hereafter called citrate-blood plasma or simply blood plasma, were decanted and analyzed for total protein concentration and cobalamin-binding characteristics.

The blood plasma samples were found to contain, depending upon the physiological states of the animals, 34–45 mg protein/ml of the fluid and each milligram of the plasma proteins was found to bind between 3 and 5 ng of B_{12}.

Proteins present in citrate-blood plasma samples were fractionated by sequential application of ammonium sulfate precipitation and gel filtration techniques. Gel filtration of ammonium sulfate-precipitated material on a precalibrated Sephadex G-150 column gave two B_{12}-binding fractions designated as A and B. The molecular weights of the proteins present in fractions A and B were estimated to be 150,000 and 30,000, respectively.

Fraction A was found to bind B_{12} preferentially from mixtures containing constant amount of B_{12} but progressively increasing amounts of the analogue cobinamide dicyanide. Under the same conditions fraction B was found to bind B_{12} exclusively. Fraction A was found to bind B_{12} with affinity constant 6×10^{11} M^{-1}, which is greater than the association constant 10^{11} M^{-1} for the human transcobalamin proteins (Allen, 1973). The binding constant of fraction B for B_{12} is around 10^{11} M^{-1}, which is an order of magnitude greater than the affinity constant of 10^{10} M^{-1} for the hog intrinsic factor.

Next, we used fraction B as a B_{12}-specific binder and developed a competitive binding radioassay for the direct determination of the vitamin in ocean waters. The detection limit of the assay is around 5 pg B_{12}/ml and standard deviation at this level is ± 1 pg (unpublished data). A typical standard curve for reading the concentration of B_{12} in unknown samples is shown in Fig. 1. Protocols for the

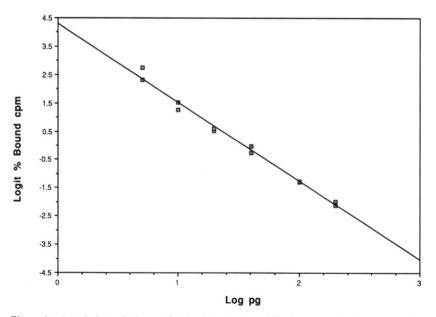

Figure 1. A typical standard curve for the determination of B_{12} in ocean water by a competitive binding radioassay. Protocol for the assay has been reported (Sharma *et al.*, 1979).

preparation of standard curves and analysis of B_{12} in unknown samples by competitive binding RID techniques have been published by several investigators (Liu and Sullivan, 1972; Sharma *et al.*, 1979; Bain *et al.*, 1982). These protocols are also supplied with B_{12}-binding radioassay kits sold by diagnostic companies for the determination of vitamin B_{12} in human blood plasma (Becton Dickinson Co., 1985; Schneider and Stroinski, 1987).

Direct gel filtration of *Limulus* citrate-blood plasma samples over a Sephadex G-150 column also furnished fraction A and fraction B cobalamin-binding proteins. In these simplified fractionation experiments, an indentation in the elution profile of fraction A proteins was invariably observed (Fig. 2). This fraction (tubes 25–60 in Fig. 2) was divided into two parts of the indentation. These subfractions were analyzed for B_{12} specificity and B_{12} binding constants. The data revealed that fraction A is a mixture of at least two types of cobalamin-binding proteins having molecular weights 200,000 and 150,000 (also see below). One protein seemed to bind B_{12} specifically and the other appeared to associate cobalamins indiscriminately. The B_{12}-binding constants of these proteins are six times greater than that of fraction B. If the B_{12}-specific protein present in fraction A could be isolated

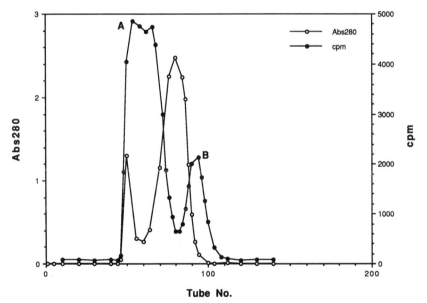

Figure 2. Gel filtration of *Limulus* citrate-blood plasma over a precalibrated Sephadex G-150 column (3 × 90 cm); sample size 10–30 ml; eluant 0.1 M ammonium bicarbonate; flow rate 40 ml/hr; fraction size 90 drops (5 ml)/tube; total fractions 140; detection 280 nm; temperature 4°C. (○) Absorbance at 280 nm. Elution profiles (●) of B_{12}-binding proteins A (mol. wt. 200,000) and B (mol. wt. 30,000) were determined using $^{57}CoB_{12}$ radioassay (Sharma *et al.*, 1985).

in a pure state, then a B_{12} radioassay with better low-end sensitivity could be developed.

As a first step toward the selection or development of further purification procedures, fractions A and B were analyzed by sodium dodecyl sulfate–polyacrylamide gel electrophoresis (SDS-PAGE). After electrophoresis, the protein bands were visualized by staining two slots of the gel with Coomassie blue (Laemmli, 1970). Bands due to B_{12}-binding proteins were located by extracting 1-cm slices of the gel slab with 0.02 M phosphate buffer (pH 8.2) and measuring cpm bound by the extracts after incubation with $^{57}CoB_{12}$. Molecular weights of the proteins corresponding to each band were estimated from a Ferguson curve prepared by plotting log mol. wt. of standard proteins against their electrophoretic mobilities on the same gel as used for the analysis of fractions A and B. The photograph of a typical gel slab is shown in Fig. 3.

Electrophoresis data revealed that fractions A and B contain, in addition to B_{12}-binding proteins, close to a dozen other polypeptides ranging in molecular weight from over 200,000 to 10,000. Fraction A B_{12}-binding proteins produced a fairly broad band stretching from mol. wt. 150,000 to mol. wt. 75,000. This observation suggests that fraction A contains more than one type of cobalamine-binding protein ranging in molecular weight from 150,000 to 75,000. The B_{12}-binding protein in fraction B was found to give a sharp band in the zone mol. wt. = 10,000. This was indeed a surprising observation because gel filtration experiments had consistently indicated that the molecular weight of fraction B

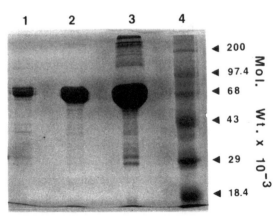

Figure 3. Gel electrophoresis of B_{12}-binding protein fractions A and B from Sephadex G-150 column. Lane 1, fraction B (20 μg); lane 2, fraction A (5 μg); lane 3, fraction A (20 μg); lane 4, standard molecular weight markers. Proteins were separated on a 12.5% SDS–polyacrylamide gel and stained with Coomassie blue according to Laemmli's (1970) method.

cobalamin-binding protein is around 30,000. This discrepancy may be rationalized by suggesting that fraction B B_{12}-binding protein is actually an oligomer (trimer) of the corrinoid-binding peptide having mol. wt = 10,000. Under SDS-PAGE conditions, the oligomer dissociates into its momomer, which migrates to the region of its true molecular weight of 10,000.

It should be pointed out that gel regions where fractions A and B cobalamin-binding proteins migrate stain faintly or not at all with Coomassie blue. This indeed is a reflection of extremely low levels of B_{12}-binding proteins in the two chromatographic fractions used in SDS-PAGE experiments. The intensely stained bands at mol. wt. = 70,000 in lanes where fractions A and B were run are due to the monomers of the 48-subunit respiratory protein hemocyanin, mol. wt. = 3.8 $\times 10^6$ (Brenowitz and Moore, 1982). Most of the Coomassie blue-stained bands between the origin and the zone corresponding to mol. wt. = 70,000 in lanes of fraction A are due to the oligomers of hemocyanin monomers.

Electronic spectra and quantitative protein determinations of fractions A and B revealed that subunits of hemocyanin and their oligomers account for over 90% of the proteins present in the two fractions. This conclusion was supported by the Sephadex G-150 gel filtration chromatogram of *Limulus* citrate-blood plasma. This chromatogram is shown in Fig. 2. The largest UV-absorbing peak in the chromatogram is due to the subunits of hemocyanin. This peak merges with peaks A and B of cobalamin-binding fractions. Because of the overlap of peaks, the B_{12}-binding fractions are expected to be significantly contaminated by the dissociation products of hemocyanin.

The dissociation of hemocyanin into lower-molecular-weight monomers and oligomers seems to have been triggered by the citrate buffer that was added to *Limulus* blood to prevent clumping and lysing of amebocytes. Citrate buffer chelates calcium ions, which are absolutely essential for keeping hemocyanin in its native quaternary state (Brenowitz and Moore, 1982).

Attempts to isolate B_{12}-binding proteins from fractions A and B by ion-exchange (Grasbeck *et al.*, 1962) and affinity chromatography (Allen and Majerus, 1972) were unsuccessful. We therefore designed a new scheme for isolating B_{12}-binding proteins from *Limulus* blood in a state of high purity. This scheme entailed six steps:

1. Collection of *Limulus* blood as neat specimens or in 3% saline solution containing 0.1% of nonionic detergent Tween 80. Tween 80 prevents clumping and lysing of amebocytes without disturbing the quaternary structure of hemocyanin.
2. Centrifugation of whole blood at 10,000 $\times$ *g* to pellet amebocytes and obtain supernatant blood plasma.
3. Ultracentrifugation of blood plasma at 30,000 $\times$ *g* to pellet hemocyanin and obtain hemocyanin-free supernatants, hereafter called serum.

4. Analysis of blood plasma and serum samples by HPLC using a precalibrated LKB gel filtration column.
5. Large-scale isolation of B_{12}-binding proteins from serum by chromatography over Sephadex G-150.
6. Final purification of proteins by HPLC using reverse-phase C-18 column.

Two important features of this scheme are noteworthy. First, dissociation of hemocyanin into various levels of subunit structures is prevented by eliminating citrate buffer from the blood collection protocol. Second, hemocyanin is removed from *Limulus* blood plasma prior to the application of fractionation procedures. These modifications should yield B_{12}-binding chromatographic fractions without contamination by subunits of hemocyanin.

Using the modified protocol outlined in step 1, new blood samples were collected from horseshoe crabs captured twice a week from New Jersey beaches during the breeding season (summer months, May–August). Blood samples were also taken from the crabs that were obtained once a month from local fishermen during the nonbreeding winter months (September–April). Analysis of these samples by steps 2–4 of the six-step isolation and purification scheme revealed that neat specimens of *Limulus* hemolymph contain different types of cobalamin-binding proteins during the breeding and nonbreeding seasons. Fraction A and fraction B cobalamin-binding proteins previously isolated from *Limulus* citrate-blood plasma samples turned out to be the dissociation products of a single oligomeric corrinoid-binding protein (see below) present in the horseshoe crab blood during the breeding season only. During nonbreeding winter months *Limulus* hemolymph contains a cobalamin-binding protein that does not dissociate into fraction A and fraction B cobalamin-binding proteins. This conclusion is based upon the analyses of a large number of *Limulus* blood plasma samples that were prepared from *Limulus* whole blood collected during the breeding and nonbreeding months of the years 1985–1987. Hereafter native B_{12}-binding proteins that are present in *Limulus* blood during breeding and nonbreeding seasons will be called S-protein (S for summer) and W-protein (W for winter), respectively.

The molecular weight of the breeding-season S-protein was estimated to be around 300,000 by HPLC gel filtration experiments (Fig. 4). This protein was found to bind vitamin B_{12} preferentially from mixtures containing a constant amount of $^{57}CoB_{12}$ but progressively increasingly amounts of the analogue cobinamide. The affinity constant of S-protein for B_{12} was estimated to be 4×10^{12} M^{-1}.

The binding of B_{12} to S-protein was found to be inhibited by seawater. This inhibition was traced to have been caused by calcium ions that are present in marine waters at concentrations ranging from 8 to 10 mmole/liter (Riley, 1975). Other major ions (Mg^{2+}, Na^+, and Cl^-) of the seawater were found not to alter the affinity of S-protein for B_{12}.

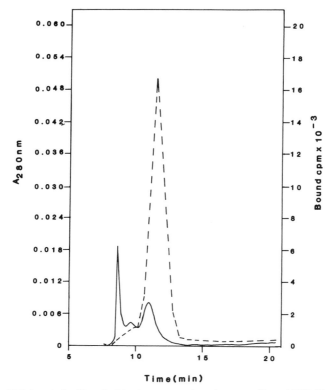

Figure 4. HPLC analysis of *Limulus* blood serum samples using a precalibrated LKB Glass Pak TSK G 3000 SW (8 × 300 mm) column. Analysis protocol: Mobile phase, a 50:50 mixture of 0.1 M KH_2PO_4 and K_2HPO_4; gradient type, isocratic; flow rate 0.6 ml/min; detection 280 nm. To monitor the elution profile of cobalamin-binding proteins, a 30-μl aliquot of serum was incubated with 30 μl of a $^{57}CoB_{12}$ solution for 10 min. A portion (30 μl) of the incubation mixture was injected into the column and 1/2-min fractions were collected up to 20 min. Radioactivity of these fractions was measured using liquid scintillation counting techniques. This figure shows the elution pattern (————) of B_{12}-binding protein (mol. wt. 300,000) present in the summer blood plasma of the animals.

Quantitative data on the inhibition produced by various levels of calcium ions to the affinity of S-protein for B_{12} are shown graphically in Fig. 5. The inhibition curve shown in this figure represents cpm bound by a substoichiometric amount of S-protein from mixtures containing a constant amount of $^{57}CoB_{12}$ (40 pg; cpm = 15,000) but progressively increasing amounts (0–20 μmole) of calcium ions. Almost identical data were obtained when aliquots of S-protein were incubated sequentially first with calcium ions and then with $^{57}CoB_{12}$. Calcium ions were also found to release B_{12} from preformed B_{12}–S-protein holocomplex. The stoichiometry of the release of B_{12} was the same as is predicted by the curve shown in Fig. 5.

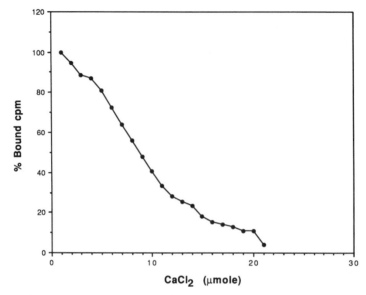

Figure 5. Inhibition of B_{12} binding to S-protein by calcium ions. In a typical assay, mixtures of 40 pg $^{57}CoB_{12}$ (cpm 15,000) with 0–2 μmole $CaCl_2$ in 1.0 ml of distilled water were incubated with 100-μl aliquots of 50-times-diluted *Limulus* blood serum for 10 min. Free B_{12} was removed by incubating with dextran-coated charcoal. After centrifugation, the supernatants were mixed with scintillation fluid (10 ml), and cpm bound by S-protein was measured in a scintillation counter.

The inhibitory effect of calcium ions was completely reversed by the chelator EDTA. These cumulative data and the sigmoid nature of the inhibition curve shown in Fig. 5 suggest that Ca^{2+} is an allosteric inhibitor of the binding of B_{12} to S-protein and, unlike other cobalamin-binding proteins thus far discovered, the S-protein is an oligomer of several corrinoid-binding subunits.

    The observation that calcium ions not only inhibit the binding of S-protein to B_{12} but also release the vitamin from the B_{12}–S-protein complex was used to estimate the amount of S-protein in *Limulus* blood that is already saturated with cobalamins. To achieve this goal, a small aliquot of 50-times-diluted *Limulus* blood plasma was incubated with excess $^{57}CoB_{12}$; the free B_{12} was removed by treatment with dextran-coated charcoal and the bound cpm in the supernatant due to $^{57}CoB_{12}$–S-protein complex was determined. In a separate experiment the diluted blood plasma was first incubated with calcium ions and the released cobalamins were removed by incubation with dextran-coated charcoal. After centrifugation, the supernatant was incubated with $^{57}CoB_{12}$ in the presence of EDTA, and after treatment with dextran-coated charcoal the cpm bound by total apo S-protein was measured. A comparison of the data from the two experiments revealed that as much as 50% of the S-protein in *Limulus* blood is already saturated with corrinoids, while the other half exists as an apoprotein.

Like all oligomeric proteins, the *Limulus* S-protein may be caused to dissociate into its corrinoid-binding constituent subunits and/or lower-molecular-weight aggregates of the subunits by modification of the solution environment. The dissociation of this protein in a pH 4.0 citrate buffer to give three cobalamin-binding polypeptides having mol. wt. around 200,000, 150,000 (protein A), and 30,000 (protein B) has already been mentioned (see Fig. 2). When a sample of *Limulus* breeding-season blood serum is titrated with dilute HCl and brought to a pH value below 4.0, the native cobalamin-binding protein dissociates to give polypeptides A and B only, i.e., the polypeptide having mol. wt. 200,00 is not obtained (Fig. 6). The kinetics of the dissociation of S-protein into peptides A and B at pH values 3.5 was studied using HPLC and B_{12}-binding radioassays. The information is presented in Fig. 7. At pH 3.5 the S-protein dissociates almost instantaneously (15 min) to give polypeptides A and B having $^{57}CoB_{12}$-binding

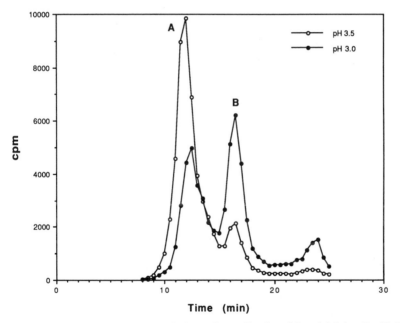

Figure 6. Dissociation of the multimeric S-protein at pH values 3.0 and 3.5 into B_{12}-binding oligomers A (mol. wt. 150,000) and B (mol. wt. 30,000). The dissociation was studied by a combination of $^{57}CoB_{12}$ radioassays and HPLC. A sample of *Limulus* summer blood serum was cooled to 4°C and adjusted to the desired pH by titrating with dilute HCl. After an incubation period of 15 min, 10 μl of the pH-adjusted blood serum was reacted with $^{57}CoB_{12}$ and the reaction mixture was analyzed by HPLC using the column and analysis protocol described in Fig. 4.

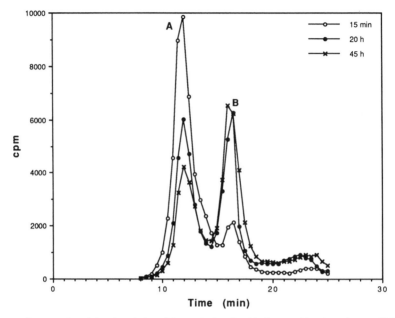

Figure 7. Kinetics of the dissociation of S-protein into B_{12}-binding peptides A (mol. wt. 150,000) and B (mol. wt. 30,000) at pH 3.5. Horseshoe crab blood serum sample (5 ml) was cooled to 4°C and adjusted to pH 3.5 with dilute HCl. After incubation periods of 15 min, 20 hr, and 45 hr, 10-μl aliquots of the pH-adjusted serum were analyzed by radioassay and HPLC as described in Fig. 6.

capacities in the ratio of 6:1. Thereafter, the concentration of protein A keeps on decreasing, while that of B keeps on increasing until, after 20 hr, the B_{12}-binding capacities of the two peptides are almost equal. After 20 hr, the concentration of A decreases further while that of B remains constant. At pH 3.0 the S-protein dissociated immediately to give peptides A and B in amounts having equal B_{12}-binding capacities (Fig. 6). In all other respects dissociation at pH 3.0 was similar to the dissociation at pH 3.5.

When a sample of *Limulus* breeding-season blood serum adjusted to pH 3.0 was brought back to neutrality and kept at 4°C for 12 hr, the peptide B was found to aggregate to give peptide A.

The S-protein was also found to dissociate into cobalamin-binding peptides A and B when breeding-season blood plasma was subjected to repeated freezing and thawing over a period of 2 months. Similar treatment of breeding-season serum samples was found to convert the S-protein into a cobalamin-binding protein having low B_{12}-binding constant (10^8 M^{-1}) and low specificity for the vitamin.

Of particular significance to the assignment of a quaternary structure for the multisubunit S-protein was the observation that the B_{12}-binding abilities of its lower-molecular-weight dissociation products, the peptides A and B, were not influenced significantly by the calcium ions. This finding suggests that in S-protein, various subunits are arranged in a fashion so as to produce an allosteric site for the binding of calcium ions. Alternatively, the B_{12}-binding subunits of S-protein are arranged around a calcium-binding regulatory protein. In either case the binding of calcium ions to a site other than the B_{12}-binding region could inhibit the binding of the vitamin in a cooperative fashion.

As pointed out earlier, during nonbreeding winter months *Limulus* blood contains a different corrinoid-binding protein, called the W-protein. This protein binds B_{12} and its analogues indiscriminately with an affinity constant 10^9 M^{-1}. This affinity constant is almost three order of magnitude lower than the binding constant 1×10^{12} M^{-1} of S-protein for B_{12}. Upon mixing the winter blood plasma of *L. polyphemus* with citrate buffer pH 4.0 or upon titration of winter blood serum samples with dilute HCl to a pH value 3.0, the W-protein dissociates to give a single peptide having molecular weight around 150,000. This peptide, unlike peptide A obtained from dissociation of S-protein, binds B_{12} and its analogues indiscriminately with an affinity constant close to 10^8 M^{-1}. In spite of these differences in the corrinoid-binding characteristics of W- and S-proteins, both proteins were found to have approximately the same molecular weight of 300,000 and B_{12}-binding abilities of both proteins were inhibited by calcium ions.

All along we were considering the possibility that *Limulus* blood plasma corrinoid-binding proteins may have been produced and excreted by amebocytes, the cells present in the hemolymph of the horseshoe crab. To test this hypothesis, amebocytes were extracted with pyrogen-free water. The extracts, commonly called *Limulus* amebocyte lysates or simply LAL (Watson *et al.*, 1987), were found to bind B_{12} and its analogues indiscriminately with affinity constant that varied from batch to batch within the range 10^9–10^{12} M^{-1}. Gel filtration of LAL over a precalibrated Sephadex G-150 column gave two types of cobalamin-binding proteins, one having molecular weight around 150,000 and the other with molecular weight approaching 40,000. The LAL protein having mol. wt. 150,000 was found to be identical with the cobalamin-binding protein obtained by treating the winter *Limulus* blood plasma with citrate buffer pH 4.2. The second LAL cobalamin-binding protein, mol. wt. 40,000, bound B_{12} and its analogues non-selectively with affinity constant around 10^9 M^{-1}. This protein is different from any of the other corrinoid-binding proteins that have been isolated from *Limulus* blood or obtained upon adjusting the pH of the *Limulus* blood serum within the range 4–3 with dilute HCl. The most general conclusion that may be drawn from this and other data that has been gathered on *Limulus* corrinoid-binding proteins is as follows:

In order to sequester, conserve, and transport vitamin B_{12} and/or its analogues, the horseshoe crab elaborates cobalamin-binding peptides having mol. wt.

around 10,000. These monomers assemble to produce aggregates (e.g., peptides A and B isolated from *Limulus* summer citrate-blood plasma) having different corrinoid-binding characteristics. These aggregates further associate to produce *Limulus* blood plasma S- and W-proteins whose B_{12}-binding abilities are inhibited by calcium ions. When *Limulus* blood plasma or blood serum samples are repeatedly thawed and frozen, the multimeric S- and W-proteins dissociate and the dissociated lower-molecular-weight oligomers reassemble to produce hybrid corrinoid-binding molecules having intermediate B_{12}-binding characteristics. This phenomenon seems to be analogous to the formation of five types of isozymes when a mixture of two types of tetrameric lactate dehydrogenases are subjected alternately to mildly denaturing and renaturing conditions. Much further work needs to be carried out to establish if this analogy is indeed valid.

Using radiolabeled B_{12} ($^{57}CoB_{12}$), several *in vitro* and *in vivo* experiments were conducted to establish the physiological roles of corrinoid-binding S- and W-proteins present in the horseshoe crab. These experiments have revealed that the B_{12} bound to S- or W-proteins is not utilized by *Limulus* amebocytes. Vitamin B_{12} bound to these proteins is ultimately picked up by the book gills of the animals. Our preliminary data seem to suggest that S- and W-proteins serve as a store of cobalamins in the horseshoe crab. They most probably do not perform B_{12}-transport functions in the animal.

2.2. Vitamin B_{12}-Binding Proteins of Marine Phytoplankton

It is not known whether horseshoe crabs satisfy their biological need for B_{12} by absorbing the vitamin directly from seawater or utilize the vitamin injested with food. In case the horseshoe crabs are not sequestering B_{12} from seawater, then these organisms should not contain B_{12}-binding proteins having the highest possible affinity constant and specificity for the vitamin. There is, however, no doubt that marine phytoplankton are among those organisms which absorb B_{12} directly from seawater. Consequently, marine phytoplankton may contain proteins which bind B_{12} with affinity and selectivity much better than that of *Limulus* corrinoid-binding proteins. To test this hypothesis, the marine phytoplankton *Thalassiosira pseudonana* was grown in a vitamin B_{12}-free medium. When the cell density in the medium reached the stationary phase (3×10^6 cells/ml), the cells and the medium were separated by centrifugation. Analysis of the supernatant growth medium revealed that it contains protein which binds B_{12} specifically with affinity constant 5×10^{11} M^{-1}. The B_{12}-binding protein(s) present in the growth medium was concentrated by ultrafiltration. The proteins present in the retentate were analyzed by HPLC using a precalibrated gel filtration column. This analysis revealed that the growth medium of *T. pseudonana* contains a single type of protein having molecular weight around 70,000. Unlike *Limulus* cobalamin-binding proteins, the *T. pseudonana* B_{12}-binding protein was found to require calcium ions to associate with the vitamin.

Aqueous extracts of *T. pseudonana* were also found to contain corrinoid-binding proteins. These proteins, however, showed low specificity and low binding constant (10^8 M^{-1}) for B_{12}. Very similar results were obtained when the growth media and cell extracts of another phytoplankton, *Monochrysis lutheri*, were analyzed for proteins which may bind B_{12} specifically, with affinity constant in excess of 10^{11} M^{-1}.

The marine phytoplankton proteins which bind B_{12} with high specificity and avidity were found to occur at an extremely low concentration in the growth media of the organisms. Consequently, marine phytoplankton cannot be a cheap direct commercial source of B_{12}-binding proteins for use in the radioassay kits sold by many diagnostic companies. Nevertheless, large-scale production of these proteins may be accomplished by using the principles of biotechnology. Nucleotide sequences which encode B_{12}-binding protein in a phytoplankton may be reverse-transcribed to generate cDNA. The cDNA may be "cloned" into a plasmid vector. The recombinant DNA thus obtained may be introduced into a microorganism which can be cultured rapidly. The transformed microorganism may produce the desired protein in larger quantities without the need for further purification. Work on this project is in progress in our laboratory.

3. LIMULUS PROTEINS FOR THE DETECTION OF ENDOTOXINS

The term endotoxins signifies a class of complex lipopolysaccharides (LPS) that are found primarily, if not exclusively, on the outer cell wall of Gram-negative bacteria. These relatively heat-stable toxins are released when the integrity of the cell is disturbed, although some Gram-negative bacteria seem to shed LPS continuously throughout their life cycles (Freidman *et al.*, 1988; Prior, 1990). According to the latest information on the structure of endotoxins, these macromolecules (mol. wt. = 400,000 to several millions) contain a phosphorylated polyacylglucosamine dissacharide, called lipid A, covalently linked to a polysaccharide moiety which contains certain deoxy hexoses not known to occur anywhere else in nature (Kotani and Takada, 1988; Schaechter *et al.*, 1989). Lipid A of endotoxins appears to be responsible for most, if not all, of the toxicity shown by LPS, whereas the polysaccharide portion determines the specific antigenicity of the molecule (Majde and Person, 1981).

The biological effects of bacterial endotoxins are numerous and depend upon the amount of LPS ingested either accidentally through the use of contaminated materials or via infection with Gram-negative bacteria. When present in sufficient amount in the bloodstream, endotoxins can trigger a cascade of lethal effects, culminating in multiple organ failure and death (Majde and Person, 1981; Friedman *et al.*, 1988; Prior, 1990). In smaller doses, endotoxins cause fever and increase antibody synthesis and inflammation. Humans appear to be among the most sensitive species to the toxic effects of endotoxins and doses as low as a few ng/kg can cause detectable pyrogenic effects.

According to Morrison *et al.* (1985), infection with Gram-negative bacteria kills up to 100,000 Americans a year. Endotoxins have been implicated as a major contributing factor in this mortality. Because endotoxins continue to present serious health problems, a rapid, sensitive, and reliable test for their detection in human body fluids and in materials used or consumed by humans should be of great value in the prevention and treatment of endotoxemia. The foundation of such a test was laid in 1965 by Levin and Bang while working on the pathophysiology of the horseshoe crab (Cohen *et al.*, 1979; Levin, 1985). The story goes as follows.

In 1885, W. H. Howell showed that the blood of *Limulus polyphemus* rapidly clotted when it was collected in a glass vessel and exposed to air. Almost 80 years later this phenomenon was investigated in greater detail by Levin and Bang (Levin, 1985). These investigators discovered that the coagulation of *Limulus* blood, as observed by Howell, was a result of the reaction of endotoxins with a protein system released by amebocytes, the cells present in the hemolymph of the animal. This conclusion was based upon the observation that aqueous extracts of washed amebocytes gel rapidly upon addition of endotoxins, while amebocyte-free *Limulus* blood plasma shows no such property. The clotting of *Limulus* amebocyte lysate (LAL) in the presence of LPS is the basis of what later came to be known as the LAL test for the detection of endotoxins in pharmaceutical preparations, water supplies, body fluids, biologically active reagents such as mitogens, erythroprotein, and cytokines, etc. (Watson *et al.*, 1987; Prior, 1990).

Since its discovery in 1965, several variations of the original LAL test have been reported (Watson *et al.*, 1987; Prior, 1990). The most widely used variation is the chromogenic assay, which is more sensitive and allows the detection of as little as 1 pg of endotoxin in 1 ml of the test sample. The chromogenic assay is based upon the biochemical mechanism for the interaction of endotoxins with a specific protein system present in LAL.

The biochemical basis of the LAL test has been investigated by several scientists (Morita *et al.*, 1985; Iwanaga *et al.*, 1986; Muta *et al.*, 1990; Prior, 1990). These investigators have shown that, similar to the human blood clotting system, the LAL clotting system is a bifurcated cascade of proteolytic reactions involving several serine zymogens and a soluble gelation protein called coagulogen (Fig. 8). One branch of the coagulation cascade is triggered by small amounts of endotoxic LPS and the other is started by minute amounts of $(1\rightarrow3)$-β-D-glucan. Endotoxins initiate the cascade by converting protein C into

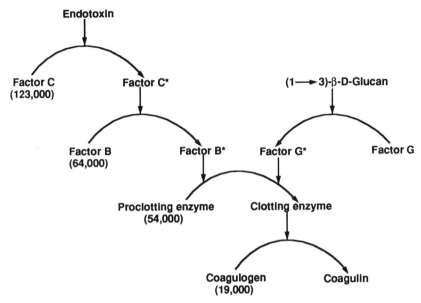

Figure 8. Schematic representation of LAL coagulation system. Numbers in parentheses are the molecular weights of various proteins.

the activated factor C* without change in its molecular weight. The activated factor C* converts the zymogen B into the proteolytic enzyme B*, which transforms the proclotting enzyme into the clotting enzyme. The clotting enzyme removes a large peptide fragment from coagulogen to give coagulin, which forms a clot or a gel. In the second clotting pathway, the proclotting enzyme is converted to the clotting enzyme with the participation of LAL factor G, which is activated to G* by (1→3)-β-D-glucan. Thus, the endotoxin and glucan pathways converge on a common sequence of final steps to form a coagulin gel.

The clotting enzyme formed in LAL gelation reactions cleaves not only coagulogen but also certain synthetic peptides having *p*-nitroanilide (PNA) as the chromogenic group. Thus, when the substrate BOC-Leu-Gly-Arg-PNA is used in the LAL test for endotoxin, the clotting enzyme hydrolyses the synthetic peptide to give BOC-Leu-Gly-Arg and PNA. The PNA gives an intense yellow color to the reaction mixture. The intensity of the yellow color is directly proportional to the concentration of the endotoxins in the test samples. This is the basis of the quantitative LAL assay for the estimation of endotoxins in materials used or consumed by humans. Principles of several other modified LAL procedures for the quantitation of endotoxins have been reviewed in the literature (Watson *et al.*, 1987; Friedman *et al.*, 1988; Prior, 1990).

Although numerous LAL tests are performed daily to detect the presence of endotoxins in radiolabeled drugs, pharmaceutical preparations, invasive medical devices, etc., several problems regarding the test remain. Most prominent among these problems are the variability of the potency of LAL preparations from batch to batch, the initiation of the clotting cascade by chemicals other than endotoxins, and the inhibition of the cascade by inhibitors present in human body fluids. Because of the last factor the use of the LAL test in clinical medicine has not been as successful as in the pharmaceutical industry (Levin, 1985; Prior, 1990). Further work to improve the LAL test for the detection of endotoxins in human body fluids is in progress in various laboratories.

4. MARINE LECTINS

Lectins are naturally occurring glycoproteins that differentially recognize and noncovalently bind sugars using the principle of stereochemical fit of lock-and-key type between complementary molecules (Cohen, 1984; Kocourek, 1986; Sharon and Lis, 1989; Marchalonis and Schluter, 1990). Although first isolated from plants, lectins are now known to occur in all unicellular and multicellular organisms, each with a particular carbohydrate-binding specificity. In their natural environments, lectins may be found attached to cellular surfaces and/or dissolved in the hemolymph of the animals.

Among several properties of lectins, the one which has received most attention is their remarkable ability to selectively agglutinate erythrocytes and bacterial and other normal and malignant cells by reversibly combining with complementary glycosyl residues on opposing cells (Cohen, 1986). As expected, the cell-agglutinating activities of lectins are competitively inhibited by saccharides of the same types as are found on the surfaces of cells being agglutinated. Besides acting as agglutinins, some lectins exhibit potent mitogenic and opsonin-like activities. Because of these properties lectins are sometimes called cell recognition molecules, with the additional potential of being important in many biological processes, such as fertilization, embryogenesis, cell migration, organ formation, immune defense, and microbial infection (Sharon and Lis, 1987). It is with this background in mind that we summarize the current information on lectins of marine organisms. We begin with the lectins of the horseshoe crab, *Limulus polyphemus*.

The hemolymph of *L. polyphemus* has been known to contain potent agglutinins for mammalian erythrocytes since the beginning of the 20th century (Noguchi, 1903). Subsequent investigations by several investigators have demonstrated the presence of multiple lectins in *Limulus* serum (Cohen *et al.*, 1984; Marchalonis and Schulter, 1990). One of these lectins agglutinates horse erythrocytes specifically. This lectin (limulin) has been purified to homogeneity and characterized (Marchalonis and Edelman, 1968; Robey and Liu, 1981; Liu *et al.*, 1982; Nguyen *et al.*, 1986).

Limulin is a large, 18-subunit protein of mol. wt. 350,000. Subunits of the lectin are identical or nearly identical, having a minimum molecular weight of approximately 20,000. The amino acid sequence of limulin exhibits considerable identity with the amino acid sequences of human and rabbit C-reactive proteins (Liu *et al.*, 1982; Nguyen *et al.*, 1986). The horseshoe crab lectin agglutinates erythrocytes by virtue of its specific binding to sialic acid (Pistole, 1982; Cohen, 1986; Pistole and Graf, 1984). Other structures recognized by limulin are phosphoryl choline, glucoronic acid, *N*-acyl-muramic acid, *N*-acyl-aminosugars, and bacterial lipopolysaccharides (Pistole, 1982; Cohen *et al.*, 1984). The specific site on the bacterial lipopolysaccharide molecule that interacts with limulin is 2-keto-3-deoxyoctonate.

Pistole and Graf (1984) have suggested that in the horseshoe crab blood, limulin acts as a specific recognition factor (opsonin) for Gram-negative bacteria. Once recognition occurs, then *Limulus* amebocytes with the participation of another, yet unknown serum component kill and phagocytize the microorganisms. Although this model for the biological role of limulin is attractive, direct experimental evidence for the participation of the lectin and the other putative component in *Limulus* defense against Gram-negative bacteria is lacking.

We have observed (M. K. Sahni, M. Z. Wahrman, H. Sanchez, and G. M. Sharma, unpublished data) that hemocyanin-free blood plasma of *L. polyphemus* contains a factor which not only inhibits the growth of the marine phytoplankton *Thalassiosira pseudonana* at an extremely low level, but also agglutinates the phytoplankton cells to produce "grapelike" structures (Fig. 9). This factor appears to have a molecular weight of less than 5000 and is present at a very low concentration in the *Limulus* hemolymph. It has yet to be ascertained if this molecule truly satisfies all the criteria required of a lectin. If it turns out to be the case, then this component would be the smallest lectin that has ever been isolated from a terrestrial or marine organism. Because of its small size, this molecule may be ideally suited to a study of the modes of interaction between the lectin and its complementary carbohydrate residue(s) by NMR spectroscopy.

Lectins that bind sialic acid residues specifically have also been found to be present in the hemolymph of the lobster *Homarus americanus* (Abel *et al.*, 1984). One of these lectins (LAg1) has been isolated in a pure state by affinity chromatography. Like limulin, the lobster lectin LAg1 is a large multimeric protein with a mol. wt. of 700,000. Subunits of this protein have a mol. wt. of 70,000. Depending upon the solution environment, the monomeric subunits of LAg1 undergo dissociation and reassociation to produce lower-molecular-weight aggregates.

Unlike limulin, which agglutinates horse red blood cells better than it does human red blood cells, LAg1 agglutinates human red blood cells better than it does

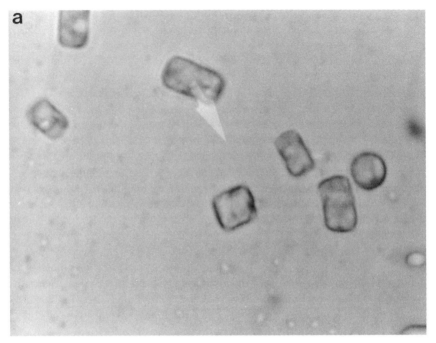

Figure 9. (a) Shapes of normal cells of the marine phytoplankton *Thalassiosira pseudonana* when viewed under a light microscope. Cells were grown in a nutrient-rich seawater medium as described by Carlucci (1973) and McLachlan (1973).

horse red blood cells. As far as carbohydrate specificity is concerned, the lobster lectin binds *N*-acetylneuraminic acid (sialic acid), *N*-acetylgalactosamine, as well as *N*-acetylglucosamine. The hemagglutin activity of LAg1 is inhibited not only by these saccharides, but also by glycoproteins such as bovine mucin, glyco-phorin, and fetuin.

The specificity of LAg1 for sialic acid has been used to separate cells which markedly differ in the sialic acid content of their extracellular surfaces. Thus, Abel *et al.* (1984) successfully used LAg1 to separate immature mouse thymocytes (low sialic acid content) from mature thymocytes (high sialic acid content). The point to be made here is that lectins with different carbohydrate specificities are excellent tools to monitor the changes that take place on the surfaces of cells as they undergo maturation and differentiation and are ready to interact with other cells to produce specific biological events. Lectins may also be used to detect the changes in the content of carbohydrates, glycoproteins, and glycolipids of body fluids when living systems are challenged by diseases or are in different metabolic states.

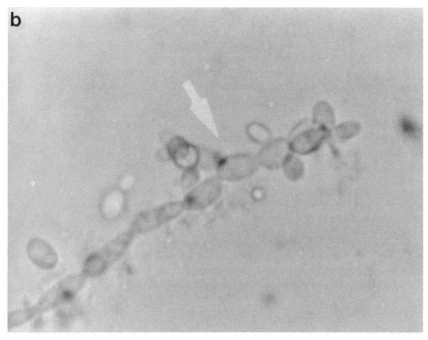

Figure 9. *(Continued)* (b) "Grapelike" structures produced when *T. pseudonana* cells were cultured in the presence of a low-molecular-weight fraction isolated by Sephadex G-150 chromatography of *Limulus* blood serum samples.

Furthermore, lectins, because of their sharp glycosyl specificities, may be coupled to Sepharose to prepare affinity columns for the purification of glycoproteins and glycolipids. The possibility that lectins present in the body fluids of certain marine invertebrates and fishes may serve as tools to detect sex-related components on cells has been suggested (Smith, 1984). There is always the possibility of exploiting the carbohydrate specificities of lectins to deliver drugs to specific tissues in the body. The biomedical significance of lectins has been emphasized in greater detail in several books and articles (Cohen, 1984; Kocourek, 1986; Sharon and Lis, 1989, and references cited therein).

It is remarkable that lectins have also been found to occur in marine sponges, the simplest of animals, where they are responsible for cell–cell recognition and aggregation (Muller and Muller, 1980; Brandley and Schnaar, 1986). Recently a lectin called bindin has been isolated from sea urchin sperm. This lectin is responsible for species-specific adhesion of sperm to the egg and plays an important role in sea urchin fertilization (Glabe *et al.*, 1982).

Although lectins are not antibodies, they seem to be active partners in the biological phenomenon related to the recognition of self from nonself. The prevailing view is that lectins are primordial cell recognition molecules. Partial support for this hypothesis is provided by the structural homology that seems to exist between proteins of the vertebrate immune system on the one hand and lectins of the marine invertebrates such as oysters, sea urchins, blue crabs, and tunicates on the other. References to this aspect of lectin research may be found in a recent review article on the origin of immunoglobulins and immune recognition molecules by Marchalonis and Schluter (1990).

5. CONCLUSION

J. Levin, the pioneer in the field of the *Limulus* amebocyte lysate test for the detection of endotoxin, concluded his paper on the history of this test (Levin, 1985) with a quote from A. V. Hill, the distinguished British physiologist. The same quote is appropriate as an opening sentence of the conclusion of this article. "By the method of comparative physiology, or of experimental biology, by the choice of suitable organ, tissue or process, in some animal far removed in evolution, we may often throw light upon some function or process in higher animals or in man" (Hill, 1960).

Many marine organisms are precursors of terrestrial plants and animals, while others are far removed in evolution. The important point is that marine organisms live, grow, and reproduce in an environment which is highly competitive, aggressive, and, therefore, very demanding. In order to grow in such an environment, marine organisms need to make proteins which are very apt at recognizing and sequestering micronutrients which exist at an extremely low concentration in natural sources. For survival, marine animals should be making proteins which, compared to similar proteins of highly evolved land organisms, may be more potent in recognizing and eliminating foreign invaders. The information provided in this chapter certainly supports this view. As pointed out in the introduction, immediate benefits of research on marine recognition proteins would be in the area of clinical diagnostics. Long-term basic research on these proteins should reveal those molecular features which make these proteins more potent in their biological activities. With this knowledge in hand, and by using the powerful new tools of genetic engineering, the biological effectiveness of these proteins may be further improved, much to the benefit of human health.

ACKNOWLEDGMENTS, The authors wish to thank the undergraduate students Barbara Mueller, Dusan Vlajovanov, Kevin Murray, and Humberto Sanchez, who were participants in some of the experimental work reported here on B_{12}-binding proteins of marine organisms. We also thank James J. Finn, President, Finn-Tech

Industries, Inc., New Jersey, for generously providing horseshoe crab blood and amebocyte lysate. We are grateful to our secretary, Margie Kelleher, for typing the manuscript. This work was financially supported by Sea Grant Agency. Additional support was provided by the School of Science and Mathematics through a NJDHE Challenge to Excellence Grant. Sea Grant publication number NJSG-91-248.

REFERENCES

Abel, C. A., Campbell, P. A., Vanderwall, J., and Hartman, A. L., 1984, Studies on the structure and carbohydrate binding properties of lobster agglutinin 1 (LAg1), A sialic acid-binding lectin, in: *Progress in Clinical and Biological Research*, Vol. 157. *Recognition Proteins, Receptors and Probes: Invertebrates* (E. Cohen, ed.), Alan R. Liss, New York, pp. 103–114.

Allen, R. H., 1973, Human Vitamin B_{12} transport proteins, *Prog. Hematol.* **9:**57–84.

Allen, R. H., and Majerus, P. W., 1972, Isolation of vitamin B_{12}-binding proteins using affinity chromatography, *J. Biol. Chem.* **247:**7695–7701.

Bain, B., Broom, G. N., Woodside, J., Litwinczuk, R. A., and Wickramasingle, S. N., 1982, Assessment of a radioisotopic assay for vitamin B_{12} using an intrinsic factor preparation with R-proteins blocked by vitamin B_{12} analogs, *J. Clin. Pathol. Lond.* **35:**1110–1113.

Becton Dickinson Co., 1985, Simultrac Radioassay Kit for Simultaneous Determination of Vitamin B_{12} and Folate Levels in Serum and Plasma, Orangeburg, New York.

Brandley, G. K., and Schnaar, R., 1986, Cell-surface carbohydrates in cell recognition and response, *J. Leukocyte Biol.* **40:**97–111.

Brenowitz, M., and Moore, M., 1982, The subunit structure of *Limulus* hemocyanin, in: *Progress in Clinical and Biological Research*, Vol. 81 (J. Bonaventura, C. Bonaventura, and S. Tesh, eds.), Alan R. Liss, New York, pp. 257–267.

Bruno, S. F., and Staker, R. D., 1978, Seasonal vitamin B_{12} and phytoplankton distribution near Napeaque Bay, New York (Block Island Sound), *Limnol. Oceanogr.* **23:**1045–1050.

Carlucci, A. F., 1973, Bioassay: Cyanocobalamin, in: *Handbook of Phycological Methods* (J. R. Stein, ed.), Cambridge University Press, Cambridge, pp. 387–394.

Carlucci, A. F., and Cuhel, R. L., 1977, Vitamins in the South Polar Seas: Distribution and Significance of dissolved and particulate vitamin B_{12}, thiamin and biotin in the Southern Indian Ocean, in: *Adaptations within Antarctic Ecosystems* (Proceedings of Third SCAR Symposium on Antarctic Biology; G. A. Llano, ed.), Gulf Printing Co., Houston, Texas, pp. 115–132.

Cohen, E. (ed.), 1984, *Progress in Clinical and Biological Research*, Vol. 157. *Recognition Proteins, Receptors and Probes: Invertebrates*, Alan R. Liss, New York.

Cohen, E., 1986, Detection of membrane receptors by arthropod agglutinins, in: *Hemocytic and Humoral Immunity in Arthropods* (A. P. Gupta, ed.), pp. 493–503.

Cohen, E., Bang, F. B., Levin, J., Marchalonis, J. J., Pistole, T. G., Prendergast, R. A., Schuster, C., and Watson, S. W. (eds.), 1979, *Progress in Clinical and Biological Research*, Vol. 29, *Biomedical Applications of the Horseshoe Crab (Limulidae)*, Alan R. Liss, New York.

Cohen, E., Vasta, G. R., Korytnyk, W., Petrie III, C. R., and Sharma, M., 1984, Lectins of the Limulidae and hemagglutination-inhibition by sialic acid analogs and derivatives, in: *Progress in Clinical and Biological Research*, Vol. 157. *Recognition Proteins, Receptors and Probes: Invertebrates* (E. Cohen, ed.), Alan R. Liss, New York, pp. 55–69.

Droop, M. R., 1982, Vitamin B_{12} and marine ecology. IV. The kinetics of uptake, growth and inhibition in *Monochrysis lutheri*, *J. Mar. Biol. Assoc. U.K.* **48:**41–46.

Enghild, J. J., Thogersen, I. B., Salvesen, G., Fey, G. H., Figler, N. L., Gonias, S. L., and Pizzo, S. V.,

1990, α-Macroglobulin from *Limulus polyphemus* exhibits proteinase inhibitory activity and participates in a hemolytic system, *Biochemistry* **29**:10070–10080.

Fiala, M., and Oriol, L., 1984, Vitamin B_{12} and phytoplankton in the Antarctic Ocean: Distribution and experimental approach, *Mar. Biol. Berl.* **79**:325–332.

Friedman, H., Klein, T. W., Nakano, M., and Nowotny, A. (eds), 1988, *Advances in Experimental Medicine and Biology*, Vol. 256. *Endotoxin*, Plenum Press, New York.

Fusetani, N., 1990, Research toward "Drugs from the Sea", *New J. Chem.* **14**:721–728.

Glabe, C. G., Grabel, L. B., Vacquier, V. D., and Rosen, S. D., 1982, Carbohydrate specificity of sea urchin sperm bindin: A cell surface lectin mediating sperm–egg adhesion, *J. Cell Biol.* **94**:123.

Grasbeck, R., Simons, K., and Sinkkonen, I., 1962, Purification of intrinsic factor and vitamin B_{12}-binders from human gastric juice, *Ann. Med. Exp. Biol. Fenn.* **40**:Suppl. 6.

Hayes, A. N., Williams, D. J., and Skelkon, D., 1985, Vitamin B_{12} (cobalamin) and folate blood levels in geriatric reference group as measured by two kits, *Clin. Biochem.* **18**:56–71.

Hill, A. V., 1960, Experiments on frogs and men, in: *The Ethical Dilemma of Science and other Writings* (A. V. Hill, ed.), Rockefeller Institute Press, New York, pp. 24–38.

Iwanaga, S., Morita, T., Miyata, T., Nakamura, T., and Aketagawa, J., 1986, The hemolymph coagulation system in invertebrate animals, *J. Protein Chem.* **5**:255.

Kaul, P. N., and Daftari, P., 1986, Marine pharmacology: Bioactive molecules from the sea, *Annu. Rev. Pharmacol. Toxicol.* **26**:117–142.

Kawano, K., Yoneya, T., Miyata, T., Yoshikawa, K., Tokunaga, F., Terada, Y., and Iwanaga, S., 1990, Antimicrobial peptide, Tachyplesin I., Isolated from hemocytes of the horseshoe crab (*Tachypleus tridentatus*), *J. Biol. Chem.* **265**:15365–15367.

Kocourek, J., 1986, Historical background, in: *The Lectins: Properties, Functions and Applications in Biology and Medicine* (I. E. Liener, N. Sharon, and I. J. Goldstein, eds.), Academic Press, pp. 1–32.

Kotani, S., and Takada, H., 1988, Structural requirements of lipid A for endotoxicity and other biological activities—An overview, in: *Advances in Experimental Medicine and Biology*, Vol. 256. *Endotoxin* (H. Friedman, T. W. Klein, M. Nakano, and A. Nowotny, eds.), Plenum Press, New York, pp. 13–43.

Laemmli, U. K., 1970, Cleavage of structural proteins during the assembly of the head of bacteriophage T_4, *Nature* **227**:680–685.

Levin, J., 1985, The history of the development of the *Limulus* amebocyte lysate test, in: *Progress in Clinical and Biological Research*, Vol. 189 (J. W. T. Cate, H. R. Buller, A. Sturk, and J. Levin, eds.), Alan R. Liss, New York, pp. 3–28.

Liu, Y. K., and Sullivan, L. W., 1972, An improved radioisotope dilution assay for serum vitamin B_{12} using hemoglobin-coated charcoal, *Blood* **39**:426–432.

Liu, T. Y., Robey, F. A., and Wang, C. M., 1982, Structural Studies on C-reactive protein, *Ann. N. Y. Acad. Sci.* **389**:251–262.

Majde, J. A., and Person, R. J. (eds.), 1981, *Progress in Clinical and Biological Research*, Vol. 62. *Pathophysicological Effects of Endotoxins at the Cellular Level*, Alan R. Liss, New York.

Marchalonis, J. J., and Edelman, G. M., 1968, Isolation and characterization of a natural hemoagglutinin from *Limulus polyphemus*, *J. Mol. Biol.* **32**:453–465.

Marchalonis, J. J., and Schluter, F., 1990, Origin of immunoglobulins and immune recognition molecules, *BioScience* **40**:758–768.

McLachlan, J., 1972, Growth media-marine, in: *Handbook of Phycological Methods* (J. R. Stein, ed.), Cambridge University Press, Cambridge, pp. 25–51.

Miyata, T., Tokunaga, T. Yonega, T., Yoshikawa, K., Iwanaga, S., Niwa, M., Takao, T., and Shimonishi, Y., 1989, Antimicrobial peptide, isolated from horseshoe crab hemocytes. Tachyplesin II., and polyphemusins I and II: Chemical structures and biological activity, *J. Biochem.* (Tokyo) **106**:663–668.

Morita, T., Nakamura, T., Miyata, T., and Iwanaga, S. 1985, Biochemical characterization of *Limulus* clotting factors and inhibitors which interact with bacterial endotoxins, in: *Progress in Clinical and Biological Research*, Vol. 189 (J. W. T. Cate, H. R. Buller, A. Sturk, and J. Levin, eds.), Alan R. Liss, New York, pp. 53–64.

Morrison, D. C., Duncan, R. L., Jr., and Goodman, S. A., 1985, *In vivo* biological activities of endotoxin, in: *Progress in Clinical and Biological Research*, Vol. 189 (J. W. T. Cate, H. R. Buller, A. Sturk, and J. Levin, eds.), Alan R. Liss, New York, pp. 81–98.

Muller, W. E. G., and Muller, I., 1980, Sponge cell aggregation, *Mol. Cell. Biochem.* **29**:131.

Muta, T., Hashimoto, R., Miyata, T., Nishimura, H., Toh, Y., and Iwanaga, S., 1990, Proclotting enzyme from horseshoe crab hemocytes, *J. Biol. Chem.* **265**:22426–22433.

Nakamura, T., Furunaka, H., Miyata, T., Tokunaga, F., Muta, T., Iwanaga, S., Niwa, M., Takao, T., and Shimonishi, Y., 1988, Tachyplesin, a class of antimicrobial peptide from the hemocytes of the horseshoe crab (*Tachypleus tridentatus*), *J. Biol. Chem.* **263**:16709–16713.

Nguyen, N. Y., Suzuki, A., Boykins, R. A., and Liu, T. Y., 1986, The amino acid sequence of *Limulus* C-reactive protein: Evidence of polymorphism, *J. Biol. Chem.* **261**:10450–10455.

Nishijima, T., and Yoshiniko, M., 1979, Seasonal variations in the concentration of dissolved vitamin B_{12}, thiamin and biotin in Susaki Harbor and Nomi Inlet Japan, *Rep. Usa. Mar. Biol. Inst. Kachi Univ.* **1**:1–8.

Noguchi, H., 1903, A study of immunization haemolysins, agglutinnins, preciptins and coagulins in cold-blooded animals, *Zentralbl. Bakteriol. Abt. Orig.* **33**:353–362.

Pistole, T. G., 1982, *Limulus* lectins: Analogues of vertebrate immunoglobulins, in: *Progress in Clinical and Biological Research*, Vol. 81 (J. Bonaventura, C. Bonaventura, and S. Tesh, eds.), Alan R. Liss, New York, pp. 283–288.

Pistole, T. G., and Graf, S. A., 1984, Bactericidal activity of *Limulus* lectins and amebocytes, in: *Progress in Clinical and Biological Research*, Vol. 157. *Recognition Proteins, Receptors and Probes: Invertebrates* (E. Cohen, ed.), Alan R. Liss, New York, pp. 71–81.

Prior, R. B. (ed.), 1990, *Clinical Applications of the Limulus Amebocyte Lysate Test*, CRC press, Boca Raton, Florida.

Riley, J. P., 1975, Analytical chemistry of sea water, in: *Chemical Oceanography* (J. P. Riley and G. Skirrow, eds.), Academic Press, pp. 193–514.

Rinehart, K. L., Holt, T. G., Fregeau, N. L., Keifer, P. A., Wilson, G. R., Perun, T. J., Sakai, R., Thompson, A. G., Stroh, J. G., Shield, L. S., and Seigler, D. S., 1990a, Bioactive compounds from aquatic and terrestrial sources, *J. Nat. Prod.* **53**:771–792.

Rinehart, K. L., Holt, T. G., Fregeau, N. L., Stroh, J. G., Keifer, P. A., Sun, F., Li, H., and Martin, D. G., 1990b, Potent antitumor agents from the Caribbean tunicate *Ecteinascidia turbinata*, *J. Org. Chem.* **55**:4512–4515.

Robey, F. A., and Liu, T. Y., 1981, Limulin: A C-reactive protein from *Limulus polyphemus*, *J. Biol. Chem.* **256**:969–975.

Schaechter, M., Medoff, G., and Schlessinger, D. (eds.), 1989, *Mechanisms of Microbial Disease*, Williams and Wilkins, Baltimore.

Scheuer, P. J., 1989, Marine natural products and biomedicine, *Med. Res. Rev.* **9**:535–545.

Schneider, Z., and Stroinski, A., 1987, *Comprehensive B_{12}; Chemistry Biochemistry, Nutrition and Medicine*, de Gruyter, Berlin, New York.

Sennett, C., and Rosenberg, E., 1982, Transmembrane transport of cobalamin in prokaryotic and eukaryotic cells, *Annu. Rev. Biochem.* **50**:1053–1086.

Sharma, G. M., Dubois, H. R., Pastore, A. T., and Bruno, S. F., 1979, Comparison of the determination of cobalamins in ocean waters by radioisotope dilution and bioassay techniques, *Anal. Chem.* **51**:196.

Sharma, G. M., Shigeura, T., and Liu, L. Y. X., 1985, Vitamin B_{12}-binding proteins of the horseshoe crab *Limulus polyphemus*, *Biochem. Biophys. Res. Commun.* **128**:241–248.

Sharon, N., and Lis, H., 1987, A century of lectin research, *Trends Biochem. Sci.* **12**:488–491.

Sharon, N., and Lis, H., 1989, Lectins as cell recognition molecules, *Science* **246**:227–233.

Smith, A. C., 1984, Sex specific agglutination of human erythrocytes by body fluids from marine invertebrates and fishes: A preliminary study, in: *Progress in Clinical and Biological Research*, Vol. 157. *Recognition Proteins, Receptors and Probes: Invertebrates* (E. Cohen, ed.), Alan R. Liss, New York, pp. 97–102.

Swift, D. G., 1980, Vitamins and phytoplankton growth, in: *The Physiological Ecology of Phytoplankton* (I. Morris, ed.), Blackwell, Oxford, pp. 329–368.

Swift, D. G., and Guillard, R. R. L., 1977, Diatoms as tools for assay of total B_{12}-activity and cyano cobalamin activity in sea water, *J. Mar. Res.* **35**:309–320.

Topham, R., Cooper, B., Tesh, S., Godette, G., Bonaventura, C., and Bonaventura, J., 1988, Isolation, purification and characterization of an iron binding protein from the horseshoe crab, *Biochem. J.* **252**:151–157.

Watson, S. W., Levin, J., and Novitsky, T. J. (eds.), 1987, *Detection of Bacterial Endotoxins with the Limulus Amebocyte Lysate Test*, Alan R. Liss, New York.

Wu, C. H., and Narahashi, T., 1988, Mechanism of action of novel marine neurotoxins on ion channels, *Annu. Rev. Pharmacol. Toxicol.* **28**:141–161.

6

Medical and Biotechnological Applications of Marine Macroalgal Polysaccharides

Donald W. Renn

1. INTRODUCTION

A considerable number of prescription drugs contain land-plant-derived ingredients. Although marine plants are not as evolutionarily advanced as land plants, when one considers the fact that over 3600 different varieties of marine macroalgae or seaweeds exist, it is not surprising that among these are some having pronounced pharmacological activities or other properties useful in the biomedical arena. What is surprising is that some of this pharmacological activity is attributed to the polysaccharides they contain, particularly those that are sulfated. Although the results of many studies ascribing different physiological properties to these sulfated polysaccharides can be found in the literature, few, if any, have ever reached commercial importance. Currently, their most important use is in biomedical research, where their inflammatory, immune-stimulating and -suppressing, thrombosis-causing, and other properties are used to induce a particular response in model systems for studying and/or screening materials for potential therapeutic value. Over the past 35 years, a number of papers, patents, and review articles have addressed these pharmacological properties in considerable detail. It is not the intent of this chapter to add another comprehensive review to the literature.

Donald W. Renn • FMC Corporation, Rockland, Maine 04841.

Marine Biotechnology, Volume 1: Pharmaceutical and Bioactive Natural Products, edited by David H. Attaway and Oskar R. Zaborsky. Plenum Press, New York, 1993.

Instead, in the first portion of this chapter, a description of the more important physiological properties will be addressed and a few leading references given.

Because diagnosis is an integral part of the medical process, and since aqueous gels of seaweed polysaccharides are essential to a variety of diagnostic assays, a section on this has been included.

The final portion of this chapter is devoted to the macroalgal polysaccharides as they are used in biotechnology to create pharmaceuticals. Most of the major advances in modern biotechnology leading to a variety of therapeutics, such as recombinant DNA insulins and tissue plasminogen activators, would not have been possible without the availability of the polysaccharides from marine macro-algae, or seaweeds (Renn, 1990).

The four major classes of macroalgae are Rhodophyta (red algae), Phaeo-phyta (brown algae), Chlorophyta (green algae), and Cyanophyta (blue-green algae). Only the red and brown macroalgae are currently sources of commercial polysaccharides of significant value. Carrageenans and agar, from which agarose is derived by purification, are obtained from red algae, but not from the same genera. Algins are obtained from a number of species of brown algae and are present in all. Alginates, the salts of alginic acid or algin, are composed of D-mannuronic and L-guluronic acid residues. "Carrageenan" is a generic term for a complex family of sulfated polysaccharides extracted from a number of different red seaweeds. They are all sulfated linear galactans, whose idealized basic structural unit, carrabiose, is an alternating D-galactose, 3,6-anhydro D-galactose disaccharide. Three major types of carrageenans, designated kappa, lambda, and iota, are available commercially. These differ in the amounts and position of ester sulfate substituents and content of 3,6-anhydro galactose. kappa-Carrageenan gels in the presence of potassium ions to form strong crisp gels, while lambda-carrageenan is nongelling but forms viscous solutions. iota-Carrageenan gels in the presence of calcium ions to form elastic gels. Because of their strongly anionic nature, these carrageenans exhibit a high degree of protein reactivity. Agars, mixtures of polysaccharides extracted from certain red seaweeds, have achieved commercial importance because of their ability to gel aqueous solutions at low concentrations. Agarobiose the idealized disaccharide repeating subunit of agar and agarose, is composed of D-galactose and 3,6-anhydro L-galactose. Other, less well-defined seaweed polysaccharides have also been found to exhibit physiologi-cal activities. As appropriate, these will be mentioned by weed source.

2. PHYSIOLOGICAL ACTIVITIES

This section is meant to be a survey which includes recent updates, when available, rather than to be a comprehensive review. In addition, emphasis will be on the commercially available polysaccharides obtained from marine algae.

2.1. Antiviral

Another chapter in this book is devoted to marine organism-derived antiviral agents, so only a brief discussion of the antiviral activities attributed to macroalgal polysaccharides will be included. The antiviral activity of the seaweed polysaccharides, particularly the sulfated ones, has been known for many years and is the subject of many reports. Good background reference papers are Ehresmann *et al.* (1979a, b), González *et al.* (1987), and Neushul (1990). The pioneering efforts of Merigan and Finkelstein (1968), which demonstrated that anionic polymers elicited interferon response as part of their antiviral activity mechanism, laid the groundwork for present-day screening.

Remedies from folk medicine are frequently tapped to obtain leads for therapeutics. The application of seaweed fronds has been a topical folk cure for "cold sores" among sailors and seacoast dwellers for some time. Screening aqueous extracts of seaweeds for antiviral activities, Deig *et al.* (1974) found that extracts from a number of red macroalgae inhibited both type 1 and 2 herpes simplex virus. Subsequent studies (Hatch *et al.*, 1979) indicated that this activity was attributable, at least in part, to the sulfated polysaccharides contained in the active extracts.

Since the advent of AIDS, the search for antiviral agents has intensified. A flurry of activity has followed the recent discovery that dextran sulfate exhibited anti-HIV activity (Ueno and Kuno, 1987; Bagasra and Lischner, 1988). Sulfated polysaccharides, including those from marine macroalgae, have been shown to inhibit replication of the human immunodeficiency virus (HIV) *in vitro* (Ueno *et al.*, 1987; Nakashima *et al.*, 1987a, b; Baba *et al.*, 1988; Neushul, 1988; De Clerq *et al.*, 1988).

Although considerable inhibitory activity toward a variety of viruses has been demonstrated, as reported by many scientists, no commercially available human antiviral therapeutics have evolved from marine algal sources, either because their efficacy is exhibited only *in vitro* or deleterious side effects have been observed *in vivo*.

2.2. Anticoagulant, Antithrombic

Heparin, one of the most effective anticoagulant therapeutics, is a sulfated amino polysaccharide. Using analogies, as is common in the search for therapeutics, it is not unusual that the sulfated polysaccharides from marine macroalgae should have been considered and tested for antithrombic activity. Although there are a number of reports substantiating this activity, particularly, but not exclusively, with carrageenan (Houck *et al.*, 1958; Hawkins and Leonard, 1963; Anderson and Duncan, 1965; Schimpf *et al.*, 1969; Kindness *et al.*, 1979; Efimov *et al.*, 1983), none of the seaweed polysaccharides matched the anticoagulant activity of the heparin controls.

2.3. Antitumor, Antimetastatic

Parish and associates (Parish *et al.*, 1987; Coombe *et al.*, 1987; Parish and Snowden, 1988) discovered that sulfated polysaccharides, including fucoidin and carrageenans, inhibit tumor mestastasis in rat test systems by inhibiting the action of the tumor cell-derived heparanases involved in membrane crossing. Nagumo (1983) discovered that sulfated polysaccharides from *Sargassum kjellmanianum* inhibited mouse S-180 tumor growth, and Matsumoto *et al.* (1984) reported that carrageenan, while not active alone, significantly potentiated the effect of mitomycin against leukemia L-1210 ascites tumor in mice. Carrageenan has also been found to stimulate lectin-dependent cell-mediated cytotoxicity against HEp-2 human epipharnyx carcinoma cells (Perl *et al.*, 1983).

2.4. Anti-Ulcer

Carrageenan interferes with the proteolytic activity of pepsin, both *in vivo* and *in vitro*, and has been useful in treating peptic ulcers. A comprehensive review of this activity is found in Stancioff and Renn (1975). Stanley (1982) reported the results of a study on the effect of carrageenan on the peptic and tryptic digestion of casein *in vitro*. This study indicated that carrageenans prolong the gastric digestive process, but do not interfere with subsequent enteric digestion.

2.5. Cholesterol-Lowering, Antilipemic

Carrageenan and other sulfated polysaccharides have been shown to have hypocholesterolemic activity (Houck *et al.*, 1957; Mookerjea and Hawkins, 1958; Murata, 1961; Ershoff and Wells, 1962; Fahrenbach and Riccardi, 1964). Unfortunately, the large doses of these that are necessary to demonstrate antilipemic activity would be difficult for humans to tolerate.

2.6. Immunoregulator, Cellular Response Modifier

Thomson and Fowler (1981) published an extensive review on the effects of carrageenan on the immune system. This includes descriptions of the immunological effects attributable to carrageenan, including the work of Calne, who used carrageenan as an immunosuppressant in renal transplant management. When coupled with their earlier review (Thomson *et al.*, 1979) on the immunopharmacology of carrageenan, a firm base is provided for understanding the immunoregulatory and cellular response modifier properties of one class of seaweed polysaccharides.

Since the time these comprehensive reviews appeared in the literature, a number of other papers have been generated by researchers testing the effects of the sulfated seaweed polysaccharides on specific cells, as well as the overall immune system. The immune adjuvant properties of carrageenans have been observed and studied (Hanazawa *et al.*, 1982; Mancino and Minucci, 1983), indicating they have immune stimulant as well as immune suppressant activity. Carrageenans have been shown to act as human T-cell mitogens (Sugawara *et al.*, 1982), induce T-cell suppressor activity in mice (Lukić *et al.*, 1983), modify lymphocyte migration via specific receptors (Brenan and Parish, 1986), induce production of a granulocyte-macrophage colony-stimulating factor (GM-CSF) (Shikita *et al.*, 1981), stimulate populations of mouse B cells (Kolb *et al.*, 1981), and stimulate the release of dialyzable helper factors from mouse spleen cells (Evelegh *et al.*, 1982).

2.7. Hemostats, Wound Dressings

Alginate fiber preparations have been used for some time to suppress bleeding associated with surgical procedures, as wound dressings, and as swabs for culture sampling. This property is more attributable to the absorptive capacity of calcium-insolubilized alginate fibers for water than to physiological activity. However, a recent paper by Attwood (1989) indicates that calcium alginate dressings accelerate split skin graft donor site healing and permit recropping to take place in half the usual time.

2.8. Cervical Dilator

Sterile dried stipes of *Laminaria* can be used as cervical dilators in obstetrics and for gynecological examinations (Stein and Borden, 1984). The stipe or "laminaria tent" is inserted in the cervix and causes it to gradually expand and, in doing so, effect dilation. Although it is probably the physical swelling of the polysaccharides contained in the stipe rather than a physiological response that enables this application, this is another example of the varied medical uses for marine algal polysaccharides.

3. MODEL SYSTEMS FOR NEW DRUG SCREENING

Before screening for discovery of new therapeutics can begin, a meaningful assay system must be developed. Macroalgal polysaccharides have played a major role in the discovery and development of new, more potent anti-inflammatory drugs.

3.1. Use of Physiological Activities to Develop Model Systems for New Drug Screening

The most important current medically oriented use for sulfated seaweed polysaccharides is in biomedical research, where their inflammatory, immune-stimulating and -suppressing, thrombosis-causing, and other properties are used to induce a particular response in model systems for studying and/or screening materials for potential therapeutic value. Stancioff and Renn (1975) reviewed the carrageenan physiology literature and distinguished between the sometimes adverse physiological properties resulting from the injection of carrageenan and the extensively proven safety of its oral ingestion as an essential food additive by comparing the vast difference between eating beef stew with the potentially disastrous results which would follow upon injecting it intravenously.

Many anti-inflammatory agents have been discovered using the carrageenan-induced, nonimmune inflammatory response, rat paw edema assay developed by Winter *et al.* (1962). The use of this model system for drug screening and studies of the mechanism of edema formation and recession are the subject of a number of publications, including representative ones by Niemegeers *et al.* (1964), Moersdorf and Anspach (1971), and Reiter *et al.* (1985). When carrageenan is infused into the lungs of mice or rats, a pleurisy-like, inflammatory reaction occurs which can be reversed by anti-inflammatory agents (Vinegar *et al.*, 1974, 1982; Mikami and Miyasaka, 1983; Ghiara *et al.*, 1986). An arthritis-like model can be generated when carrageenan is injected into the synovial fluid of animals (Gardner, 1960; Marroquin and Ajmal, 1970). This model has been used not only to study the mechanism of the onset and development of arthritis, but also to screen for arthritis-specific anti-inflammatory agents.

Selye (1965) reported that thrombohemorrhagic lesions could be induced by subcutaneous infiltration of carrageenan into rats and then subjecting them to cold stress, with the mechanism of action being localized thrombosis formation. Bekemeier and associates (Bekemeier *et al.*, 1984, 1985) discovered that if kappa-carrageenan was injected into rats, thromboses resulted, particularly if the intravenous route was used. After finding that this activity could be reversed by a variety of therapeutic agents, they proposed that this be used as a model for studying the factors influencing thrombosis formation and testing antithrombic substances. Whitehouse and Rainsford (1985) observed that irreversible tail necrosis occurred when rats that were primed for rat paw edema and injected with acidic nonsteroidal anti-inflammatory drugs were subjected to chilling. They have proposed that this be used as a model for developing drugs to control microvascular disturbances in humans.

4. BIOMEDICAL ASSAY APPLICATIONS

Appropriate medical treatment requires correct diagnosis. Agar and agarose, polysaccharides from marine macroalgae, are essential gel-forming components for a variety of diagnostic assays.

4.1. Serum Protein Electrophoresis

Blood not only carries nutrients to cells, tissues, and organs throughout the body, but also collects cellular wastes and metabolites for disposal and/or reprocessing. It is this function that makes the cell-free plasma or serum such an important fluid for diagnosing a variety of diseases and dysfunctions. One of the simplest indicators is the protein separation pattern obtained using agarose gel electrophoresis (Killingsworth *et al.*, 1980a). Agarose gels containing the appropriate buffers provide excellent media for separation of polyelectrolytes, particularly proteins and nucleic acids and their derivatives, by charge and/or mass using an electric potential. Separations by charge are based on differential rates of migration of charged particles toward the oppositely charged electrode when an electric potential is applied across the gel. Electrophoretic separation by mass or molecular size depends on the relative ability of the particles to migrate through the pores of the gel matrix. The smaller the molecule, the less is the restriction, and therefore the faster the movement. This type of electrophoretic sorting is frequently termed "molecular sieving electrophoresis." Since serum proteins reflect various charge-to-mass ratios, agarose gel electrophoresis is routinely used to identify protein abnormalities, including enzyme variations, in serum and plasma (Jeppsson *et al.*, 1979; Killingsworth, 1979), as well as those in other biological fluids, such as urine (Killingsworth *et al.*, 1980b) and cerebrospinal fluid (Killingsworth *et al.*, 1980c).

4.2. Immunological Assays

Applications of agarose for the immunological detection and study of various antigenic disease agents and/or markers and their specific antibodies are numerous. As a bit of background, *antigens* are any substance not recognized as self that give rise to an immunological response. *Antibodies*, or immunoglobulins, are specific proteins formed by specialized animal cells. They are synthesized in response to an antigenic stimulus and will combine specifically with that antigen to neutralize it. Many of these antigen–antibody complexes are insoluble; thus, if an antigen and the antibody specific to it diffuse separately through an agarose gel, the position where they come together will be marked by a cloudy or white so-called precipitin band. These reactions can be amplified for detection by using a variety of techniques and tags. Because of the macroporosity of the agarose gel matrix, its relative chemical neutrality, and its high clarity, agarose is an ideal medium for immunological reactions. A good introduction to a number of the immunoprecipitin-in-gel techniques and their general applications can be found in the *Scandinavian Journal of Immunology*, Vol. 4, Supplements 1 and 2 (1975), and Vol. 17, Supplement 10 (1983).

Many diseases and other physiological aberrations have a specific marker or indicator associated with them which is antigenic. In addition, one of the body's defense mechanisms is to produce specific antibodies to invading microparticulates. Agarose gel-based immunoassays can be used to detect either the antibodies or the antigenic agent, and therefore are very important in medical diagnostics. A review of the various methods available and leading references can be found in the Agarose Monograph (FMC BioProducts, 1988a).

4.3. Microbial Growth

In 1882, Dr. Robert Koch formally announced the use of agar, extracted from seaweed, as a new solid culture medium for microorganisms, following his now-famous experiments on tuberculosis bacteria. Amazingly, this, the most significant use of nonfood and pharmaceutical agars, has not changed in over 100 years. It is still the medium of choice for general microbiological growth and identification, and thus is an important medium for use in detecting the presence of culturable disease-causing microbes in body fluids and/or lesions. Although agar has long been the standard medium for microorganism and cell culture, even the bacteriological-grade agars vary considerably from lot to lot and contain varying proportions of unknown entities, some of which are reported to be toxic to microorganisms and plant and animal cells. This leads to slower or no growth of sensitive cells and microorganisms. Agarose, because of its higher degree of purity and consistency, is being used increasingly by scientists for critical cultures (FMC BioProducts, 1988b). Not only are agar and agarose used for microbial disease detection, but also in determining which therapeutics will be useful in treatment, an example being the use of antibiotic sensitivity discs. With the availability of lower-gelling, lower-melting-temperature agarose derivatives (Guiseley, 1976, 1987), cells and other heat-labile substances can be incorporated into gelling media at considerably lower temperatures than before.

4.4. Genetic Disorder Detection by Restriction Fragment Analysis

Some genetic disorders can be detected by studying the number and shape of chromosomes. Other genetic disorders can only be detected by examining the DNA contained in the chromosomes. Variations in the nucleotide composition of DNA can be determined by cutting the gene or gene fragment with restriction enzymes. By choosing the appropriate enzyme that differentiates between nucleotide sequences in the desired variable region, then determining the number and relative lengths of the restricted, or cut, DNA fragments by separation on gel electrophoresis, genetic diseases, or a predisposition toward them, can be detected. The range of sieving or separation obtained by using different types and/or concentrations of agarose gel as the electrophoresis medium is ideal for this determination. Although it is desirable for this process, the defective gene itself does not have to have been isolated. Genetic markers, or segments of the DNA having unusual nucleotide sequences which occur in the neighborhood of the defective gene, can also be used. A variation of this technique is the basis for DNA fingerprinting.

5. BIOTECHNOLOGY-ORIENTED APPLICATIONS

The term *biotechnology* means something different to each of us—from modifying and culturing living systems to create useful products, to test-tube babies, to the human clones of science fiction. As a general definition, *biotechnology* is the manipulation and/or use of all or part of a specific biological system to generate a desired product. Biotechnology is not a science unto itself, but a family of tools and techniques which can be used to solve problems and create products and/or processes. These tools include genetic engineering or recombinant DNA technology, plant and animal cell and tissue culture, enzymes, fermentations, immobilized bioreactors, biochemistry, and immunology. Biotechnology is nothing new, and applications of it have existed for thousands of years, as in winemaking and the use of other fermentation products. Recent understandings and breakthroughs are what have catapulted biotechnology to the prominent position it holds today.

For separation of genes and gene fragments, agarose gel electrophoresis has been indispensable. Recombinant insulin, tissue plasminogen activator, Factor VIII, interferons, and many other human therapeutics have all been developed using agarose gels in at least one step of their discovery and development. It is questionable whether the recent great strides in cancer research, particularly the discovery and understanding of oncogenes, and in AIDS research would have been possible without agarose and/or its derivatives. With the discovery of the polymerase chain reaction technique, it has been possible to take a single copy of a particular DNA and amplify it a millionfold within an hour and then use this to identify a particular pathogen. Separation of the amplified fragments from the reactants is done on agarose. Agar continues to be the medium of choice for cultivation and selection of the transformed microbial hosts in genetic engineering. More efficient bioconversions are the result of microbial encapsulation technology using agar and agarose, as well as the insoluble salts of algin and carrageenan.

5.1. Media for Gene Fragment Separations

Enzymes called restriction endonucleases that cleave DNA between certain defined nucleotide sequences have been discovered in nature. The DNA sequence coding for a specific protein, such as a human therapeutic polypeptide or protein, can be cut from a gene. Agarose gel electrophoresis is used to separate the desired DNA fragment. Concurrently, as a "vector" or foreign gene-introducing agent, a bacterial plasmid composed of circular DNA can be isolated and the same or an analogous enzyme used to cut the vector DNA. When a plasmid is used, the isolated DNA fragment is spliced into the plasmid DNA replacing the portion removed and is glued there by another enzyme called a "ligase." The altered plasmid is reintroduced into the host and codes for the specific protein desired. By this process *Escherichia coli* or other microbial hosts can produce, for example, human insulin by fermentation. A variety of vectors, including viruses, can be used to introduce the foreign genes into their hosts, which can be anything from bacteria to plant or animal cells, to ultimately human beings, as will be in the case of vaccines and gene therapy.

5.2. Media for Gene Mapping

Each human cell nucleus contains 46 chromosomes, each containing many gene segments coding for specific polypeptides or proteins. Some segments are nearly identical in all individuals. Others vary, and in a number of cases, these are indicators of genetic disorders or susceptibility to a variety of diseases. An ambitious project is underway to map all of the human chromosomes, that is, to determine which chromosomal segments are responsible for what function. Human genome mapping is being approached in a variety of ways, but all use, at some point, a step requiring electrophoretic separation of DNA fragments on agarose gels. Because charge densities are essentially equal in genomic DNA and smaller restriction enzyme-cleaved fragments, all migrate according to size in electric fields. By appropriate choice of agarose, concentration, buffers, and electrophoresis conditions, DNA ranging in size from 10 base pairs to chromosomal-fragment-size megabase pairs can be separated. It is also possible to separate particles such as phages, viruses, and capsids using even lower concentrations of agarose.

5.2.1. Chromosome Fragment Separation

Small chromosomes and large chromosomal DNA fragments, greater than 40 kilobases, created by restriction enzyme treatment can only be separated on agarose gels using a technique called pulsed field gel electrophoresis, or PFGE. In PFGE, an electric current is alternately imposed at a predetermined angle to the direction of electrophoretic migration. A number of modifications of this technique, first described by Schwartz, Cantor, and associates (Schwartz *et al.*, 1983), have been developed, many for human disease-, genetic disorder-specific chromosomal fragment separations.

5.3. Cell Immobilization

After a host cell has been altered to contain a foreign gene coding for a desired bioactive protein by genetic technology manipulations, the cells must be propagated in order to produce the desired therapeutic in large quantities. Occasionally, it is beneficial to immobilize the cells by encapsulation. Using one of the gel-forming seaweed polysaccharides—agar, agarose, kappa-carrageenan, or algin—as the encapsulating agent is an effective way of doing this. This topic is the subject of a recent review by Guiseley (1989).

5.3.1. Artificial Organs

Because experimental implants of insulin and other hormone-producing cells and tissues have been shown to reverse the physiological effects caused by the lack of these hormones or bioregulators, methods are being devised to implant the living cells or tissues, yet keep them isolated to prevent immune rejection and other deleterious effects. Cages of agarose, its low-gelling-temperature hydroxy-ethyl derivatives (Guiseley, 1976, 1987), and calcium alginate are the algal polysaccharides cited as being compatible with this application. Encapsulated insulin-producing Islets of Langerhans cells have been used as model systems with considerable success (Lim and Sun, 1980; Howell *et al.*, 1982; Sun *et al.*, 1982; Bouhaddioui *et al.*, 1985; Goosen *et al.*, 1985, 1989; Gin *et al.*, 1987; Iwata *et al.*, 1988). Yarmush *et al.* (1988) reported that rat hepatocytes retained their enzymatic detoxifying activity when immobilized in alginate droplets. In another report (Adaniya *et al.*, 1987), preimplantation mouse embryos were kept alive by encapsulation in an artificial womb of calcium alginate.

6. PROJECTIONS FOR THE FUTURE

Although no marine macroalgal polysaccharides are now commercially available for therapeutic use, as seen in the preceding discussions, this has not reduced their importance to medical science. To attempt to predict whether or not any will ever find a niche as a human therapeutic agent would be mere speculation. There is no doubt in my mind that the seaweed polysaccharides, their salts, and/or derivatives will play an increasingly important role in the search for causes, diagnosis, and treatment of diseases, aberrant physiological conditions, and genetic disorders. Known seaweed polysaccharides will continue to be screened for therapeutic value as new assays are developed, and new ones will be discovered and characterized. An ever-increasing number of gene fragments correlating with genetic disorders and susceptibility to various diseases, such as specific types of malignancies and cardiovascular diseases, will be discovered during the human genome effort, using agarose gel media. As knowledge of these genes evolves, gene therapy will follow where applicable, as will vaccines using specific gene expression products. Modifications of current and future therapeutics with polysaccharide tails will alter both the response and the excretion times. Effective controlled-release drug-delivery systems will be developed using seaweed polysaccharides. Artificial organs, using therapeutic-producing cells and tissues encapsulated in algal polysaccharides, will become medical realities, rather than laboratory curiosities. In the future, as now, marine macroalgal polysaccharides will be an integral part of the discovery process of medical science, even if not in a therapeutic role.

Polysaccharides from seaweeds have enabled scientists to develop modern biotechnology tools and techniques leading to new therapeutics and other important molecules. These same techniques can be applied advantageously to seaweeds and their biosynthetic systems. Seaweeds are living organisms and thus contain chromosomes and other nucleic acid components that are subject to genetic manipulation through conventional crossbreeding and/or the emerging new techniques of biotechnology. Examples of these techniques as they are applied to land plants are numerous. In contrast, reports of genetic manipulations with marine macroalgae are relatively few, but some excellent pathfinding work has already been done.

Potentially productive applications of biotechnology to seaweeds can be accomplished in a number of ways. These include: (1) use of algae as hosts for foreign genes to produce desired products, including therapeutics, (2) transfer of algal genes coding for the enzymes that produce polysaccharides or other desired materials into bacteria or other microorganisms, and (3) manipulation of algal genes to enhance the productivity of desired products, such as polysaccharides, by increasing copy numbers of limiting biosynthetic enzymes, or alteration of photosynthetic partitioning so that less energy goes into cell division and more goes into polysaccharide or other desired entity productivity. Use of these and/or other more conventional techniques, such as mutant selection and selective mutation, should enable scientists to customize marine macroalgae to better meet future needs. The momentum has begun to build, the potential is great, and the continued dedicated, interdisciplinary efforts of scientists can make it happen.

REFERENCES

Adaniya, G. K., Rawlins, R. G., Miller, I. F., and Zaneveld, L. J. D., 1987, Effect of sodium alginate encapsulation on the development of preimplantation mouse embryos, *J. In Vitro Fertil. Embryo Transf.* **4**(6):343–345.

Anderson, W., and Duncan, J. G. C., 1965, The anticoagulant activity of carrageenan, *J. Pharm. Pharmacol.* **17**:647–654.

Attwood, A. I., 1989, Calcium alginate dressing accelerates split skin graft donor site healing, *Br. J. Plastic Surgery* **42:**373–379.

Baba, M., Snoeck, R., Pauwels, R., and De Clerq, E., 1988, Sulfated polysaccharides are potent and selective inhibitors of various enveloped viruses, including herpes simplex virus, cytomegalovirus, vesicular stomatitis virus, and human immunodeficiency virus, *Antimicrob. Agents Chemother.* **1988:**1742–1745.

Bagasra, O., and Lischner, H. W., 1988, Activity of dextran sulfate and other polyanionic polysaccharides against human immunodeficiency virus, *J. Infect. Dis.* **158**(5):1084–1087.

Bekemeier, H., Hirschelmann, R., and Giessler, A. J., 1984, Carrageenin-induced thrombosis in the rat and mouse as a test model of substances influencing thrombosis, *Biomed. Biochim. Acta* **43:** 347–350.

Bekemeier, H., Hirschelmann, R., and Giessler, A. J., 1985, Carrageenin-induced thrombosis in rats and mice: A model for testing antithrombic substances? *Agents Actions* **16**(5):446–451.

Brenan, M., and Parish, C. R., 1986, Modification of lymphocyte migration by sulfated polysaccharides, *Eur. J. Immunol.* **16:**423–430.

Bouhaddioui, N., Crespy, N., and Orsetti, A., 1985, Functional study of rat Islets of Langerhans cultured in agarose cells, *C. R. Soc. Biol.* **179:**21–26.

Coombe, D. R., Parish, C. R., Ramshaw, I. A., and Snowden, J. M., 1987, Analysis of the inhibition of tumour metastasis by sulfated polysaccharides, *Int. J. Cancer* **39:**82–88.

De Clerq, E. D. A., Ito, M., Shigata, S., and Baba, M., 1988, Therapeutic and prophylactic application of sulfated polysaccharides against AIDS, European Patent Application No. 0 293 826.

Deig, E. F., Ehresmann, D. W., Hatch, M. T., and Riedlinger, D. J., 1974, Inhibition of herpesvirus replication by marine algae extracts, *Antimicrob. Agents. Chemother.* **6:**524–525.

Efimov, V. S., Usov, A. I., Ol'skaya, T. S., Baliunis, A., and Roskin, M. Y., 1983, Comparative study of anticoagulant activity of sulfated polysaccharides obtained from red sea algae, *Farmakol. Toksikol.* **46**(3):61–67.

Ehresmann, D. W., Deig, E. F., Hatch, M. T., DiSalvo, L. H., and Vedros, N. A., 1979a, Antiviral substances from California marine algae, *J. Phycol.* **13:**37–40.

Ehresmann, D. W., Deig, E. F., and Hatch, M. T., 1979b, Antiviral properties of algal polysaccharides and related compounds, in: *Marine Algae in Pharmaceutical Science* (H. A. Hoppe, T. Levring, and Y. Tanaka, eds.), de Gruyter, New York, pp. 294–302.

Ershoff, B. H., and Wells, A. F., 1962, Effects of gum guar, locust bean gum and carrageenan on liver cholesterol of cholesterol-fed rats, *Proc. Soc. Exp. Biol. Med.* **110:**580–582.

Evelegh, M. J., Clark, D. A., and McCandless, E. L., 1982, Carrageenan stimulates the release of dialyzable helper factors, *Immunol. Lett.* **5:**247–252.

Fahrenbach, M. J., and Riccardi, B. A., 1964, Method of reducing cholesterol levels, U.S. Patent No. 3,148,114.

FMC BioProducts, 1988a, The Agarose Monograph, in: *FMC BioProducts Source Book*, FMC Corp., Rockland, Maine, pp. 88–94.

FMC BioProducts, 1988b, The Agarose Monograph, in: *FMC BioProducts Source Book*, FMC Corp., Rockland, Maine, pp. 95–101.

Gardner, D. L., 1960, Production of arthritis in the rabbit by the local injection of the mucopolysaccharide carragheenin, *Ann. Rheum. Dis.* **19:**369–376.

Ghiari, P., Bartalini, M., Tagliabue, A., and Boraschi, D., 1986, Anti-inflammatory activity of IFN-β in carrageenan-induced pleurisy in the mouse, *Clin. Exp. Immunol.* **66:**606–614.

Gin, H., Dupuy, B., Baquey, C., Ducassou, D., and Aubertin, J., 1987, Agarose encapsulation of Islets of Langerhans: Reduced toxicity *in vitro*, *J. Microencapsulation* **4:**239–242.

González, M. E., Alarcón, B., and Carrasco, L., 1987, Polysaccharides as antiviral agents: Antiviral activity of carrageenan, *Antimicrob. Agents Chemother.* **1987:**1388–1393.

Goosen, M. F. A., O'Shea, G. M., Gharapetian, H. M., Chou, S., and Sun, A. M., 1985, Optimization of microencapsulation parameters: Semipermeable microcapsules as a bioartificial pancreas, *Biotechnol. Bioeng.* **27**:146–150.

Goosen, M. F. A., O'Shea G. M., and Sun, A. M., 1989, Microencapsulation of living tissue and cells, U.S. Patent No. 4,806,355.

Guiseley, K. B., 1976, Modified agarose and agar and method of making same, U. S. Patent 3,956,273.

Guiseley, K. B., 1987, Natural and synthetic derivatives of agarose and their use in biochemical separations, in: *Industrial Polysaccharides: Genetic Engineering, Structure/Property Relations and Applications* (M. Yalpani, ed.) Elsevier, Amsterdam, pp. 139–147.

Guiseley, K. B., 1989, Chemical and physical properties of algal polysaccharides used for cell immobilization, *Enzyme Microb. Technol.* **11**:706–716.

Hanazawa, S., Ishikawa, T., and Yamaura, K., 1982, Comparison of the adjuvant effect of antibody response of three types of carrageenan and the cellular events in the induction of the effect, *Int. J. Immunopharmac.* **4**(6):521–527.

Hatch, M. T., Ehresmann, D. W., and Deig, E. F., 1979, Chemical characterization and therapeutic evaluation of anti-herpesvirus polysaccharides from species of Dumontiaceae, in: *Marine Algae in Pharmaceutical Science* (H. A. Hoppe, T. Levring, and Y. Tanaka, eds.), de Gruyter, New York, pp. 344–363.

Hawkins, W. W., and Leonard, V. G., 1963, The antithrombic activity of carrageenan in human blood, *Can. J. Biochem. Physiol.* **41**:1325–1327.

Houck, J. C., Morris, R. K., and Lazaro, E. J., 1957, Anticoagulant, lipemia clearing, and other effects of anionic polysaccharides extracted from seaweed, *Proc. Soc. Exp. Biol. Med.* **96**:528–530.

Howell, S. L., Ishaq, S., and Tyhurst, M., 1982, Possible use of agarose gels as encapsulating media for transplantation of Islets of Langerhans, *J. Physiol.* **324**:20P.

Iwata, H., Amemiya, H., Matsuda, T., Takano, H., and Akutsu, T., 1988, Microencapsulation of Langerhans Islets in agarose microbeads and their application for a bioartificial pancreas, *J. Bioact. Compat. Polym.* **3**(4):356–369.

Jeppsson, J. O., Laurell, C. B., and Franzen, B., 1979, Agarose gel electrophoresis, *Clin. Chem.* **25**: 629–638.

Killingsworth, L. M., 1979, Plasma proteins in health and disease, in: *Critical Reviews in Clinical Laboratory Sciences* (J. Batsakis and J. Savory, eds.), CRC Press, Cleveland, Ohio, pp. 1–30.

Killingsworth, L. M., Cooney, S. K., and Tyllia, M. M., 1980a, Protein analysis: The closer you look, the more you see, *Diagn. Med.* **1980**(January/February):3–7.

Killingsworth, L. M., Cooney, S. K., and Tyllia, M. M., 1980b, Protein analysis: Finding clues to disease in urine, *Diagn. Med.* **1980**(May/June):1–5.

Killingsworth, L. M., Cooney, S. K., Tyllia, M. M., and Killingsworth, C. E., 1980c, Protein analysis: Deciphering cerebrospinal fluid patterns, *Diagn. Med.* **1980**(March/April):1–7.

Kindness, G., Long, W. F., and Williamson, F. B., 1979, Enhancement of antithrombin III activity by carrageenans, *Thromb. Res.* **15**(1/2):49–60.

Kolb, J.-P. B., Quan, P. C., Poupon, M.-F., and Desaymard, C., 1981, Carrageenan stimulates populations of mouse B cells mostly nonoverlapping with those stimulated with LPS or dextran sulfate, *Cell. Immunol.* **57**:348–360.

Lim, F., and Sun, A. M., 1980, Microencapsulated islets as bioartificial endocrine pancreas, *Science* **210**:908–910.

Lukić, M. L., Vukmanović, S., Ramić, Z., and Mostarica-Stojković, M., 1983, Carrageenan induces T-cell suppressor activity in mice, *Periodicum Biologorum* **85**(Suppl. 3):41–43.

Manicino, D., and Minucci, M., 1983, Adjuvant effects of ι, κ, and λ carrageenans on antibody production in BALB/c mice, *Int. Arch. Allergy Appl. Immunol.* **72**:359–361.

Marroquin, A., and Ajmal, M., 1970, Carrageenin-induced arthritis in the specific pathogen-free pig, *J. Comp. Pathol.* **80**:607–611.

Matsumoto, T., Taguchi, A., Okawa, Y., Mikami, T., Suzuki, K., Suzuki, S., and Suzuki, M., 1984, Growth-inhibition of leukemia L-1210 ascites tumor by combination of carrageenan and mitomycin C *in vivo*, *J. Pharm. Dyn.* **7**:465–471.

Merigan, T. C., and Finkelstein, M. S., 1968, Interferon stimulating and *in vivo* antiviral effects of various synthetic anionic polymers, *Virology* **35**:363–374.

Mikami, T., and Miyasaka, K., 1983, Effects of several anti-inflammatory drugs on the various parameters involved in the inflammatory response in rat carrageenin-induced pleurisy, *Eur. J. Pharmacol.* **95**:1–12.

Moersdorf, K., and Anspach, K., 1971, Potency of various antiinflammatory agents in carrageenan induced inflammation in comparison to other inflammation models, *Arch. Int. Pharmacodyn. Ther.* **192**(1):111–127.

Mookerjea, S., and Hawkins, W. W., 1958, The antilipaemic activity of *Laminaria* sulfate, *Can. J. Biochem. Physiol.* **36**:261–268.

Murata, K., 1961, Effects of carrageenan on serum lipids and atherosclerosis in rabbits, *Nature* **191**: 189–190.

Nagumo, T., 1983, Antitumor polysaccharides from seaweeds, *Jpn. Kokai Tokkyo Koho*, JP 58,174,329 (CA 100:12641f, 1984).

Nakashima, H., Kido, Y., Kobayashi, N., Motoki, Y., Neushul, M., and Yamamoto, N., 1987a, Antiretroviral activity in a marine red alga: Reverse transcriptase inhibition by an aqueous extract of *Schizymenia pacifica*, *J. Cancer Res. Clin. Oncol.* **113**:413–416.

Nakashima, H., Kido, Y., Kobayashi, N., Motoki, Y., Neushul, M., and Yamamoto, N., 1987b, Purification and characterization of an avian myeloblastosis and human immunodeficiency virus reverse transcriptase inhibitor, sulfated polysaccharides extracted from sea algae, *Antimicrob. Agents Chemother.* **1987**:1524–1528.

Neushul, M., 1988, Method for the treatment of AIDS virus and other retroviruses, U.S. Patent No. 4,783,446.

Neushul, M., 1990, Antiviral carbohydrates from marine red algae, *Hydrobiologia* **204/205**:99–104.

Niemegeers, C. J. E., Verbrugger, F. J., and Janssen, 1964, Effect of various drugs on carrageenan-induced oedema in the rat hind paw, *J. Pharm. Pharmacol.* **16**:810–816.

Parish, C. R., and Snowden, J. McK., 1988, Sulphated polysaccharides having metastatic and/or anti-inflammatory activity, PCT Application No. WO 88/05301.

Parish, C. R., Coombe, D. R., Jakobsen, K. B., Bennett, F. A., and Underwood, P. A., 1987, Evidence that sulphated polysaccharides inhibit tumor metastasis by blocking tumor-cell-derived heparanases, *Int. J. Cancer* **40**:511–518.

Perl, A., Gonzalez-Cabello, R., and Gergely, P., 1983, Stimulation of lectin-dependent cell-mediated cytotoxicity against adherent HEp-2 cells by carrageenan, *Clin. Exp. Immunol.* **54**:567–572.

Reiter, M. J., Schwartzmiller, D. H., Swingle, K. F., Moore, G. I., Goldlust, M. B., Heghinian, K., DeVore, D. P., Choy, B., and Weppner, W. A., 1985, Comparison of anti-inflammatory compounds in the carrageenan induced paw edema model and the reversed passive arthus model utilizing the same animal, *Life Sci.* **36**:1339–1346.

Renn, D. W., 1990, Seaweeds and biotechnology—Inseparable companions, *Hydrobiologia* **204/205**:7–13.

Schimpf, K., Lenhard, J., and Schaaf, G., 1969, The influence of the polysaccharide carrageenin on the clotting mechanism of the blood in the rabbit *in vitro* and *in vivo*, *Thrombosis Diathesis Haemorrhagia* **21**(3):524–533.

Schwartz, D. C., Saffran, W., Welsh, J., Haas, J., Goldenberg, M. M., and Cantor, C. R., 1983, New techniques for purifying large DNAs and studying their properties and packaging, *Cold Spring Harbor Symp. Quant. Biol.* **47**:189–195.

Selye, H., 1965, Induced hypersensitivity to cold, *Science* **149**:201–202.

Shikita, M., Tsuneoka, K., Hagiwara, S., and Tsurufuji, S., 1981, A granulocyte-macrophage colony-stimulating factor (GM-CSF) produced by carrageenin-induced inflammatory cells of mice, *J. Cell. Physiol.* **109:**161–169.

Stancioff, D. J., and Renn, D. W., 1975, Physiological effects of carrageenan, in: *Physiological Effects of Food Carbohydrates* (A. Jeanes and J. Hodge, eds.), American Chemical Society, Washington, D.C., pp. 282–295.

Stanley, N. F., 1982, The effect of carrageenan on peptic and tryptic digestion of casein, *Prog. Food Nutr. Sci.* **6:**161–170.

Stein, J. R., and Borden, C. A., 1984, Causative and beneficial algae in human disease conditions: A review, *Phycologia* **23**(4):485–501.

Sugawara, I., Ishizaka, S., and Möller, G., 1982, Carrageenans, highly sulfated polysaccharides and macrophage-toxic agents: Newly found human T lymphocyte activator, *Immunobiology* **163:**527–538.

Sun, A. M., O'Shea, G., van Rooy, H., and Goosen, M., 1982, Microencapsulation D'îlots de Langerhans et pancréas artificiel, *J. Annu. Diabetol. Hotel Dieu* **1982:**161–168.

Thomson, A. W., and Fowler, E. F., 1981, Carrageenan: A review of its effects on the immune system, *Agents Actions* **11**(3):265–272.

Thomson, A. W., Fowler, E. F., and Pugh-Humphreys, R. G. P., 1979, Immunopharmacology of the macrophage-toxic agent carrageenan, *Int. J. Immunopharmacol.* **1:**247–261.

Ueno, R., and Kuno, S., 1987, Dextran sulphate, a potent anti-HIV agent *in vitro* having synergism with zidovudine (Letter), *Lancet* **1:**1379.

Ueno, R., Ueno, R., Kuno, S., and Tabata, A., 1987. Treatment of diseases caused by retroviruses, European Patent Application No. 0 240 098.

Vinegar, R., Truax, J. F., and Selph, J. L., 1974, Carrageenan-induced pleurisy as a quantitative model for anti-inflammatory drug testing, *Pharmacologist* **16**(2):291.

Vinegar, R., Truax, J. F., Selph, J. L., and Voelker, F. A., 1982, Pathway of onset, development, and decay of carrageenan pleurisy in the rat, *Fed. Proc.* **41:**2588–2595.

Whitehouse, M. W., and Rainsford, K. D., 1985, A model of peripheral microvascular injury: Irreversible caudal necrosis induced in carrageenan-inflamed rats treated with anti-inflammatory drugs and mild chilling: A pluricausal thrombo-haemorrhagic phenomenon, *Int. J. Tissue React.* **7**(2):127–131.

Winter, C. A., Risley, E. A., and Nuss, G. W., 1962, Carrageenan-induced edema in the hindpaw of the rat as an assay for antiinflammatory drugs, *Proc. Soc. Exp. Biol. Med.* **111:**544–547.

Yarmush, M. L., Tomkins, R. G., Carter, E. A., and Burke, J. F., 1988, Enzymatic function of alginate-immobilized rat hepatocyte, *FASEB J.* **2**(4):A738.

7

Antitumor and Cytotoxic Compounds from Marine Organisms

Francis J. Schmitz, Bruce F. Bowden, and Stephen I. Toth

1. INTRODUCTION

An earlier review of antitumor and cytotoxic compounds from marine organisms which covered the literature into early 1986 was published in 1987 by Munro *et al.* (1987). The current review is intended to provide a comprehensive review of the field from the beginning of 1986 to early 1991. The primary aim was to include all the marine natural products reported to have any type of cytotoxic or antitumor activity. In addition to compounds reported to be toxic to a variety of cultured cancer cell lines, we have included compounds that show activity in the brine shrimp assay or which inhibit development of fertilized sea urchin or starfish eggs, simple assays which correlate to some extent with cytotoxicity. A considerable literature has developed regarding some of the most promising marine antitumor agents, such as didemnin B, the bryostatins, and the dolastatins. Information on some of the pharmacologic and mechanistic studies of these compounds has been included. The chapter is organized according to structural type, although in some cases a given compound could be assigned equally well to different categories.

The reports cited in this chapter were retrieved from computer literature

Francis J. Schmitz and Stephen I. Toth • Department of Chemistry and Biochemistry, University of Oklahoma, Norman, Oklahoma 73019. *Bruce F. Bowden* • Department of Chemistry and Biochemistry, James Cook University, Townsville, Queensland 4811, Australia.

Marine Biotechnology, Volume 1: Pharmaceutical and Bioactive Natural Products, edited by David H. Attaway and Oskar R. Zaborsky. Plenum Press, New York, 1993.

searches using *Chemical Abstracts* (CA) and also NAPRALERT. Terms such as marine, antitumor, anticancer, antineoplastic, cytotoxic, and phylla names were used. The NAPRALERT search, which was conducted after CA searching, retrieved articles not picked up by the latter. This seemed to be due to the lack of inclusion of biological data in the abstracts or titles of articles. Undoubtedly some relevant articles have been missed, but we hope the number is small. In this age of computerized searching it is important that reference to any biological activity be included in the abstract to facilitate retrieval by interested researchers.

The review by Munro *et al.* (1987) lists approximately 185 compounds of marine origin for which some type of cytotoxic activity had been reported. In contrast, 434 compounds are described in this chapter, which covers only 5½ years. Only about 30 of these compounds are also found in the earlier review. This indicates the increase in efforts directed to discovering potential antitumor drugs and reflects the fact that more investigators have adopted a bioassay-guided strategy in their search for new natural products.

The terminology used in this chapter follows the definitions recommended by the National Cancer Institute (Suffness and Douros, 1982). Cytotoxicity refers to toxicity to tumor cells in culture. The terms antitumor and antineoplastic are used for *in vivo* activity in experimental models and are not appropriate for describing *in vitro* results. Anticancer is the term reserved for describing data from clinical trials in humans.

The cytotoxicity data reported most frequently are from mouse lymphocytic leukemia cell cultures (P388 and L1210) and human epidermoid carcinoma of the mouth cell cultures (KB). Results are expressed in terms of the dose which inhibits cell growth to 50% of the control growth ($ED_{50} = ID_{50} = IC_{50}$). These are usually expressed in $\mu g/ml$. According to the National Cancer Institute's criteria, an active *extract* is one with an ED_{50} of <20 $\mu g/ml$ and a *pure compound* is active if the $ED_{50} <4$ $\mu g/ml$ (Geran *et al.*, 1972; Suffness and Pezzuto, 1991). A few results are given as LD_{50}, which refers to the dose that kills 50% of the cells.

The most common *in vivo* results reported are those for PS ($=$ P388) leukemia in mice, L1210 lymphoid leukemia, and B16 melanoma. These results are given in percent as a ratio of mean survival time of the test group compared to that of a control group (T/C). A significant level of activity in the PS test is 20% increase in survival time (T/C $=$ 120%). For the other tests the criterion for activity is T/C 125%. A few results are given for human tumor xenografts (in athymic mice) and the T/C value here refers to the change in weight of the treated tumor $\times 100 \div$ change in weight of control tumors. T/C values of $\leq 20\%$ are considered significant and T/C values of $\leq 10\%$ are biologically important (Suffness and Douros, 1982; Suffness *et al.*, 1989).

Although only a few different types of bioassays were used for detection of the cytotoxic compounds reported in this chapter, many more assays related to cancer drug discovery are available. Suffness and Pezzuto (1991) have compiled and discussed an extensive list of such assays.

A number of review articles describing antitumor and other bioactive compounds have appeared in recent years. Many of these are limited in scope. References to these are collected under a "Review Articles" section at the end of the chapter references.

2. POLYKETIDES

2.1. Fatty Acid Metabolites

A novel azacyclopropene, dysidazirine (**1**), isolated from the sponge *Dysidea fragilis*, was found to have an IC_{50} of 0.27 μg/ml against L1210 (Molinsky and Ireland, 1988). Ficulinic acids A (**2**) and B (**3**) from the sponge *Ficulina ficus* were also found to inhibit L1210 cell growth (ID_{50} 10–12 μg/ml) (Guyot *et al.*, 1986).

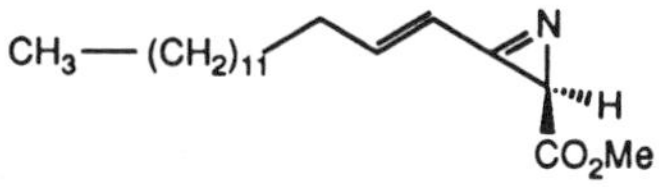

(**1**)

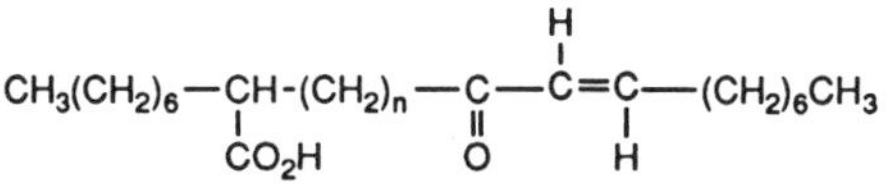

(**2**) Ficulinic acid A: n = 7
(**3**) Ficulinic acid B: n = 9

2.2. Long-Chain Acetylenes

Numerous aliphatic acetylenic compounds have been isolated from sponges and a number of these have been reported to be cytotoxic. Five monoacetylenic alcohols (**4–8**) from the sponge *Cribrochalina vasculum* collected in Belize were toxic to the mouse P388 cell line (IC_{50} 1.0, 1.3, 1.1, 0.2, 0.1 μg/ml, respectively) and they also showed *in vitro* immunosuppressive activity in lymphocyte reaction tests (Gunasekera and Faircloth, 1990). This appears to be the first report of branched-chain aliphatic acetylenic compounds from marine organisms.

Duryne (**9**), isolated from the Caribbean sponge *Cribrochalina dura*, is toxic

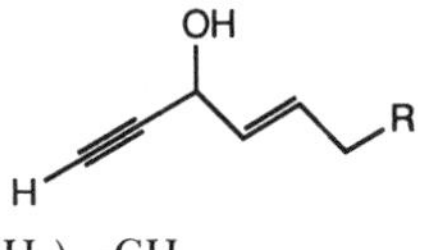

(**4**) R = $(CH_2)_{13}$-CH_3
(**5**) R = (E)-$(CH_2)_8$-CH=CH-$(CH_2)_5$-CH_3
(**6**) R = $(CH_2)_9$-CH(CH_3)-$(CH_2)_3$-CH_3
(**7**) R = $(CH_2)_{12}$CH(CH_3)$_2$
(**8**) R = (E)-$(CH_2)_8$-CH=CH-$(CH_2)_4$-CH(CH_3)$_2$

$$H-C\equiv C-CHOH-CH\overset{E}{=}CH-(CH_2)_9-CH=CH-(CH_2)_9-\overset{E}{CH}=CH-CHOH-C\equiv C-H$$

(9)

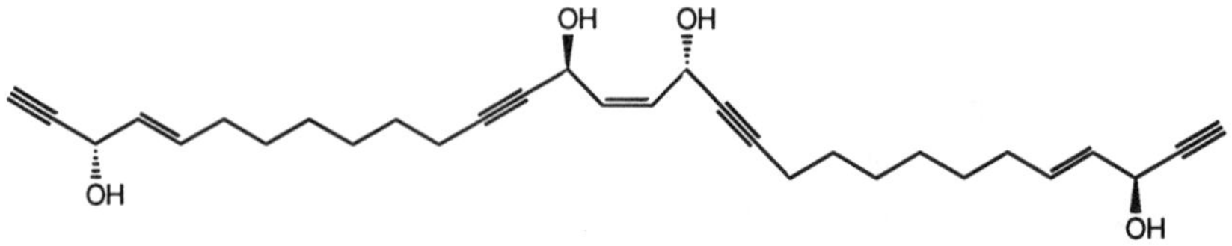

(10)

to murine leukemia cells (IC_{50} 0.07 μg/ml) and also colon, lung, and mammary cell lines, with MIC (minimum inhibitory concentration) of 0.1 μg/ml (Wright *et al.*, 1987a). Petrosynol (**10**), from a *Petrosia* sp. of sponge, shows antibiotic activity and is also active in the starfish egg assay at 1 μg/ml (Fusetani *et al.*, 1987c). Cimino *et al.* (1990) have described a number of C_{46} polyacetylenes, typified by **11–16**, that were active in the brine shrimp assay (IC_{50} 0.002–0.12

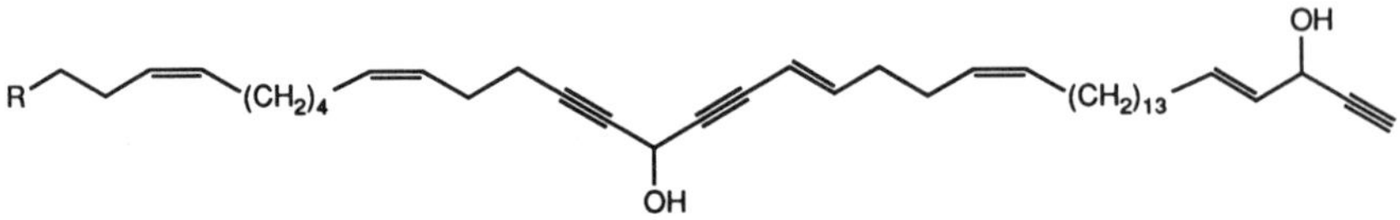

(11) R = A

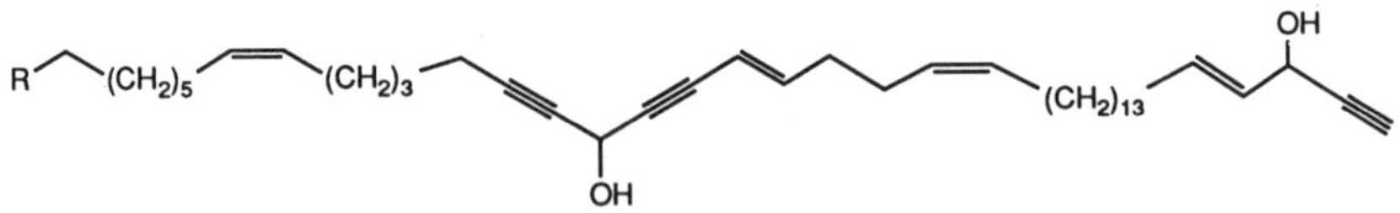

(12) R = A

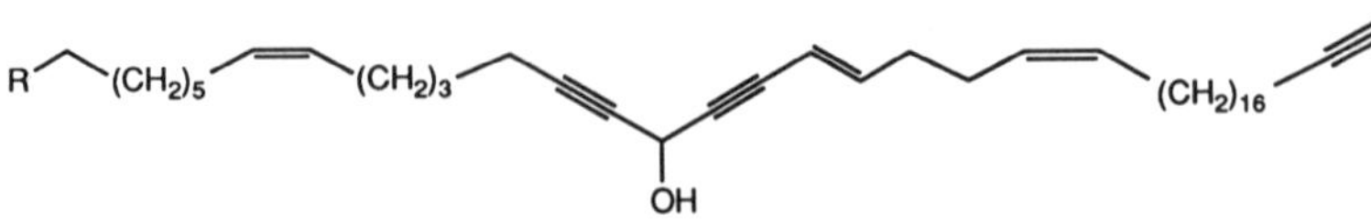

(13) R = A
(14) R = B

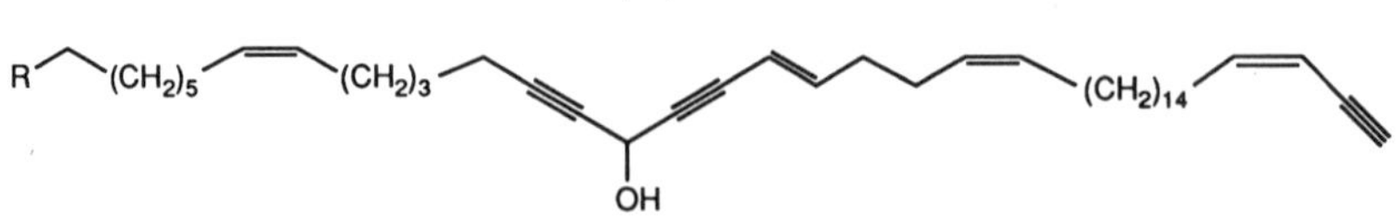

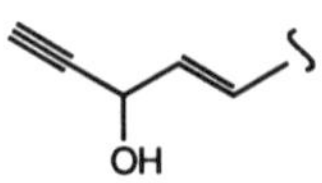

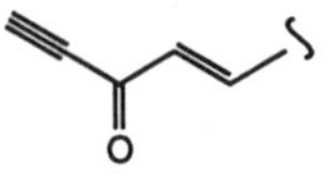

(15) R = A
(16) R = B

A

B

μg/ml) and also the sea urchin egg assay (LC$_{50}$ 1–50 μg/ml). These acetylenes were isolated from the sponge *Petrosia ficiformis* and the activity level is reported to be the highest ever recorded for compounds in the brine shrimp assay.

2.3. Aliphatic Ester Peroxides

A new sponge (order: Listhistida; family: Theonellidae), collected at a depth of 800 m using a submersible, yielded a group of homologous branched-chain fatty acids (**17–21**) which contain a peroxide group in the chain. Three of these

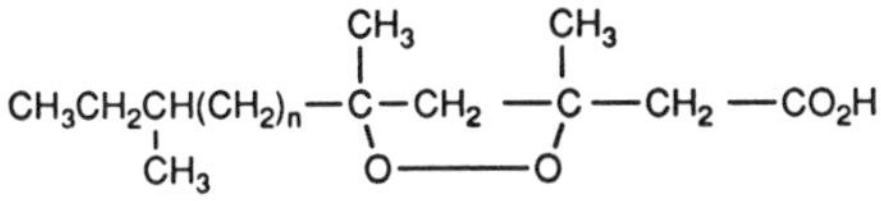

$$\text{CH}_3\text{CH}_2\text{CH(CH}_2)_n-\overset{\overset{\text{CH}_3}{|}}{\underset{|}{\text{C}}}-\text{CH}_2-\overset{\overset{\text{CH}_3}{|}}{\underset{\diagdown}{\text{C}}}-\text{CH}_2-\text{CO}_2\text{H}$$

(**17**) n = 9   (**19**) n = 11
(**18**) n = 10 (**20**) n = 12
            (**21**) n = 13

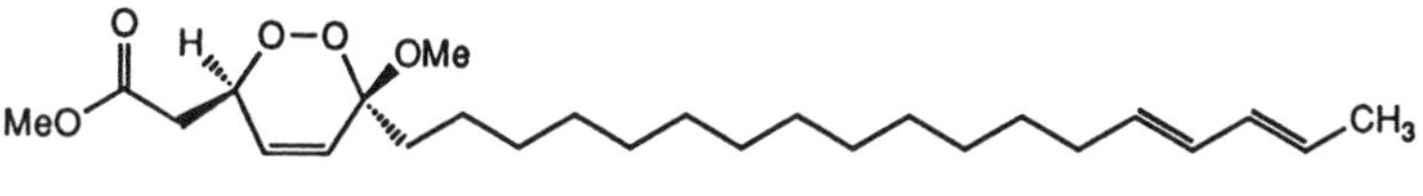

(**22**)

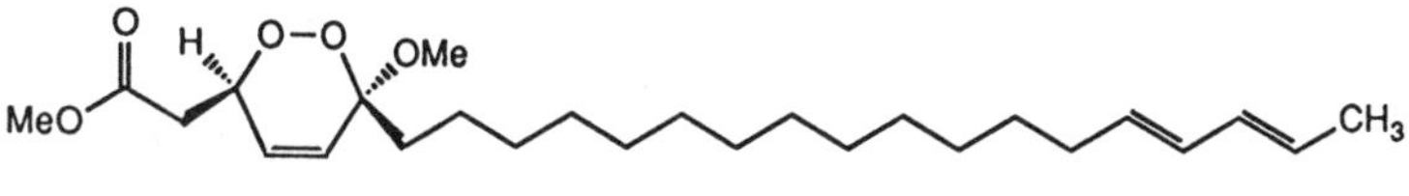

(**23**)

were cytotoxic to the P388 cell line (ED$_{50}$ 0.5 μg/ml) and also showed anti-microbial activity (Patil, 1989). Quinoa *et al.* (1986b) isolated xestins A and B (**22, 23**), from a *Xestospongia* sp. and reported that these esters were toxic against P388 cells (ID$_{50}$ 0.3 and 3 μg/ml, respectively). Ester **22** also showed *in vitro* activity against three different human tumor cell lines (A549, lung; HTC-8, colon; MDAMB, mammary). Similar esters of shorter chain length obtained from the sponge *Plakortis lita* (**24–27**) showed slightly greater toxicity to P388 cells (IC$_{50}$

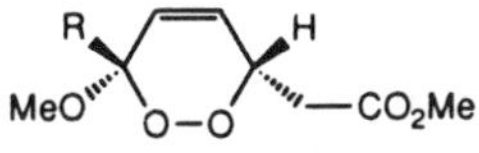

(**24**) R =   —(CH$_2$)$_{11}$—CH$_3$
(**25**) R =   —(CH$_2$)$_9$—CH$_3$
(**26**) R =   —(CH$_2$)$_7$—CH$_3$
(**27**) R =   —(CH$_2$)$_9$—CH$_3$ / CH$_3$

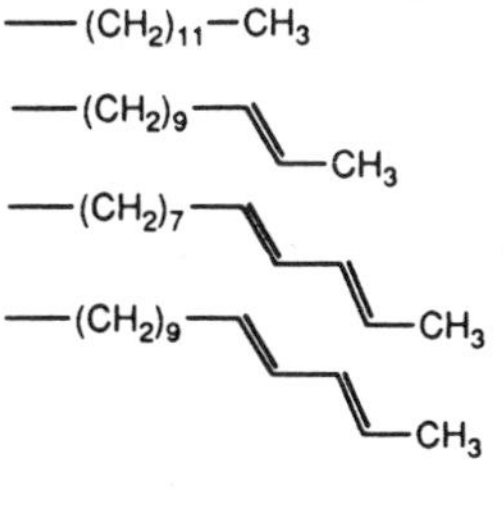

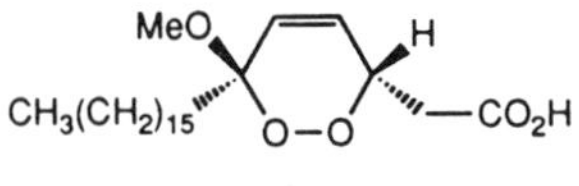

(28)

0.05–0.01 µg/ml), while the acid chondrillin (28) was slightly less active (IC_{50} 5 µg/ml) (Sakemi *et al.*, 1987). The much more highly branched cyclic peroxide acids 29 and 31 isolated from the sponge *Plakortis angulospiculatis*, collected in Venezuela, as well as the derived esters 30 and 32, also inhibited the growth of P388 cells (IC_{50} 0.2–0.9 µg/ml) (Gunasekera *et al.*, 1990a).

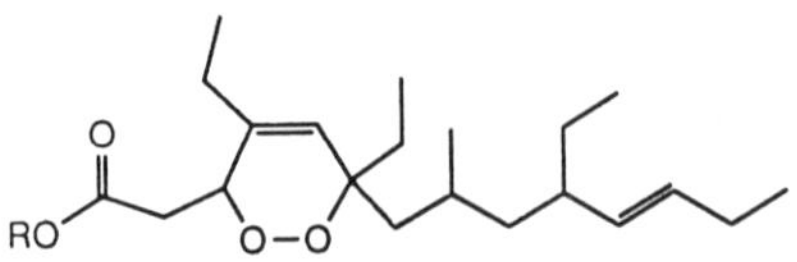

(29) R = H
(30) R = CH₃

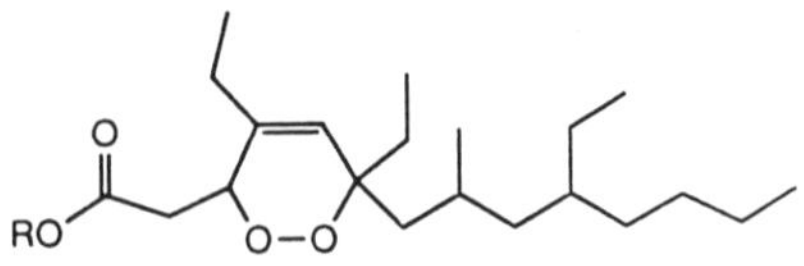

(31) R = H
(32) R = CH₃

2.4. Prostanoids

The initial discovery of prostaglandins in a gorgonian (Weinheimer and Spraggins, 1969) helped draw attention to marine organisms as a source of novel natural products. Prostanoids and other eicosanoids are discussed more extensively in the chapter by Gerwick and Bernart in this book. Some of the marine prostaglandins have shown interesting levels of cytotoxicity. Munro *et al.* (1987) reviewed the activity of the clavulones, punaglandins, and chlorovulone-I. More recently, two different syntheses of (7E)- and (7Z)-punaglandin-4 (33, 34) have

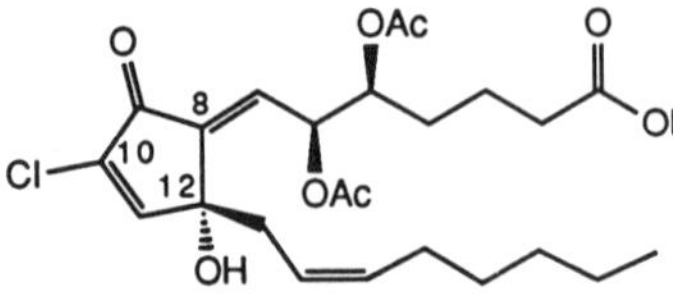

(33) 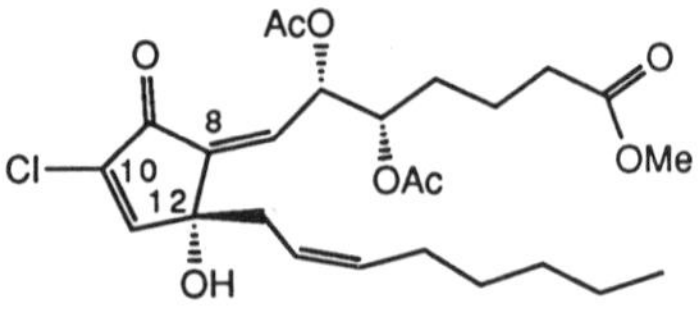   (34)

been reported which led to a minor structural revision of these isomers (stereochemistry at C-12) (M. Suzuki *et al.*, 1988; Mori and Takeuchi, 1988). This same revision applies to (7E)- and (7Z)-punaglandin-3. Interestingly, the stereoisomeric (7E)- and (7Z)-punaglandin-4 showed virtually identical toxicity to L1210 cells (IC_{50} 0.07 and 0.06 µg/ml, respectively).

A series of C-10 halogenated prostaglandins have been described. The first of these, chlorovulone-I (35), from the stolonifer *Clavularia viridis*, showed activity against human promyelocytic leukemia (HL-60) cells *in vitro* (ED_{50} 0.01 µg/ml)

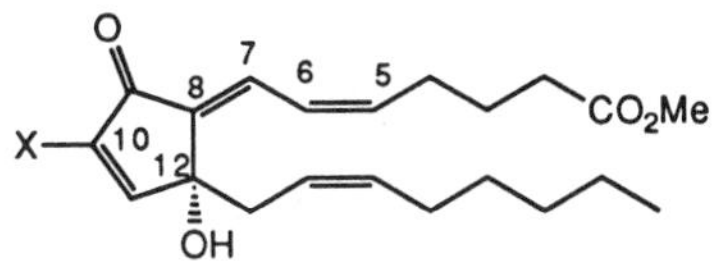

(**35**) X = Cl
(**36**) X = Br
(**37**) X = I

(Iguchi *et al.*, 1985; Nagaoka *et al.*, 1986). Slightly later the very novel bromo and iodo analogs were discovered, i.e., bromovulone-I (**36**) and iodovulone-I (**37**) (Iguchi *et al.*, 1986). These had virtually the same level of activity against HL-60 cells (IC$_{50}$ 0.025 and 0.03 µg/ml, respectively) as chlorovulone-I. 10,11-Epoxychlorovulone-I (**38**), isolated from the same source as the other chloro-vulones and the clavulones, namely *Clavularia viridis*, was also very active against HL-60 cells *in vitro* (IC$_{50}$ 0.04 µg/ml) (Iguchi *et al.*, 1987).

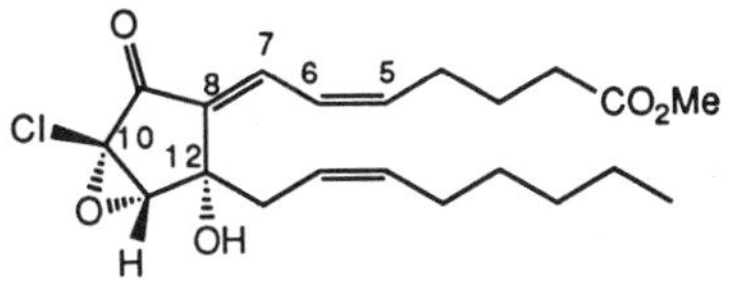

(**38**)

As a result of various studies (Honda *et al.*, 1985, 1987, 1988a,b) some structure–activity relationships among the prostanoids have been noted. The order of antiproliferative and cytotoxic activity found is chlorovulone-I > bromovulone-I = iodovulone-I > clavulones-I and II > PGA$_2$. The alkylidenecyclopentanone structure was required for activity, but the epoxy prostanoids, which lack the cross-conjugated cyclopentenone system, were also active. A hydroxyl group at C-12 enhanced activity, but the stereochemistry at C-12 was not important. Halogens at C-10 clearly potentiated activity (Cl > Br = I > H), as did the presence of the diene moiety at C-5,6 and C-7,8. Bivariate DNA/bromodeoxyuridine analysis using a flow cytometer showed that chlorovulone-I transiently arrests cell cycle progression from G1 to S after 24 hr exposure to nontoxic concentrations, and caused lasting blockage of leukemia cells at G1 at the cytotoxic concentration.

2.5. Complex Polyketides

Higa *et al.* (1987) isolated a novel array of substituted cyclohexenes from a new species of acorn worm, and three of these metabolites (**39–41**) were reported to inhibit the growth of P388 cells, with **41** being the most active (IC$_{50}$ 0.010 µg/ml). A total synthesis of clavularin A (**42**) has been reported (Still and Shi, 1987). This diketone was reported to have a lethal effect on cultured cells

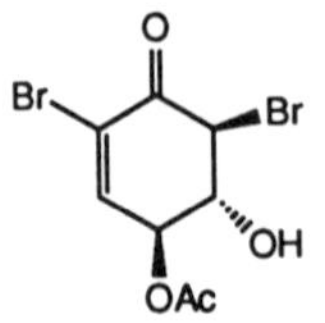

(39) 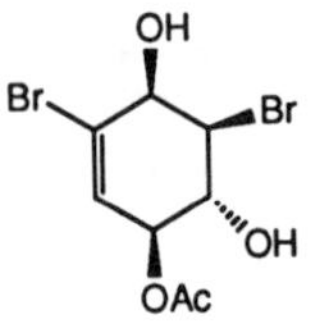   (40) 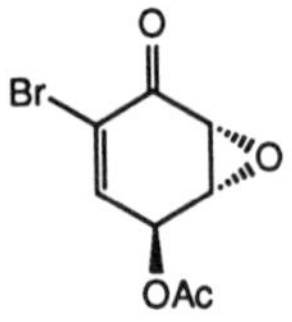   (41)

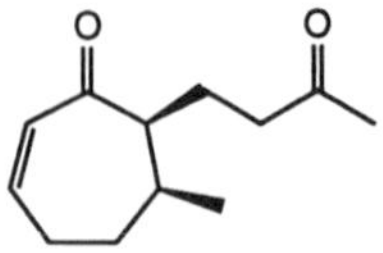

(42)

transformed with polynoma virus (PV$_1$) cells). Two cytotoxic styrylchromones have been reported from marine algae. Hormothamnione (**43**) was obtained from *Hormothamnion enteromorphoides* Grunow and was found to be active against both P388 cells (IC$_{50}$ 0.0046 μg/ml) and HL-60 cells (0.0001 μg/ml) (Gerwick *et al.*, 1986). Hormothamnione (**43**) appears to exert its cytotoxic action by

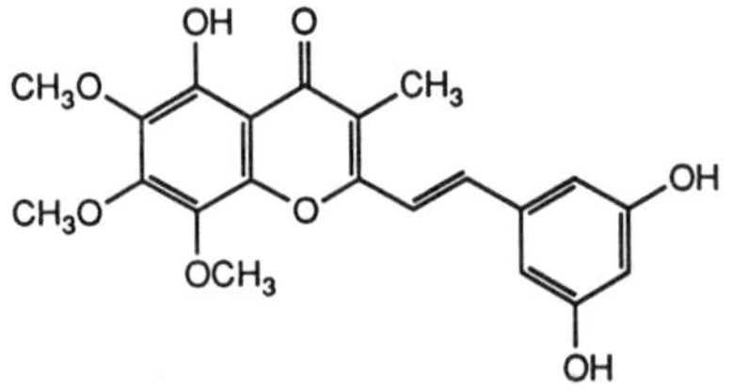

(43)

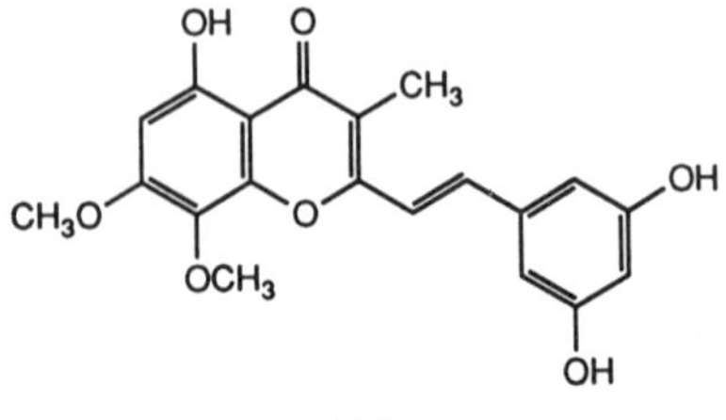

(44)

inhibiting RNA synthesis. 6-Desmethoxyhormothamnione (**44**), which displayed an IC$_{50}$ against KB cells of 1 μg/ml, was isolated from *Chrysophaeum taylori* (Gerwick, 1989). Two syntheses of hormothamnione have also been reported (Ayyanger *et al.*, 1988; Alonso and Brossi, 1988).

A marine hydroid, *Abietinaria* sp., was the source of abietinarins A and B (**45, 46**), both of which were toxic to L1210 leukemia cells (ED$_{50}$ < 10 μg/ml) (Pathirana *et al.*, 1990). Schmitz and Bloor (1988) reported that the known

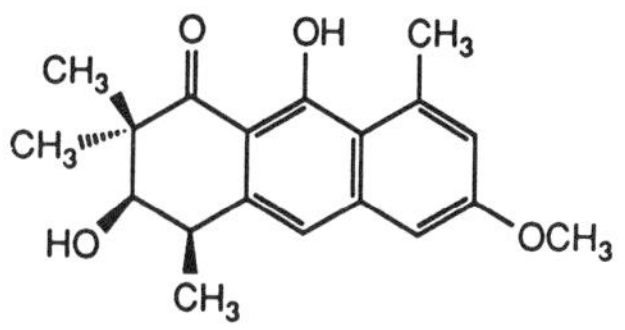

(45) 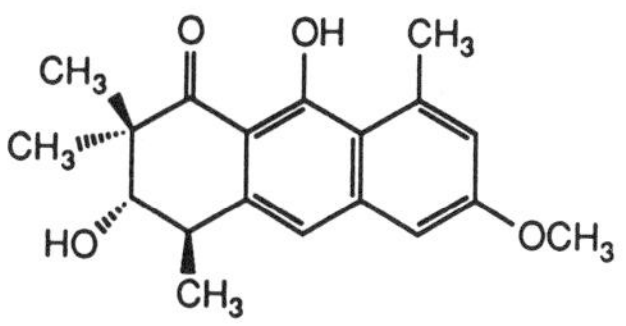   (46)

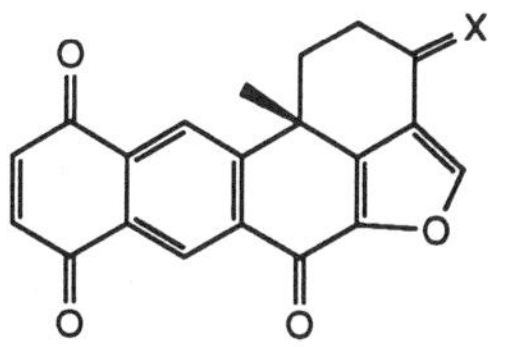

(47) X = H,H   Xestoquinone
(48) X = O   Halenaquinone

xestoquinone (47) but not the related ketone halenaquinone (48) was mildly cytotoxic (ED_{50} 2.9 μg/ml vs. P388). These authors reported the occurrence of three new related metabolites, one of which, adociaquinone B (49) has the same level of activity as 47. Harada *et al.* (1988) carried out a total synthesis of (+)-halenaquinone and its corresponding hydroquinol form and proved the absolute configuration of these compounds to be that shown by formula 48.

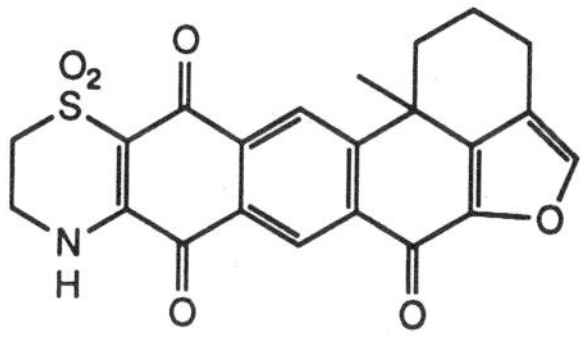

(49) Adociaquinone B

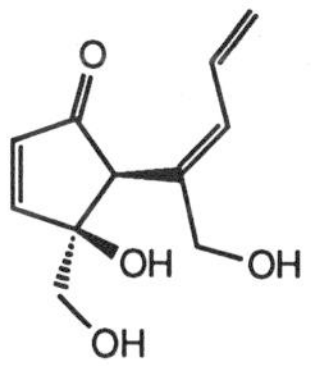

(50) 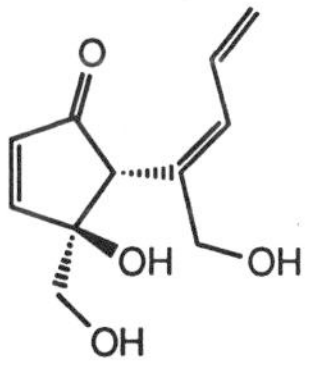   (51)

The stereoisomeric didemnenones C and D (50, 51) were found to be equally toxic to L1210 leukemia cells, with an IC_{50} 4.5 μg/ml. These novel metabolites were both obtained by Lindquist *et al.* (1988) from tunicates collected from opposite sides of the world, *Didemnum voeltzkowski* from Fiji and *Trididemnum* cf. *cyanophorum* from the Bahamas. Hemibrevetoxin B (52), was isolated by Prasad and Shimizu (1989) from a red tide organism, *Gymnodinium breve* found in

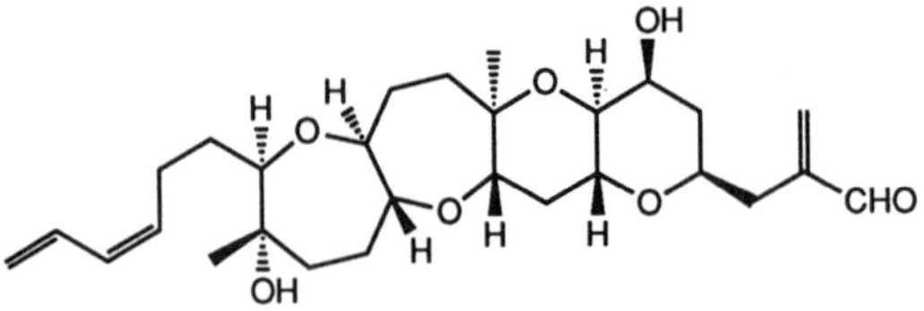

(52)

the Gulf of Mexico. This polyether comprises one-half of the carbon skeleton of the well-known red tide toxin brevetoxin B (Lin *et al.*, 1981) and is reported to be toxic at 5 μg/ml to mouse neuroblastoma cells. *In vitro* cytotoxicity to B16-F10 murine melanoma cells (IC_{50} 0.0072 μg/ml) and HCT-16 human colon tumor cells (IC_{50} 0.5 μg/ml) has been reported for the eight-membered ring lactone octalactin A (53) (Tapiolas *et al.*, 1991). Interestingly, octalactin B, which differs from

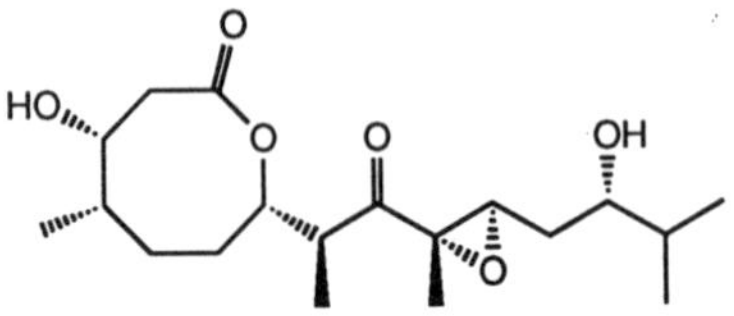

(53)

octalactin A only by having a double bond in place of the epoxide moiety, was inactive in these tests. These novel lactones were obtained from a marine bacterium, *Streptomyces* sp., which was isolated from the surface of a gorgonian, *Pacifigorgia* sp., collected in the Sea of Cortez.

The strongly cytotoxic polyether carboxylic acids acanthifolicin (54) and okadaic acid (55) were reported in contiguous communications in 1981 (Schmitz

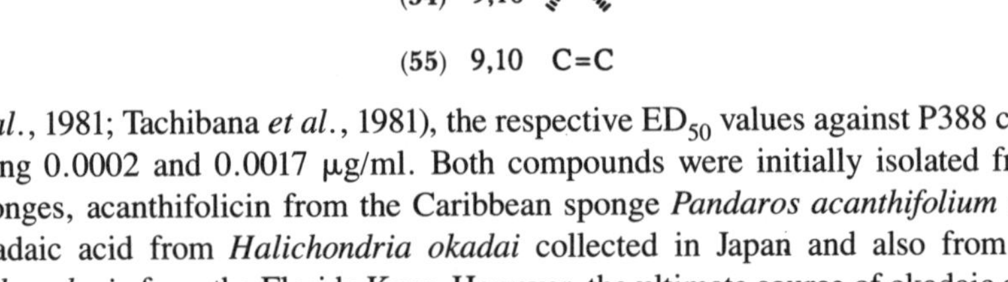

(54) 9,10

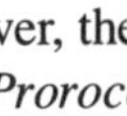

(55) 9,10 C=C

et al., 1981; Tachibana *et al.*, 1981), the respective ED_{50} values against P388 cells being 0.0002 and 0.0017 μg/ml. Both compounds were initially isolated from sponges, acanthifolicin from the Caribbean sponge *Pandaros acanthifolium* and okadaic acid from *Halichondria okadai* collected in Japan and also from *H. melanodocia* from the Florida Keys. However, the ultimate source of okadaic acid has been proven to be a dinoflagellate, *Prorocentrum lima* (Murakami *et al.*,

1982), and presumably acanthifolicin is also produced by a dinoflagellate. Okadaic acid was subsequently found to be a tumor promoter (Suganuma *et al.*, 1988) and also has been shown to be a potent and selective inhibitor of protein phosphatases type I and IIa (Bialojan and Takai, 1988). This latter property has led to extensive use of okadaic acid as a pharmacologic and biochemical tool and numerous papers reporting its use have been published (Cohen *et al.*, 1990). A total synthesis of okadaic acid has also been reported (Isobe *et al.*, 1986). As might be expected, acanthifolicin has been shown to inhibit protein phosphatase I and IIa to about the same extent as okadaic acid (Holmes *et al.*, 1990; Nishiwaki *et al.*, 1990). Although the report of cytotoxic and antitumor activity of these compounds predates this review, the intense interest in okadaic acid as a biochemical and pharmacologic tool prompted a brief review here.

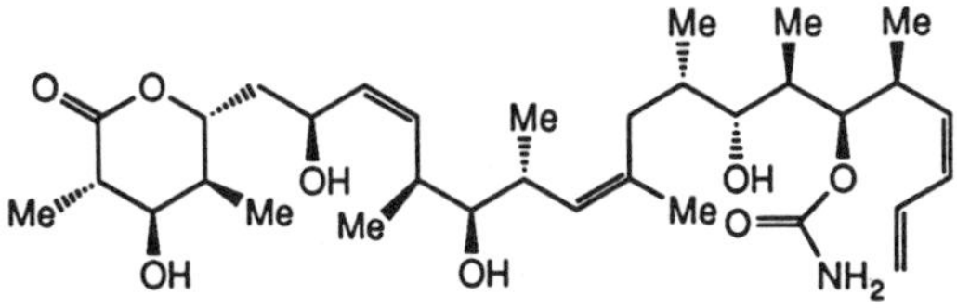

(56)

Discodermolide (**56**), obtained from the Caribbean sponge (*Discodermia dissoluta* (Gunasekera *et al.*, 1990b), shows significant cytotoxicity (IC$_{50}$ 0.5 μg/ml vs. P388) and also immunosuppressive activity. The latter property has made this an interesting molecule for further studies (Schreiber, 1991).

2.6. Macrolides

2.6.1. Assorted Macrolides

The macrolides are a class of metabolites to which many novel members have been added in the past few years. Latrunculin (**57**) was reported by Kakou *et al.*

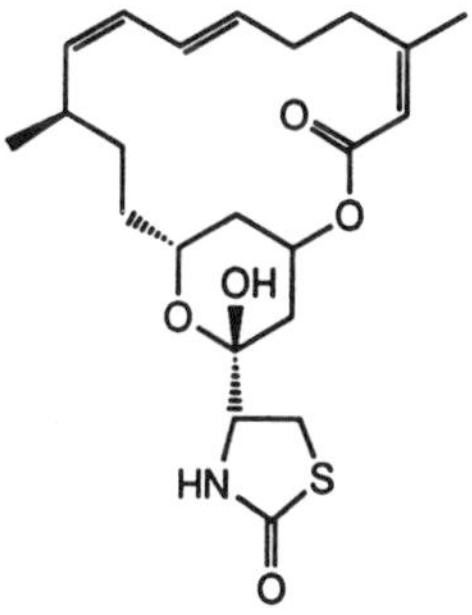

(57)

(1987) to be toxic to HEP-2 and MA-104 cells at 0.072 and 0.23 μg/ml, respectively. Latrunculin was first reported in 1980, at which time only its ichthyotoxicity and ability to induce cell morphology were described (Kashman *et al.*, 1980). Five macrolides have been reported from a cultured dinoflagellate, *Amphidinium* sp., obtained from Okinawa. Amphidinolide A (**58**) was reported as toxic to two types of leukemia cells, L1210 (IC$_{50}$ 2.4 μg/ml) and L5178Y (IC$_{50}$ 3.9

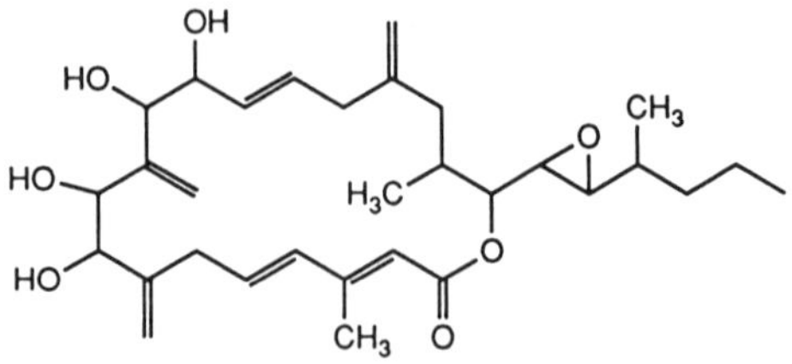

(58)

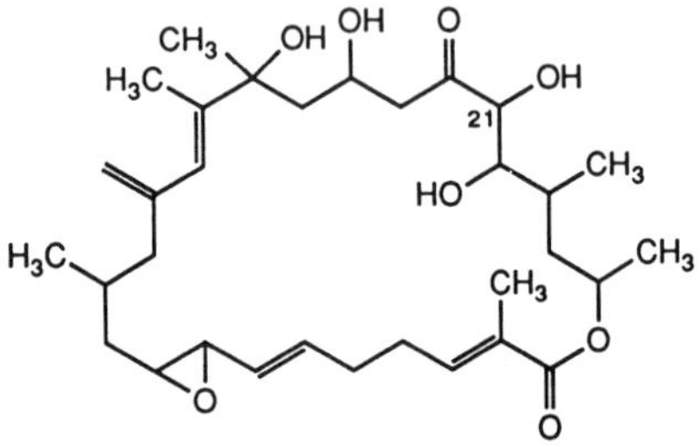

(**59**) C-21 stereoisomer of (**60**)
(**60**) C-21 stereoisomer of (**59**)

μg/ml) (J. Kobayashi *et al.*, 1986). Amphidinolides B and D (**59, 60**), which only differ in stereochemistry (stereoisomeric at C21), appear to have profoundly different activities, with amphidinolide B (**59**) (IC$_{50}$ 0.00014 μg/ml vs. L1210) being more than 100 times as active as amphidinolide D (IC$_{50}$ 0.019 μg/ml) (Ishibashi *et al.*, 1987b; J. Kobayashi *et al.*, 1989a). Amphidinolide C (**61**) is also

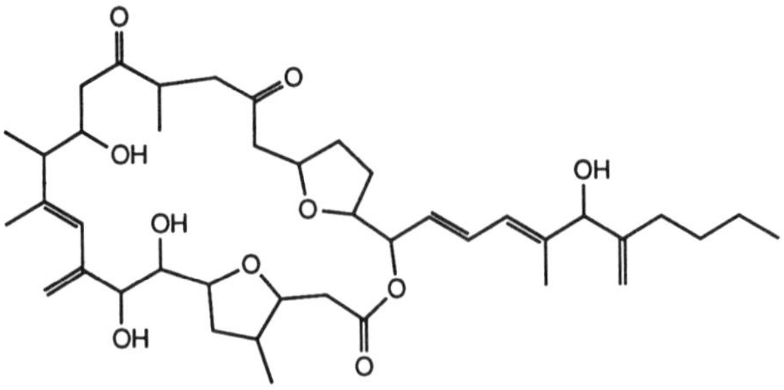

(61)

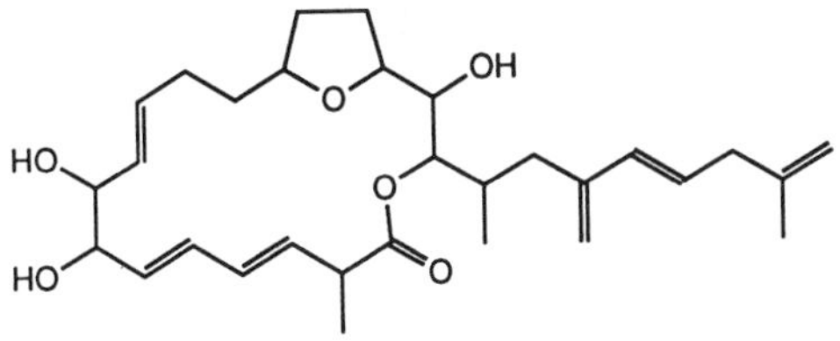

(62)

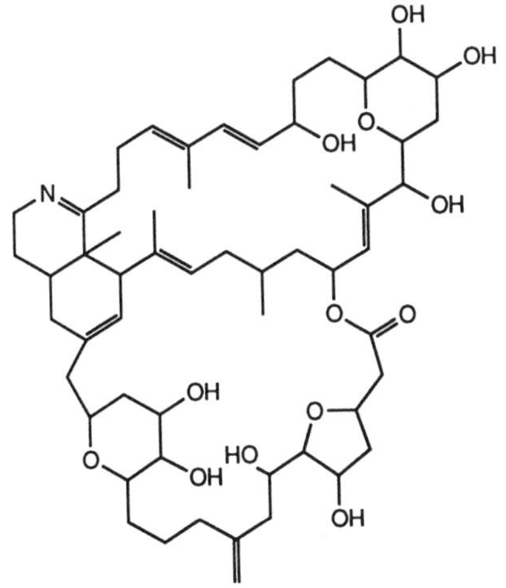

(63)

quite cytotoxic (IC_{50} 0.0058 µg/ml vs. L1210) (J. Kobayashi *et al.*, 1988e), while amphidinolide E (62) is less so (IC_{50} 2 µg/ml) (J. Kobayashi *et al.*, 1990a). Prorocentrolide (63) (IC_{50} 20 µg/ml vs. L1210) was obtained from the same dinoflagellate species from which okadaic acid was initially isolated (Torigoe *et al.*, 1988).

Iejimalides A and B (64, 65) are antileukemic lactones extracted from the Okinawan tunicate *Eudistoma* cf. *rigida* (IC_{50}, respectively, 0.062 and 0.032 µg/ml vs. L1210 and 0.022 and 0.001 µg/ml vs. L5178Y leukemia cells) (J.

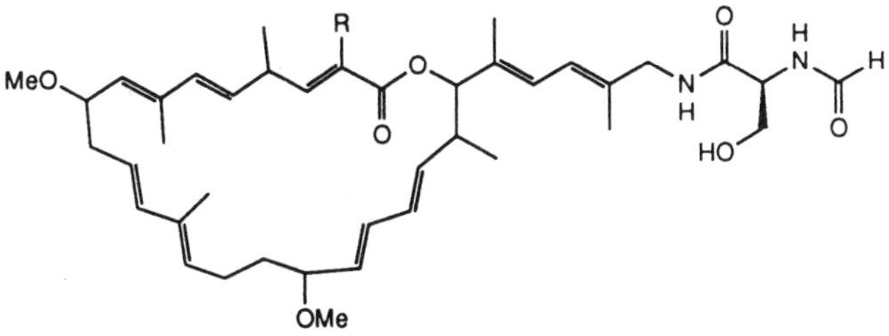

(64) Iejimalide A: R = H
(65) Iejimalide B: R = CH₃

Kobayashi *et al.*, 1988b). In their continuing work with marine microorganisms, Fenical's group has obtained a group of unsaturated macrolides from an unidentified deep-sea marine bacterium (Gustafson *et al.*, 1989). One of these, macrolactin A (**66**), was reported to be cytotoxic (IC_{50} 3.5 μg/ml vs. B16-F10 murine melanoma cells) and also showed antiviral activity.

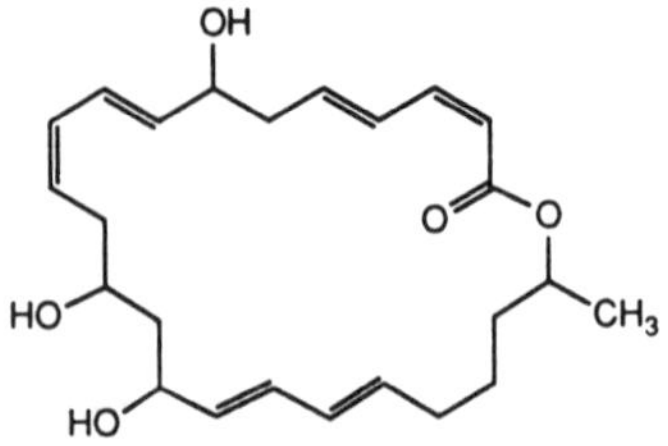

(**66**)

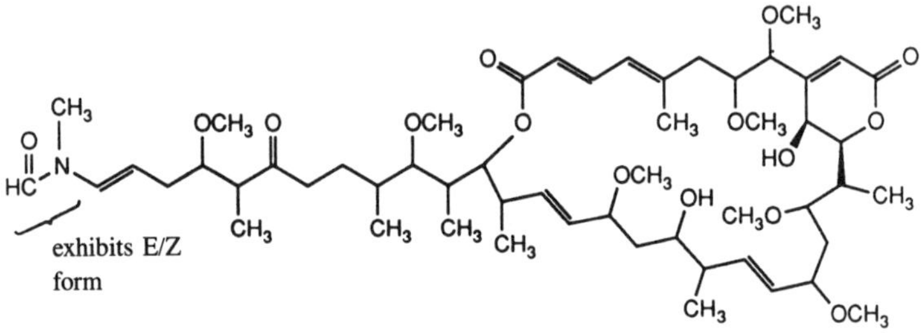

(**67**)

Sphinxolide (**67**) is a 26-membered macrolide that has been reported to be highly active against KB cells (IC_{50} 35 $\times$ 10^{-6} μg/ml). It was isolated from an unidentified Pacific nudibranch (Guella *et al.*, 1989). In back-to-back papers, two different groups each reported pairs of macrolides that appear to differ only in stereochemistry at a few centers (see structures). Quinoa *et al.* (1988) reported fijianolides A and B (**68, 69**) from the Vanuatuan sponge *Spongia mycofijiensis** and reported IC_{50} for fijianolide A of 9 μg/ml against P388 and 11 μg/ml vs. HT-29 human colon tumor cells. The diacetate of fijianolide B was tested against the same cell lines: IC_{50} 6 μg/ml vs. P388 and 0.5 μg/ml vs. HT-29. Fijianolide B (**69a**) was also isolated from the nudibranch *Chromodoris lochi*, which is often found attached to *S. mycofijiensis*. Corley *et al.* (1988a) discovered **68b** (isolaulimalide) and **69b** (laulimalide) in the extracts of a sponge (*Hyatella* sp.) and its nudibranch

---

*Now assigned the new name *Leiosella lavis* as a result of a reclassification (P. Crews, personal communication).

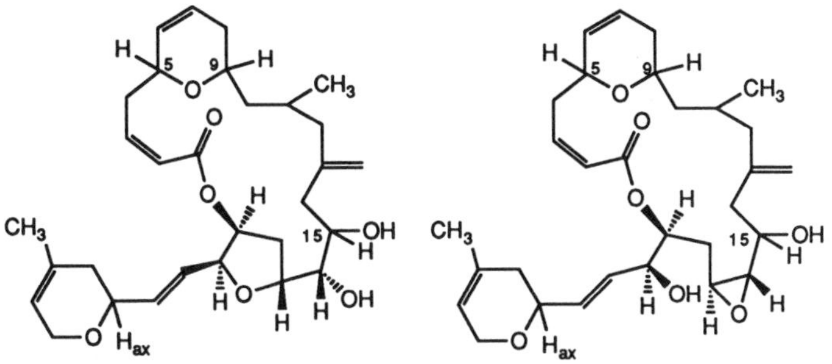

(68a) Fijianolide A
(68b) Isolaulimalide

(69a) Fijianolide B
(69b) Laulimalide

(68a) and (69a): H-5 (ax), H-9(eq), C-15(S*)
(68b) and (69b): H-5(eq), H-9(ax), C-15(unspecified)

predator (*Chromodoris lochi*). These lactones were tested against KB cells: IC_{50} laulimalide, 0.015 µg/ml; isolaulimalide, > 0.2 µg/ml.

The gelatinous egg masses of nudibranchs have been observed to be generally free of predation and this led Roesener and Scheuer (1986) to investigate the egg masses of *Hexabranchus sanguineus* from Hawaii. Two potent antibacterial macrolides, ulapualides A (**70**) and B (**71**) were isolated, which were also potent inhibitors of L1210 cell growth (IC_{50} 0.01–0.03 µg/ml). Simultaneously, Fusetani's group was isolating a series of closely related lactones from *Hexabranchus* egg masses collected in the Ryukyus. The first of these to be reported was kabiramide C (**74**) (Matsunaga *et al.*, 1986). Subsequently, Fusetani's group

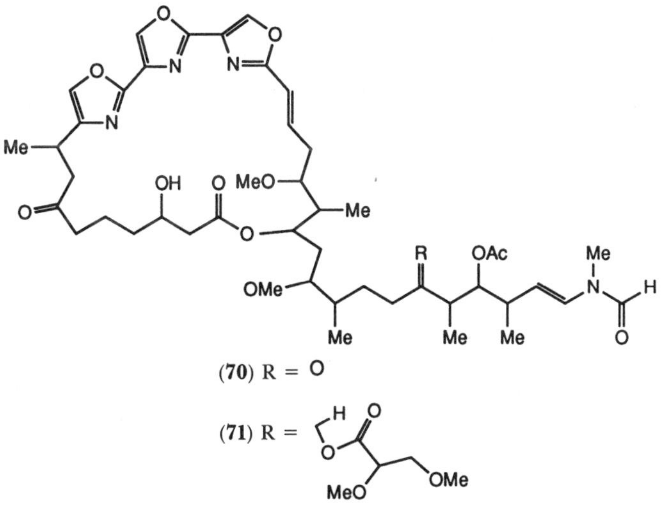

(**70**) R = O

(**71**) R =

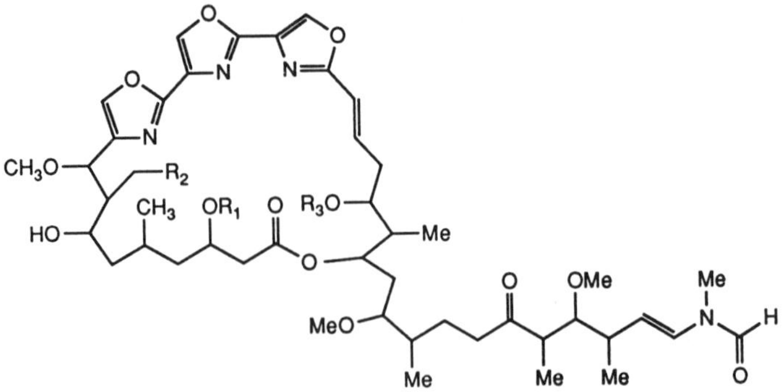

(72) Kabiramide A: R_1 = $CONH_2$, R_2 = OH, R_3 = CH_3
(73) Kabiramide B: R_1 = $CONH_2$, R_2 = R_3 = H
(74) Kabiramide C: R_1 = $CONH_2$, R_2 = H, R_3 = CH_3
(75) Kabiramide D: R_1 = R_2 = H, R_3 = CH_3
(76) Kabiramide E: R_1 = $COCH_3$, R_2 = H, R_3 = CH_3

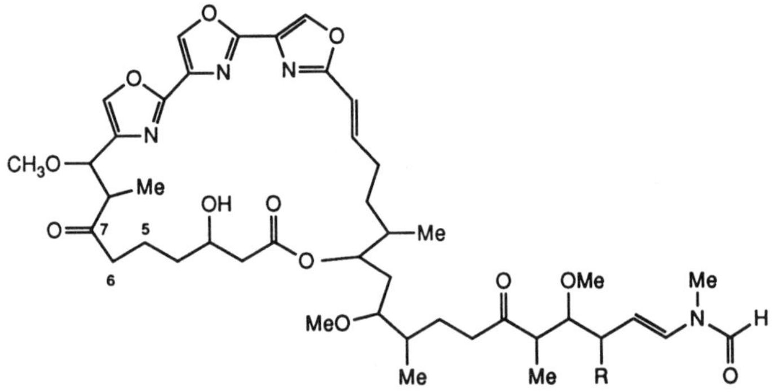

(77) R = H
(78) R = CH_3
(79) R = H, $\Delta^{5,6}$

Cytotoxicity: Sea urchin egg assay, L1210 cells

		IC_{99}	IC_{50}
72	Kabiramide **A**	1 μg/ml	0.03 μg/ml
73	Kabiramide **B**	0.2 μg/ml	0.03 μg/ml
74	Kabiramide **C**	0.2 μg/ml	0.01 μg/ml
75	Kabiramide **D**	0.2 μg/ml	0.02 μg/ml
76	Kabiramide **E**	0.2 μg/ml	0.02 μg/ml
77	Dihydrohalichondramide	0.5 μg/ml	0.03 μg/ml
78	33-Methyldihydrohalichondramide	0.5 μg/ml	0.05 μg/ml

described kabiramides A (**72**), B (**73**), D (**75**), and E (**76**) plus dihydrohalichondra-mide (**77**) and 33-methyldihydrohalichondramide (**78**) from *Hexabranchus* egg masses (Matsunaga *et al.*, 1989b, and references therein). Interestingly, all of these compounds have virtually the same level of activity in the sea urchin egg assay and against L1210 cells *in vitro* (see data under the structures). Ulapualide-B (**71**) was isolated along with **72–78**. Kabiramides B (**73**) and C (**74**) were subsequently isolated from a Palauan sponge *Halichondria* sp. and halichon-dramide (**79**) was obtained from a *Halichondria* sp. collected at Kwajelein (Kernan and Faulkner, 1987).

Three other tris-isoxazole-containing macrolides are mycalolides A–C (**80–82**), which were found in a sponge of the genus *Mycale*. Although **80–82** are highly cytotoxic (IC$_{50}$ 0.0005–0.001 μg/ml vs. B16), they have not shown promising results *in vivo*, due to their high toxicity (Fusetani *et al.*, 1989b).

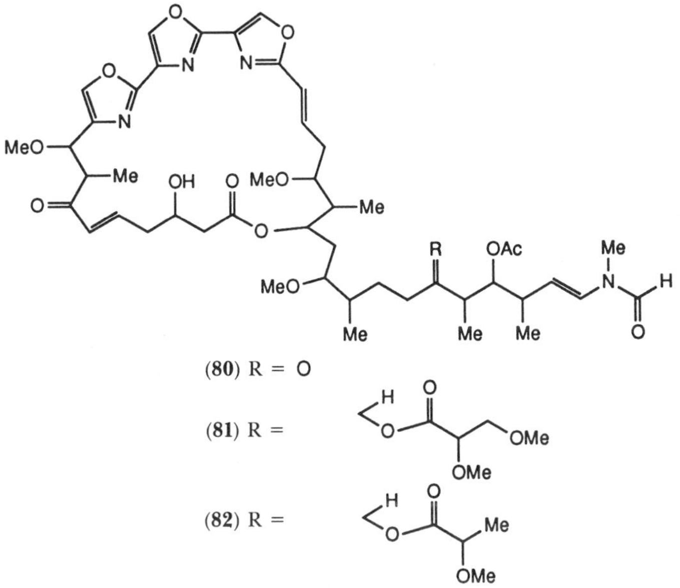

(**80**) R = O

(**81**) R =

(**82**) R =

The tunicate *Lissoclinum patella*, collected from various localities, has been the source of an extensive list of cyclic peptides (see Section 4.13). In addition to these, three novel, highly cytotoxic macrolides have been reported from this organism by two different groups in back-to-back papers. Patellazoles A–C (**83–85**) were isolated from a Fijian collection of *L. patella* by Zabriskie *et al.* (1988), while patellazole B (**84**) was also obtained from *L. patella* collected in Guam (Corley *et al.*, 1988b). The former group reported that **83–85** are very active in the National Cancer Institute's human cell line protocol (mean IC$_{50}$ of 10^{-3}–10^{-6}

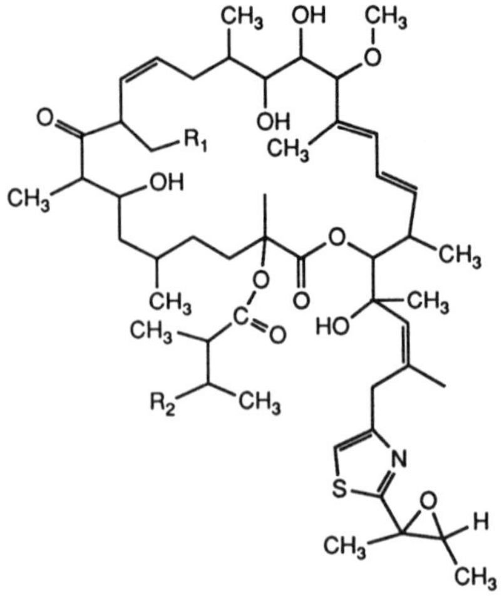

(83) $R_1 = R_2 = H$
(84) $R_1 = H, R_2 = OH$
(85) $R_1 = R_2 = OH$

μg/ml), while the latter group reported that **84** was active against KB cells, with an IC_{50} of 0.0003 μg/ml.

In addition to yielding okadaic acid (**55**), the Japanese sponge *Halichondria okadai* has been the source of a group of very complex and biologically active macrolides, halichondrins B (**86**) and C (**87**), norhalichondrins A (**88**), B (**89**), and C (**90**), and homohalichondrins A (**91**), B (**92**), and C (**93**) (Hirata and Uemura, 1986). Data on *in vitro* activity against B16 melanoma are given under the structures. Halichondrin B (**86**) showed good *in vivo* activity against B16 melanoma in mice [T/C values of 203–244%, depending on dose (5–20 μg/kg) and regimen], against P388 leukemia in mice (T/C 323% @ 10 μg/kg), and

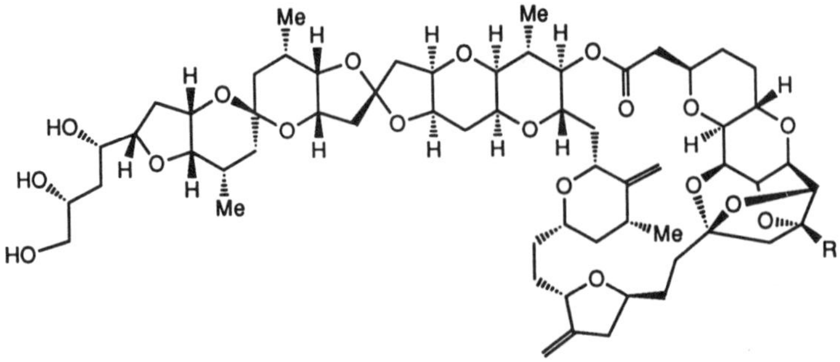

(86) Halichondrin B: R = H
(87) Halichondrin C: R = OH

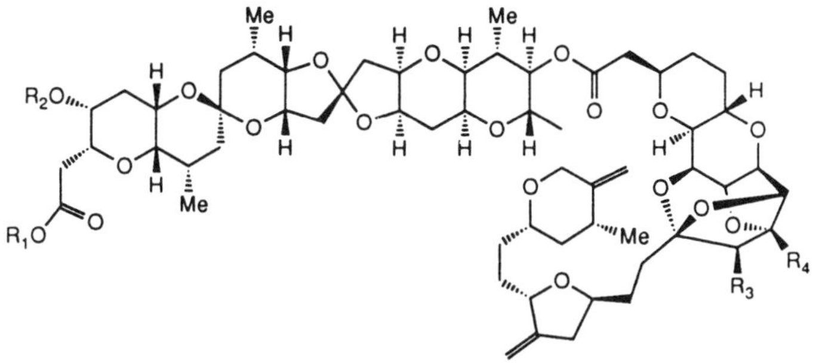

(88) Norhalichondrin A: $R_1 = R_2 = H$, $R_3 = R_4 = OH$
(89) Norhalichondrin B: $R_1 = R_2 = R_3 = R_4 = H$
(90) Norhalichondrin C: $R_1 = R_2 = R_3 = H$, $R_4 = OH$

against L1210 in mice (T/C 207–375% with doses of 50–100 µg/kg under various injection schedules). Homohalichondrin B (92) is comparable in activity *in vivo* to halichondrin B (86), which is about 50 times as active as norhalichondrin A (88). No cytotoxicity data were given for 89, 90, and 93; they were described as having activities "inferior to halichondrin B" (86). The investigators concluded that it is important for antitumor activity that the tricyclic ring be relatively lipophilic and that the terminal group have two or three hydroxyls, but not a carboxylate. They speculate that the tricyclic ring of these molecules may get into the lipid bilayer of the biomembrane.

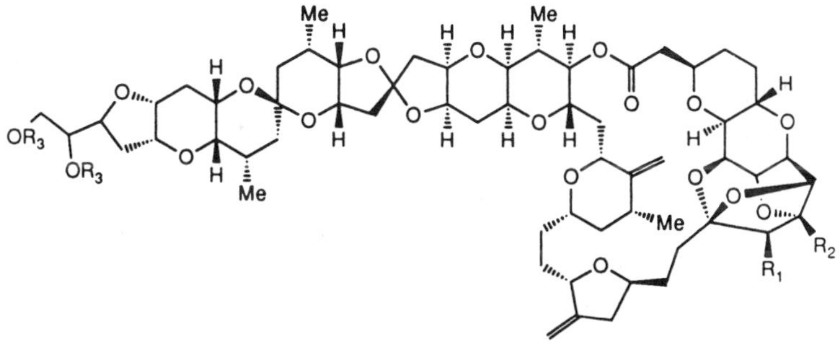

(91) Homohalichondrin A: $R_1 = R_2 = OH$, $R_3 = H$
(92) Homohalichondrin B: $R_1 = R_2 = R_3 = H$
(93) Homohalichondrin C: $R_1 = R_3 = H$, $R_2 = OH$

88	Norhalichondrin A	0.0052 µg/ml
86	Halichondrin B	0.000093 µg/ml
91	Homohalichondrin A	0.00026 µg/ml
87	Halichondrin C	0.00035 µg/ml
92	Homohalichondrin B	0.0001 µg/ml

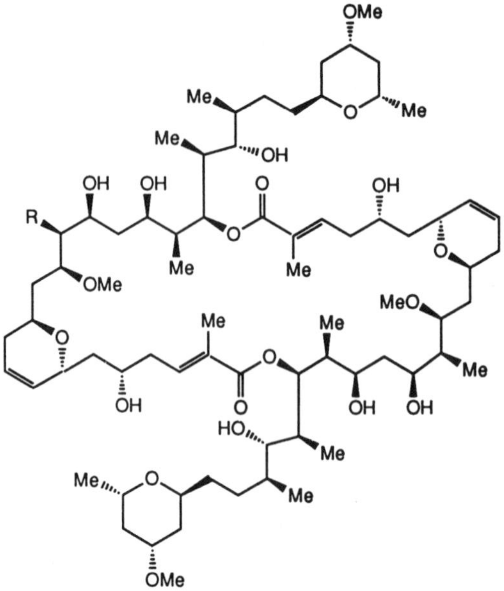

(94)   Misakinolide A: R = CH₃
        (Bistheonelllide A)
(95)   Bistheonellide B: R = H

The dimeric macrolide misakinolide A (94), obtained from a *Theonella* sp. of sponge, was initially reported by Sakai *et al.* (1986a) as a monomeric structure. It was very cytotoxic against several cell lines: IC_{50} for P388 was 0.01 μg/ml and for each of the human tumor cell lines HCT-8, A549, and MDA-MB-231 was 0.0005–0.005 μg/ml. Slightly later it was shown that misakinolide A was identical to bistheonellide A (94), which was isolated by another group from a *Theonella* sp. of sponge, and that the correct structure is the dimeric one shown in this formula (Kato *et al.*, 1987a,b). The desmethyl analog of 94, bistheonellide B (95), was isolated along with 94. Misakinolide A (94) and bistheonellide B (95) inhibit development of starfish embryos at 0.1–0.2 μg/ml (Kato *et al.*, 1987a). From the Okinawan sponge *Theonella swinhoei*, Kitagawa's group found additional members of this dimeric lactone group. Swinholide A (96) was reported first (M. Kobayashi *et al.*, 1989), followed by swinholides B (97) and C (98) and isoswinholide A (99) (M. Kobayashi *et al.*, 1990a). Subsequently, the absolute stereostructure of swinholide A was determined by a combination of chemical behavior and x-ray crystallographic analysis (Kitagawa *et al.*, 1990; M. Kobayashi *et al.*, 1990b). The absolute stereochemistry of misakinolide A (= bistheonellide A) (94) has been determined by chemical correlation with swinholide A (Tanaka *et al.*, 1990). Swinholides A–C (96–98) show approximately the same levels of activity against KB cells ($IC_{50} \sim 0.05$ μg/ml), while isoswinholide A (99) is less active

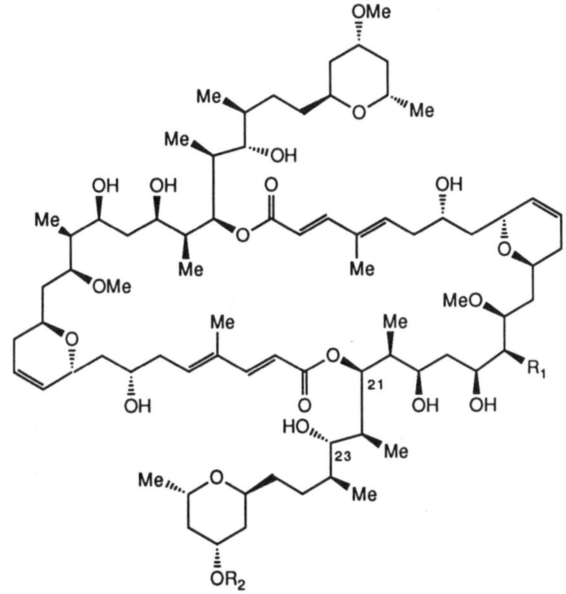

(96) Swinholide A: $R_1 = R_2 = CH_3$
(97) Swinholide B: $R_1 = H, R_2 = CH_3$
(98) Swinholide C: $R_1 = CH_3, R_2 = H$

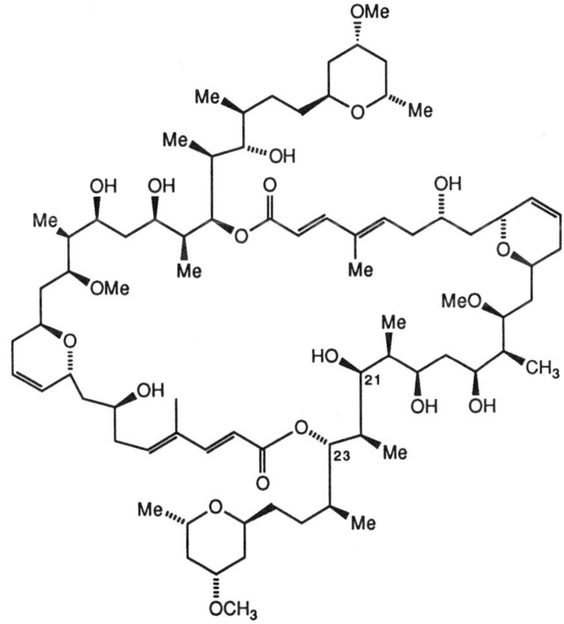

(99) Isoswinholide A

(IC$_{50}$ 1.1 μg/ml). These results suggest that the lactone ring size and conformation are important factors for cytotoxicity.

2.6.2. Bryostatins

Bryostatins 1–15 (**100–114**), isolated primarily from the bryozoan *Bugula neritina*, are a group of macrolides remarkable both for their novel structures and their biological activity. The first 13 of these were reviewed earlier by Munro *et al.* (1987), but all of the structures are reproduced here along with cytotoxicity data (see Table I) to give a comprehensive overview. A recent in-depth review of all the

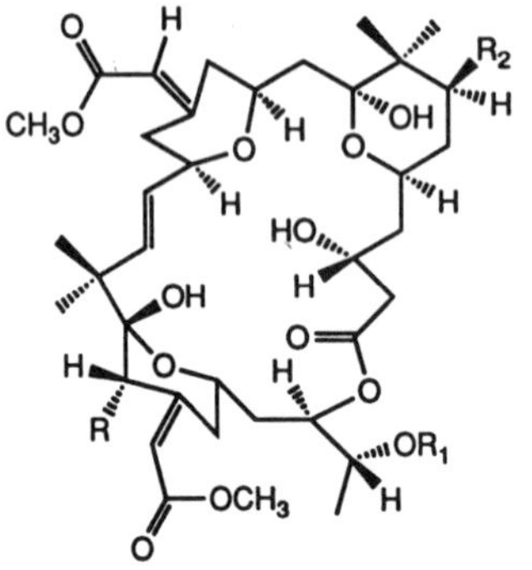

		R	R$_1$	R$_2$
(**100**)	Bryostatin 1	B	H	A
(**101**)	Bryostatin 2	B	H	OH
(**103**)	Bryostatin 4	D	H	C
(**104**)	Bryostatin 5	A	H	C
(**105**)	Bryostatin 6	A	H	D
(**106**)	Bryostatin 7	A	H	A
(**107**)	Bryostatin 8	D	H	D
(**108**)	Bryostatin 9	D	H	A
(**109**)	Bryostatin 10	H	H	C
(**110**)	Bryostatin 11	H	H	A
(**111**)	Bryostatin 12	B	H	D
(**112**)	Bryostatin 13	H	H	D
(**113**)	Bryostatin 14	OH	H	C
(**114**)	Bryostatin 15	E	H	A

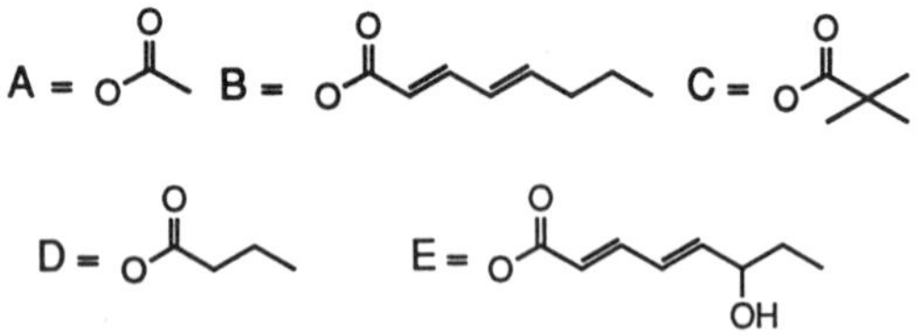

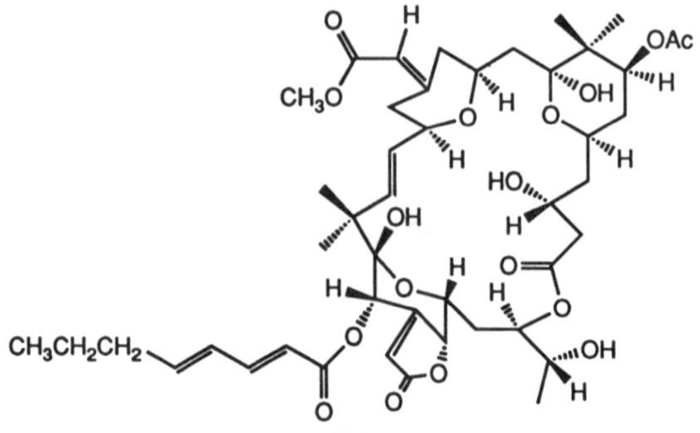

(**102**)

Table I. Origin and P388 Activity of the Bryostatins

Compound	Bryostatin	Organism[a,b]	P388 Lymphocytic leukemia	
			In vivo	*In vitro*
			Percent life extension (dose, μg/kg)	ED_{50}, (μg/ml)
100	1	1A	52–96 (10–70)	8.9×10^{-1}
101	2	1A	60 (30)	
102	3	1A	63 (30)	
103	4	1BCD, 2C	62 (42)	10^{-3}–10^{-4}
104	5	1BCD, 2C	88 (185)	1.3×10^{-3}–2.6×10^{-4}
105	6	1BCD, 2C	82 (185)	3.0×10^{-3}
106	7	1BCD	77 (92)	2.6×10^{-5}
107	8	1AC, 2C	74 (110)	1.3×10^{-3}
108	9	1BC	40 (80)	1.2×10^{-3}
109	10	1BC	—	7.6×10^{-4}
110	11	1BC	64 (92.5)	1.8×10^{-5}
111	12	1A	47–68 (30–50)	1.4×10^{-2}
112	13	1A	—	5.5×10^{-3}
113	14	1C	—	3.3×10^{-1}
114	15	1A	—	1.4

[a]1, *Bugula neritina*; 2, *Amathia convoluta*.
[b]A, California; B, Gulf of California (Mexico); C, Gulf of Mexico (Florida); D, Gulf of Sagami (Japan).

bryostatins has been compiled by Pettit (1991). The latest two members of this group, bryostatins 14 (**113**) and 15 (**114**), are described in a paper which summarizes the structure of most of the bryostatins and gives leading references to the compounds and some of their key activities (Pettit *et al.*, 1991a, and references therein). The absolute configurations of the bryostatins has recently been established (Pettit *et al.*, 1991b) and is that shown by structures **100–114**. Bryostatins 4 (**103**) and 5 (**104**) have also been isolated from the sponge *Lissodendoryx isodictyalis* along with bryostatins A and B, for which structures have not yet been reported (Pettit *et al.*, 1986a). Extracts of the ascidian *Aplydium californicum* have also been reported to yield bryostatins 4 (**103**) and 5 (**104**) (Pettit *et al.*, 1986b). Bryostatin 8 (**107**) was obtained from a collection of specimens of the marine bryozoan *Amathia convoluta* with which some *Bugula neritina* was associated (Pettit *et al.*, 1985). A symbiotic association between *B. neritina* and the sponge *L. isodictyalis* and *A. californicum* is believed to exist.

Bryostatins have been found to stimulate human hematopoietic cells (May *et al.*, 1987). This suggests the potential of using bryostatins for treating neoplastic bone marrow failure states. Bryostatin 1 also triggered activation and differentiations of peripheral blood cells from β-chronic lymphocytic leukemia patients (Drexler *et al.*, 1989). Bryostatin 1 activates protein kinase C (PKC), as do phorbol esters; however, it induces only a subset of the biological responses induced by the phorbol esters (Drexler *et al.*, 1990). Bryostatin 1 competes with phorbol esters for binding to PKC, but in contrast to PKC, it inhibits tumor promotion in SENCAR mouse skin (Hennings *et al.*, 1987) and blocks the effects of phorbol esters on differentiation in HL-60 promyelocytic leukemia cells and the Friend erythro-leukemia cells (Kraft *et al.*, 1989; Dell'Aquila *et al.*, 1987; Stone *et al.*, 1988). Bryostatin 1 also blocks phorbol ester-induced arachidonic acid metabolite release in C3H 10T1/2 mouse fibroblasts (Dell'Aquila *et al.*, 1988). Some of the other bryostatins cause variable arachidonic acid metabolite release.

Wender *et al.* (1988) have attempted to model the bryostatins to the phorbol ester pharmacophore via study of the binding affinities of naturally occurring and semisynthetic bryostatins for protein kinase C. These studies plus computer modeling suggested that the C-21, C-1, and C-19 oxygens of the bryostatins give an excellent spatial correlation with the proposed critical elements of the phorbol ester pharmacophore.

Bryostatins 1 and 2 greatly enhanced the efficiency of recombinant inter-leukin-2 in initiating development of *in vivo* primed cytotoxic T lymphocytes during *in vitro* incubation. This suggests that the use of bryostatins with recombi-nant interleukin-2 may lead to the use of lower concentrations of recombinant interleukin-2 to avoid the undesirable effects of high doses of the latter (Trenn *et al.*, 1988).

3. TERPENES

3.1. Monoterpenes

Cytotoxicity has been reported for a few marine monoterpenes in the past few years. Two groups have reported cytotoxic halogenated monoterpenes isolated from the sea hare *Aplysia kurodai*. Miyamoto *et al.* (1988), isolated aplysia-terpenoid A (**115**), which inhibited the growth of L1210 cells (IC$_{50}$ 10 μg/ml) and also showed insecticidal activity against German cockroaches. Kusumi *et al.* (1987) described four new halogenated monoterpenoids, aplysiapyranoids A–D (**116–119**), from *A. kurodai*. These terpenoids exhibited very mild cytotoxicity against Vero, MDCK, and B16 cells (IC$_{50}$ 19–96 μg/ml), the most active being aplysiapyranoid D (**119**) (IC$_{50}$ against Moser human tumor cells 14 μg/ml).

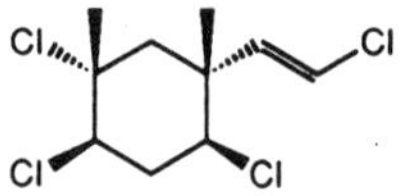

(115)

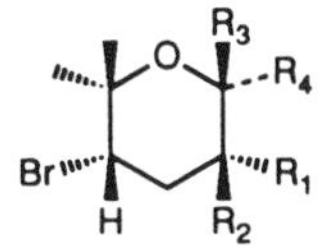

(116) R_1 = H, R_2 = Br, R_3 Me, R_4 = (E)-chlorovinyl
(117) R_1 = H, R_2 = Br, R_3 = (E)-chlorovinyl, R = Me
(118) R_1 = Cl, R_2 = H, R_3 = Me, R_4 = (E)-chlorovinyl
(119) R_1 = Cl, R_2 = H, R_3 = (E)-chlorovinyl, R_4 = Me

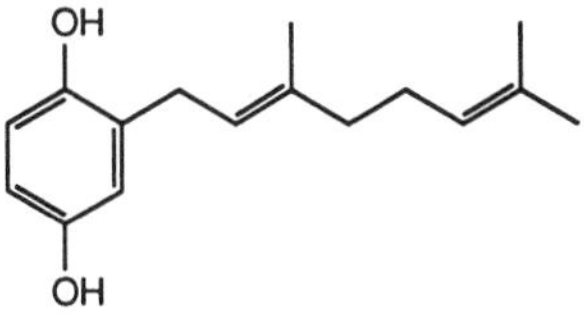

(120) 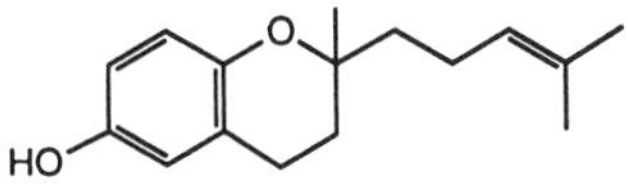   (121)

Geranyl hydroquinone (120) and a related pyran, cordiachromene A (121), isolated from the ascidian *Aplydium antillense*, were active against both KB and P388 cells (for 120, IC_{50} was 4.3 and 0.035 μg/ml vs. KB and P388, respectively; for 121, IC_{50} was 36 and 0.5 μg/ml vs. KB and P388, respectively) (Benslimane *et al.*, 1988). Hydrallmanol A (122), a phenylated *p*-menthane derivative, was obtained from a Nova Scotian hydroid, *Hydrallmania falcata*, and its structure was confirmed by synthesis (Pathirana *et al.*, 1989). The synthetic racemic material was mildly toxic to L1210 cells (ED_{50} 30 μg/ml).

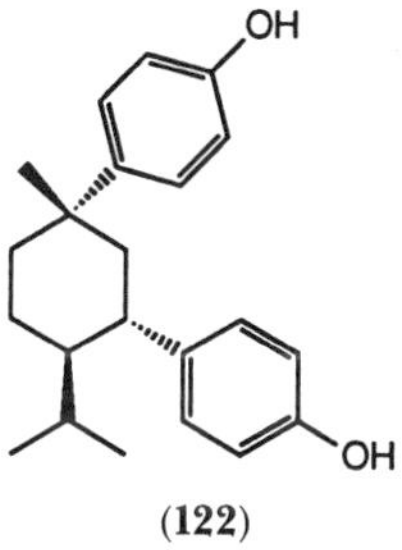

(122)

3.2. Sesquiterpenes

The glycosylated metabolite moritoside (123), isolated by Fusetani *et al.* (1985) from the gorgonian *Euplexaura* sp., inhibits division of fertilized starfish eggs (*Asterina pectinifera*) at 1 μg/ml. The unusual terpenoids anthoplalone (124) and noranthoplone (125) show modest toxicity to B16 murine melanoma cells (IC_{50} 22 and 16 μg/ml) (Zheng *et al.*, 1990a). They were obtained from an anemone, *Anthopleura pacifica* Uchida, and have rearranged isoprenoid skeletons. Curcu-

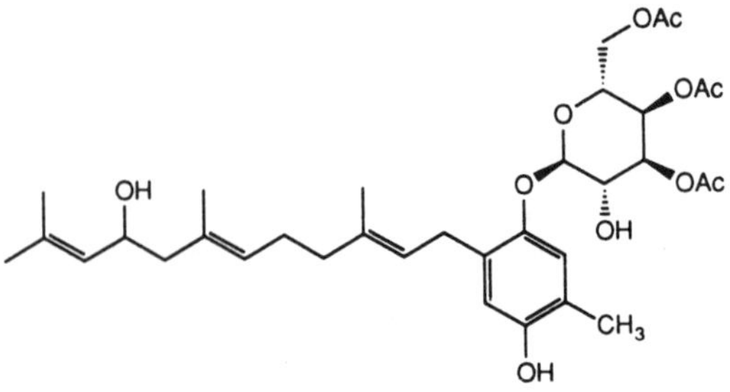

(123)

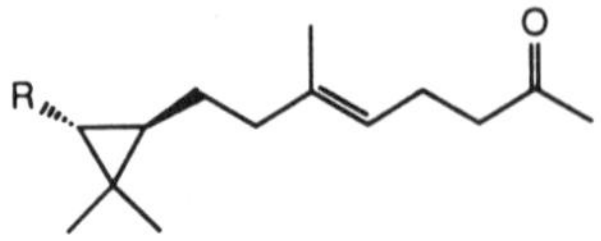

(124) R = CHO
(125) R = H

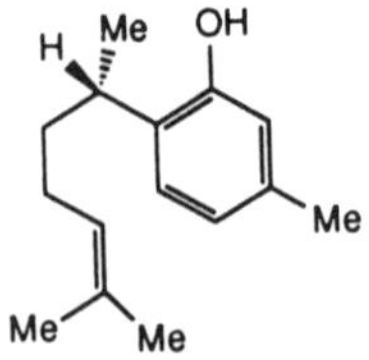

(126)

phenol (126), which was extracted from the sponge *Didiscus flavus* collected in both shallow and deep waters in the Bahamas and Belize, inhibits growth of several cell lines [IC_{50} 7 μg/ml vs. P388; MIC for human cell lines: A-549 (lung) 10 μg/ml; HCT-8 (colon) 0.1 μg/ml; MDAMB (mammary) 0.1 μg/ml] (Wright *et al.*, 1987b). Nephtheoxydiol (127), a hydroperoxy germacrane from an Okinawan soft coral, *Nephthea* sp., was found to be active against B16 melanoma cells IC_{50} 0.1 μg/ml (Kitagawa *et al.*, 1987a). Metachromins A (128) and B (129), isolated from the sponge *Hippospongia* cf. *metachromia*, were found to be toxic to L1210 cells (IC_{50} 2.4 and 1.62 μg/ml, respectively) (Ishibashi *et al.*, 1988).

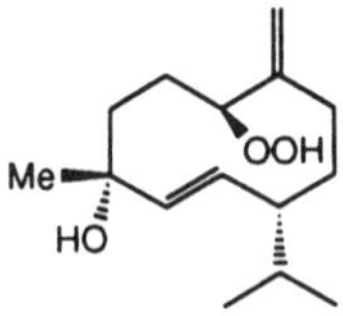

(127

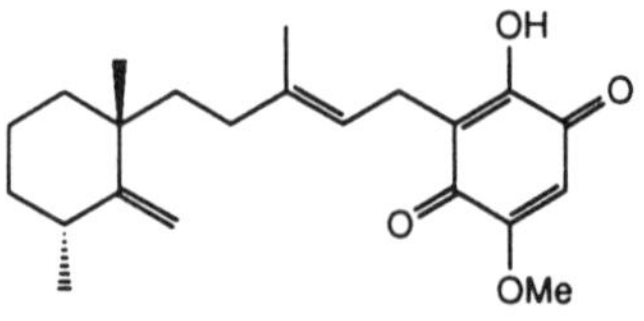

(128)

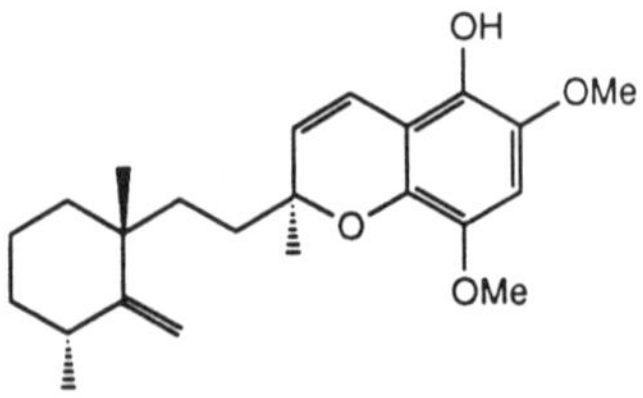

(129)

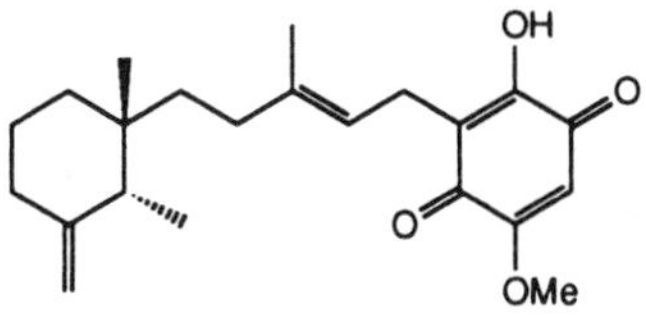

(130)

Metachromin C (**130**) is also active against L1210 cells (IC_{50} 2.0 μg/ml) as well as L5178Y cells at 0.92 μg/ml (J. Kobayashi *et al.*, 1989b). All three compounds also showed coronary vasodilating effects and inhibited potassium chloride-induced contraction of the rabbit isolated coronary artery.

The halogenated chamigrene derivative **131**, obtained from the digestive gland of the sea hare *Aplysia dactylomela*, was reported in a patent to be cytotoxic (Snader and Higa, 1987). Its configuration and biogenesis have been discussed by Sakai *et al.* (1986c). A variety of oxygenated derivatives (**132–137**) of the sesquiterpene lepidozene were isolated from the anemone *Anthopleura pacifica* Uchida and found to inhibit growth of murine melanoma cells. The IC_{50} range was 0.7–4.5 μg/ml; the hydroperoxy metabolites were among the most active (Zheng *et al.*, 1990b).

Sakemi and Higa (1987) reported that linderazulene (**138**), 2.3-dihydrolinder-

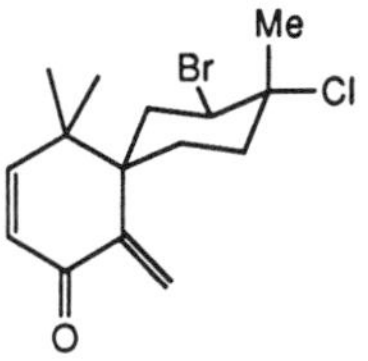

(131)

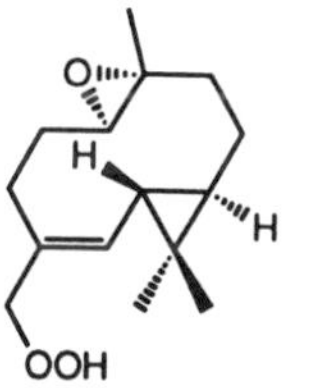

(132)

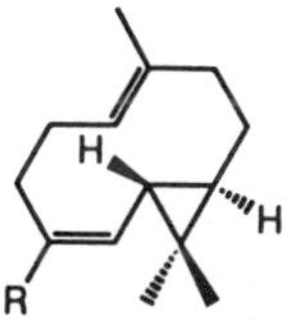

(133) R = CHO
(134) R = CH$_2$OH

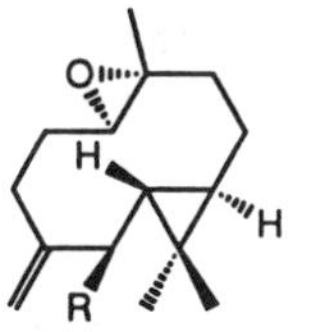

(135) R = OOH

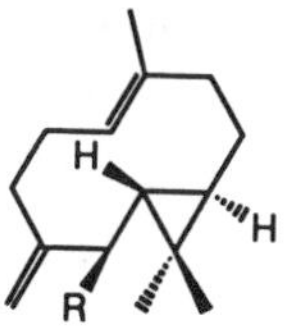

(136) R = OOH
(137) R = OH

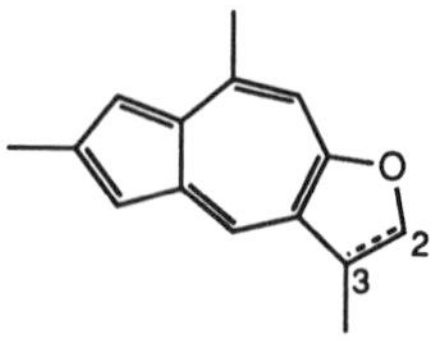

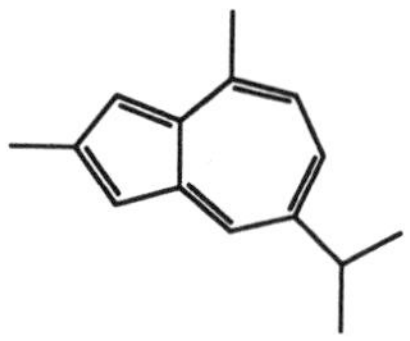

(138) Δ^2 (139)
(140) 2,3 dihydro

azulene (140), and guaiazulene (139), pigments isolated from the gorgonian *Acalycigorgia* sp., showed "moderate" activity against P388 and that the linderazulenes displayed immunostimulatory activity at "low concentrations." Cardellina and Barnekow (1988) found eight oxidized nakafuran sesquiterpenes in the extract of the sponge *Dysidea etheria*, of which one (141) showed weak activity against KB cells (ED_{50} 22 μg/ml) and another (142) was active in the brine shrimp assay (LD_{50} 38 μg/ml). Dendrolasin (142a) is another sesquiterpene furan for

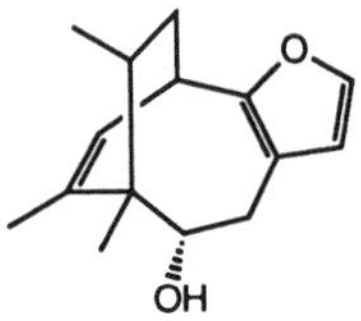

(141) 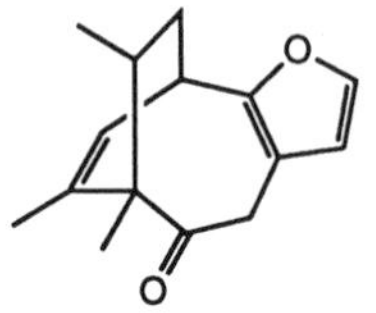   (142)

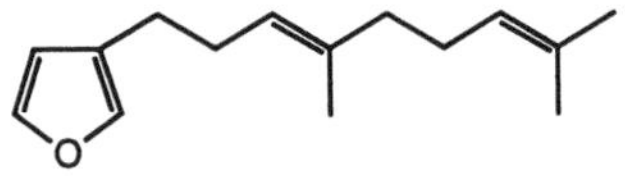

(142a)

which some cytotoxicity was reported (completely toxic to HEP-2 cells at 24 μg/ml, but inactive against MA-104 cells) (Kakou *et al.*, 1987). This furan was isolated by these workers from the Fijian sponge *Spongia microfijiensis** and an associated nudibranch, *Chromodoris lochi*.

Studies on the mechanism of action of avarol (143), obtained from the sponge *Dysidea avara*, indicate that it interferes with the mitotic processes, thus preventing telophase formation. It has been suggested that inhibition of growth may be due to changes of the intracellular pools and/or alterations of the permeability properties of the cell membranes for the precursors (Mueller *et al.*, 1985). Avarol diacetate exhibited cytotoxicity similar to that of avarol. Avarol monoacetate (144) was isolated from the sponge *Dysidea avara*, collected near Naples, Italy, and was

*Now assigned the new name *Leiosella lavis* as a result of a reclassification (P. Crews, personal communication).

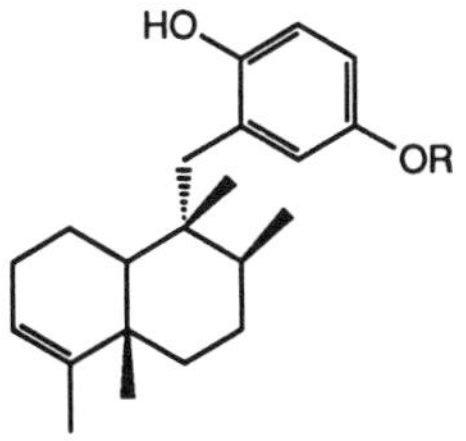

(143) R = H
(144) R = Ac

found to be more toxic in the brine shrimp bioassay than avarol itself (143, LC_{50} 0.18 μg/ml; 144, LC_{50} 0.09 μg/ml) (Crispino *et al.*, 1989).

A number of metabolites with the general avarol structure but with oxidation or substituent differences in the aromatic ring have been isolated from the sponge *Smenospongia* sp. (Kondracki and Guyot, 1987; Kondracki and Guyot, 1989a,b; Kondracki *et al.*, 1989a,b). All of these quinones and hydroquinones (145–149) showed similar levels of activity against L1210; see the data given with the structures.

Kohmoto *et al.* (1987a) isolated the known puupehenone (150) from a deep-

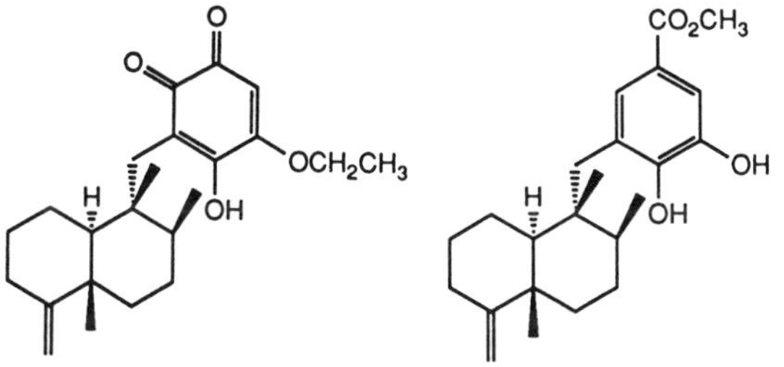

(145) Smenorthoquinone
(L1210, ID_{50} 1.5 μg/mL)

(146) Smenospondiol
(L1210, ID_{50} 4 μg/mL)

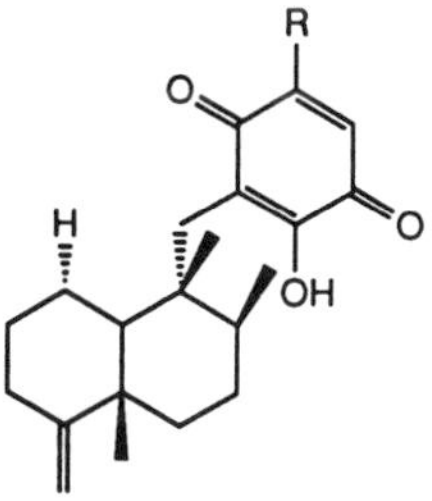

			Cytotoxicity	L1210 cells
(147)	Smenoquinone:	R = OH	Smenoquinone:	ID_{50} 2.5 μg/ml
(148)	Smenospongine:	R = NH_2	Smenospongine:	ID_{50} 1.5 μg/ml
(149)	Smenospongiarine:	R = $NHCH_2CH_2CH(CH_3)_2$	Smenospongiarine:	ID_{50} 4.0 μg/ml

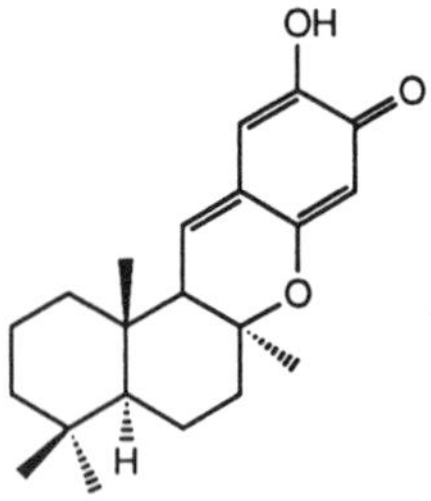

(150)

water sponge, *Strongylophora hartmani*, and established that it inhibits the growth of a number of tumor cell lines (IC_{50}: P388, 1 μg/ml; A549 human lung, 0.1–1 μg/ml; HCT-8 human colon, 1–10 μg/ml; MCF-7 human mammary, 0.1–1 μg/ml). Puupehenone showed very modest *in vivo* effects (P388 19% increase in lifetime @ 25 mg/kg for 9 days). The halogenated sesquiterpenes (**151–153**) isolated from the green alga *Neomeris annulata* were active in the brine shrimp bioassay (LD_{50} 9–16 μg/ml), but were inactive against KB cells ($ED_{50} > 20$ μg/ml) (Barnekow *et al.*, 1989).

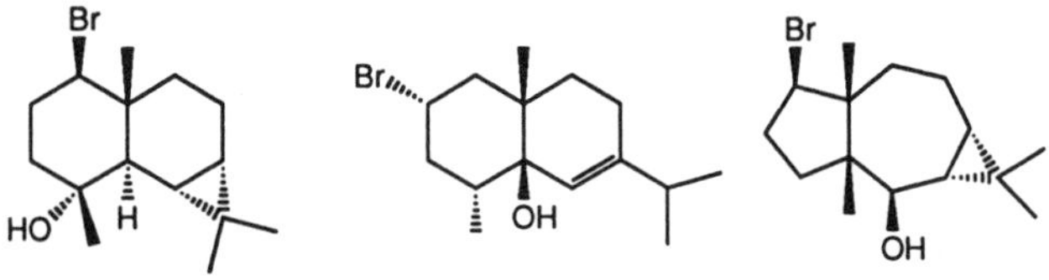

(151)   (152)   (153)

3.3. Diterpenes

The previous review on marine antitumor compounds by Munro *et al.* (1987) included a number of linear diterpenes characterized by a 1,4-diacetoxybutadiene moiety. This is a common feature of green algae metabolites (Faulkner, 1991, and earlier reviews cited therein), which Fenical and Paul (1984) pointed out is frequently associated with high biological activity. Three of these types of diterpenes not included in the Munro *et al.* (1987) review are **154–156**. All three compounds completely inhibited cell division of fertilized sea urchin eggs at 8

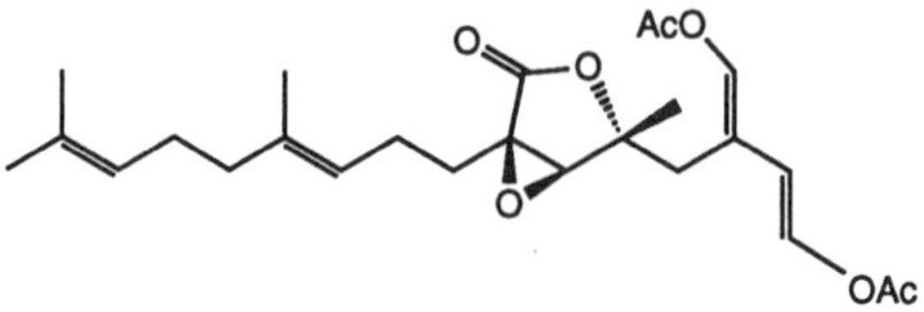

(154)

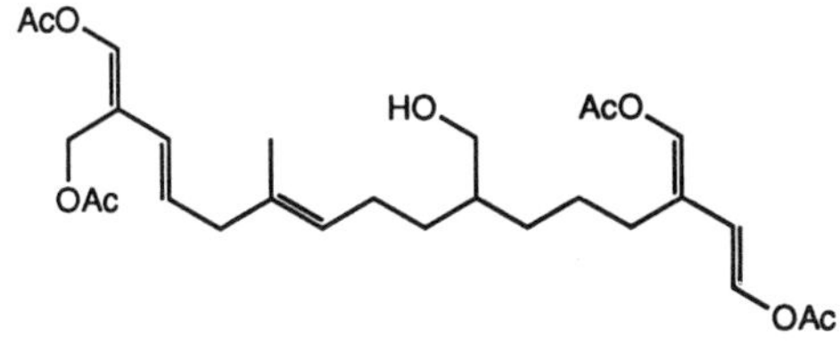

(155)

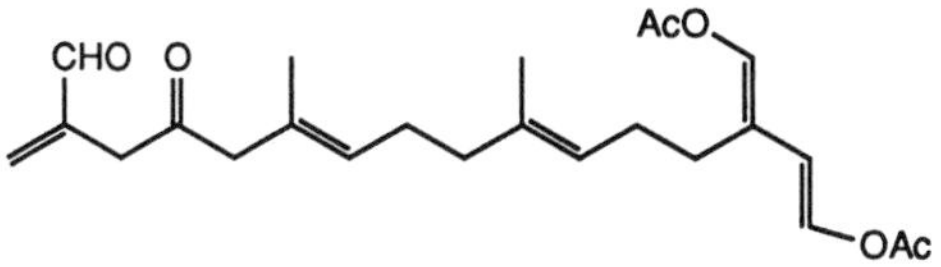

(156)

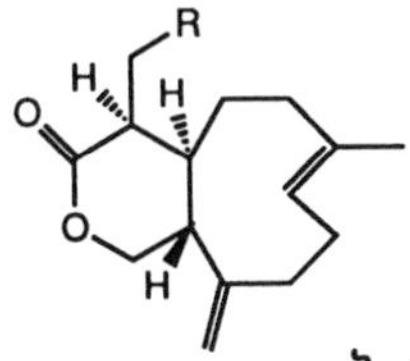

(157) Acalycixeniolide A: R =

(158) Acalycixeniolide B: R =

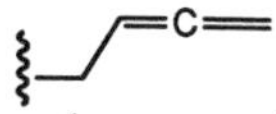

µg/ml. The sources of these diterpenoids were three species of green algae (Chlorophyta; order Caulerpales) from Guam: *Udotea argtentea*, *Tidemania expeditionis*, and *Chlorodesmis fastigiata* (Paul and Fenical, 1985).

Five norditerpenes with the xenicane skeleton (**157–161**), four of which have the uncommon allene functionality, have been isolated by Fusetani's group from several gorgonians of the genus *Acalycigorgia* (Hokama *et al.*, 1988; Fusetani *et al.*, 1987a, 1989a). These compounds all interfered with cell division of fertilized starfish eggs (*Hemicentrotus pulcherrimus*) (1–50 µg/ml range) and **159**, **160**, and **161** displayed toxicity to P388 cells (IC$_{50}$ 0.27, <2.5, and 2.5 µg/ml, respectively). From the brown alga *Dictyota dichotoma* collected in Okinawa the

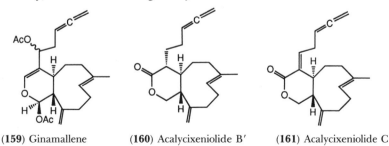

(159) Ginamallene   **(160)** Acalycixeniolide B'   **(161)** Acalycixeniolide C

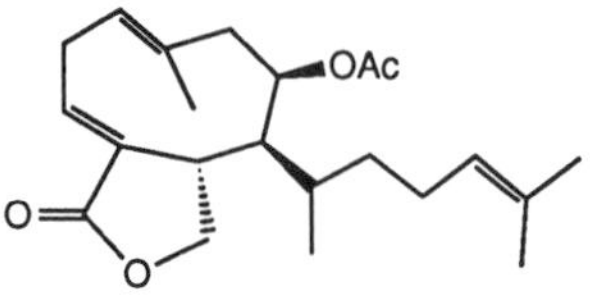

(162) 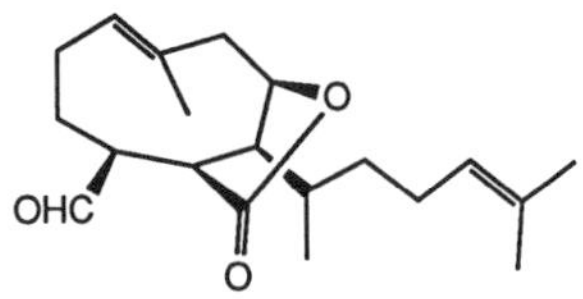   (163)

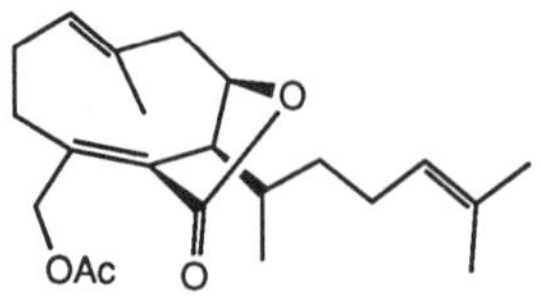

(164) 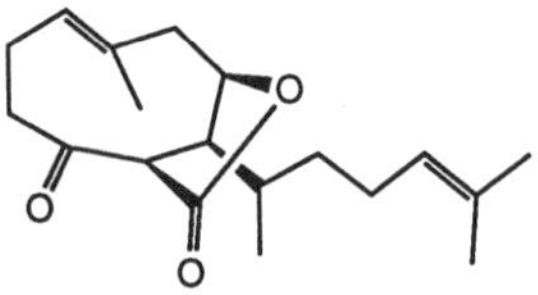   (165)

four lactones acetoxydictyolactone (**162**), dictyotalide A (**163**), dictyotalide B (**164**), and nordictyotalide (**165**) were obtained, all of which were toxic to B16 melanoma cells (IC_{50} 1.57, 2.57, 0.58, and 1.58 μg/ml, respectively) (Ishitsuka *et al.*, 1988).

Munro *et al.* (1987) reviewed a number of cembranoids isolated from marine sources that have displayed cytotoxicity. The majority of these possess the α-methylene lactone functionality, a feature which is associated with the cytotoxic activity of numerous natural products (Hoffmann and Rabe, 1985). Additional examples of α-methylene γ-lactone cembranolides have been reported in recent years. Kericembranolides A–E (**166–170**) were extracted from the soft coral

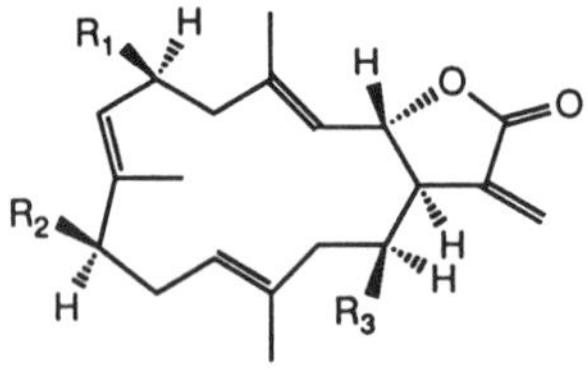

	R₁	R₂	R₃
(**166**)	OAc	H	H
(**167**)	H	OAc	H
(**168**)	OAc	H	OAc
(**169**)	OAc	H	OH
(**170**)	OH	H	OH

Clavularia koellikeri and all had very similar activity against B16 melanoma cells (IC_{50} 3.8, 2.5, 1.3, 1.2, and 1.8 μg/ml, respectively) (M. Kobayashi *et al.*, 1986). The cembranolides **171** and **172** and the peroxide-containing metabolite denticula-tolide (**173**), reported from the soft coral *Sinularia mayi*, had similar degrees of toxicity to B16 cells ("cytotoxicity at" 8.4, 2.1, 3.6 μg/ml) (Kusumi *et al.*, 1988a). Denticulatolide (**173**) had been reported earlier as an ichthyotoxin obtained

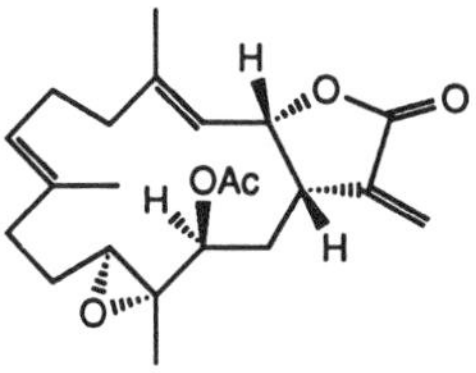

(171) 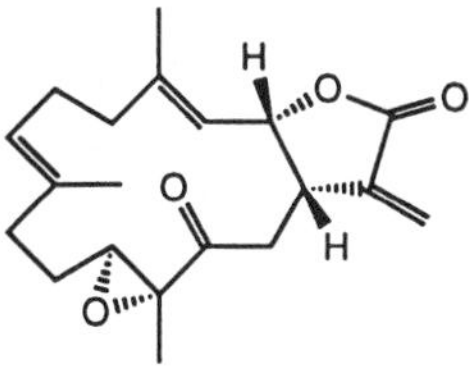　　　　　　　(172) 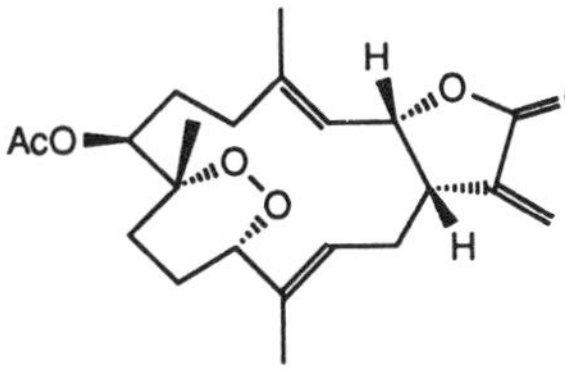　　　　　　　(173)

from another soft coral, *Lobophytum denticulatum* (Uchio *et al.*, 1985). The highly functionalized bipinnatins a–d (174–177), isolated from the gorgonian *Pseudopterogorgia bipinnata*, were active against P388 cells (IC$_{50}$ 0.9, 3.2, 46.6, and 1.5 μg/ml) (Wright *et al.*, 1989). The α,β-unsaturated carbonyl functionality of the oxidized isopropyl residue is clearly required for the best activity.

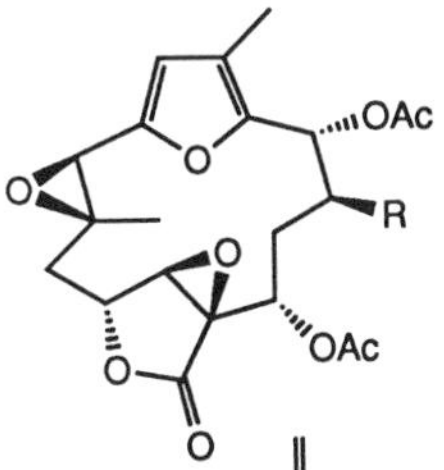

(174) R = CO$_2$Me

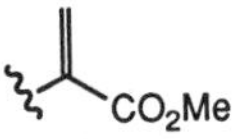

(175) R = CHO

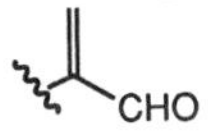

(176) R = CH$_3$

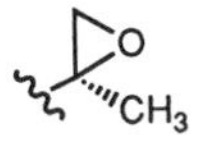

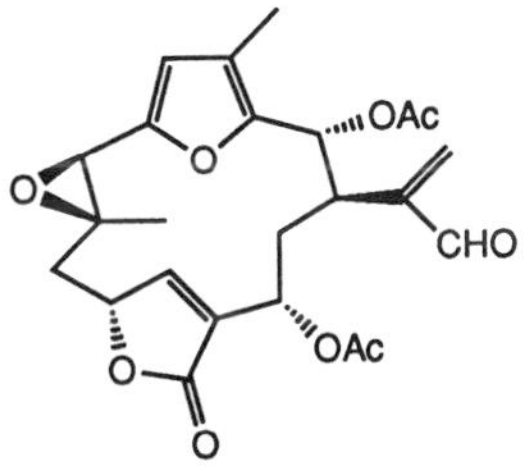

(177)

Two very unusual biscembranoids from the soft coral *Sarcophyton glaucum*, methyl sarcophytoate (178) and methyl chlorosarcophytoate (179), were reported by Kusumi *et al.* (1990). These were active against B16 cells at 7.5 and 12.0 μg/ml, respectively.

Fusetani *et al.* (1989d, 1990) determined that the isonitriles kalihinene (180) and isokalihinol B (181) were cytotoxic to P388 cells (IC$_{50}$ 1.2 and 0.8 μg/ml) and

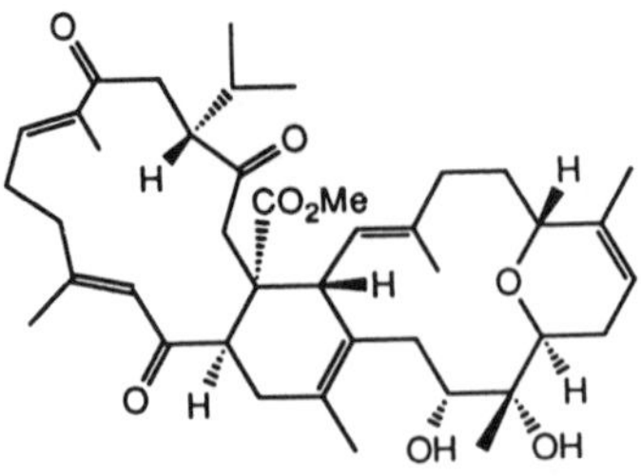

(178) 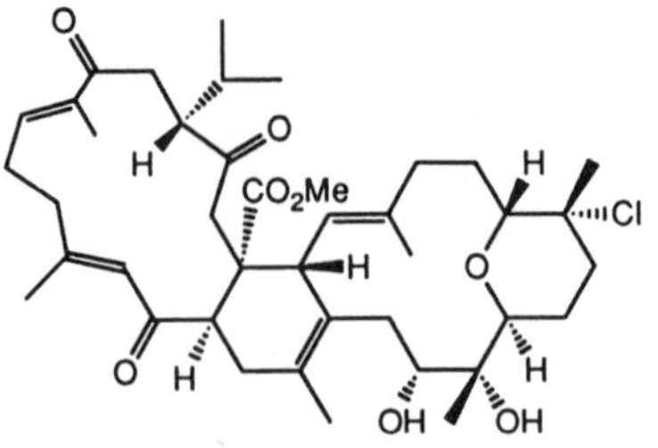   (179)

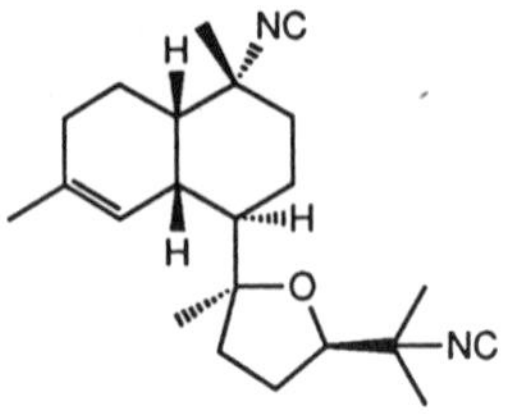

(180) 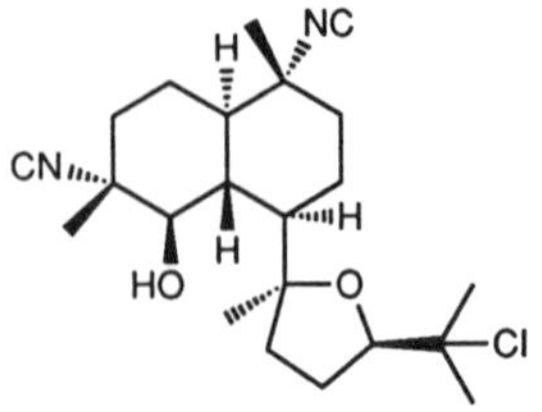   (181)

also showed antifungal activity. The source of **180** and **181** was the sponge *Acanthella klethra* collected at Kuchinoerabu Island, Japan.

A group of 11 rearranged diterpenoids called mediterraneols was isolated from the brown alga *Cystoseira mediterranea* collected on the French Mediterranean coast. Four of these, mediterraneols A–D (**182–185**), inhibited the motility of sea urchin sperm and cell division of sea urchin eggs (ED_{50} 2 μg/ml). Mediterraneols A and B (**182, 183**) show modest *in vivo* activity versus P388 (28% increase in lifetime @ 32 mg/kg) (Francisco *et al.*, 1986).

Two diterpenes with rearranged spongiane skeletons (**186, 187**) isolated by Rudi and Kashman (1990) from a Red Sea *Dysidea* sp. of sponge, were active

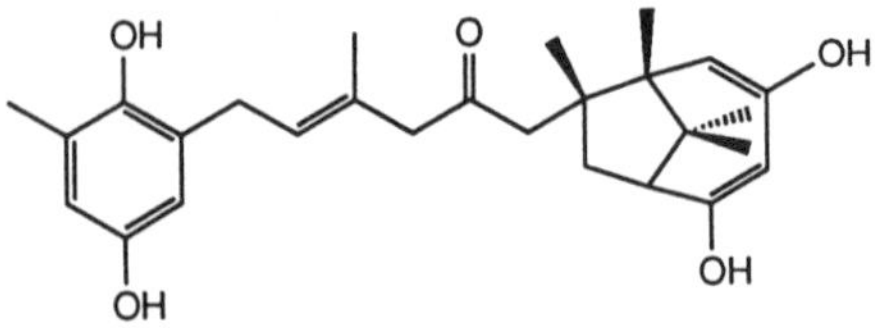

(182)

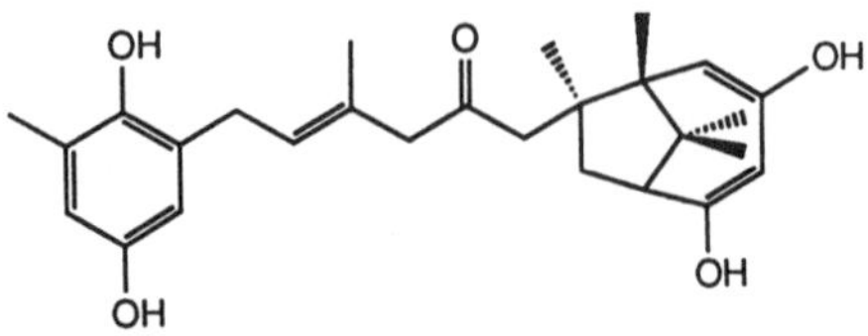

(183)

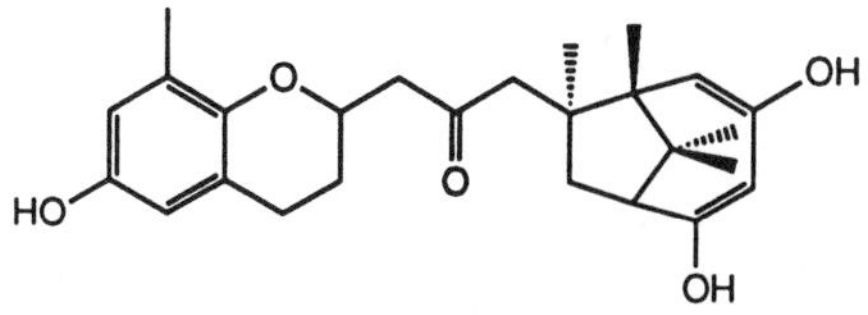

(184)

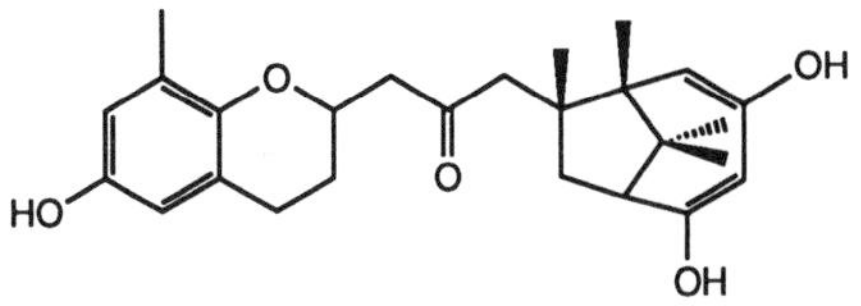

(185)

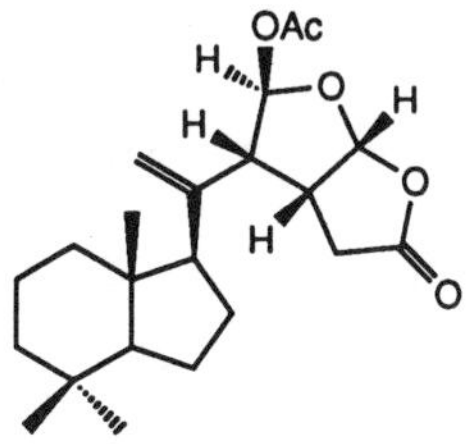

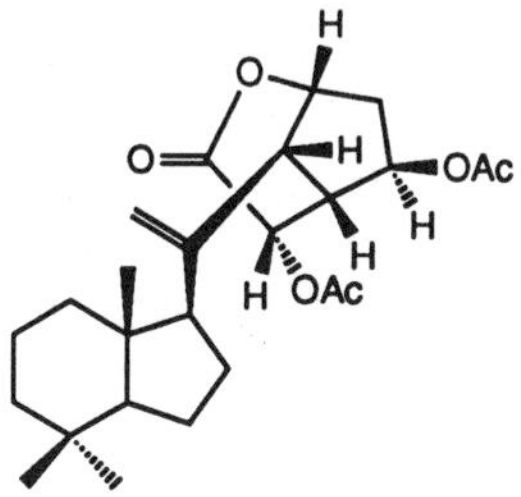

(186)   (187)

against P388 (IC_{50} 1.5 and 1.2 µg/ml). Chromodorolide A (188), which possesses a different rearranged version of the spongiane skeleton, was isolated from the Indian Ocean nudibranch *Chromodoris cavea*. It also showed P388 *in vitro* (ED_{50} 20 µg/ml) as well as some *in vivo* activity, with a T/C of 125% at 4 µg/ml (Dumdei *et al.*, 1989).

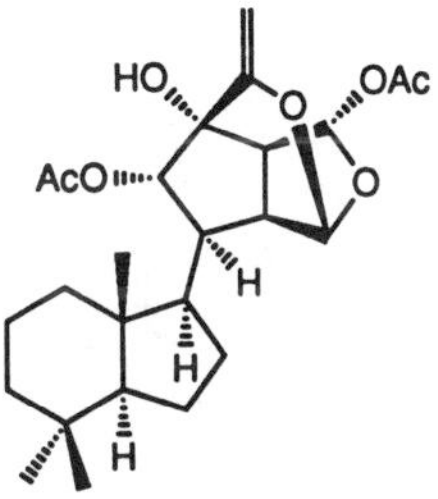

(188)

Agelasimines A and B (189, 190) were reported to be cytotoxic (ED_{50} ~2–4 µg/ml vs. L1210), but were also evaluated for other pharmacologic activity (Fathi-Afshar and Allen, 1988). In the range ~3–14 µM they relax rabbit gut smooth muscle and inhibit nucleoside transport into rabbit erythrocytes. It is speculated that they may act as Ca^{2+} channel antagonists as well as α-adrenergic blockers.

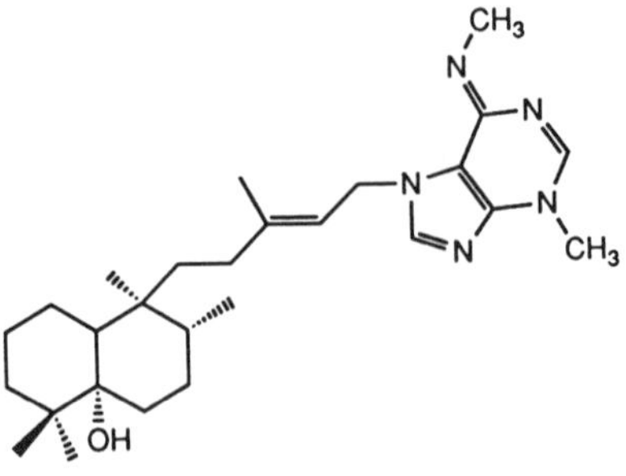

(189) 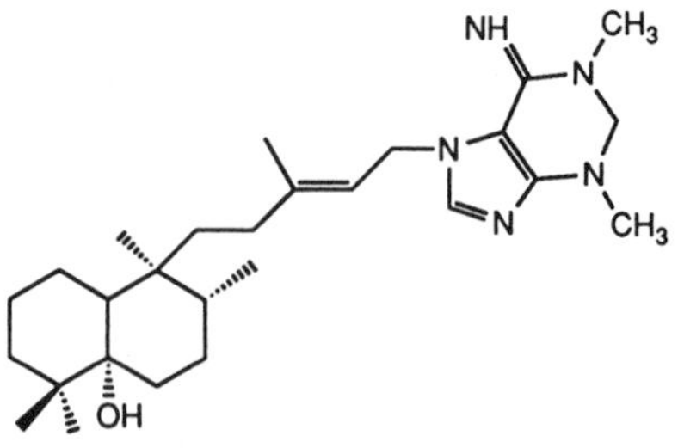   (190)

These diterpene-adenine compounds caused accumulation of L1210 cells at the G1 stage of the cell cycle *in vitro*, but had no *in vivo* activity against P388 in mice (Fathi-Afshar *et al.*, 1989). The source of these metabolites was the sponge *Agelas mauritiana* collected at Enewetak Atoll.

Stolonidiol (191) and its monoacetate (192), new diterpenes with a novel skeleton, which were obtained from an Okinawan soft coral, *Clavularia* sp., were found to be quite cytotoxic (IC$_{50}$ 0.015 μg/ml vs. P388) and were also toxic to

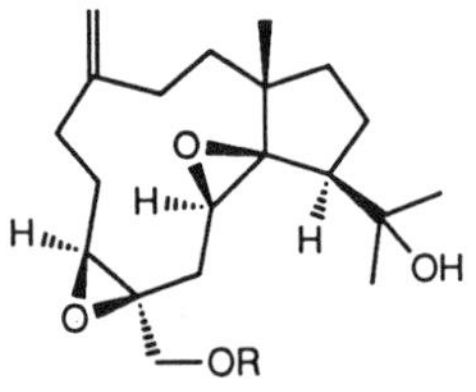

(191) R = H
(192) R = Ac

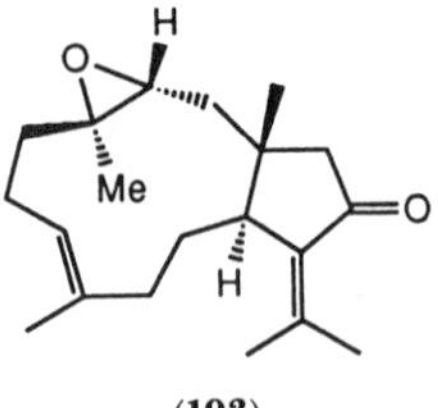

(193)

killifish at 10 and 17 μg/ml, respectively (Mori *et al.*, 1987a,b, 1988). The closely related claenone (193) was found to also inhibit division of fertilized sea urchin eggs, but was not active against P388 cells (Mori *et al.*, 1988).

Several representatives of the eunicellin class of diterpenes have been found to cytotoxic. Sclerophytin A (194), from the soft coral *Sclerophytum capitalis* collected at Enewetak, was quite active against L1210 (IC$_{50}$ 0.001 μg/ml) (Sharma and Alam, 1988).

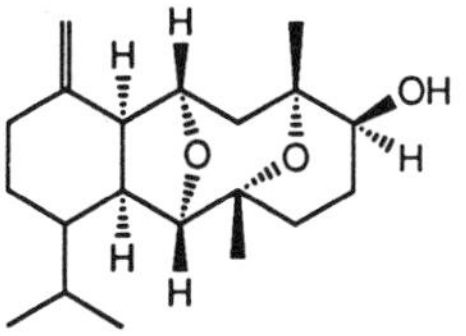

(194)

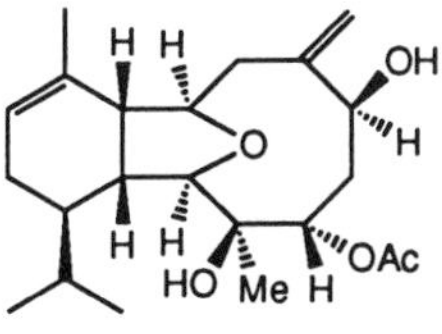

(195)

Alcyonin (195), from the soft coral *Sinularia flexibilis*, was active against Vero cells at 55 μg/ml (Kusumi *et al.*, 1988b). Fusetani *et al.* (1989e) reported that astrogorgin (196) and ophirin (197), isolated from a gorgonian of the *Astrogorgia* sp., inhibit cell division of fertilized starfish eggs, each at 10 μg/ml.

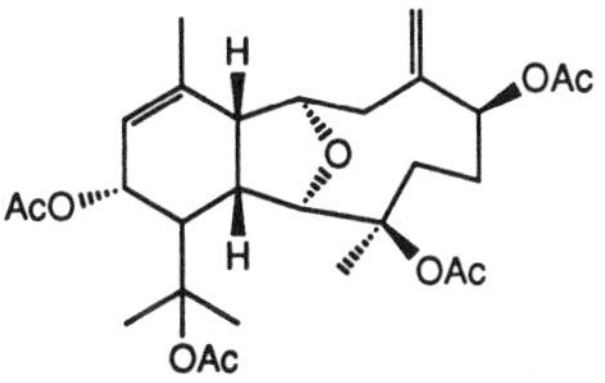

(196)

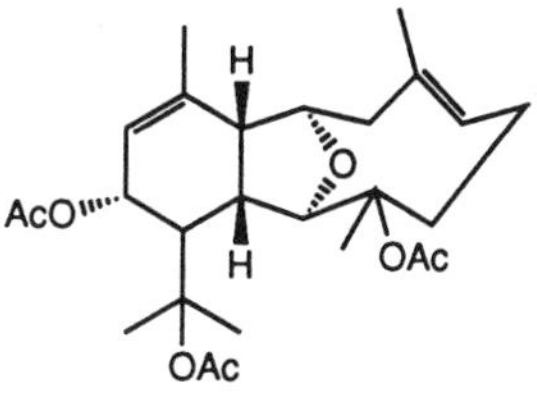

(197)

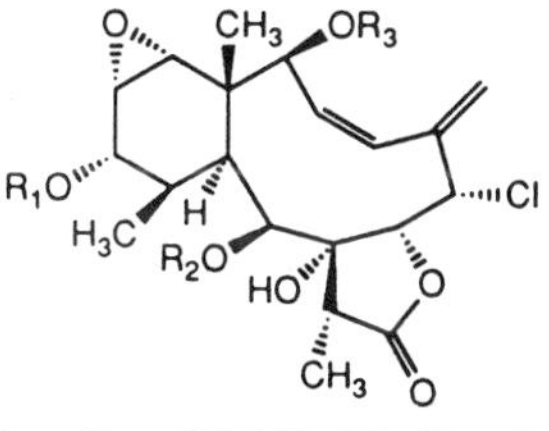

(198) $R_1 = R_3 = COCH_2CH_2CH_3$, $R_2 = Ac$
(199) $R_1 = R_2 = R_3 = Ac$

Another common marine diterpene skeleton is illustrated by briantheins V (198) and Z (199) reported by Coval *et al.* (1988). These lactones showed modest activity against P388 cells (ED_{50} 13 and 10 μg/ml, respectively). They were isolated from the Caribbean gorgonian *Briareum asbestinum*, which has been the source of many other briarein-type compounds. Tubiporein (200), also of the diterpenoid class, was isolated by Natori *et al.* (1990) from a Japanese soft coral,

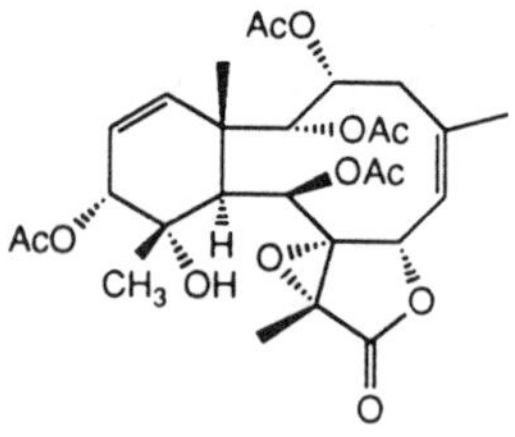

(200)

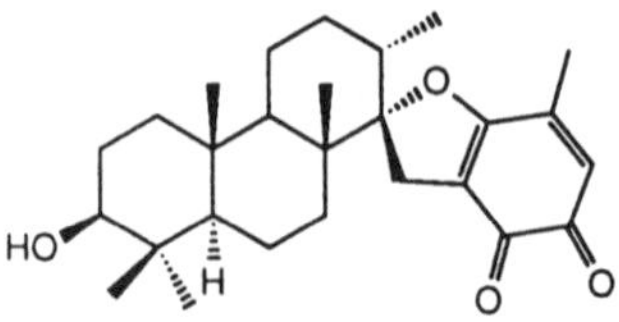

(201)

Tubipora sp., and was shown to be active against B16 melanoma cells with an IC_{50} of 2 µg/ml.

The mechanism of action of stypoldione (201), which is known to inhibit the division of fertilized sea urchin eggs and cultured mammalian cells, has been studied further (O'Brien *et al.*, 1986). This ortho-quinone is believed to add to sulfhydryl groups at or near the sensitive reaction sites on cellular proteins, and/or add to the sulfhydryl groups of glutathione, thereby affecting the function of sensitive sulfhydryl-dependent proteins directly.

Kohmoto *et al.* (1987b) found three cytotoxic spongiadiols in a deep-water Caribbean sponge, *Spongia* sp.: spongiadiol (202), epispongiadiol (203), and isospongiadiol (204) with IC_{50} values of 0.5, 12.5, and 2 µg/ml, respectively.

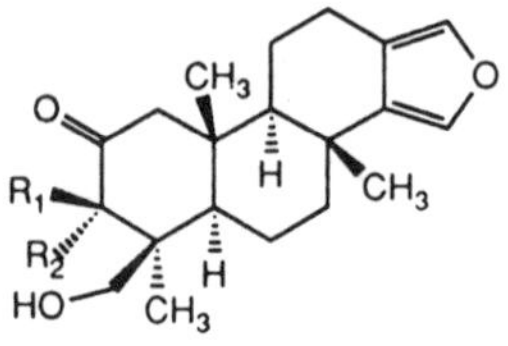

(202) R_1 = H, R_2 = OH
(203) R_1 = OH, R_2 = H

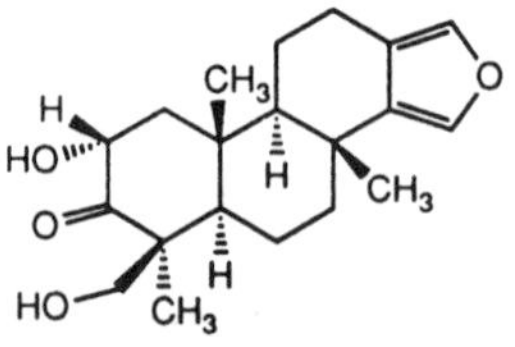

(204)

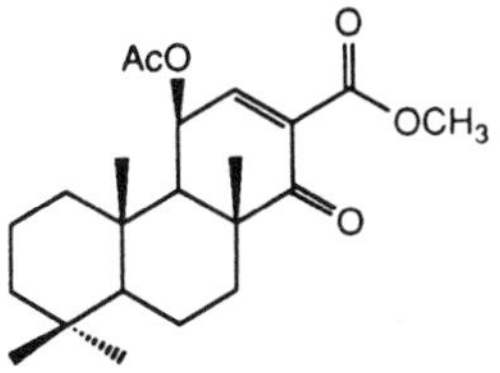

(205)

These diols also showed antiviral activity against herpes simplex (HSV-1) at the same concentrations. Toth (1990) found a new sponge metabolite, aplypallidenone (205) from an *Aplysilla* sp., to be active against P388 (IC_{50} 0.01 μg/ml).

Schmitz *et al.* (1982) initially reported the structure and cytotoxicity (ED_{50} 3.8 μg/ml) of parguerol (206), one member of a group of brominated diterpenes isolated from the sea hare *Aplysia dactylomela* collected in Bimini, Bahamas.

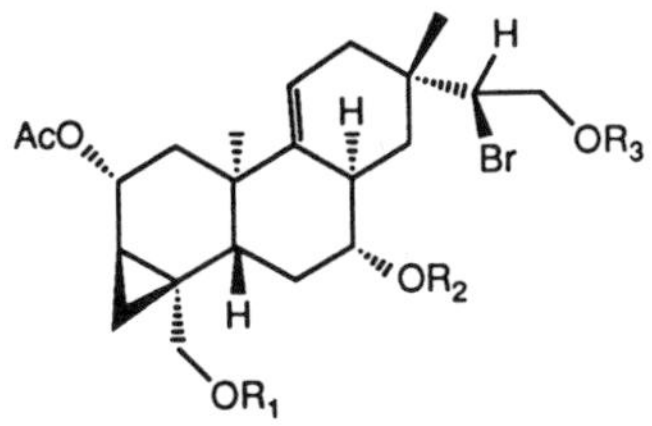

	R_1	R_2	R_3
(206)	H	H	H
(207)	Ac	Ac	Ac
(208)	Ac	H	Ac

T. Suzuki *et al.* (1989) have established by x-ray analysis the absolute configuration (shown in drawing) of parguerol peracetate (207) obtained from the red alga *Laurencia obtusa*. This group also isolated the 16,19-diacetyl analog of parguerol (208) and demonstrated that it is cytotoxic (P388, IC_{50} 3.5 μg/ml). From the same alga a new parguarene (209) was obtained in which the cyclopropane ring was opened (Takeda *et al.*, 1990a). Diterpene 209 is less active than most other parguarenes against P388 (IC_{50} 25 μg/ml), but is quite toxic to B16 cells (IC_{50} 0.78 μg/ml). Testing of a variety of compounds of the parguarene series with variations

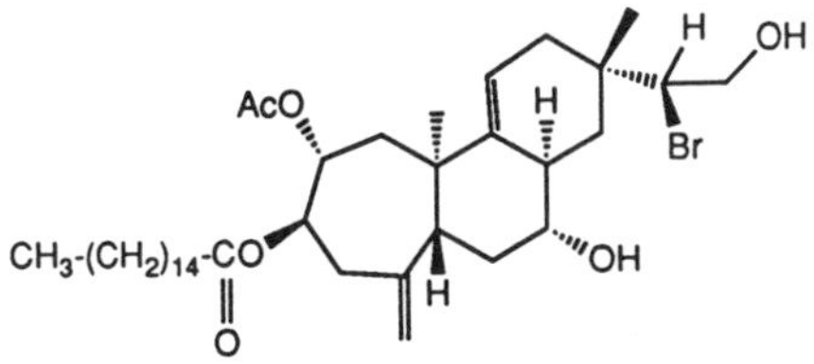

(209)

in the number and position of acetyl groups suggested that the OAc at C-2 and bromine at C-15 are indispensable for cytotoxic activity, but acetyl groups (except for that at C-2) are not always necessary (Takeda *et al.*, 1990b).

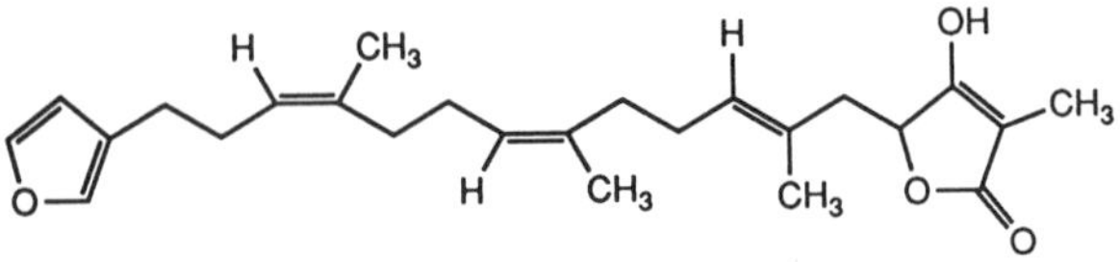

(210)

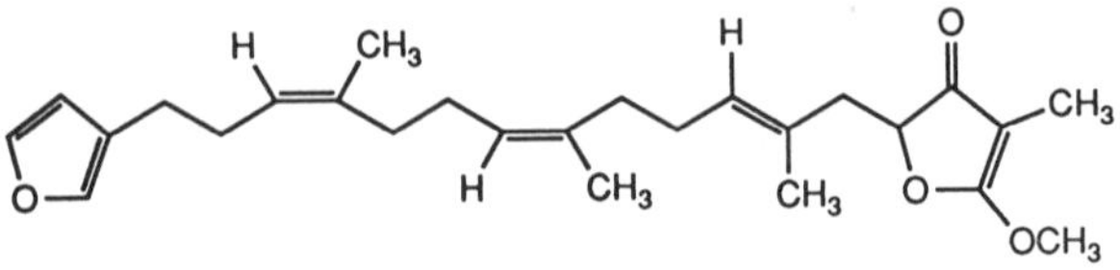

(211)

3.4. Sesterterpenes

A number of cytotoxic furanosesterterpenes have been obtained from a variety of sponges. Palominin (**210**) from a Caribbean *Ircinia* sp., and the synthetic derivative **211** were both found to be toxic to brine shrimp at 1 μg/ml (Garcia and Rodriguez, 1990). Variabilin (**212**) and the related sesterterpene tetronic acids (**213–216**), from an *Ircinia* sp. of sponge, were all described as being cytotoxic to

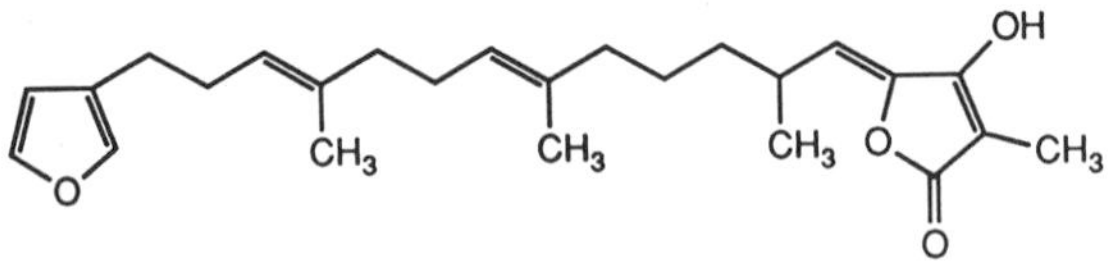

(212)

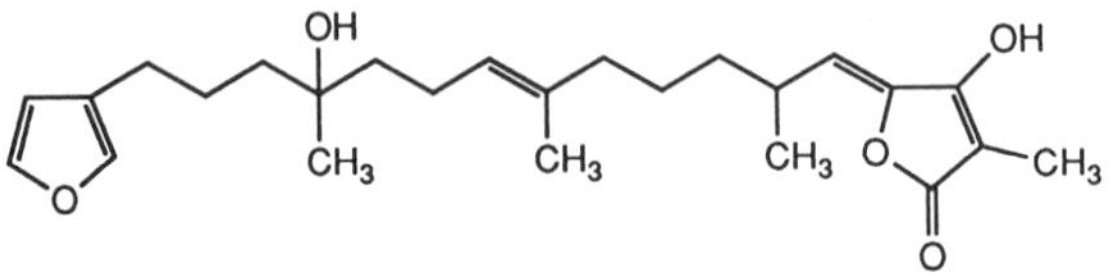

(213)

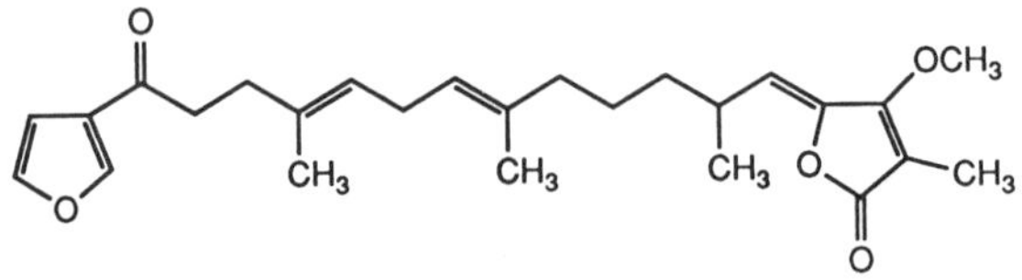

(214)

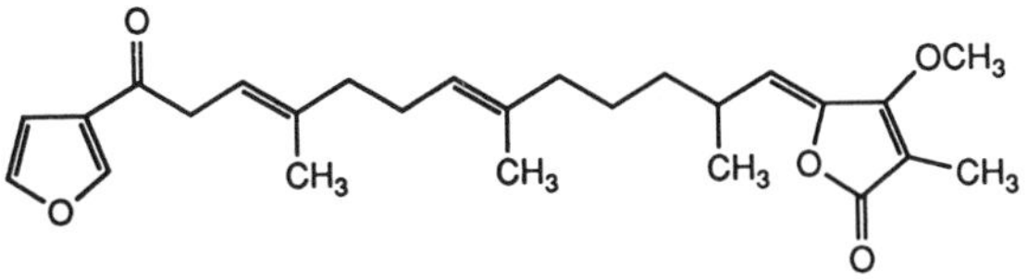

(215)

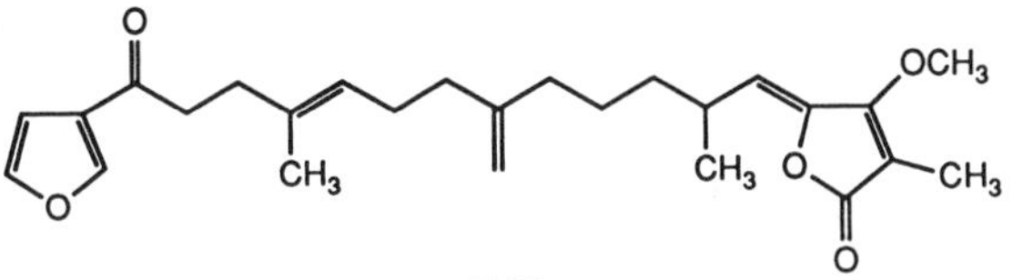

(216)

host BSC cells at 2 μg/ml in an antiviral assay (Barrow *et al.*, 1988). DeGiulio *et al.* (1989) found the difurans (217–220) obtained from a north Adriatic sample of the sponge *Spongia officianalis*, to be toxic to brine shrimp at 0.09–1.6 μg/ml. Okinonellins A and B (221, 222), from a *Spongionella* sp., were reported to inhibit division of fertilized starfish eggs at 5 μg/ml (Kato *et al.*, 1986c).

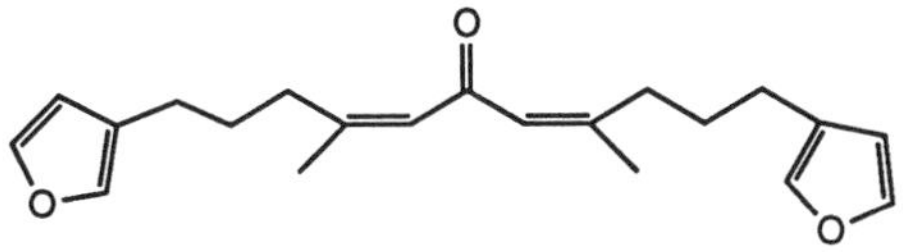

(217)

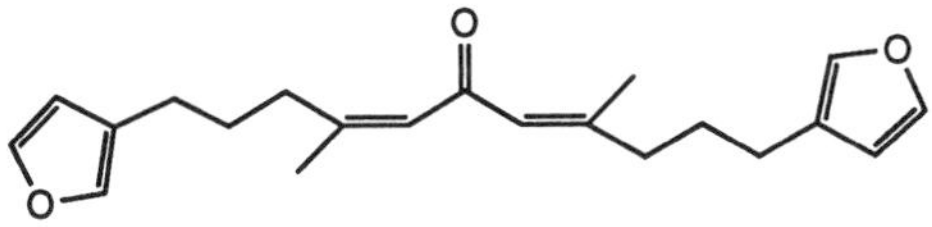

(218)

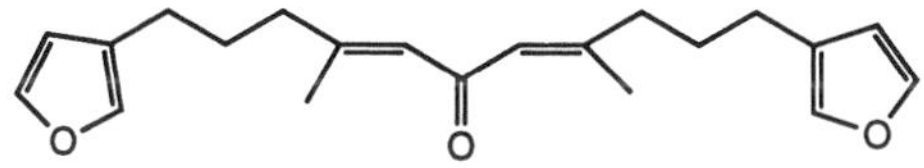

(219)

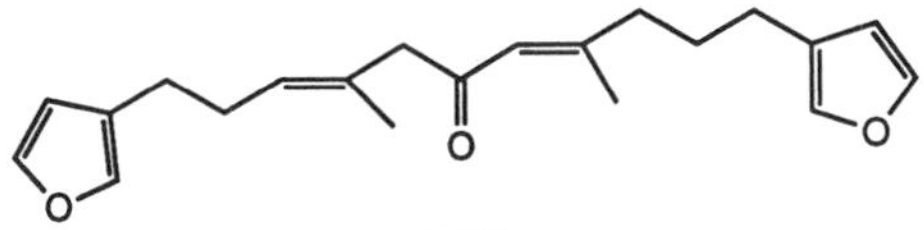

(220)

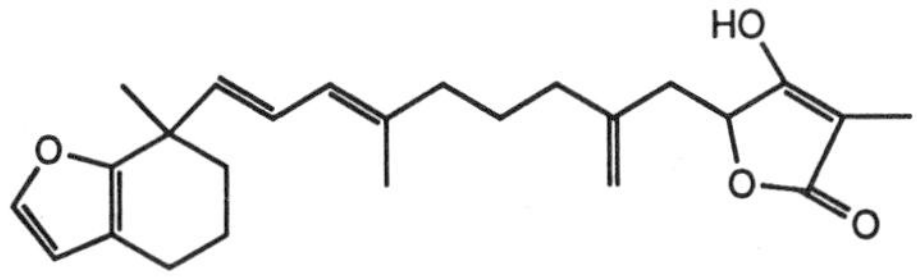

(221)

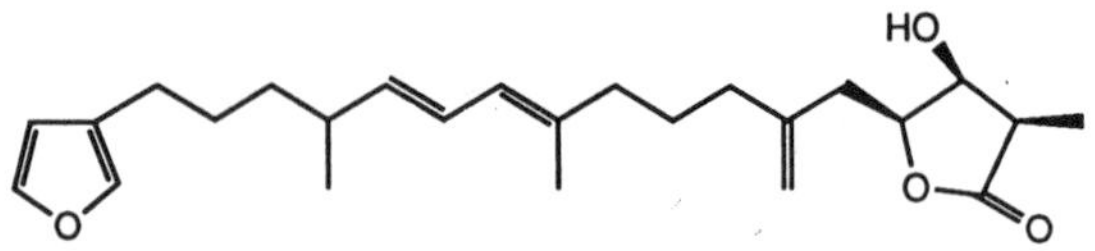

(222)

Cacospongionolide (**223**), from the marine sponge *Cacospongia mollior*, has been reported to be active in the brine shrimp assay (LD_{50} 0.1 μg/ml) and cause 75% inhibition in the crown-gall potato disk assay. It is also ichthyotoxic at 1 μg/ml (De Rosa *et al.*, 1988). The bishomo scalarene sesterterpene phyllofolia-spongin (**224**) from *Phyllospongia foliascens* inhibited P388 cell growth at 5 μg/ml (Kitagawa *et al.*, 1989). Another activity noted for this compound was an antithrombocytic inhibitory effect on ADP-induced and collagen-induced aggregation of rabbit platelets *in vitro*.

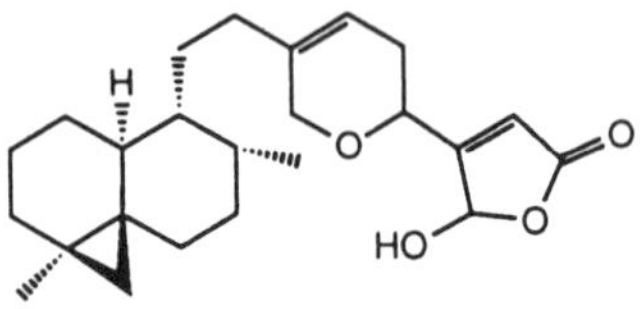

(223)

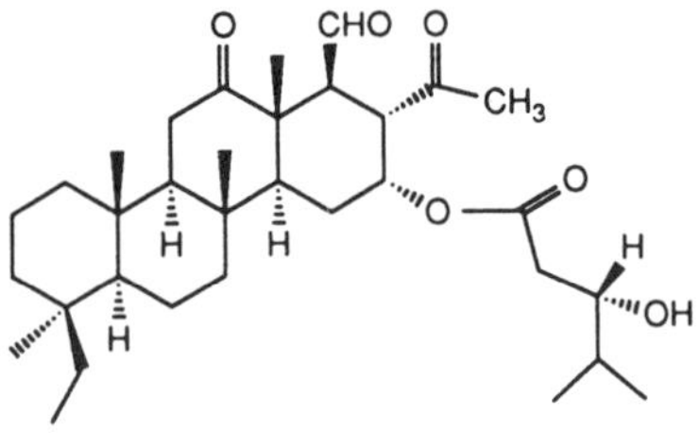

(224)

3.5. Triterpenes

The degraded squalene metabolite turbinaric acid (**225**), obtained from the brown alga *Turbinaria ornata* by Asari *et al.* (1989), showed antiproliferative effects against both murine melanoma cells (ED_{50} 26 μg/ml) and human colon carcinoma (ED_{50} 12.5 μg/ml). The thyrsiferyl family of polyether sesterterpenes (**226–232**), obtained from the Japanese red alga *Laurencia obtusa*, all show cytotoxicity, with the initially discovered member of the group, thyrsiferyl acetate (**226**), being remarkably active; see the data given with the structures (T. Suzuki *et al.*, 1985; 1987a,b). The New Zealand group initially isolated thyrsiferyl from the red alga *L. thyrsifera* Hook (Blunt *et al.*, 1978). (+)-Thyrsiferol has been synthesized by M. Hashimoto *et al.* (1988).

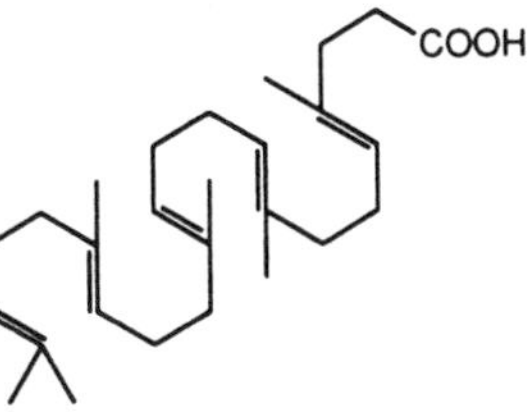

(225)

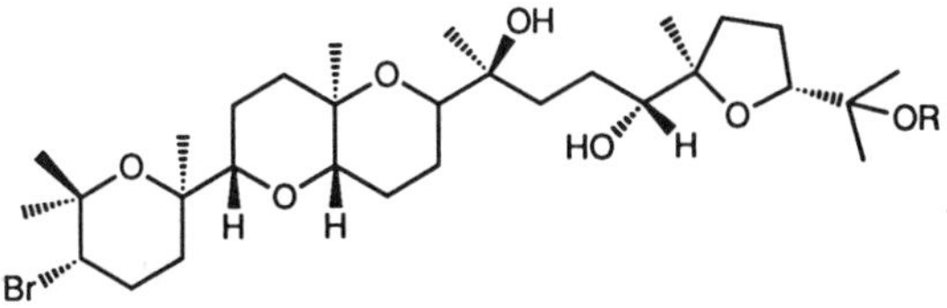

(226) Thyrsiferyl acetate: R = Ac
(227) Thyrsiferol: R = H

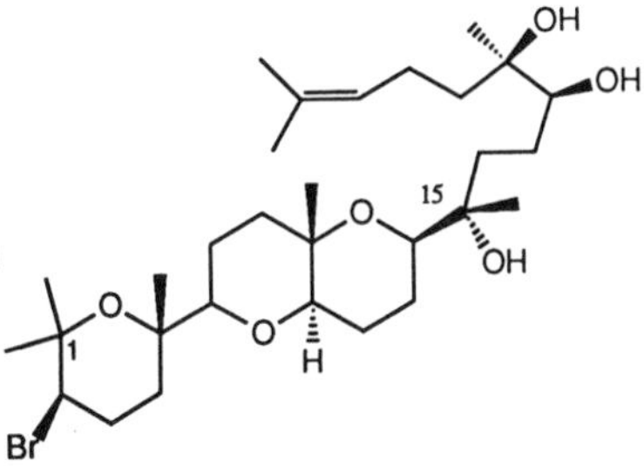

(228) Magireol A

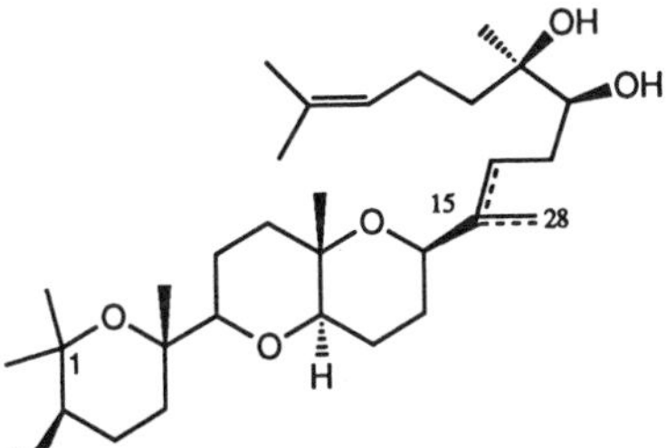

(229) Magireol B, $\Delta^{15,28}$
(230) Magireol C, Δ^{15}

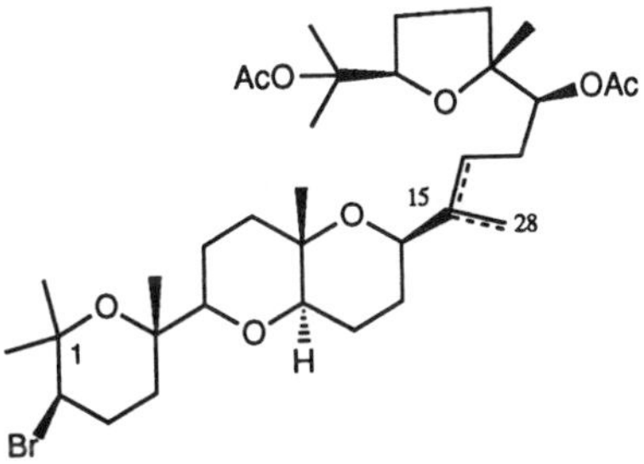

(231) 15(28)-Anhydrothyrsiferol
diacetate, $\Delta^{15,28}$
(232)
15-Anhydrothyrsiferol diacetate, Δ^{15}

	P388 ED_{50} (μg/mL)
226	0.0003
227	0.01
228	0.03
229	0.03
230	0.03
231	0.05
232	0.10

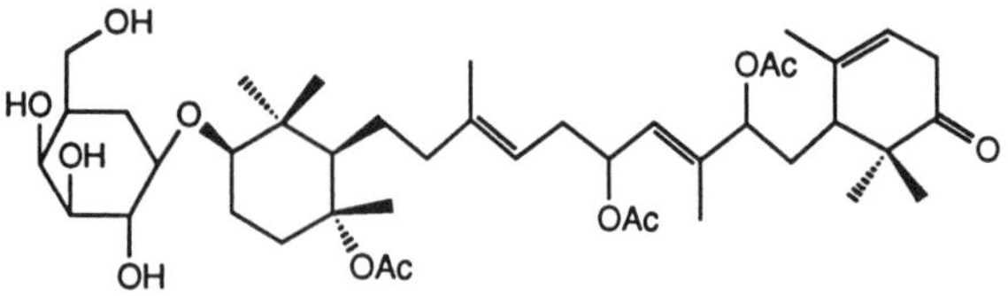

(233)

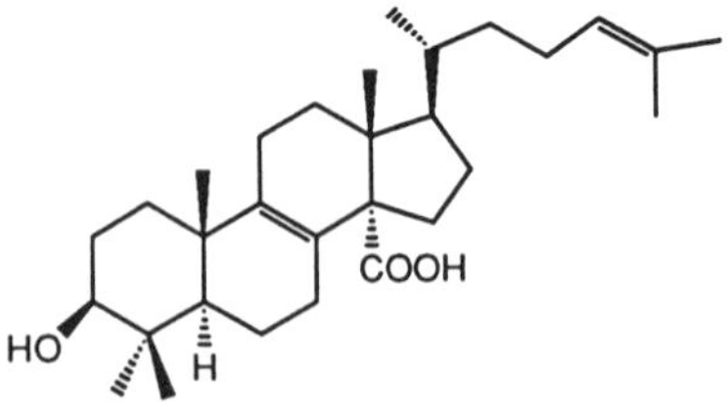

(234)

Pouoside A (**233**), a metabolite from an *Asteropus* sp. of sponge collected in Truk Lagoon, inhibited P388 cell growth with an ED_{50} of 1.5 µg/ml (Ksebati *et al.*, 1988, 1989). The Okinawan sponge *Penares* sp. was the source of penasterol (**234**), which was active against L1210 cells with an ED_{50} of 3.6 µg/ml (Cheng *et al.*, 1988a).

3.6. Sterols

3.6.1. Oxygenated Sterols and Saponins

A number of polyoxygenated sterols and glycosylated sterols have been reported to display cytotoxicity. Sarasinoside A_1 (**235**), a saponin containing amino sugars, exhibited an ED_{50} of 2.8 µg/ml against P388 cells. This saponin was isolated by Schmitz *et al.* (1988) from an *Asteropus* sp. of sponge from Truk and Guam Islands and by Kitagawa *et al.* (1987c) from the sponge *Asteropus sarasinosum* from Palau. The seco sterol astrogorgiadiol (**236**) inhibited fertilized starfish eggs, IC_{50} 50 µg/ml (Fusetani *et al.*, 1989e). This unusual sterol was isolated from a gorgonian, *Astrogorgia* sp. From another gorgonian, *Anthoplexaura dimorpha*, from Japan, Fusetani *et al.* (1987b) isolated dimorphosides A and B (**237, 238**), which also inhibit development of fertilized sea urchin eggs (IC_{50} 6 µg/ml). Eryloside A (**239**), a 4-methyl sterol glycoside extracted from the Red Sea sponge *Erylus lendenfeldi*, has been reported to have both cytotoxic (IC_{50} 4.2 µg/ml vs. P388) and antifungal activity against *Candida albicans* (MIC 15.6 µg/ml) (Carmely *et al.*, 1989a). The epoxy sterol **240** obtained from the mollusk *Planaxis sulcatus* has also been reported to be cytotoxic, but no data were given (Alam *et al.*, 1988).

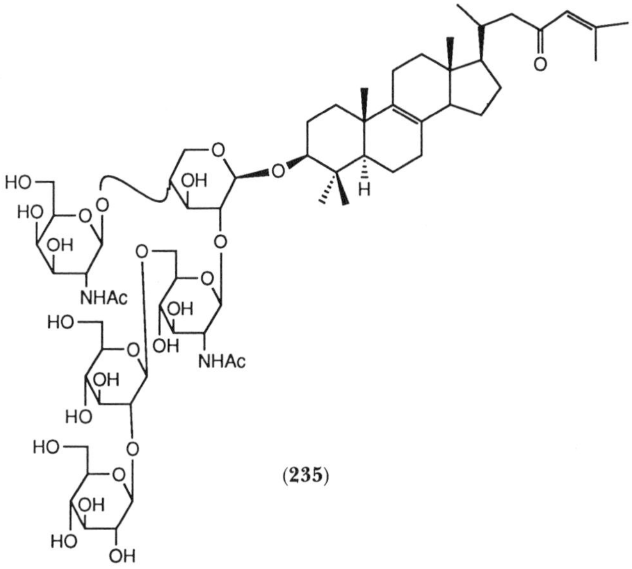

(235)

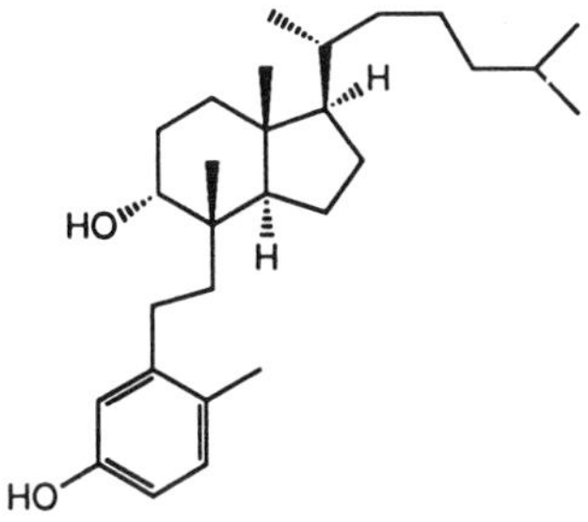

(236)

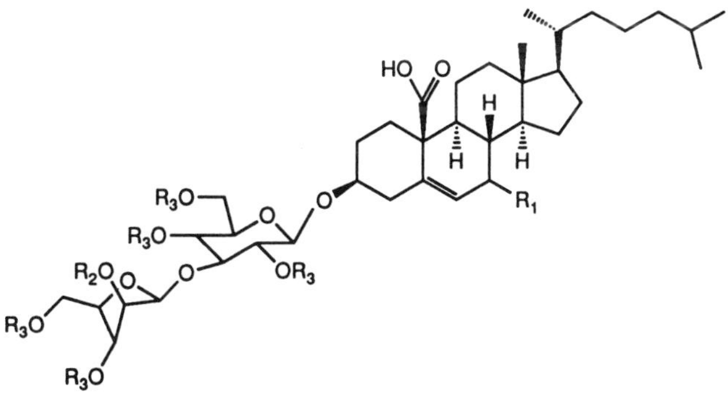

(237) Dimorphoside A: R_1 = OH, R_2 = R_3 = H
(238) Dimorphoside B: R_1 = R_3 = H, R_2 = Ac

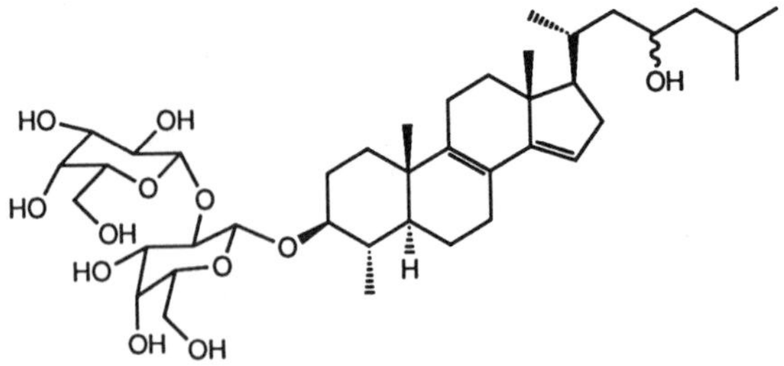

(239)

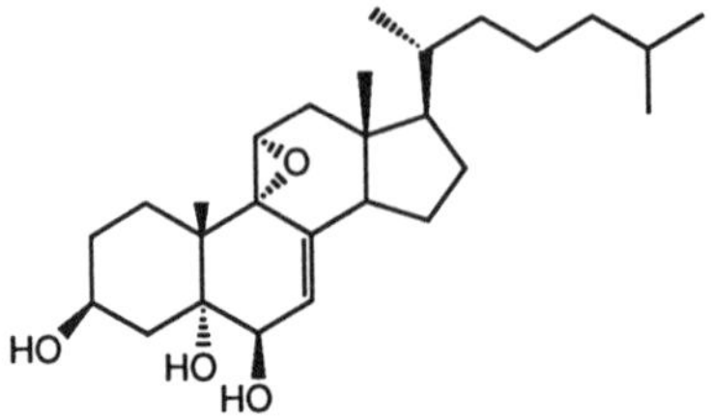

(240)

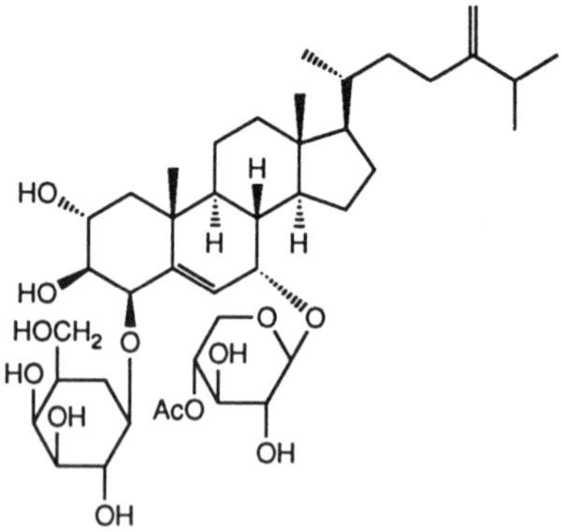

(241)

A sponge of the genus *Pachastrella* collected in Kamagi Bay, Japan, yielded pachastrelloside A (241). This compound was found to inhibit the cell division of fertilized starfish eggs, but did not affect nuclear divisions and hence allowed the formation of multinucleated unicellular embryos (Hirota *et al.*, 1990a). Higuchi *et al.* (1988) tested four highly oxygenated sterols isolated from the starfish *Asterina pectinifera* against L1210 and KB cells. All were very weakly active, but only 242 was active at less than 20 μg/ml (i.e., 14 μg/ml) against L1210. West and Cardellina (1988) reported the sterol 243 from the sponge *Dysidea etheria* to be cytotoxic to KB cells (ED_{50} 4.7 μg/ml) and also to be lethal to brine shrimp (LD_{50} 18 μg/ml). Kitagawa *et al.* (1986a) described a series of new sterols isolated from a

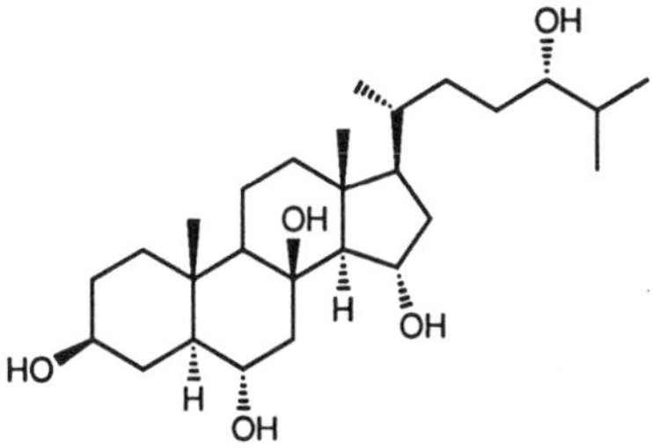

(242)

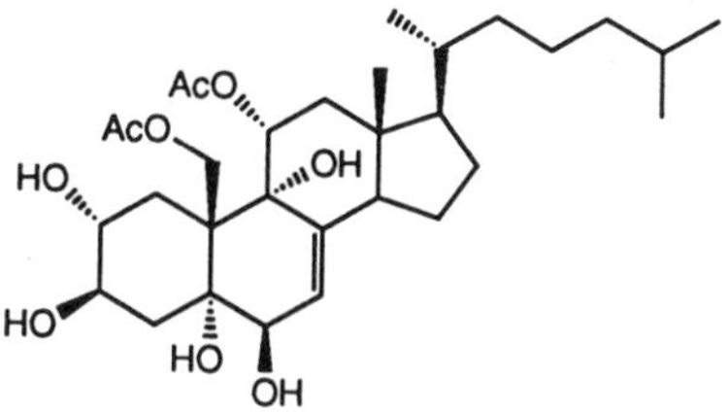

(243)

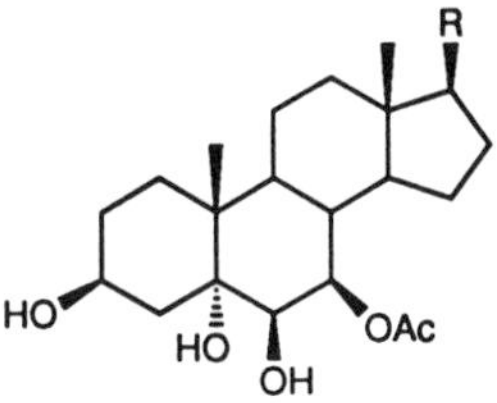

(244) Xeniasterol-a: R =

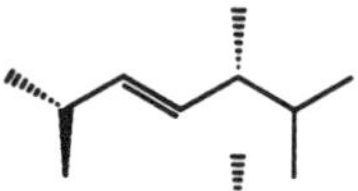

(245) Xeniasterol-b: R =

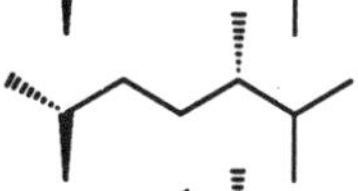

(246) Xeniasterol-d: R =

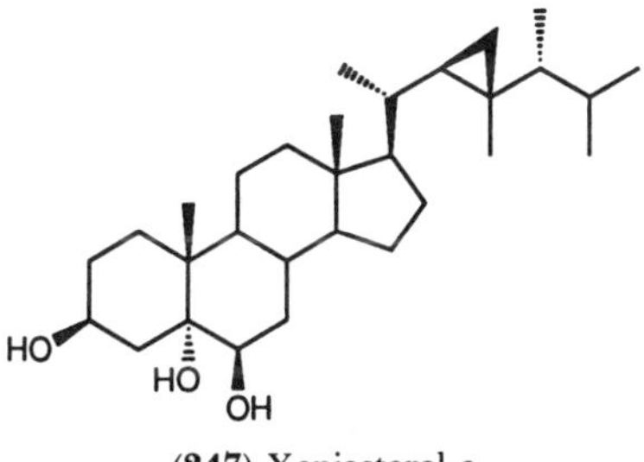

(247) Xeniasterol-c

Xenia sp. of soft coral and reported that a mixture of **244–247** was active against B-16 cells (IC_{50} 5 µg/ml). The sterol sulfate **248**, obtained from three Pacific ophiuroids, was reported to be active against mouse T-cell lymphoma cells, but no data were given (D'Auria *et al.*, 1987). The sulfated sterol glycosides **249** and **250** (imbricatosides A and B) were found to inhibit cell division of fertilized sea urchin eggs (Bruno *et al.*, 1990). These glycosides were isolated from the starfish *Dermasterias imbricata*. Toxicity to L1210 and KB cells was reported for the

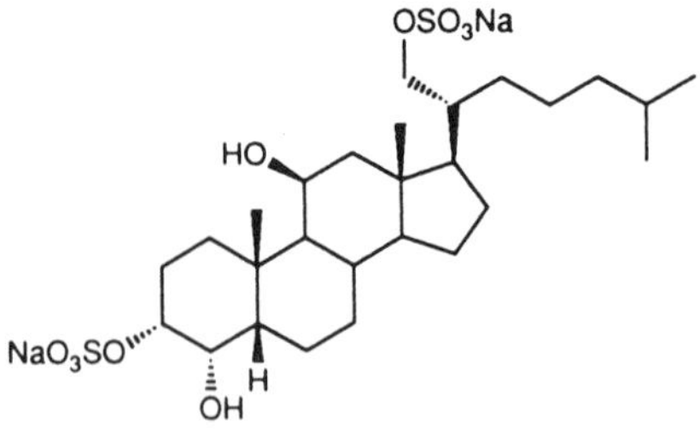

(248)

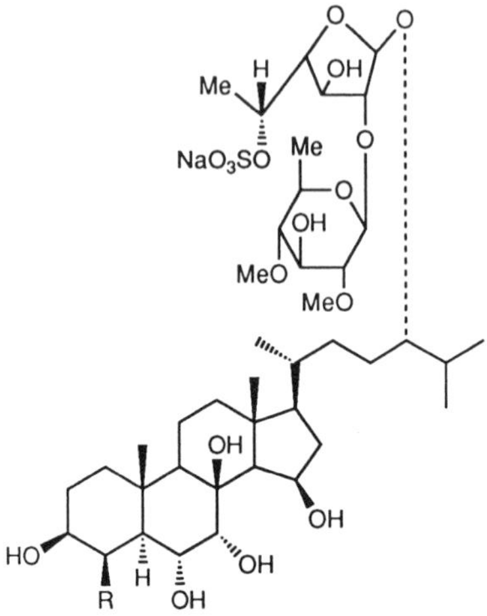

(249) R = OH
(250) R = H

sulfated steroid glycosides named pectiniosides A (**251**), C (**252**), and E (**253**), which were isolated from the starfish *Asterina pectinifera* (see the activity data given with the structures) (Dubois *et al.*, 1988). Andersson *et al.* (1989) tested the biological activity of 16 saponins and saponin-like compounds against human lymphoma cells (JURCAT); significant cell growth inhibition at 5 μg/ml or lower was noted for three polyhydroxysterols (**254–256**). Six new triterpenoid glycoside sulfates (**257–265**) from the sea cucumber *Cucumaria echinata* and three aglycones derived therefrom were active against L1210 and KB cells (see the data given with the structures) (Miyamoto *et al.*, 1990).

3.6.2. Cephalostatins

The most remarkable metabolites in the sterol category are a series of dimeric sterols designated cephalostatins (**266–271**). All of these unusual compounds have been obtained from a marine worm, *Cephalodiscus gilchristi*, collected in the

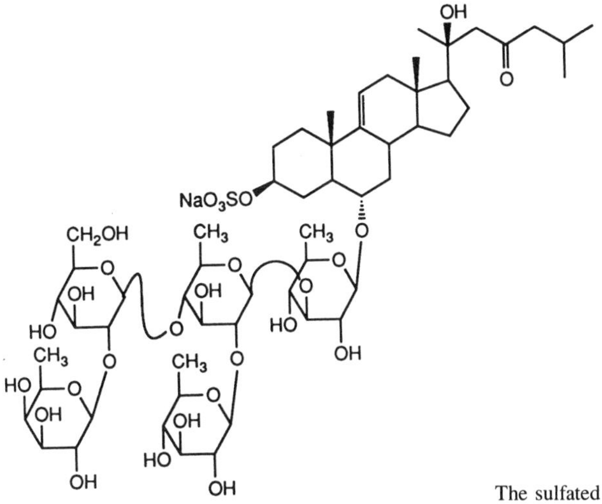

(251)

The sulfated steroid moiety
is the same for 251–253

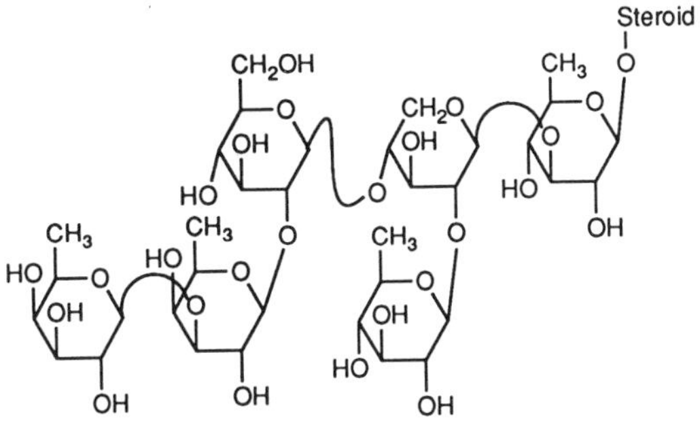

(252)

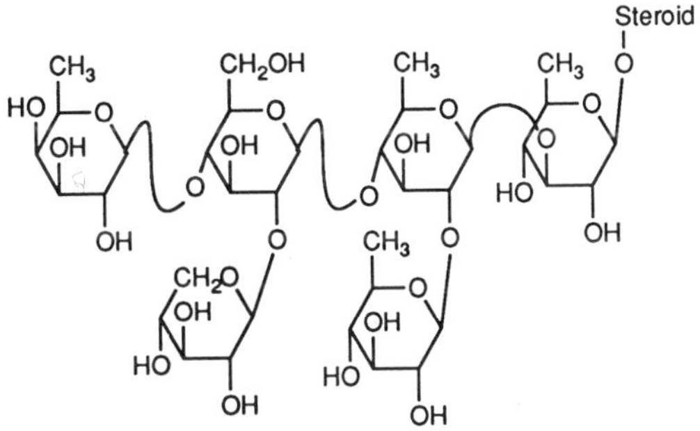

(253)

	LB1210 IC$_{50}$ (μg/mL)	KB IC$_{50}$ (μg/mL)
251	10.0	10.0
252	11.0	10.8
253	8.8	11.5

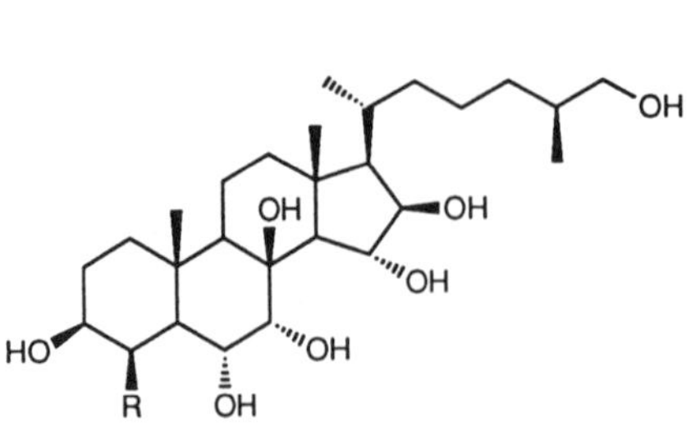

(**254**) R = H
(**255**) R = OH

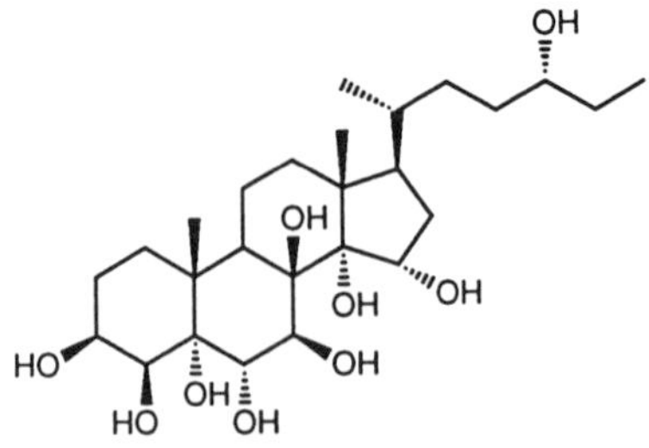

(**256**)

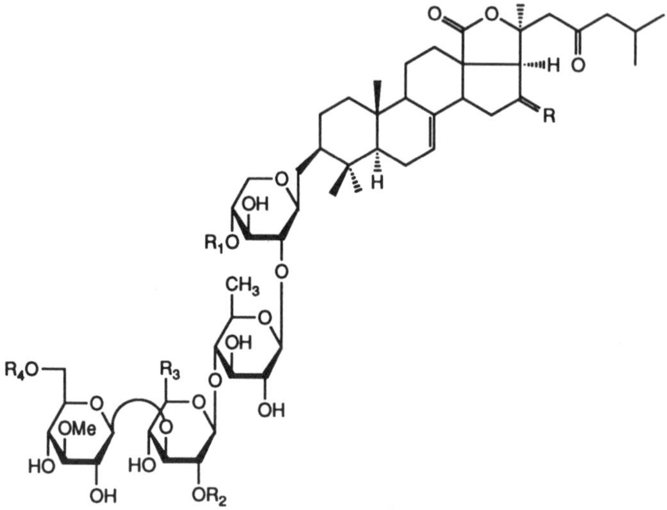

Compound	R	R_1	R_2	R_3	R_4	IC$_{50}$ (μg/mL) L1210	KB
257	O	SO$_3^-$	H	CH$_2$OSO$_3^-$	H	1.7	4.0
258	O	SO$_3^-$	SO$_3^-$	H	H	2.9	6.3
259	H$_2$	SO$_3^-$	H	CH$_2$OSO$_3^-$	H	2.8	4.0
260	O	SO$_3^-$	H	CH$_2$OSO$_3^-$	SO$_3^-$	8.4	7.6
261	O	SO$_3^-$	SO$_3^-$	H	SO$_3^-$	20.0	36.0
262	H$_2$	SO$_3^-$	H	CH$_2$OSO$_3^-$	SO$_3^-$	2.7	3.6
263	O	H	H	CH$_2$OH	H	0.34	1.2
264	O	H	H	H	H	0.32	0.7
265	H$_2$	H	H	Ch$_2$OH	H	0.26	1.1

Indian Ocean in the area of southeast Africa. The first of these to be reported, cephalostatins 1 (**266**) (Pettit *et al.*, 1988c) and 2–4 (**267–269**) (Pettit *et al.*, 1988a), with ED$_{50}$ vs. P388 in the range of 10^{-7}–10^{-9} μg/ml, are extraordinarily potent cytotoxins. A minor structure correction for cephalostatins 2–4 appeared later in 1988 (Pettit *et al.*, 1988b). Cephalostatins 5 (**270**) and 6 (**271**), while less

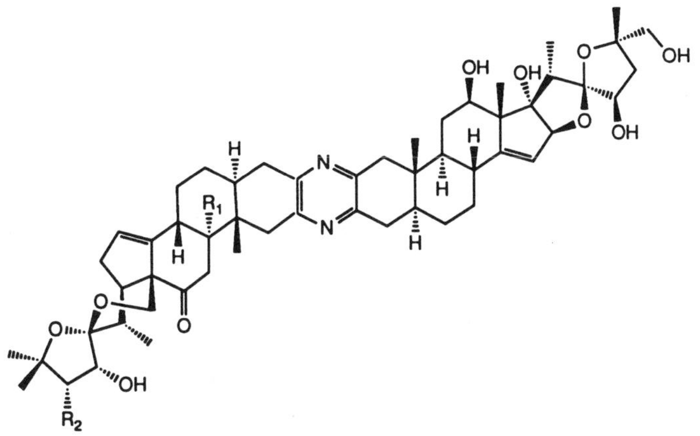

(266) Cephalostatin 1: $R_1 = R_2 = H$
(267) Cephalostatin 2: $R_1 = OH$, $R_2 = H$
(268) Cephalostatin 3: $R_1 = OH$, $R_2 = CH_3$

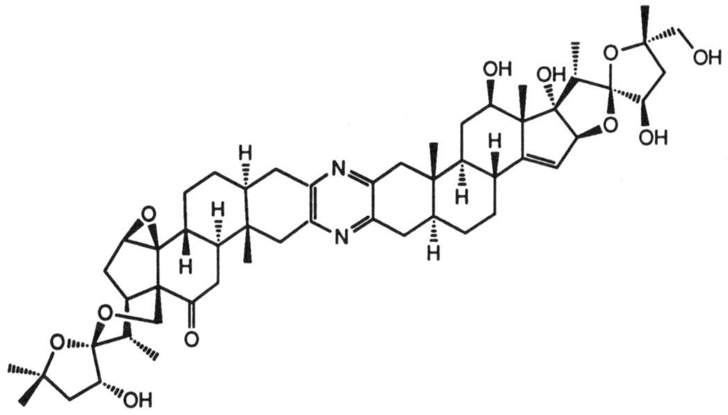

(269) Cephalostatin 4

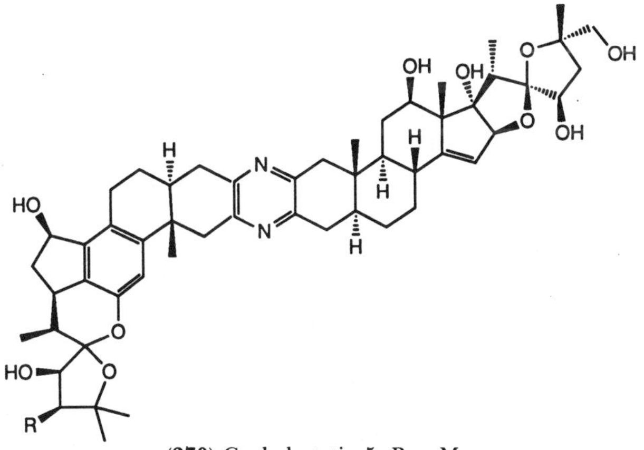

(270) Cephalostatin 5: $R = Me$
(271) Cepahlostatin 6: $R = H$

cytotoxic than the other four members of this group, are still quite active (ED_{50} 10^{-3} and 10^{-2} µg/ml vs. P388) (Pettit *et al.*, 1989e; Kamano *et al.*, 1988).

4. NITROGEN-CONTAINING COMPOUNDS

4.1. Amides of Fatty Acids

Symbioramide (**272**) is a simple amide derivative which is mildly cytotoxic to P388 cells (IC_{50} 9.5 µg/ml), but also activates SR Ca^{2+}-ATPase activity (J. Kobayashi *et al.*, 1988d). Its source was a cultured dinoflagellate, *Symbiodinium*

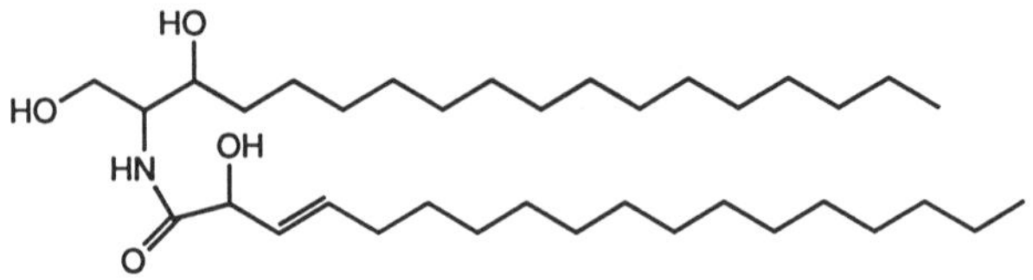

(272)

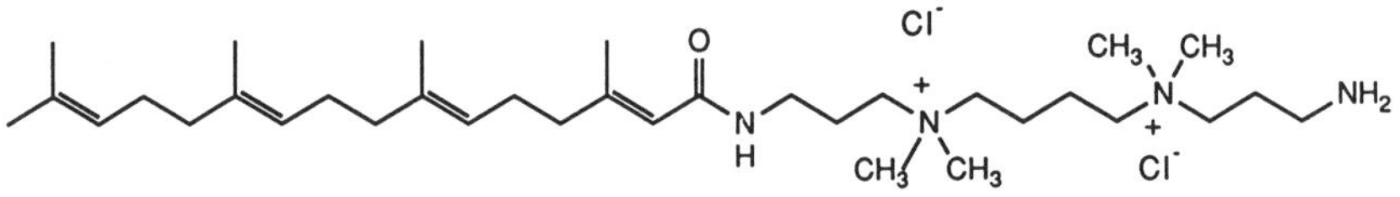

(273)

sp. The quaternized spermidine derivative (**273**), isolated from a *Sinularia* sp. of soft coral, is toxic to L1210 cells (IC_{50} 3.1 µg/ml) (T. Hashimoto *et al.*, 1988). It is one of several spermine or spermidine derivatives that have been isolated from soft corals.

4.2. Tyrosine-Based Metabolites

Tyrosine-based compounds were among the earliest metabolites isolated from sponges. One of these, aeroplysinin-I (**274**), has more recently been reported to have cytostatic activity (Kreuter *et al.*, 1989). The (+) form of **274** isolated from *Aplysina* (formerly *Verongia*) *aerophoba* was found to be cytotoxic to L5178Y mouse lymphoma cells (ED_{50} 0.3 µM), Friend erythroleukemia cells (ED_{50} 0.7 µM), human mammary carcinoma cells (ED_{50} 0.3 µg/ml), and human colon

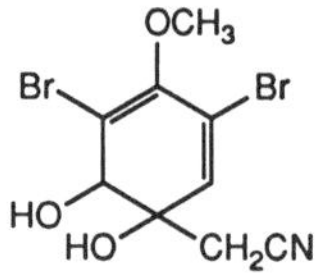

(274)

carcinoma cells (ED_{50} 3.0 μg/ml). Interestingly, aeroplysinin-I caused preferential inhibition of [^{3}H]thymidine (dThd) incorporation rates in L5178Y mouse lymphocytes compared to murine spleen lymphocytes. Aeroplysinin-I also displayed *in vivo* antileukemic activity using the L5178Y cell/NMRI mouse system. The T/C value determined was 338 at a dose of 50 mg/kg for five consecutive days. Aeroplysinin was neither a direct mutagen nor a premutagen in the *umu/Salmonella typhimurium* test system.

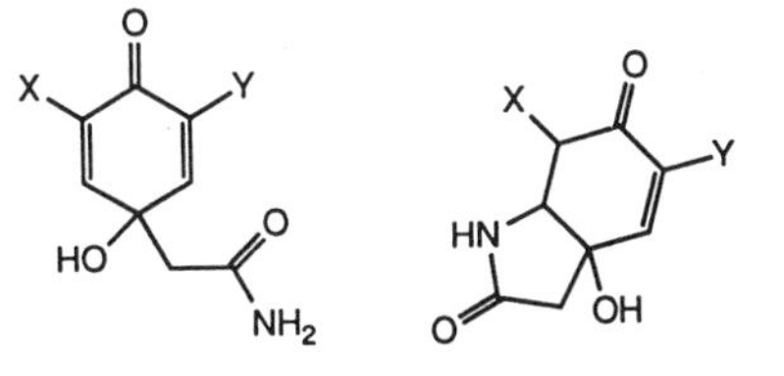

(275) X = Y = Br   (278) X = Y = Br
(276) X = Br, Y = Cl   (279) X = Br, Y = Cl
(277) X = Y = Cl   (280) X = Y = Cl

Tyrosine metabolites (275–280) isolated from *Aplysina fistularia* were all reported to have antineoplastic activity, but no data were given (Goo, 1985). Quinoa and Crews (1987a) reported that the disulfide 281 from the sponge *Psammaplysilla* sp. from Tonga was cytotoxic (IC_{50} 0.3 μg/ml) to P388 cells, but Arabshahi and Schmitz (1987) did not observe significant levels of P388 activity

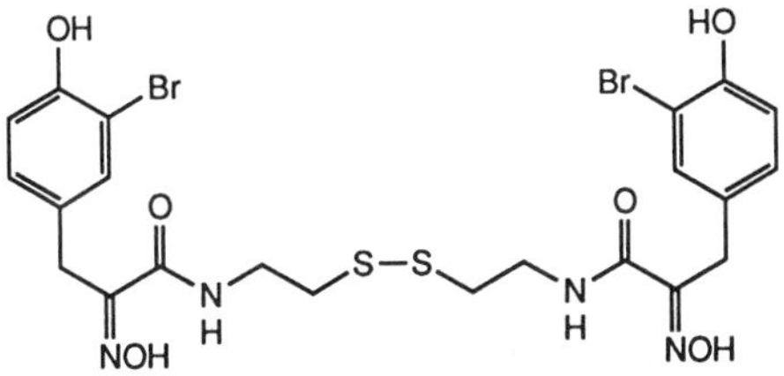

(281)

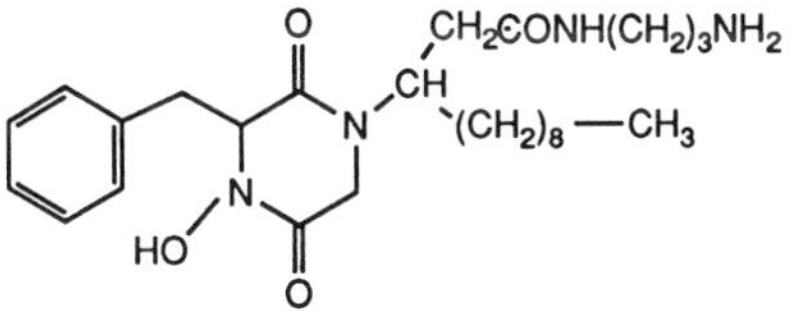

(282)

for this same compound. The diketopiperazine 282, etzionin, from a Red Sea tunicate was active against P388 at 10 μg/ml (Hirsch *et al.*, 1989). Several members of the bastadin series of cyclic amides (283–285) isolated from the sponge *Ianthella basta*, were found by Pordesimo and Schmitz (1990) to inhibit P388 cell growth (ED_{50} 2–4 μg/ml).

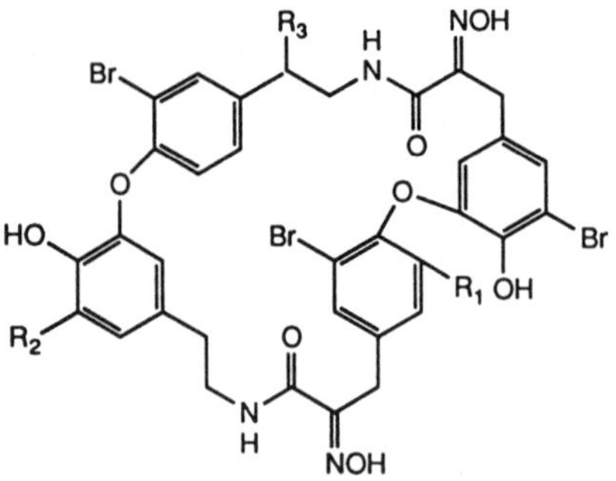

(283) Bastadin 8: R_1 = Br, R_2 = Br, R_3 = OH
(284) Bastadin 9: R_1 = Br, R_2 = H, R_3 = H

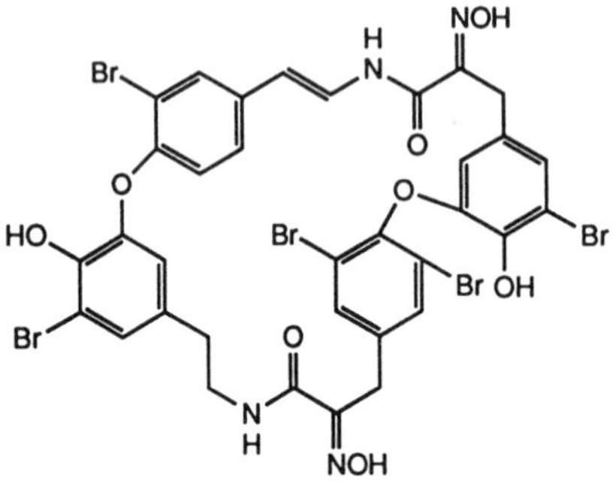

(285)

4.3. Other Amides

Bistratenes A and B (286, 287) are unusual cyclic amide-ethers that were isolated from the tunicate *Lissoclinum bistratum* collected at Heron Island, Australia. These compounds were reported to be toxic to both MRC5CV1 fibroblast cells and T24 bladder carcinoma cells, with IC_{50} of 0.07–0.09 µg/ml (Degnan *et al.*, 1989b). Mycalamides A (288) and B (289) were obtained from a New Zealand sponge, *Mycale* sp. (Perry *et al.*, 1988a, 1990), and showed antiviral and cytotoxic activity. The authors pointed out that the heterocyclic portion of 288–290 is very similar to pederin, an insect toxin known to have significant biological activities. In a more in-depth paper on activity, Burres and Clement (1989) reported that 288 and 289 are toxic to several cell lines (see the data given with the structures). These amides showed modest *in vivo* activity against P388, but mycalamide A (288) was much more active against several solid tumors. The mycalamides block DNA and protein synthesis (P388 system), with lesser effects on RNA synthesis. Onnamide A (290), from a *Theonella* sp. of sponge, shares a good portion of its structure with the mycalamides (Sakemi *et al.*, 1988). Although

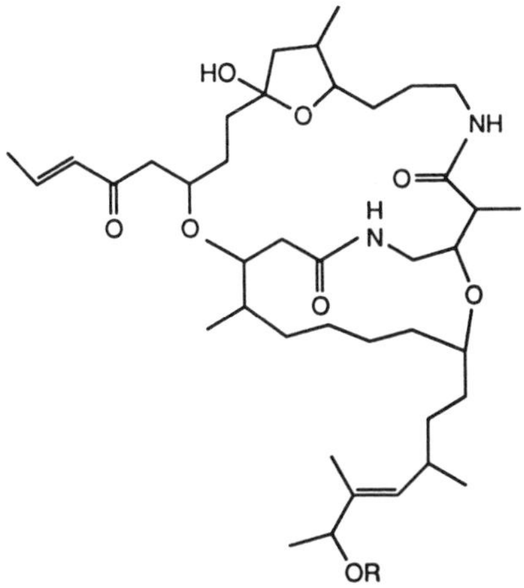

(286) Bistratene A: R = H
(287) Bistratene B: R = Ac

onnamide A inhibited P388 cell growth (IC_{50} 2.4 μg/ml), it does not show toxicity to as broad a spectrum of cell lines as do the mycalamides and it was inactive *in vivo* against P388 (Burres and Clement, 1989).

Bengamide A (291) was discovered in the extracts of a Jaspidae sponge using a bioassay for anthelminthic activity (Quinoa *et al.*, 1986a). Subsequently it was confirmed that 291 is active against both P388 and KB cells *in vitro* (IC_{50} 0.4 and 0.04 μg/ml, respectively) (P. Crews, personal communication).

Calyculins A–D (292–295) are a group of very unusual sponge metabolites described over a period of several years from the sponge *Discodermia calyx* (Kato *et al.*, 1986a,b, 1988b). Cytotoxicity data are shown under the structures. Calyculins A–D also inhibited cell division of both starfish and sea urchin eggs in the 10^{-2} μg/ml range. Calyculin A (292) exhibited *in vivo* activity against Erlich and P388 leukemia in mice (T/C 245 and 144%, respectively). Calyculin A inhibits uptake of [³H]thymidine, [³H]uridine, and [³H]leucine in L1210 murine leukemia cells. This suggested to the authors that the primary target of calyculin A is a system other than macromolecular synthesis (Kato *et al.*, 1988a,b).

An ascidian, *Diazona chinensis*, from Siquijor Island, Philippines, was the source of two very unique cytotoxins, diazonamide A (296) and B (297) (Lindquist *et al.*, 1991). Diazonamide A shows *in vitro* activity against HCT-116 human colon carcinoma and B16 murine melanoma cell lines, with IC_{50} values of less than 0.05 μg/ml. Diazonamide B is "less active."

Moore and Entzeroth (1988) found that the marine blue-green algal metabolites majusculamide D (298a) and dehydroxymajusculamide D (298b) from *Lyngbya majuscula*, collected at Enewetak Atoll, are both toxic to CCRF-CEM cells at 0.2 μg/ml.

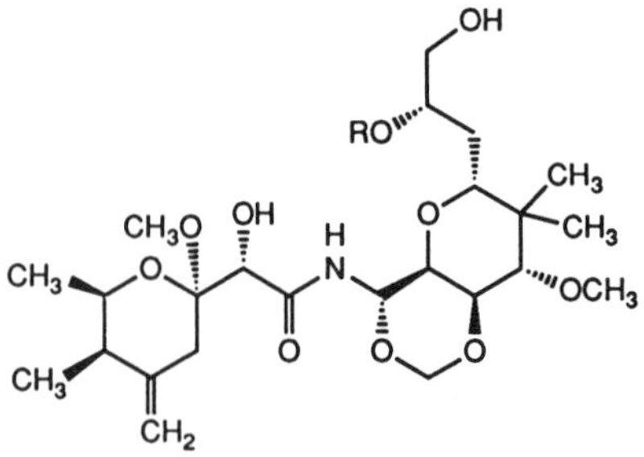

(288) R = H
(289) R = Me

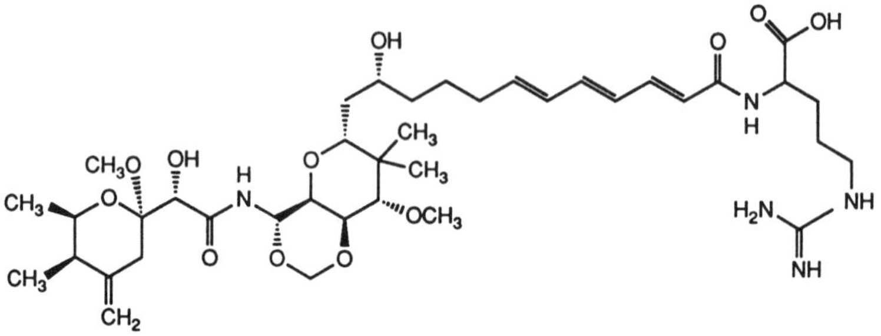

(290)

| | | IC$_{50}$ (nM) | | |
		P388	**HL-60**	**HT-29**	**A549**
288	Mycalamide A	5.2	3.0	2.8	3.6
289	Mycalamide B	1.3	1.5	1.5	0.6
290	Onnamide	2.4	25	180	170

In vivo, **P388**
288 T/C 140 (10 μg/Kg)
289 T/C 150 (2.5 μg/Kg)
290 T/C 115 (40 μg/Kg), **inactive**

Solid tumors, 288

B16 (melanoma)	T/C 245 (30 μg/Kg), 40% cures
M5076 (ovarian carcinoma)	T/C 233 (60 μg/Kg), 40% cures
Colon 26 carcinoma	T/C 149 (60 μg/Kg), 20% cures

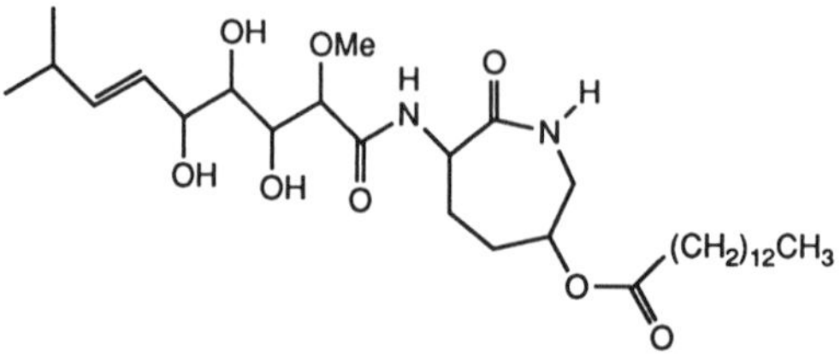

(291)

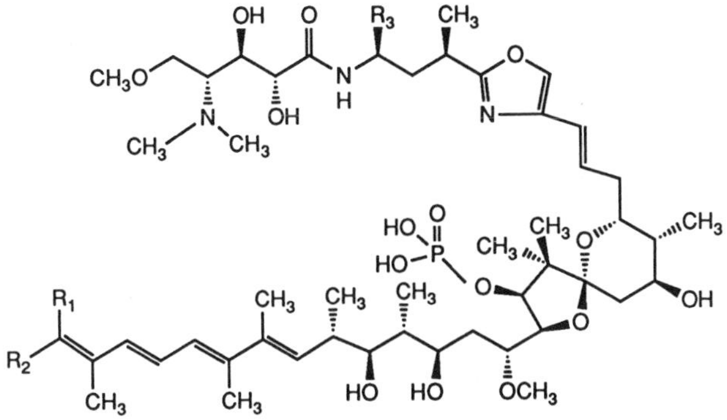

(292) Calyculin A: R_1 = CN, R_2 = R_3 = H
(293) Calyculin B: R_1 = R_3 = H, R_2 = CN
(294) Calyculin C: R_1 = CN, R_2 = H, R_3 = CH$_3$
(295) Calyculin D: R_1 = H, R_2 = CN, R_3 = CH$_3$

	L1210 IC$_{50}$ (μg/mL)		**L1210 IC$_{50}$** (μg/mL)
292	7.4×10^{-4}	294	8.6×10^{-4}
293	8.8×10^{-4}	295	1.5×10^{-3}

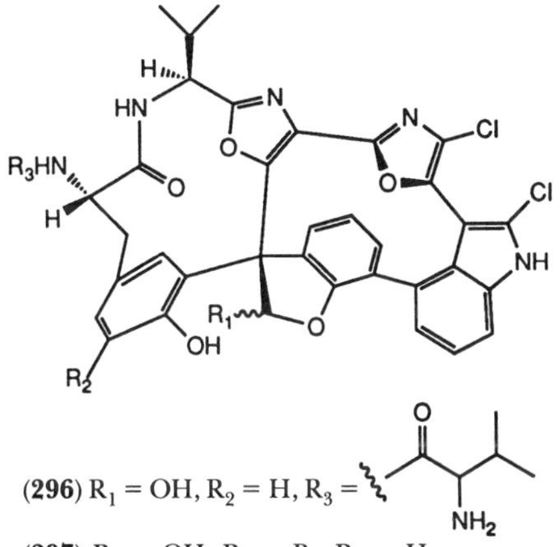

(296) R_1 = OH, R_2 = H, R_3 =
(297) R_1 = OH, R_2 = Br, R_3 = H

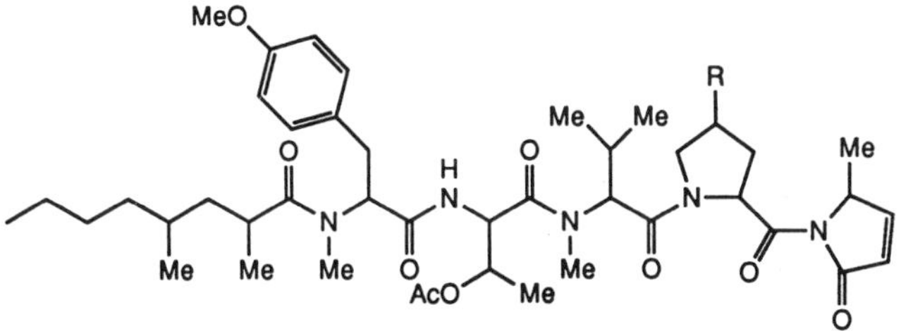

(298a) R = OH
(298b) R = H

4.4. Pyrroles

The pentabromophenylpyrrole **299** (pentabromo pseudeline), which is produced by several marine bacteria, was found to be active *in vitro* against leukemia
and melanoma cells, but was inactive *in vivo* (Laatsch and Pudleiner, 1989). Mild
cytotoxicity to L1210 cells (IC$_{50}$ 10 µg/ml) was reported for alkaloid **300** (Wright

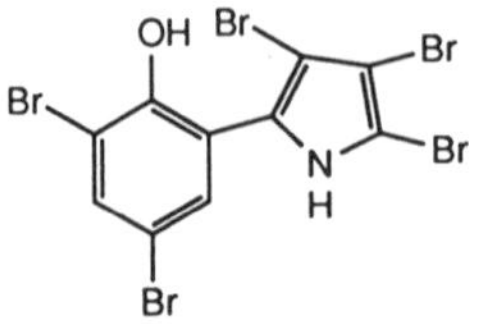

(299) 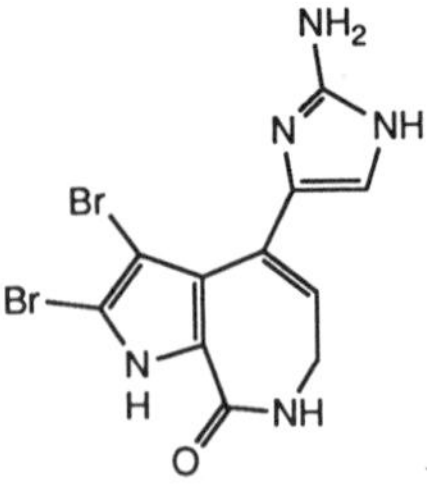   (300)

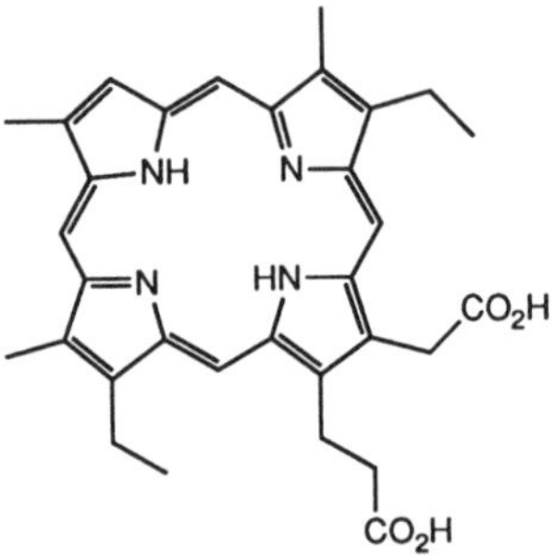

(301)

and Thompson, 1987). It was isolated from the sponges *Teichaxinella morchella*
and *Ptilocaulis walpersi*. The porphyrin corallistin A (**301**), isolated from the
demosponge *Corallistes* sp. from the Coral Sea, is toxic to KB cells (IC$_{50}$ 10
µg/ml), but inactive against leukemia and solid tumor cultures and also inactive
in vivo (D'Ambrosio *et al.*, 1989).

4.5. Imidazoles

The sponge *Pseudaxinyssa cantharella* was the source of girolline (**302**),
which is active against P388 at 0.001–1 µg/ml, and this activity was confirmed *in
vivo* in mice (P388 at 1 mg/kg doses (Ahond *et al.*, 1988). Two unusual stereoisomeric 1,2,3-trithiane derivatives (**303**, **304**) were reported from a New Zealand
ascidian, *Aplydium* sp. (Copp *et al.*, 1989). Both displayed mild activity against
P388 (IC$_{50}$ 12–13 µg/ml) and it is suspected that they may be precursors to similar

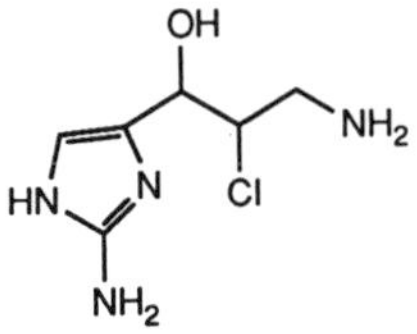

(302)

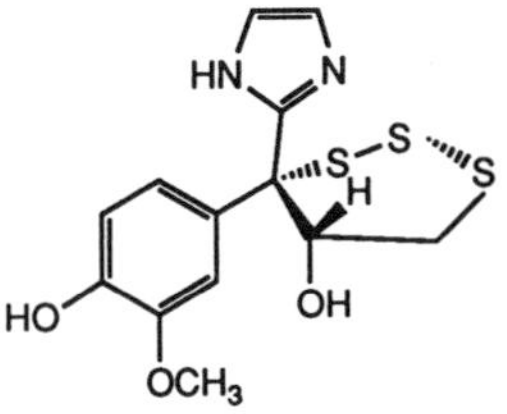

(303) 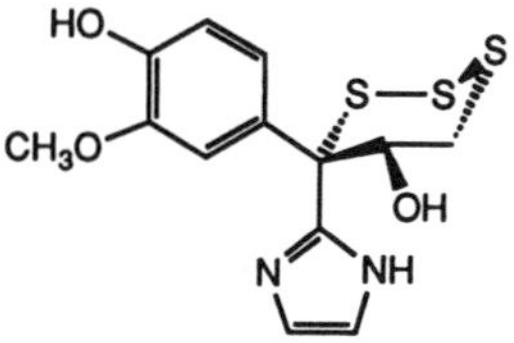   (304)

products reported by Arabshahi and Schmitz (1988). Pyronaamide (**305**), obtained from a *Leucetta* sponge from Saipan and Guam, was toxic to KB cells (MIC 5 µg/ml) (Akee *et al.*, 1990). A series of 2-amino imidazole alkaloids called naamidines (e.g., **306**) were obtained by Carmely *et al.* (1989b) from the marine sponge *Leucetta chagosensis*. The cytotoxicity of compounds in this series is 2–10 µg/ml against P388 (Y. Kashman, personal communication).

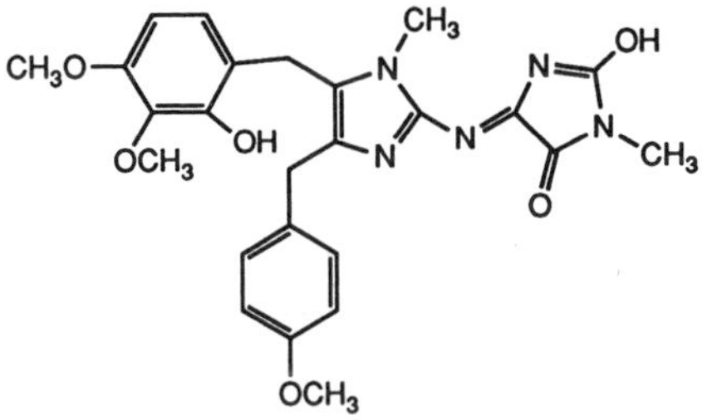

(305)

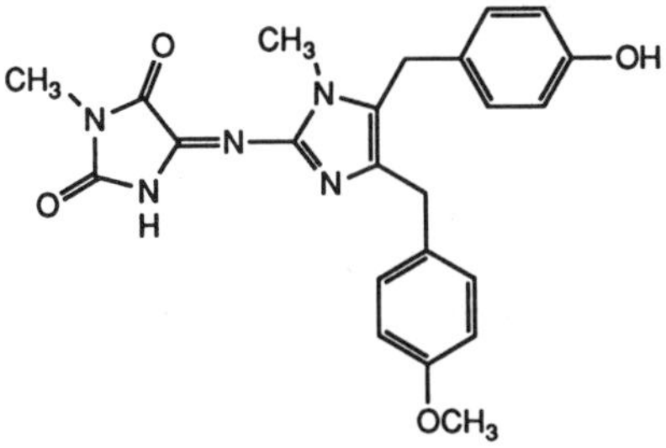

(306)

4.6. Indoles

Herbindoles A–C (**307–309**) are all cytotoxic (KB; MIC 5, >10, and 10 μg/ml, respectively) and were also found to have fish antifeedant activity (Herb *et al.*, 1990). Their source was a sponge, *Axinella* sp., from Western Australia.

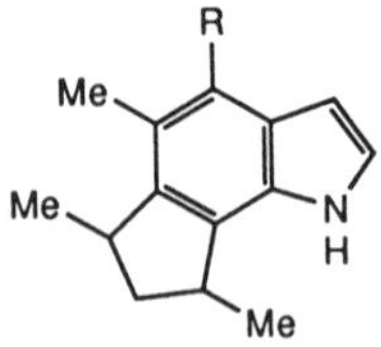

(**307**) Herbindole A: R = Me
(**308**) Herbindole B: R = Et
(**309**) Herbindole C: R = CH=CH-Et

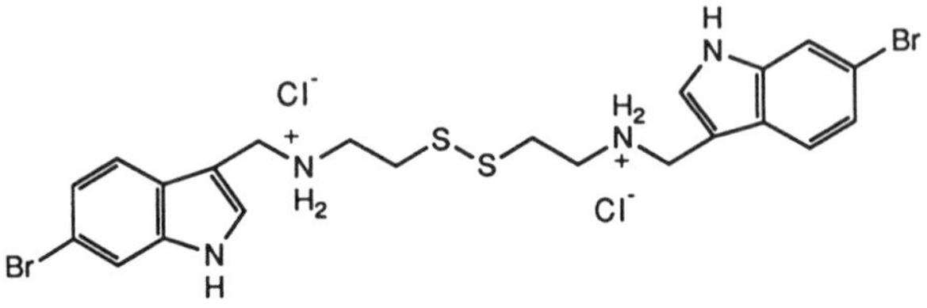

(**310**)

Citorellamine (**310**) is a symmetrical disulfide from the tunicate *Polycitorella mariae* from Fiji, which was initially reported as a monomeric structure (Moriarty *et al.*, 1987; Roll and Ireland, 1985). It is active against L1210 cells (IC$_{50}$ 3.7 μg/ml) and showed antibiotic activity against several microorganisms. A deep-water sponge, *Dragmacidon* sp., afforded dragmacidin (**311**), which is toxic to

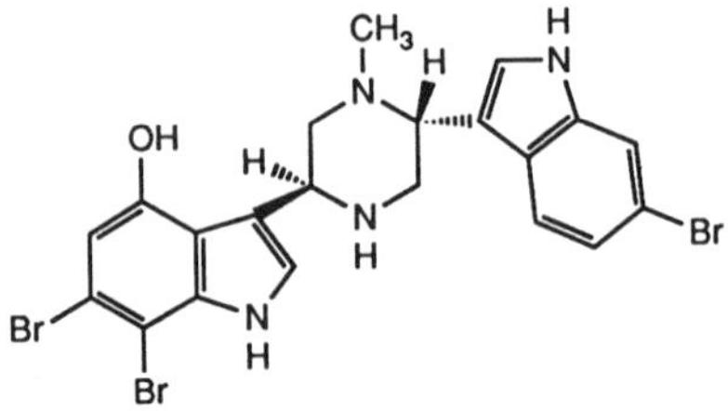

(**311**)

P388 cells (IC$_{50}$ 15 μg/ml) and also to A549 human lung, KCT-8 human colon, and MDAMB human mammary cells, all with IC$_{50}$ of 1–10 μg/ml (Kohmoto *et al.*, 1988). Morris and Andersen (1989) reported the closely related dragmacidon A (**312**); it showed cytotoxicity against L1210 cells (ED$_{50}$ 10 μg/ml) similar to that of **311**. The topsentins were discovered by two different groups at almost the same time. A group in Belgium made the initial report, which described topsentins A

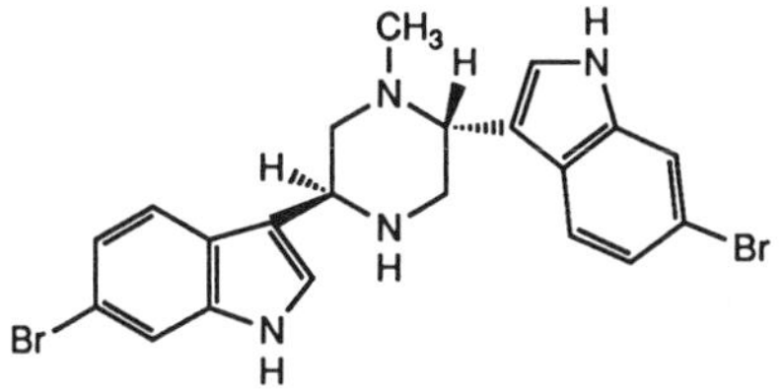

(312)

(313), B1 (314), and B2 (315) obtained from the sponge *Topsentia genitrix* (Bartik *et al.*, 1987; Braekman *et al.*, 1987a). They reported toxicity to fish at 15–20 mg/liter for these bis-indoles. Subsequently, Tsujii and Rinehart (1988) also reported 313 (designed deoxytopsentin), 314 (designated topsentin), and 315 (designated bromotopsentin), in addition to the new metabolite dihydrodeoxy-

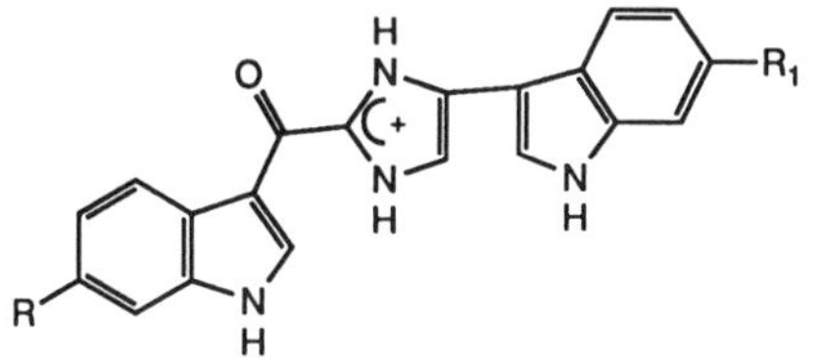

(313) Topsentin A (= deoxytopsentin): R = R_1 = H
(314) Topsentin B1 (= topsentin): R = OH, R_1 = H
(315) Topsentin B2 (= bromotopsentin): R = OH, R_1 = Br

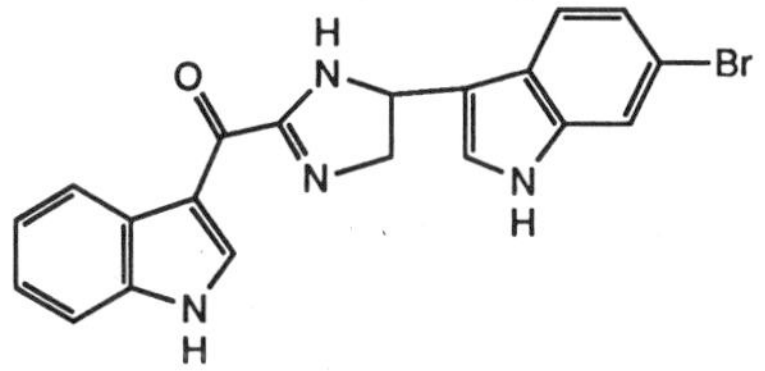

(316)

	P388 (IC_{50}) (*in vitro*) (μg/mL)	P388 (*in vivo*)
313	12.0	
314	2.0	T/C 132 at 75 mg/Kg
315	7.0	T/C 126 at 75 mg/Kg
316	4.0	

bromotopsentin (316), from a deep-sea sponge of the family Halichondriidae. In addition to *in vitro* activities against P388, very weak *in vivo* activities against P388 were reported (see the data given with the structures).

An antimicrobial pigment designated fascaplysin (317), which was obtained from a Fijian sponge, *Fascaplysinopsis* sp., killed L1210 cells (LD_{50} 0.2 μg/ml)

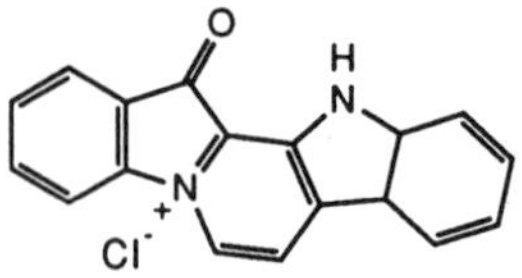

(317)

and also showed antibiotic activity against four different microorganisms (Roll *et al.*, 1988).

The eudistomins were initially reported as antiviral agents, but J. Kobayashi *et al.* (1990b) reported recently that eudistomins B (**318**), C (**319**), and D (**320**) are toxic to L1210 cells (IC$_{50}$ 3.4, 0.36, and 2.4 μg/ml, respectively) and L5178Y

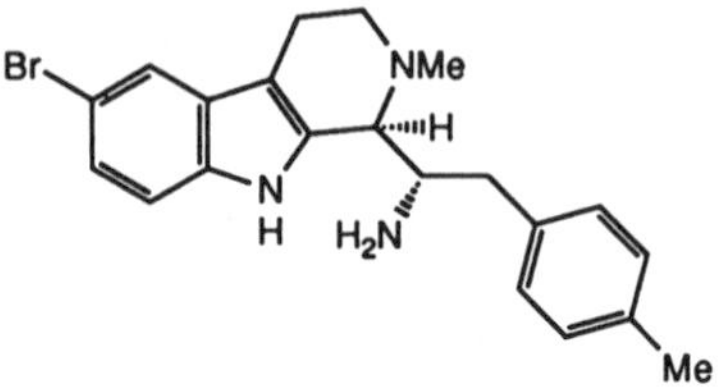

(318)

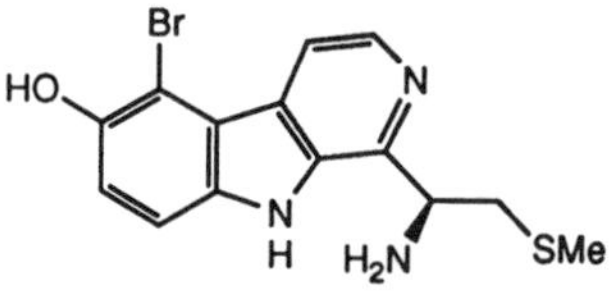

(319)

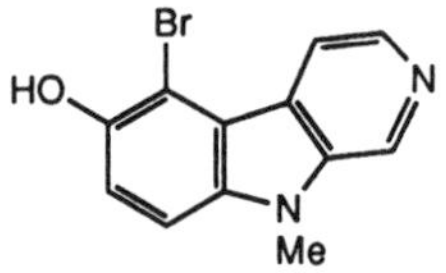

(320)

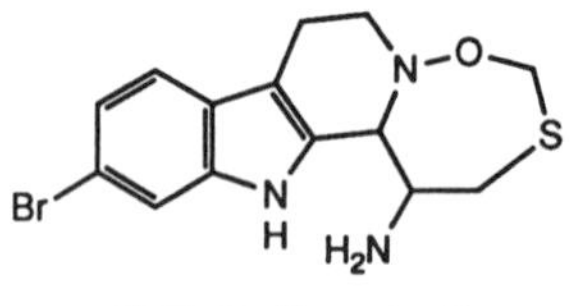

(321) Eudistomin K

cells (IC$_{50}$ 3.1, 0.42, and 1.8 μg/ml, respectively). Eudistomin K (**321**), obtained from *Riterella sigillinoides*, is described in a patent as being "very effective in inhibiting growth of L1210, P388, A549, and HCT-8 cells at varying concentrations" (Blunt *et al.*, 1988).

Grossularine-1 (**322**) and -2 (**323**) were isolated from the tunicate *Dendrodoa grossularia* (Moquin-Pattey and Guyot, 1989). Both were toxic to L1210 cells (IC$_{50}$ 4–6 μg/ml), but were reported to be more toxic to solid tumor cell lines (colon and breast down to 0.001 μg/ml). The authors report that grossularine-2 appears to act on DNA as a mono-intercalating agent.

One of the more complex alkaloid structures is that of the manzamines. Manzamine A (**324**) was reported as its hydrochloride salt from a *Haliclona* sp. of

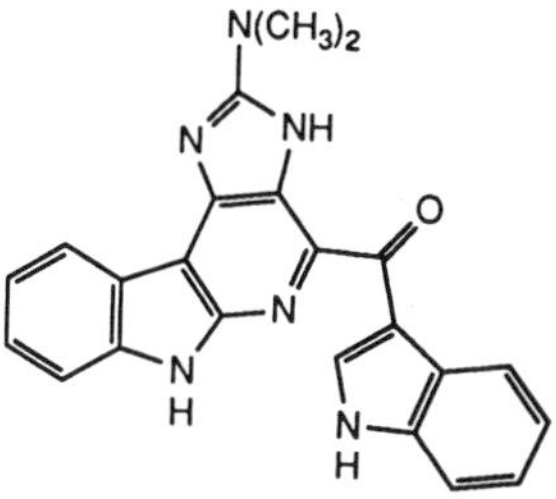

(322)

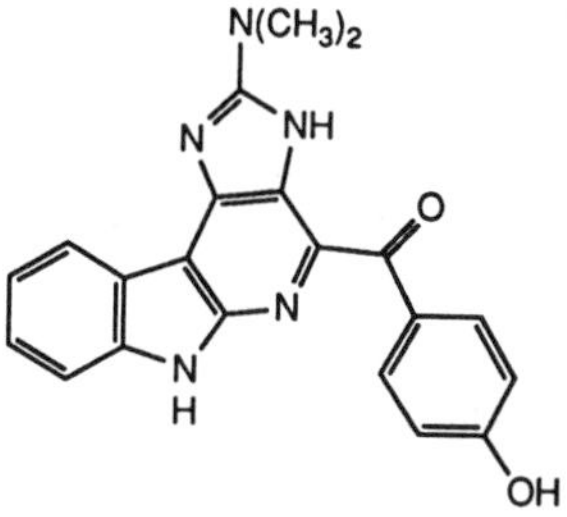

(323)

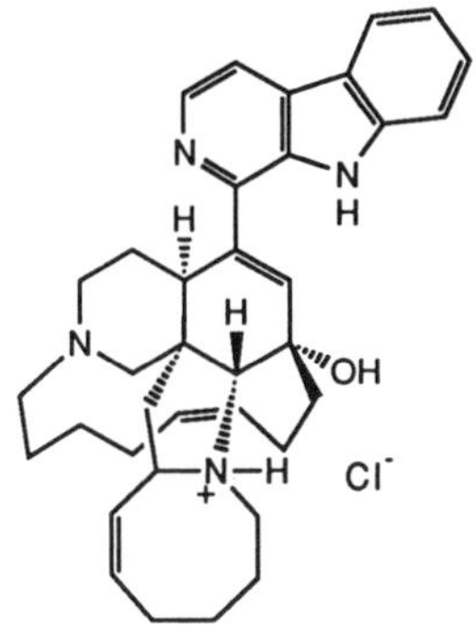

(324)

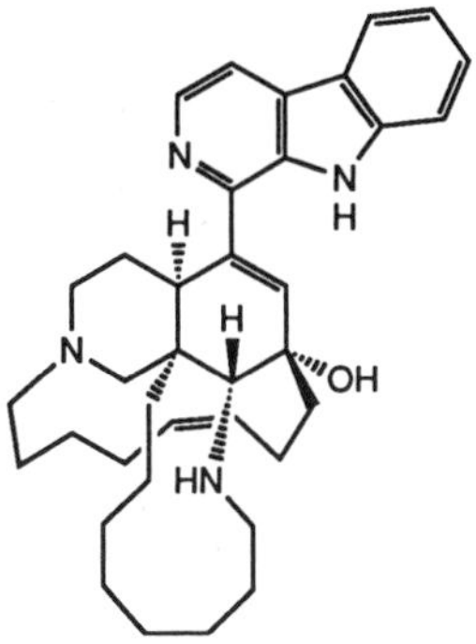

(325)

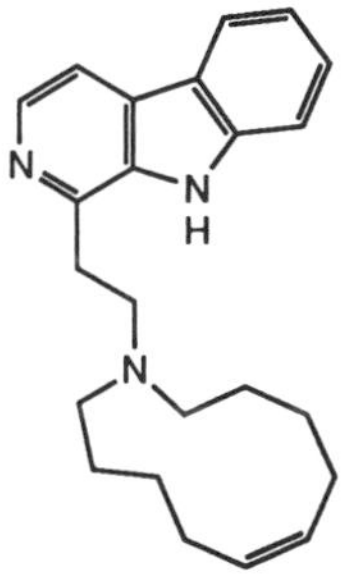

(326)

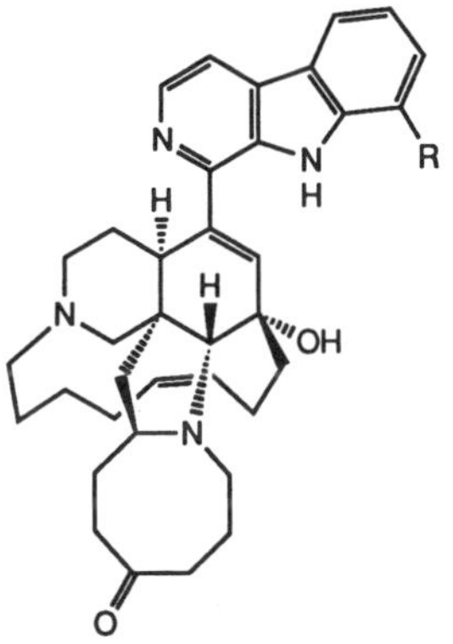

(327) R = H
(328) R = OH

P388 IC$_{50}$
(μg/mL)

324	0.07
325	6
326	3
327	5
328	5

sponge from Okinawa with an IC_{50} of 0.07 μg/ml against P388 cells *in vitro* (Sakai *et al.*, 1986b). This was followed by disclosure of manzamines B (**325**) and C (**326**) from the same sponge (Sakai *et al.*, 1987). Manzamines E (**327**) and F (**328**) were reported later from a sponge of the genus *Xestospongia* (Ichiba *et al.*, 1988). Cytotoxicity data for the manzamines are shown near the structures. Nakamura *et al.* (1987a) also reported the alkaloid **324** from a *Pellina* sp. of sponge, but assigned it the name keramine A. These authors also reported a keramine B, but the structure disclosed was incorrect and should be that of Manzamine F.

4.7. *Pyridines*

In addition to the swinholide macrolides (**96–99**), the pyridine alkaloids theonelladins A–D (**329–332**) have been isolated from the sponge *Theonella swinhoei* (J. Kobayashi *et al.*, 1989c). Cytotoxicities are shown under the struc-

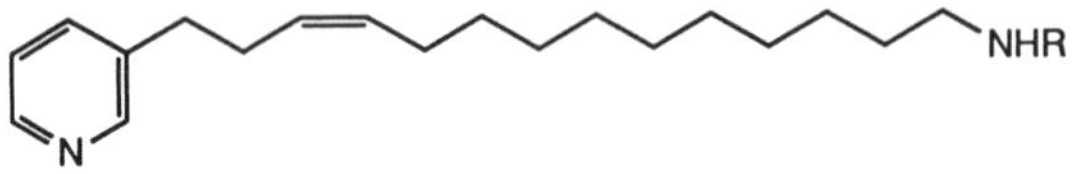

(**329**) Theonelladin A: R = H
(**330**) Theonelladin B: R = CH₃

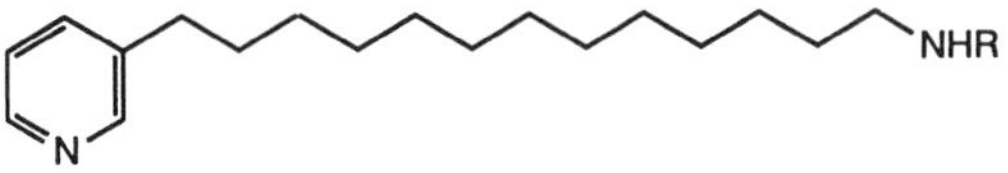

(**331**) Theonelladin C: R = H
(**332**) Theonelladin D: R = CH₃

	L1210 IC_{50} (μg/mL)	KB ED_{50} (μg/mL)
329	4.7	10
330	1.0	3.6
331	3.6	10
332	1.6	5.2

tures. These compounds are also reported to be 20 times more powerful than caffeine in causing release of Ca^{2+} from sarcoplasmic reticulum. The related pyridine alkaloids niphatynes A (**333**) and B (**334**), from a *Niphates* sp. of sponge collected in Fiji, were cytotoxic to P388 cells (IC_{50} 0.5 μg/ml) (Quinoa and Crews,

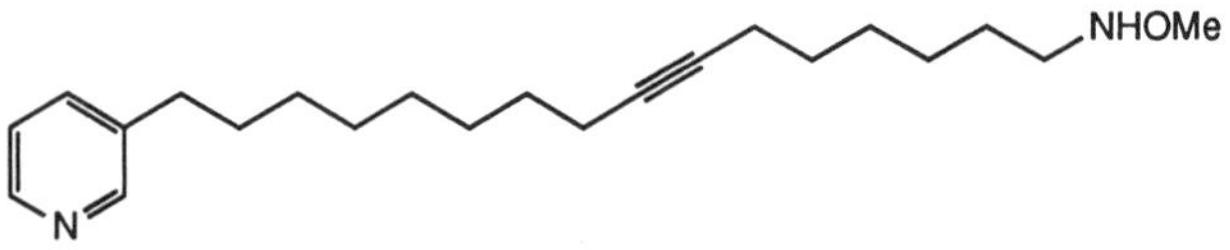

(**333**)

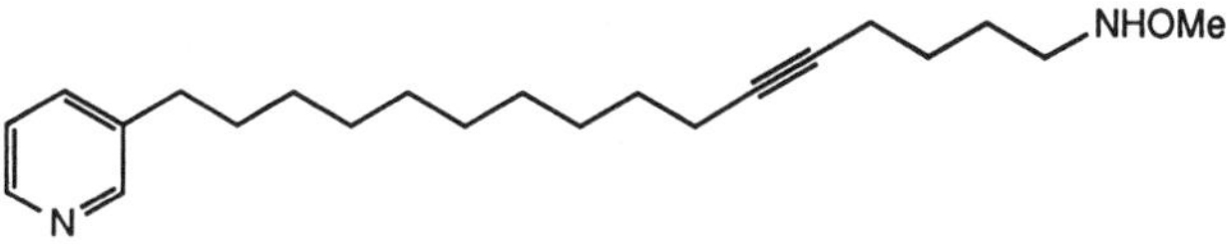

(**334**)

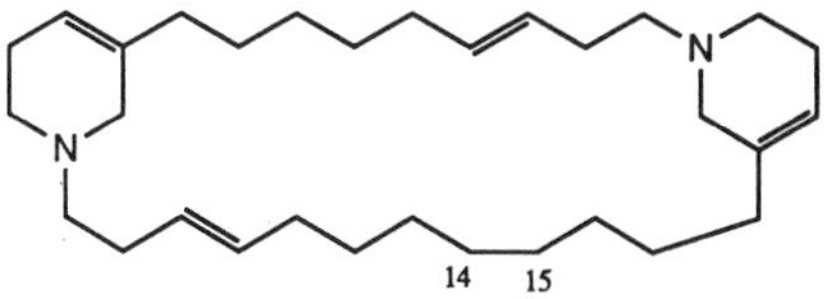

14   15

(**335**) Haliclamine A
(**336**) Haliclamin B, Δ^{14}

	P388 IC_{50} ($\mu g/mL$)	**L1210** IC_{50} ($\mu g/mL$)
335	0.75	1.5
336	0.39	0.9

Inhibition of cell division of eggs
of sea urchin *Hemicentrotus pulcherrimus*

335	5 $\mu g/mL$
336	10 $\mu g/mL$

1987b). The macrocyclic alkaloids haliclamines A (**335**) and B (**336**) were isolated from a sponge of the genus *Haliclona* and reported to be active in various cell assays; see the structures for data (Fusetani *et al.*, 1989c). These alkaloids would appear to be biogenetically related to halitoxin (Schmitz *et al.*, 1978).

The isomeric (E) and (Z) piperidine alkaloids pseudodistomins A (**337**) and B (**338**) were more active against L5178Y cells (IC_{50} for each. 0.4 $\mu g/ml$) than against L1210 cells (IC_{50} for each, 2.4 $\mu g/ml$) (Ishibashi *et al.*, 1987a). The source of these alkaloids was the Okinawan tunicate *Pseudodistoma kanoko*.

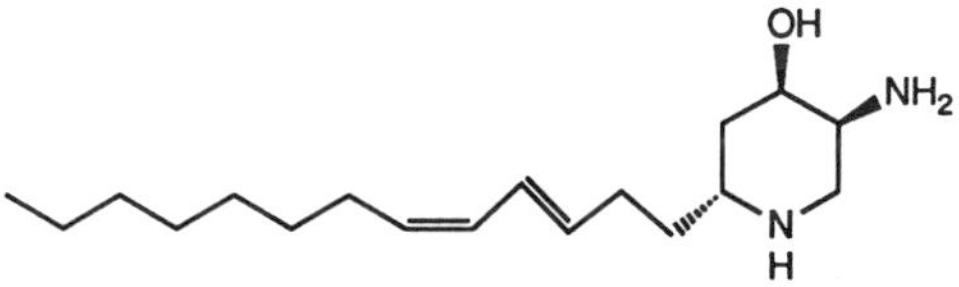

(**337**)

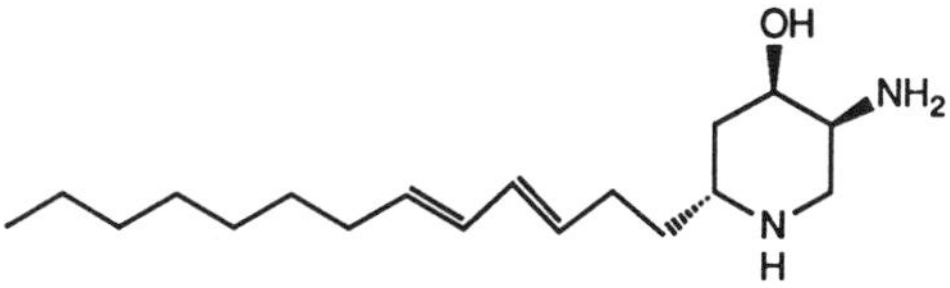

(**338**)

4.8. Quinolines and Isoquinolines

Although aaptamine (**339**) has been known for some time, recently this alkaloid and some derivatives thereof have been reported to have some *in vitro* and *in vivo* cell inhibitory activity. Compounds **340** and **341** were tested for antitumor

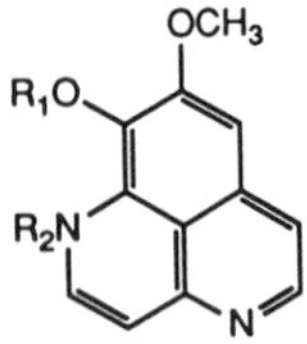

	R_1	R_2
(**339**)	CH$_3$	H
(**340**)	H	H
(**341**)	H	CH$_3$

activity against Ehrlich ascites tumors in mice. A 95% inhibition was reported in the cases of mice inoculated with Ehrlich ascites tumor cells pretreated with **340** or **341** at 25 μg/ml (Fedoreev *et al.*, 1988). These authors isolated **339–341** from a sponge of the genus *Suberites*. Nakamura *et al.* (1987b) reported **340** and **342**, isolated from the sponge *Aaptos aaptos*, to be toxic to HeLa cells, with ED$_{50}$

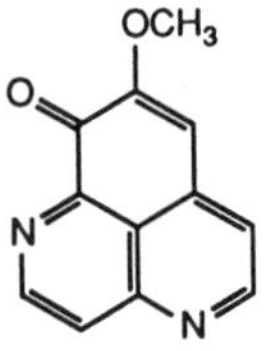

(**342**)

values of 2 and 0.87 μg/ml, respectively. A total synthesis of aaptamine has been reported (Kelly and Maguire, 1985).

A series of pyrroloquinoline alkaloids called isobatzellines A–D (**343–346**) were found in extracts of the Caribbean sponge *Batzella* sp. (Sun *et al.*, 1990). In addition to the cytotoxicity noted under the structures, these compounds have

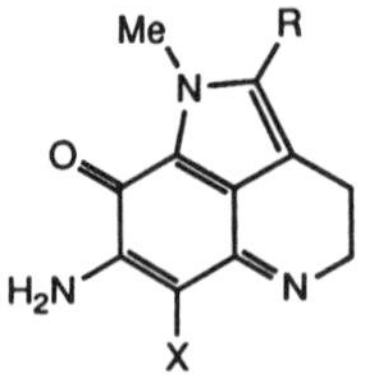

(**343**) Isobatzelline A: R = SMe, C = Cl   (**346**) Isobatzelline D
(**344**) Isobatzelline B: R = SMe, X = H
(**345**) Isobatzelline C: R = H, X = Cl

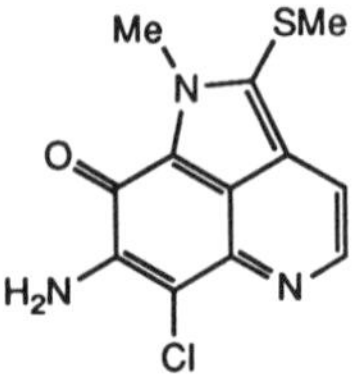

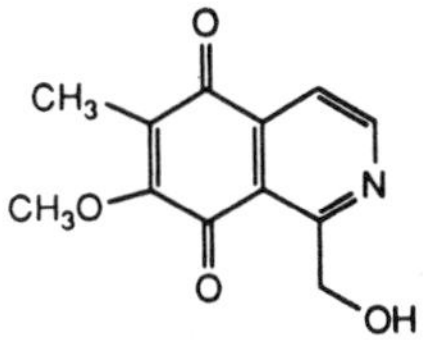

(347)

shown antifungal activity against *C. albicans*. Renierol (**347**), obtained from the Fijian sponge *Xestospongia caycedoi*, inhibited the growth of L1210 cells (IC_{50} 3 $\mu g/ml$) (McKee and Ireland, 1987).

Many secondary metabolites are suspected of being used by organisms for chemical defense. Therefore, field observations of aversion responses may guide a collector to an organism which possesses interesting biologically active metabolites. An example is the isolation of the alkaloid imbricatine (**348**). It has been

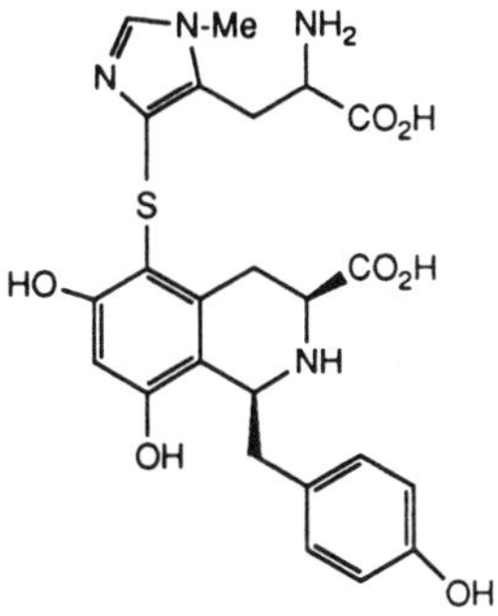

(348)

found that the presence of the starfish *Dermasterias imbricata* induces the anemone *Stomphia coccinea* to release itself from its holdfast and "swim" away to a new location. Pathirana and Andersen (1986) found that the alkaloid imbricatine (**348**) isolated from the starfish causes this aversion response. Imbricatine was also toxic to L1210 (<1 $\mu g/ml$) and *in vivo* P388 with T/C 139 at 0.5 mg/kg.

In one of the early reports of tumor-inhibiting effects by extracts of marine invertebrates, it was noted that extracts of the Caribbean tunicate *Ecteinascidia turbinata* caused dramatic increases in the lifetimes of mice inoculated with P388 cells (Sigel *et al.*, 1970). The active compounds proved quite difficult to isolate, but two groups whose work was reported in adjacent papers succeeded in isolating and characterizing a group of very complex alkaloids designated ecteinascidins (Et, followed by a number indicating the molecular weight). These compounds are very cytotoxic (see data under structures), but more importantly, a number of the compounds show excellent *in vivo* activity. Wright *et al.* (1990) described **349** (Et 729) and **350** (Et 743) while Rinehart *et al.* (1990a) reported metabolites **349**,

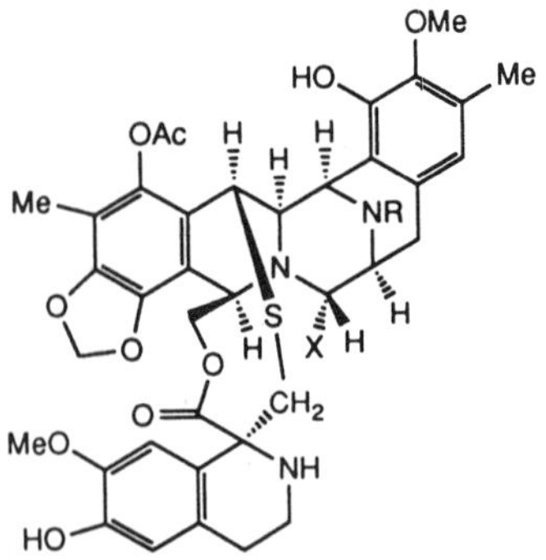

(349) Et 729: R = H, X = OH
(350) Et 743: R = Me, X = OH
(351) Et 745: R = Me, X = H
(352) Et 770: R = Me, X = CN

	In vivo activity	*In vitro* (μg/ml)
P388		
349	(3.8 μg/Kg), T/C = 214%	0.00093 (P388)
350	(15 μg/Kg), T/C = 167%	0.0013 (P388)
351	(250 μg/Kg), T/C = 111%	0.00088 (L1210)
B16 melanoma		
349	(10 μg/Kg), T/C = 246%	

350, **351** (Et 745), and **352** (Et 770). More recently Rinehart's group has described additional members of this group, Et 736 (**353**) and Et 722 (**354**), which feature a tetrahydrocarboline unit in place of one of the tetrahydroisoquinoline units (Rinehart in Schmitz and Yasumoto, 1991). The newer ecteinascidins have *in vivo*

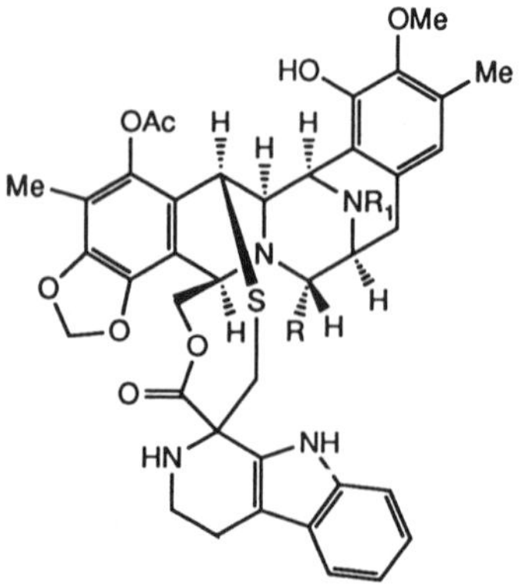

	R	R_1
(353) Et 736:	OH	CH_3
(354) Et 722:	OH	H

antitumor activities comparable to the earlier ones, but are more active against some tumors and less active against others. Based on the high levels of *in vivo* activity cited, one would hope that an effective drug might arise from among the ecteinascidins.

4.9. Quinolizidines and Indolizidines

Clavepictines A and B (**355, 356**), isolated from the tunicate *Clavelina picta*, were found to be active against P388 and three human solid tumors (A-549, U-251, and SN12K1), with an IC_{50} of 1.8–8.5 μg/ml (Raub *et al.*, 1991). The compounds effectively kill each cell line, LC_{50} 10–25 μg/ml, under conventional

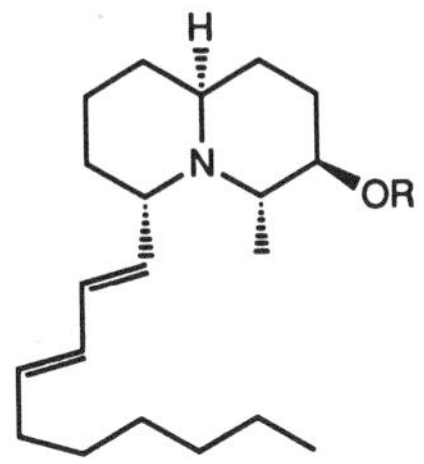

(**355**) Clavepictine A: R = AC
(**356**) Clavepictine B: R = H

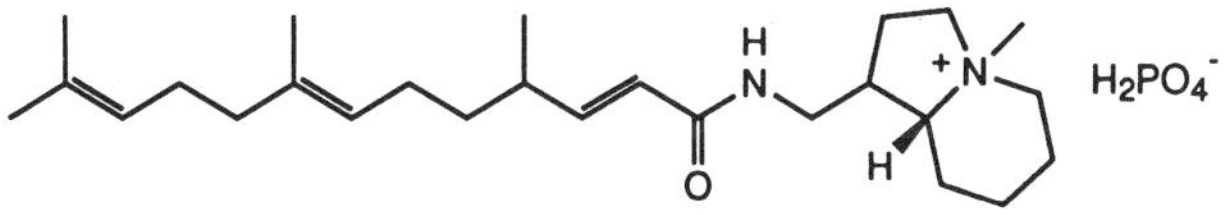

(**357**)

culture conditions. The indolizidine stellettamide A (**357**) from a sponge, *Stelleta* sp., shows antifungal activity and also inhibits K562 epithelium cell growth, IC_{50} of 5.1 μg/ml (Hirota *et al.*, 1990b).

4.10. Prianosins/Discorhabdins

Two different groups independently isolated sets of unusual, but closely related sulfur-containing alkaloids called prianosins and discorhabdins. The first of these to be reported was discorhabdin C (**360**) (Perry *et al.*, 1986). The remaining discorhabdins (**358, 359, 361** = discorhabdins A, B, D, respectively) were described in two subsequent papers (Perry *et al.*, 1988b,c). The discorhabdins were isolated from the sponges *Latrunculia brevis* and *Prianos* sp. Cytotoxicity data are shown with the structures. Only discorhabdin D (**361**) showed any *in vivo* activity in the P388 model and that was modest, T/C 132 at 20 mg/kg. Prianosin A (**362**), from the Okinawan sponge *Prianos melanos* (J. Kobayashi

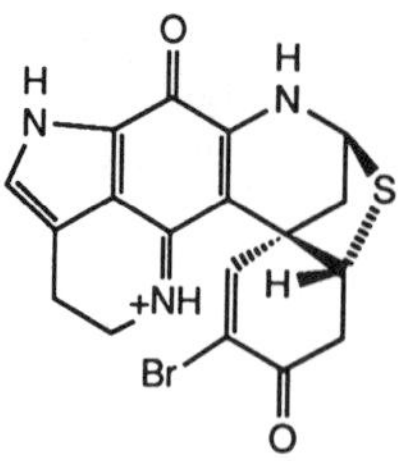

(358) 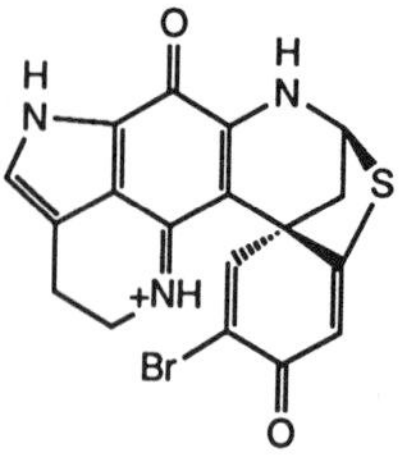   (359) 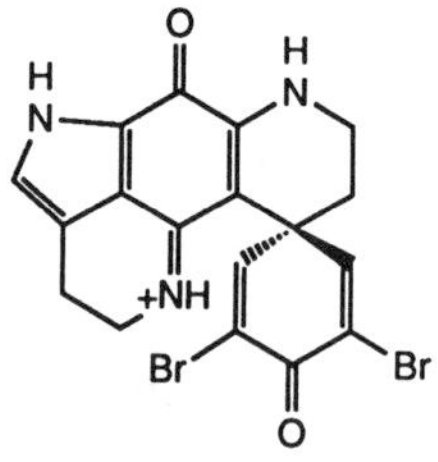   (360)

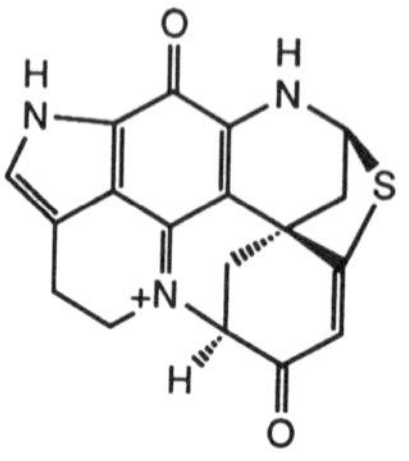

P388 IC$_{50}$
(μg/mL)

358	0.05
359	0.1
360	0.03
361	6

358–360 inactive *in vivo*; P388 model:
T/c < 120% (toxic dose 2 mg/kg)
361 active *in vivo*; P388 model:
T/C 132 % at 20 mg/kg dose

(361)

et al., 1987), is the nonprotonated form of discorhabdin A (**358**). The remaining prianosins B–D (**363–365**) were reported in 1988 (Cheng *et al.*, 1988b). Prianosin D (**365**) and discorhabdin D (**361**) are a hydroquinone/quinone pair. In addition to the activities noted with the structures, prianosin D (**365**), but not the others, induced Ca^{2+} release from sarcoplasmic reticulum, with a potency ten times that of caffeine.

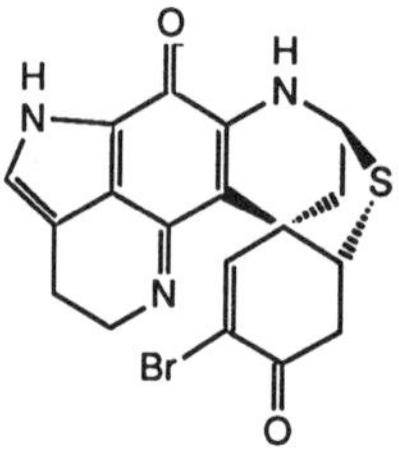

(362) 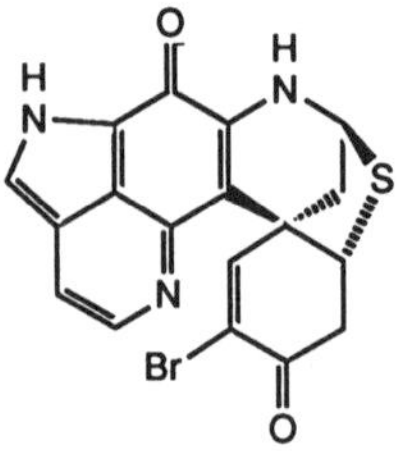   (363) 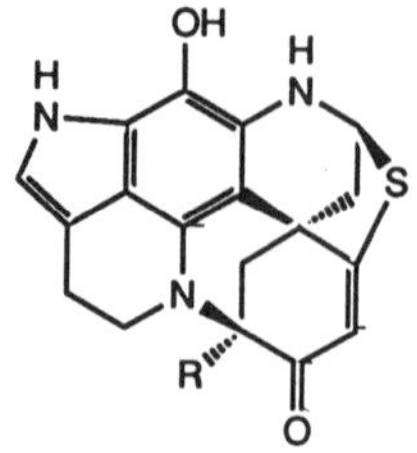   (364) R = OH
                                                  (365) R = H

	L1210 IC$_{50}$ (μg/mL)	**L5178Y IC$_{50}$** (μg/mL)	**KB IC$_{50}$** (μg/mL)
362	0.037	0.014	0.073
363	2.0	1.8	>5
364	0.15	0.024	0.57
365	0.18	0.048	0.46

4.11. Polycyclic Aromatic Alkaloids: Acridine Alkaloids

A group of polycyclic aromatic alkaloids isolated from ascidians and sponges and having in common a tetracyclic moiety that includes the acridine ring system have emerged in the past few years. The first of these to be described was amphimedine (366), isolated from a sponge from Guam, *Amphimedon* sp., which was active against P388 *in vitro* with an ED_{50} of 0.4 μg/ml, but proved inactive *in vivo* (Schmitz *et al.*, 1983). Bloor and Schmitz (1987) reported 2-bromoleptoclinidin-

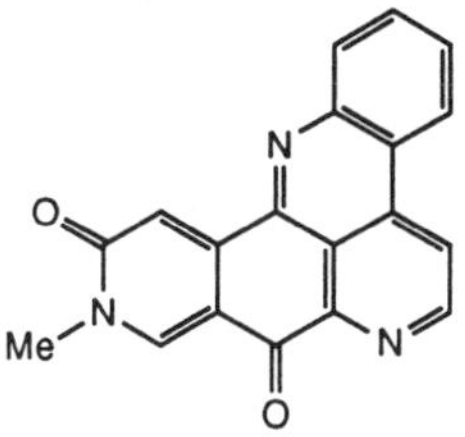

(366) 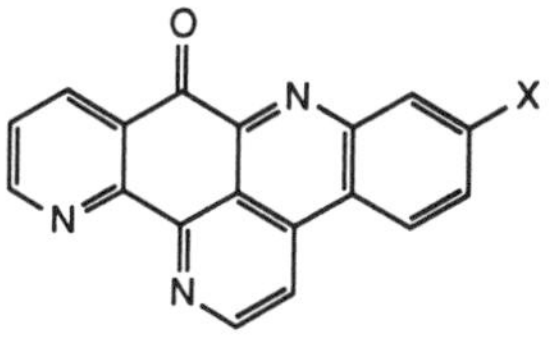   (367) X = Br
                (368) X = H

one from a *Leptoclinidines* sp. of ascidian collected in Truk, but the structure was revised later to that of 367 (DeGuzman and Schmitz, 1989) on the basis of a chemical correlation with its debromo analog, ascididemin (368), which was isolated from an Okinawan tunicate *Didemnum* sp. (J. Kobayashi *et al.*, 1988a). 2-Bromoleptoclinidinone (= 2-bromoascididemin, 367) was active against P388 cells (ED_{50} 0.4 μg/ml), while ascididemin was reported to be active against L1210 cells (IC_{50} 0.39 μg/ml). Ascididemin (368) was also reported to be seven times as active as caffeine in causing Ca^{2+} release in the sarcoplasmic reticulum.

Dercitin (369), from a deep-water species of *Dercitus* sponge, showed *in vitro* (see data with structure) and *in vivo* activity (T/C 170 at 5 mg/kg) in the P388 model (Gunawardana *et al.*, 1988; see Gunawardana *et al.*, 1992, for structure

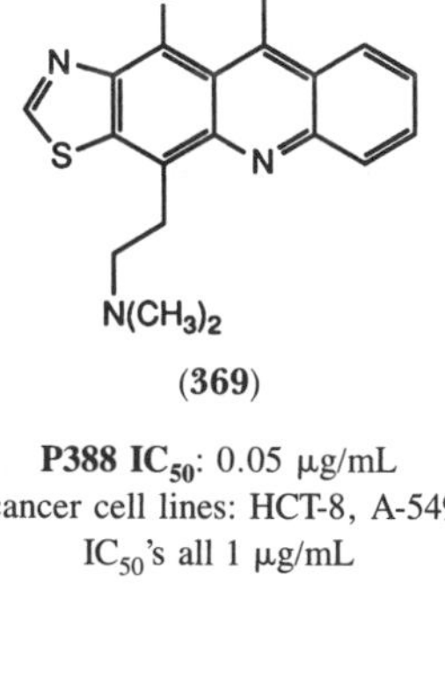

(369)

P388 IC_{50}: 0.05 μg/mL
Human cancer cell lines: HCT-8, A-549, T47D
IC_{50}'s all 1 μg/mL

revision). In addition, dercitin is described as having immunosuppressive and antiviral activity. Mechanism-of-action studies indicate that dercitin disrupts macromolecular synthesis (DNA, RNA, and protein) in the P388 system by binding to DNA and inhibiting nucleic acid synthesis. It is an effective inhibitor of DNA nick-translation at concentrations that disrupt the superhelical density of DNA. It relaxes covalently closed supercoiled ΦX174DNA, indicating intercalation as the mode of binding (Burres *et al.*, 1989). This latter paper also confirmed that dercitin shows slight *in vivo* activity against B16 tumors in mice, T/C 125 at

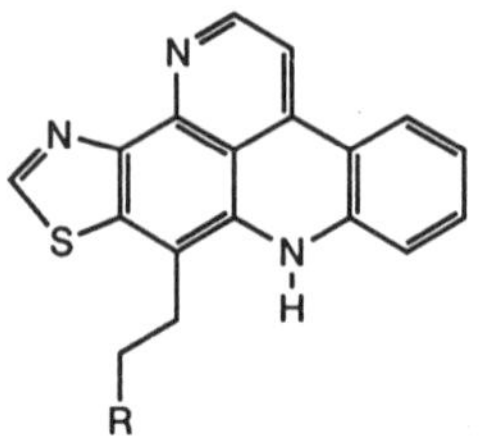

(**370**) Nordercitin: R = N(Me)$_2$
(**371**) Dercitamine: R = NHMe
(**372**) Dercitamide: R = NHCOCH$_2$CH$_3$

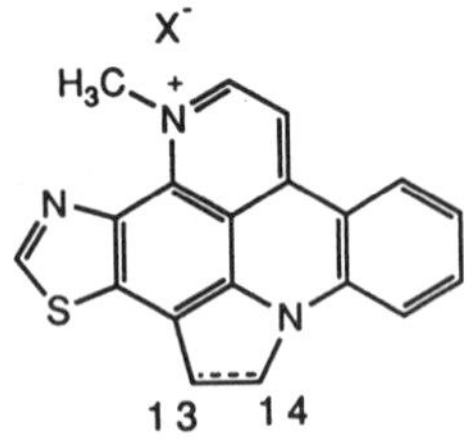

(**373**) Cyclodercitin (13,14 dihydro)
(**374**) Cyclodercitin (Δ^{13})

P388 IC$_{50}$

370	4.79 μM
371	26.7 μM
372	12.0 μM
373	1.9 μM
374	9.89 μM

1.25 mg/kg. Other members of the dercitin family (**370–373**) have been isolated from two deep-water sponges of the family Pachastrellidae (Gunawardana *et al.*, 1989; see Gunawardana *et al.*, 1992, for structure revision). In addition to the cytotoxicity data shown with the structures, nordercitin (**370**), dercitamine (**371**), and dercitamide (**372**) show immunosuppressive activity. Cyclodercitin (**373**) readily undergoes air oxidation to give **374**, which is also cytotoxic.

Carroll and Scheuer (1990) found shermilamine B (**375**) and kuanoniamines A–D (**376–379**) in extracts of a tunicate and also in a mollusk that feeds on it, *Chelynotus semperi*. Modest cytotoxicity to KB cells was determined for these alkaloids (see data with the structures). Kuanoniamine C (**378**) and dercitamide (**372**) are identical.

J. Kobayashi *et al.* (1988c) isolated cystodytins A, B, and C (**380–382**) from an Okinawan tunicate, *Cystodytes dellechiajei*. Cystodytins A (**380**) and C (**382**) were reported to inhibit L1210 cell growth (IC$_{50}$ 0.2 μg/ml for each) and also induce Ca^{2+} release in sarcoplasmic reticulum (36 and 13 times more potent than caffeine, respectively). Closely related to the cystodytins are diplamine (**383**),

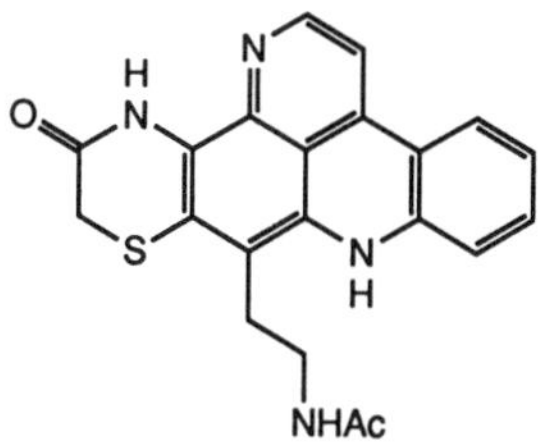

(375)

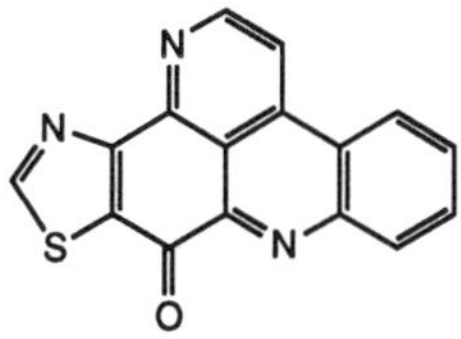

(376)

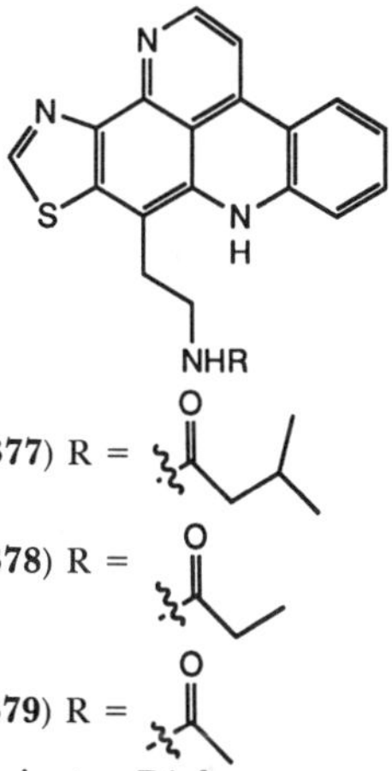

		KB IC$_{50}$
		(μg/mL)
	375	5
	376	1
	377	>10
	378	5

from a Fijian tunicate, *Diplosoma* sp. (Charyulu *et al.*, 1989), and varamines A and B (**384**, **385**), from another Fijian tunicate, *Lissoclinum vareau* (Molinski and Ireland, 1989). These three alkaloids all were active against L1210 cells, with IC$_{50}$ values of 0.02–0.05 μg/ml.

A slightly different variation in the acridine alkaloids is seen in the plakinidines, which have been isolated by two different groups. Inman *et al.* (1990), using an antiparasite bioassay, isolated plakinidines A (**386**) and B (**387**) from a *Plakortis* sp. of sponge from Fiji; subsequently it was discovered that plakinidine A inhibited reverse transcriptase activity at 1 μg/ml. Shortly thereafter West *et al.* (1990) described plakinidines A (**386**), B (**387**), and C (**388**) from the Fijian sponge *Plakortis* sp. and reported IC$_{50}$ values of 0.1, 0.3, and 0.7 μg/ml, respectively, for these compounds against L1210 cells.

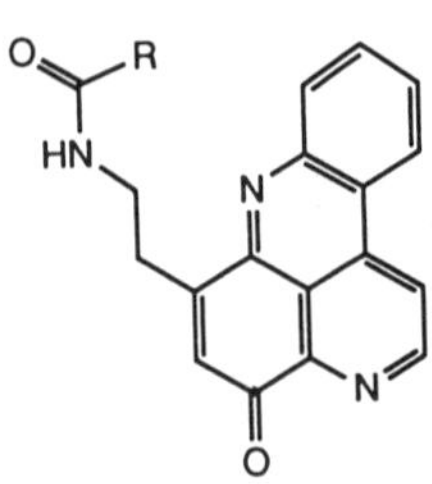

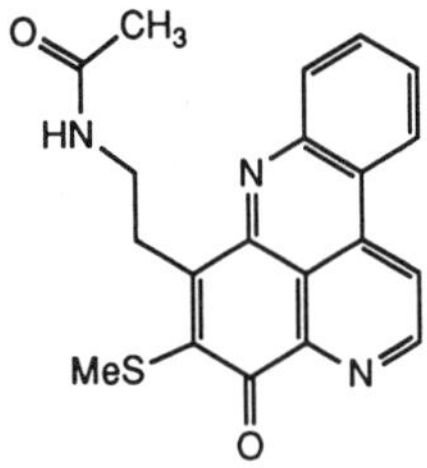

(380) R = (CH₃, CH₃)

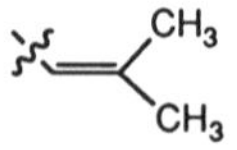

(381) R =

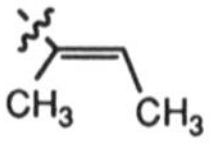

(382) R =

(383)

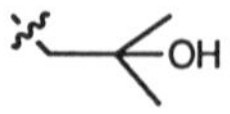

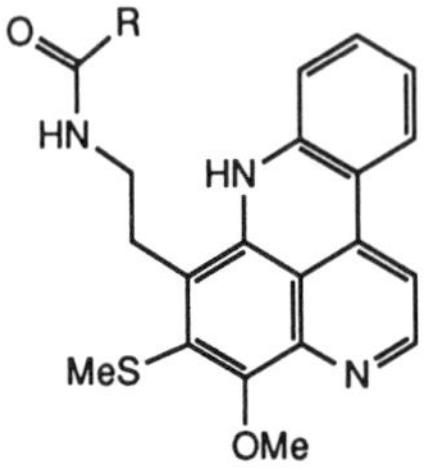

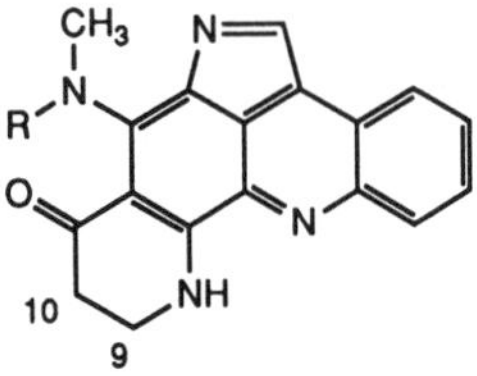

(384) R = Me
(385) R = H

(386) R = H
(387) R = CH₃
(388) R = H,Δ⁹

4.12. Guanidines

A very novel polycyclic guanidine alkaloid, ptilomycalin A (**389**), has been reported by Kashman's group (Kashman *et al.*, 1989) from a Caribbean sponge, *Ptilocaulis spiculifer*, and a Red Sea sponge, *Hemimycale* sp. Toxicity to P388 cells was confirmed for **389** (IC_{50} 0.1 μg/ml) along with antifungal (*C. albicans*) and antiviral activity (HSV @ 0.2 μg/ml).

4.13. Peptides and Depsipeptides

Cyclic peptides and depsipeptides have emerged as very important classes of bioactive compounds in marine natural products. Particular attention has accrued to these compounds because didemnin B (**391**) has been in clinical trials for several years and is the first marine natural product to attain this status. Also, the dramatic cytotoxicity of some of the dolastatins compels attention.

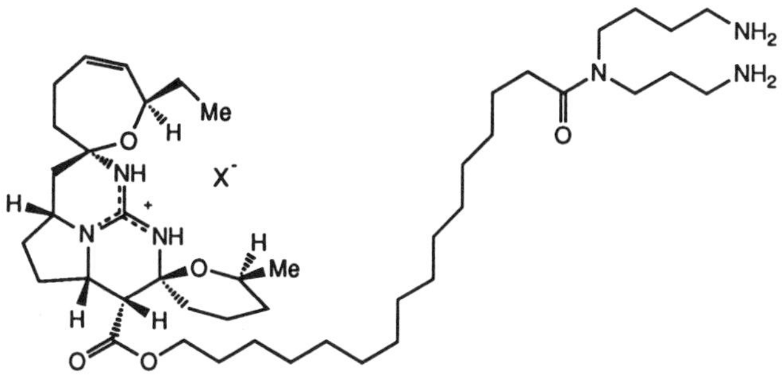

(389)

4.13.1. Didemnins

The didemnins consist of an extensive family of depsipeptides, the first members of which were reported in 1981. They have been reviewed by Munro *et al.* (1987) and also recently by Rinehart *et al.* (1990c). In addition to antitumor activity, antiviral (Rinehart *et al.*, 1983; Canonico *et al.*, 1982) and immunosuppressive activities (Montgomery and Zukoski, 1985) have also been reported for these compounds. Didemnin B (**391**) is the most prominent member of this group; it showed the best antitumor activity and was advanced to clinical trials (Chun *et al.*, 1986) and is now in phase II of these tests (Abbruzzese *et al.*, 1988; Motzer *et al.*, 1990). Some activity data for the didemnins are summarized in Table II.

Table II. In Vitro and in Vivo Inhibitory Activity of the Didemnins

Compound	Didemnin	*In vitro*[a] L1210 IC$_{50}$, μg/ml	*In vivo* P388 Percent life extension (dose, μg/kg, per inj)
390	A	0.014	25–99 (0.030–1.0)[b]
391	B	0.0025	9–40 (0.25–8.0)[b]
392	C	0.011	—
393	D	0.0050	—
394	E	0.0030	—
395	G	0.006	—
396	X	0.0044	—
397	Y	0.0056	—
	Nor-B[d]	0.0078	—
398	Dehydro-B	—	110 (0.16)[c] [118 (0.16 vs. B16)]

[a]Rinehart *et al.* (1990c).
[b]Rinehart *et al.* (1981b).
[c]Rinehart, in Schmitz and Yasumoto (1991).
[d]Ethyl group of isostatine unit replaced by methyl.

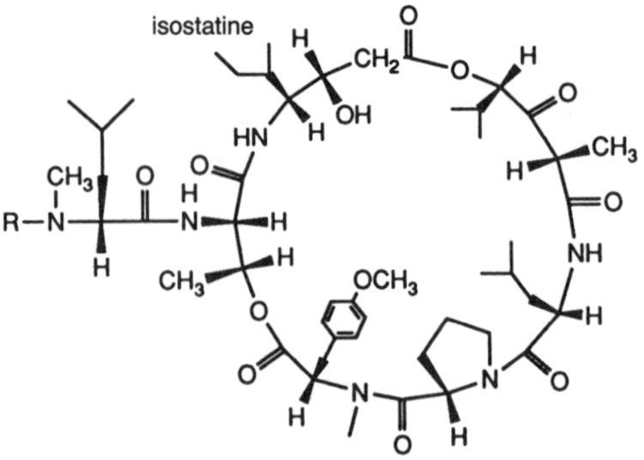

(**390**) Didemnin A: R = H

(**391**) Didemnin B: R = CH₃CHOHC—N——C—
                          (L-Lac)   (L-Pro)

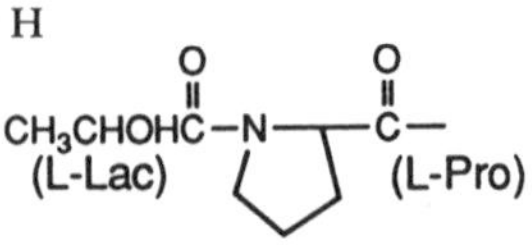

(**392**) Didemnin C: R = L-Lac
(**393**) Didemnin D: R = L-pGlu-(L-Gln)₃-L-Lac-L-Pro-
(**394**) Didemnin E: R = L-pGlu-(L-Gln)₂-L-Lac-L-Pro-
(**395**) Didemnin G: R = CHO
(**396**) Didemnin X: R = Hydec-(L-Gln)₃-L-Lac-L-Pro-
(**397**) Didemnin Y: R = Hydec-(L-Gln)₄-L-Lac-L-Pro-

Hydec = n'- C_7H_{15}

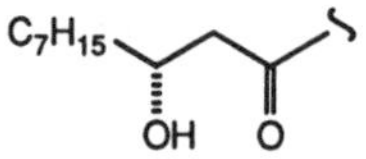

(**398**) Dehydrodidemnin B: R = CH₃C—C—N——C—
                                                (L-Pro)

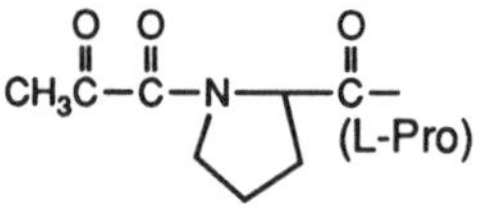

The presence of the unusual amino acid unit $3S,4R,5S$-isostatine was confirmed by synthetic studies (Rinehart *et al.*, 1987) and an x-ray analysis (Hossain *et al.*, 1988), after having been initially proposed as a statine unit. The didemnins have been isolated primarily from the Caribbean tunicate *Trididemnun solidum*, but are also present in *T. cyanophorum* and *T. palmae*. At about the same time the statine to isostatine correction was being made, Guyot *et al.* (1987) reported isodidemnine-1, which they obtained from *T. cyanophorum* collected in Guadaloupe. These workers formulated the presence of the isostatine unit and NMR studies confirmed that isodidemnine-1 and didemnin B were identical (Banaigs *et al.*, 1989). Although most of the didemnins had been reported by 1981 (Rinehart *et al.*, 1981a,b) others have been added quite recently. Two of these are didemnins X (**396**) and Y (**397**) (Rinehart *et al.*, 1990b,c). Another recent discovery is dehydrodidemnin B (**398**), which was found in *Aplydium albicans*, a Mediterranean tunicate that is in a different taxonomic family from that of *T. solidum*

(Rinehart, 1989; K. L. Rinehart, in Schmitz and Yasumoto, 1991). *In vivo* testing suggests that dehydrodidemnin B (**398**) is three to five times as active as didemnin B (**391**) and is effective against P388 leukemia (T/C 210), B16 melanoma (T/C 218), and lung tumors (T/C 0.00), all at 160 μg/kg/inj in mice. A total synthesis of nordidemnin B, a minor cytotoxic product also isolated from the tunicate *T. cyanophorum*, has been completed (Jouin *et al.*, 1989). (Nordidemnin B = didemnin B with the ethyl group of isostatine replaced by a methyl group.) Hamada *et al.* (1988) have synthesized didemnins A and B, and Li *et al.* (1990) have synthesized didemnins A, B, and C.

Some of the early mechanistic studies on didemnin B have been summarized by Chun *et al.* (1986). Preliminary studies showed that didemnin B inhibits protein, DNA, and RNA synthesis (Crampton *et al.*, 1984; Li *et al.*, 1984). Legrue *et al.* (1988) have reported that didemnin B inhibits T-lymphocyte proliferation, but not lymphokine stimulation. They concluded that didemnin B functions as an antiproliferative agent by inhibiting lymphokine-induced protein and DNA synthesis. Didemnin B inhibits Ca^{2+}/calmodulin-dependent phosphorylation of eukaryotic elongation factor 2 *in vitro* and the biological effects of the phorbol ester TPA on mouse skin *in vivo* (Geschwendt *et al.*, 1989).

Didemnin B proved not to be efficacious in a phase II trial in patients with advanced renal carcinoma (Motzer *et al.*, 1990). Didemnin B has also undergone phase II clinical trials for colorectal cancer (Abbruzzese *et al.*, 1988), but at the doses used in this preliminary study, no favorable responses were documented.

Didemnins A and B show antiviral activity against herpes simplex, Rift Valley fever, rhino- and parainfluenza, and dengue viruses, but not against rabies (Canonico *et al.*, 1982).

4.13.2. Dolastatins

The dolastatins are a collection of cyclic and linear peptides and depsipeptides which display some remarkable cell growth inhibitory action *in vitro*, and three of the dolastatins are reported to show impressive *in vivo* activities at very low doses (see Table III). The saga of the isolation of the dolastatins by Pettit's group in Arizona spans 20 years to date (Pettit *et al.*, 1981, 1987b). All the dolastatins have been isolated from the sea hare *Dolabella auricularia* collected in the Indian Ocean, but the yields are very low, ~1 mg/100 kg wet weight, and workup of large collections, e.g., 1600–1700 kg wet weight, has been carried out to produce the milligram quantities of individual compounds reported thus far. In the early 1980s Pettit *et al.* (1981) described the isolation of dolastatins 1–9 and reported cytostatic activity data for some of these. The structure of only one of these, dolastatin 3 (**399**), has been reported; however, structures for several additional members of this family, dolastatins 10–15 (**400–405**), have been described.

Table III.   In Vitro and in Vivo Activity of the Dolastatins

Compound	Dolastatin	In vitro P388 ED_{50}, µg/ml	In vivo Tumor	Percent life extension	Dose
399	3 (natural)	1×10^{-4}–1×10^{-7a}	—	—	—
	3 (synthetic)	0.16	—	—	—
400	10 (natural)	4.6×10^{-5}	B-16	40–138	1.44–11.1 µg/kg
	10 (synthetic)	10^{-4}	PS	69–102	1–4 µg/ml per kg
401	11	2.7×10^{-3}	PS	25	300–600 µg/kg
402	12	7.5×10^{-2}	—	—	—
403a	13	0.013	17–67% curative response vs. NCI human melanoma xerograph @ 3.25–26 µg/kg		
403b	Dehydro 13	"Marginally inactive"	—	—	—
404	14	0.022	—	—	—
405	15	0.0024	—	—	—

[a]The stronger inhibition reported for the natural product by Pettit *et al.* (1982) was theorized to be due to a sensitivity difference in cell lines used or trace contamination with an undetected and considerably more powerful dolostatin (Pettit *et al.* 1987a).

The first structure proposed for dolastatin 3 in 1982, which was based on ~1 mg of material using the best tools available at the time, proved to be incorrect, as were the original proposals for patellamides A–C (see below), which also contain thiazole amino acid units. The dolastatin 3 structure was revised about 5 years later to **399** via synthesis by Pettit and by others (Pettit *et al.*, 1987a, and references

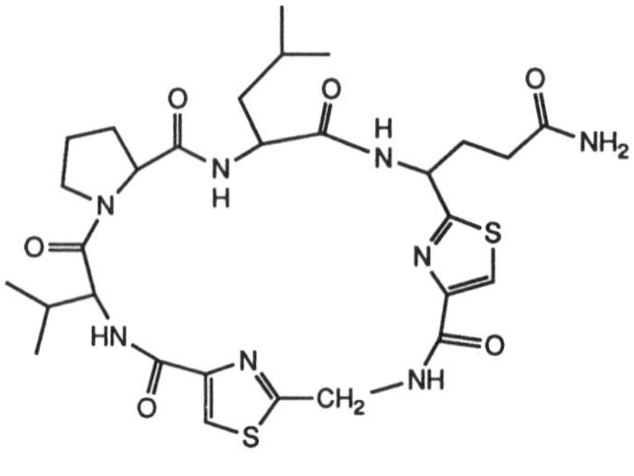

Cyclo[L-Val-L-Pro-L-Leu-L-(gln)Thz-(gly)Thz]

(**399**) Dolastatin 3

therein; Holzapfel *et al.*, 1990). Considerable effort was expended on synthesizing alternative structural possibilities, but interestingly, none of the alternative compounds which have been tested have proved active *in vitro* against the P388 cell line (Pettit *et al.*, 1987a). Thus, the synthetic efforts directed at the "wrong target" have provided some structure–activity information.

Improvements in NMR instrumentation and experiments, especially inverse detection probes and the H/H and H/C relayed experiments, have made it possible to trace the one-dimensional structures of the more recently described dolastatins with much greater reliability. High-resolution collision-activated-decomposition (MS/MS) has also been a vital tool in elucidating these structures.

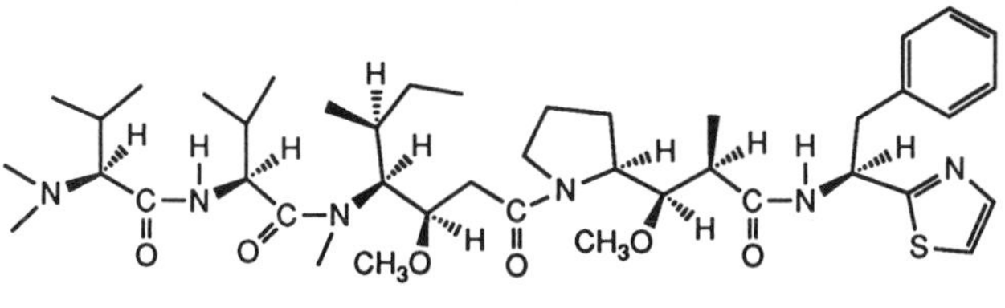

(400) Dolastatin 10

Dolastatin 10 (400), the next member of this group to be elucidated, was reported to be the most active (i.e., lowest dose) antineoplastic substance known up to that time, giving very significant *in vivo* results with microgram doses (see Table III) (Pettit *et al.*, 1987b). A synthesis which confirmed the structure and established the absolute configuration was reported a couple of years later by Pettit *et al.* (1989d). Hamada *et al.* (1991) have also reported a stereoselective synthesis of 400. Although there are 512 possible stereoisomeric structures, earlier work on dolastatin 3 suggested that the amino acids would all be of the L series, and this proved correct. Mechanistic studies revealed that dolastatin 10 (400) inhibits tubulin polymerization and tubulin-dependent GTP hydrolysis (Bai *et al.*, 1990a). Further studies by Bai *et al.* (1990b) showed that dolastatin 10 is a noncompetitive inhibitor of vincristine binding to tubulin, and these authors propose that dolastatin 10 binds to tubulin near to, but distinct from, the exchangeable nucleotide and vinca alkaloid sites.

Dolastatins 11 and 12 (401, 402) (Pettit *et al.*, 1989c; Kamano *et al.*, 1987) differ in structure only slightly from majusculamide C, isolated from a blue-green

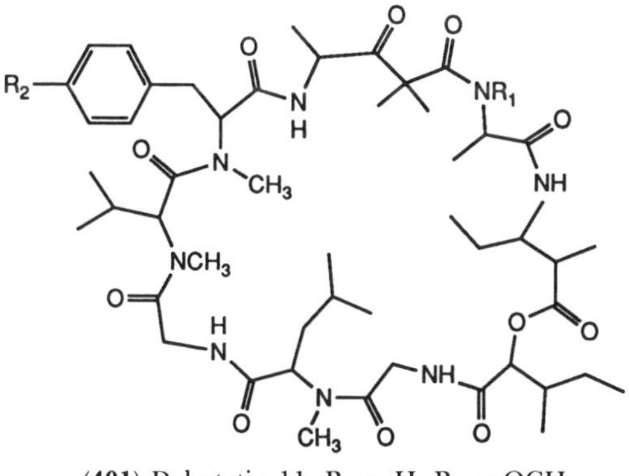

(401) Dolastatin 11: R_1 = H, R_2 = OCH_3
(402) Dolastatin 12: R_1 = CH_3, R_2 = H

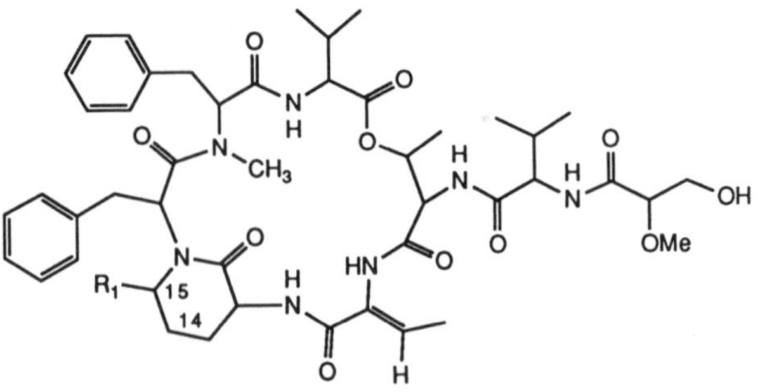

(**403a**) Dolastatin 13: R_1 = OH
(**403b**) Dehydrodolastatin 13: R_1 = H; $\Delta^{14,15}$

alga, *Lyngbya majuscula* (Carter *et al.*, 1984), and this correlation strengthens the supposition of an algal source for some or all of the dolastatins. Dolastatin 13 (**403a**) contains an unusual cyclic masked aldehyde unit and a rare dehydro amino acid. In terms of structure–activity relationships, it is noteworthy that dehydro-dolastatin 13 (**403b**), a "minor" component (Pettit *et al.*, 1989b; Kamano *et al.*, 1989), is much less active ("marginally inactive," i.e., ED_{50} 10–20 µg/ml) than **403a**. Dolastatin 14 (**404**) (Pettit *et al.*, 1990b) incorporates a long-chain hydroxy acid within its cyclodepsipeptide structure, while dolastatin 15 (**405**) (Pettit *et al.*, 1989a) is a novel linear depsipeptide that contains a pyrrolidone methyl vinyl ether group that has been noted in some other glue-green algal constituents, such as

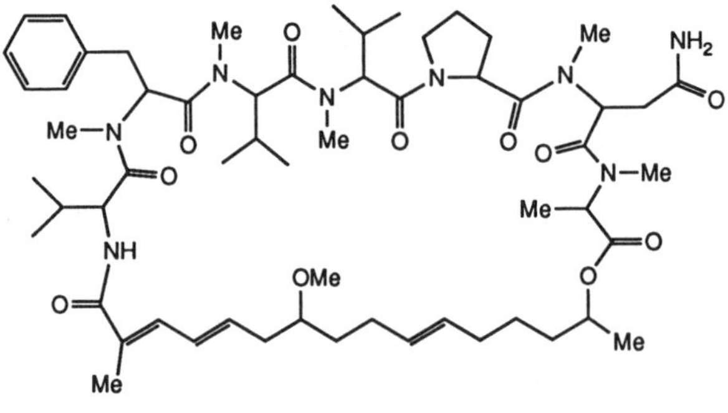

(**404**) Dolastatin 14

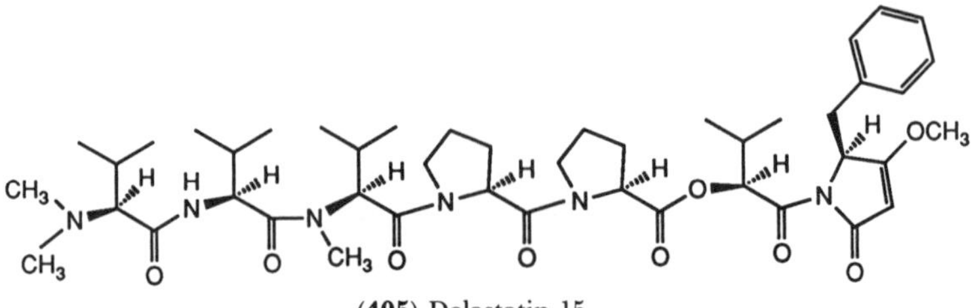

(**405**) Dolastatin 15

malyngamide A (Cardellina *et al.*, 1979) and the pukelemides (Cardellina and Moore, 1979).

4.13.3. Lissoclinum Peptides

The tunicate *Lissoclinum patella* collected from various locations has been the source of a large family of cyclic peptides that show cytotoxic (see Table IV) and other pharmacologic properties. Ulicyclamide (**413**) and ulithiacyclamide (**406**) were the first of these lipophilic cyclic peptides to be reported (Ireland and

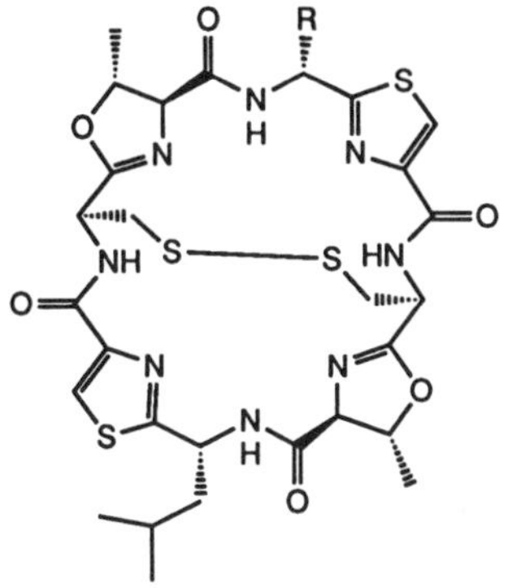

(**406**) Ulithiacyclamide: R =

(**407**) Ulithiacyclamide B: R =

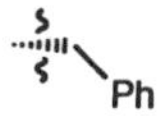

Scheuer, 1980). Subsequent work with the same species from different locations uncovered related compounds. Further structural work and synthetic efforts revealed that some of the initially proposed structures were incorrect and led to the revised structures shown here for ulicyclamide (**413**) and patellamides A–C (**409–411**); see below. The revised structures are summarized in a paper by Ireland's group (Sesin *et al.*, 1986). Total syntheses with correct structures have been reported for patellamide A (**409**) (Hamada *et al.*, 1985a), patellamide B (**410**) (Hamada *et al.*, 1985b; Schmidt and Greisser, 1986), and patellamide C (**411**) (Hamada *et al.*, 1985b). Ascidiacyclamide (**408**) was reported from an unidentified species of ascidian. It has been synthesized (Hamada *et al.*, 1985c) and its solid-state conformation determined by x-ray analysis (Ishida *et al.*, 1987). The solution conformation deduced from NMR analysis and energy minimization calculations (Ishida *et al.*, 1988, 1992) compares favorably with the solid-state conformation. Ishida *et al.* (1989) have also studied the conformation of ulithiacyclamide (**406**) by NMR and molecular modeling calculations and concluded that the solid-state and solution conformations are very similar.

Table IV. Activity Data for Lissoclinum patella Metabolites

Compound	Metabolite	IC$_{50}$, μg/ml					
		L1210	T24	MRC5CV1	P388	Lymphocytes	Other
406	Ulithiacyclamide	0.35	0.10–0.15	0.04–0.22	—	—	0.01 CEM
407	Ulithiacyclamide B	—	—	—	—	—	0.017 KB
408	Ascidiacyclamide	—	—	—	—	—	—
409	Patellamide A	3.9	—	—	—	—	0.028 CEM
410	Patellamide B	2.0	—	—	—	—	—
411	Patellamide C	3.2	—	—	—	—	—
412	Patellamide D	—	—	—	11	—	—
413	Ulicyclamide	7.2	—	—	—	—	—
414	Lissoclinamide 1	>10	—	—	—	—	—
415	Lissoclinamide 2	>10	—	—	—	—	—
416	Lissoclinamide 3	>10	—	—	—	—	—
417	Lissoclinamide 4	—	0.8	0.8	10	—	—
418	Lissoclinamide 5	—	16	15	12	20	—
419	Lissoclinamide 6	—	—	—	6.9	—	—
420	Lissoclinamide 7	—	0.6	0.04	—	0.08	—
421	Lissoclinamide 8	—	6	1	—	8	—

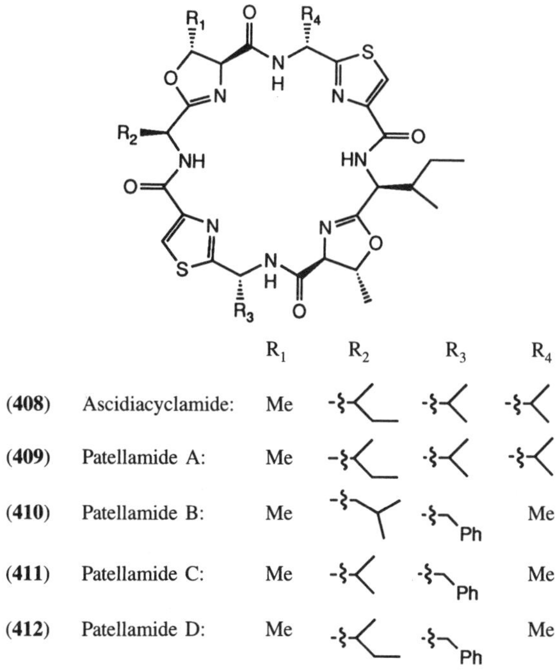

		R₁	R₂	R₃	R₄
(408)	Ascidiacyclamide:	Me			
(409)	Patellamide A:	Me			
(410)	Patellamide B:	Me			Me
(411)	Patellamide C:	Me			Me
(412)	Patellamide D:	Me			Me

Patellamide D (**412**) was described by Schmitz *et al.* (1989) along with lissoclinamides 4–6 (**417–419**), all of which were isolated from a Great Barrier Reef collection of *L. patella*. Simultaneously, Australian workers (Degnan *et al.*, 1989a) reported patellamide D and lissoclinamides 4 and 5 from the same tunicate collected at Heron Island. The Australian group has also found lissoclinamides 7 and 8 in the extracts of *L. patella* (Hawkins *et al.*, 1990). Moore's group has found ulithiacyclamide B (**407**) in *L. patella* collected in Pohnpei, the same tunicate from which this group isolated the macrolide patellazole (**84**) (D. E. Williams *et al.*, 1989). These authors found that ulithiacyclamide (**406**), ulithiacyclamide B (**407**), and patellamides B (**410**) and C (**411**) did not show any selective toxicity against solid tumor cell lines. They concluded that these compounds are general cytotoxins, probably too toxic to be active against solid tumors *in vivo*. Patellazole B (**84**) is about 100 times more cytotoxic than the ulithiacyclamides.

Shioiri *et al.* (1987) tested 35 cyclic peptides of marine origin or derivatives thereof against L1210 cells and found that ulithiacyclamide was the most active. Structure–activity considerations led the authors to conclude that the oxazoline ring was essential for cytotoxicity. Even some small peptides containing this structural moiety were cytotoxic, a result which led the authors to suggest that the macrocyclic feature may not be essential for activity. Kohda *et al.* (1989) found that ulicyclamide inhibited both DNA and RNA synthesis.

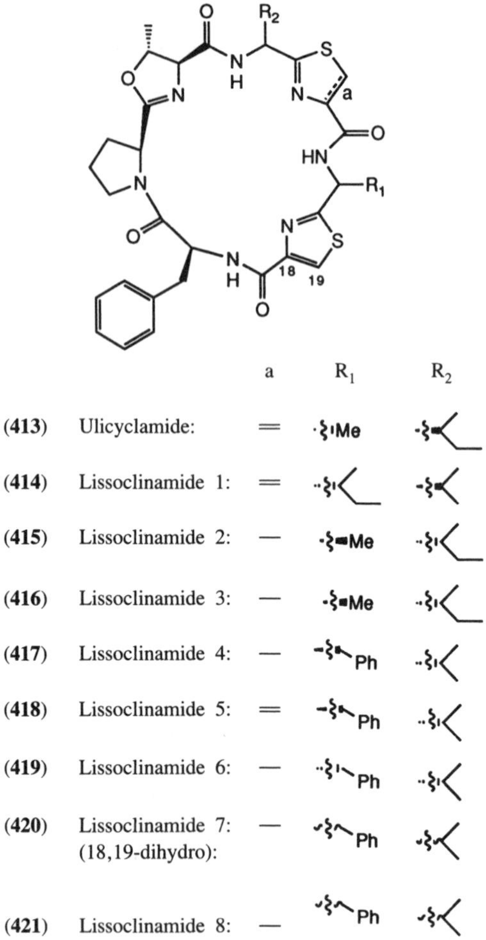

		a	R_1	R_2
(413)	Ulicyclamide:	=	·⌇ⁱMe	-⌇⊲
(414)	Lissoclinamide 1:	=	··⌇ⁱ⊲	-⌇⊲
(415)	Lissoclinamide 2:	—	-⌇⊶Me	··⌇ⁱ⊲
(416)	Lissoclinamide 3:	—	-⌇⊶Me	··⌇ⁱ⊲
(417)	Lissoclinamide 4:	—	-⌇⌒Ph	-⌇ⁱ⊲
(418)	Lissoclinamide 5:	=	-⌇⌒Ph	··⌇ⁱ⊲
(419)	Lissoclinamide 6:	—	··⌇ⁱ⌒Ph	··⌇ⁱ⊲
(420)	Lissoclinamide 7: (18,19-dihydro):	—	⌄⌇⌒Ph	-⌇⌒⊲
(421)	Lissoclinamide 8:	—	⌄⌇⌒Ph	⌄⌇⊲

Patellamide D (412) has shown some interesting multiple-drug resistance activity, the level being equal to that of verapamil, which is a standard drug used in this bioassay (Jacobs *et al.*, in this volume; R. S. Jacobs, in Schmitz and Yasumoto, 1991).

4.13.4. Other Peptides

Geodiamolides A (422) and B (423) were initially found in a Caribbean sponge, *Geodia* sp., and reported as novel cyclodepsipeptides, but no cytotoxic activity was reported (Chan *et al.*, 1987). Later these same peptides plus geodiamolides C–F (424–427) were found in a *Pseudaxinyssa* sp. of sponge collected in

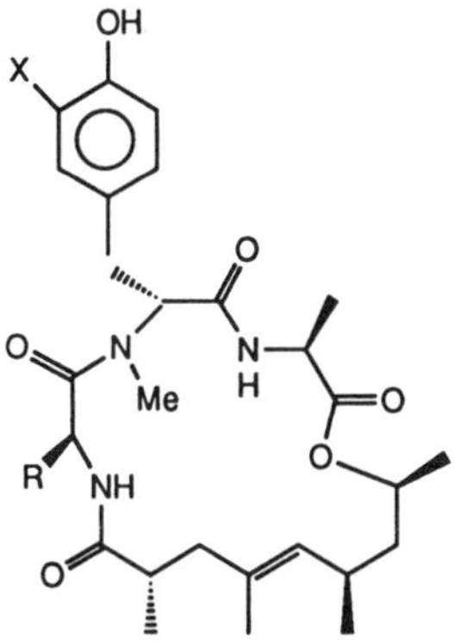

	Compound	X	R	**L1210** IC$_{50}$ (μg/mL)
(422)	Geodiamolide A	I	Me	0.0032
(423)	Geodiamolide B	Br	Me	0.0026
(424)	Geodiamolide C	Cl	Me	0.0025
(425)	Geodiamolide D	I	H	0.0039
(426)	Geodiamolide E	Br	H	0.014
(427)	Geodiamolide F	Cl	H	0.006

New Guinea and the cytotoxicity data shown under the structure were reported (Dilip de Silva *et al.*, 1990). Geodiamolide A (422) has been synthesized (White and Amedio, 1989), as has geodiamolide B (423) (Grieco and Perez-Medrano, 1988).

Two groups independently discovered the depsipeptide 428, called jaspla-kinolide by one group (Crews *et al.*, 1986) and jaspamide by the other (Zabriskie *et al.*, 1986). In both cases the source of 428 was a *Jaspis* sp. of sponge and both groups noted its potent activity against the fungus *Candida albicans* but lack of activity toward a variety of bacteria. The former group reported the compound's cytotoxicity against larynx epithelial carcinoma (IC$_{50}$ 0.32 μg/ml) and a lung cell

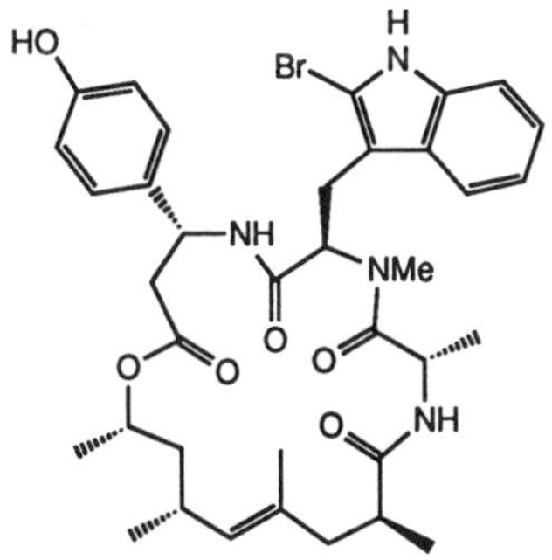

(428)

line (IC$_{50}$ 0.01 μg/ml), and the latter group noted that it was a potent insecticide. In a subsequent full paper, Inman and Crews (1989) discussed the conformational analysis of this peptide, reported cytotoxicity data, and identified the source of their peptide as *J. johnstoni*. Jasplakinolide (**428**) was found to be active against 36 solid tumor cell cultures in the National Cancer Institute human tumor panel as well as cytotoxic to P388 (IC$_{50}$ of 0.01–0.04 μg/ml). In tests with *C. albicans*, jasplakinolide reduced the rate of [^{3}H]thymidine incorporation, thus implying a suppressed rate of DNA synthesis.

Jasplakinolide was also isolated by Braekman *et al.* (1987b), who found it to be ichthyotoxic. Grieco *et al.* (1988) have synthesized (+)-jasplakinolide.

Theonella species of sponges appear to be rich sources of bioactive compounds. The macrolides misakinolide (**94**) and the swinholide series of macrolides (**96**) have been described above. Five depsipeptides named theonellapeptolides Ia–Ie, which inhibited development of fertilized sea urchin eggs, have also been isolated from a *Theonella* sp. from Japan by Kitagawa *et al.* (1986b). Inhibitory concentrations ranged from 2 to 50 μg/ml. Only the structures of the major metabolite, theonellapeptolide Id (**429**), and one minor one, theonellapeptolide Ie (**430**) (Kitagawa *et al.*, 1987b), have been reported. Nakamura *et al.* (1986)

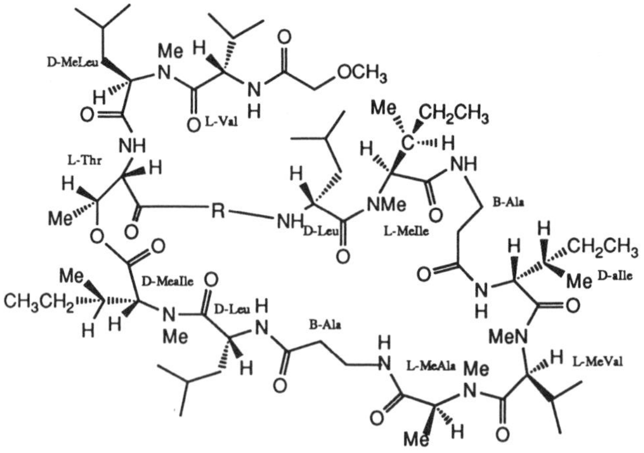

(**429**) R = β-Ala
(**430**) R = N-Me-β-Ala

independently isolated compound **429** and deduced its structure and assigned it the name theonellamine B. These latter workers reported that **429** inhibited Na, K-ATPase activity.

Theonellamide F (**431**) is an antifungal peptide isolated from a *Theonella* sp. from Japan that also shows activity against L1210 and P388 cells (IC$_{50}$ 3.2 and 2.7 μg/ml, respectively) (Matsunaga *et al.*, 1989a). Although the name theonellamide

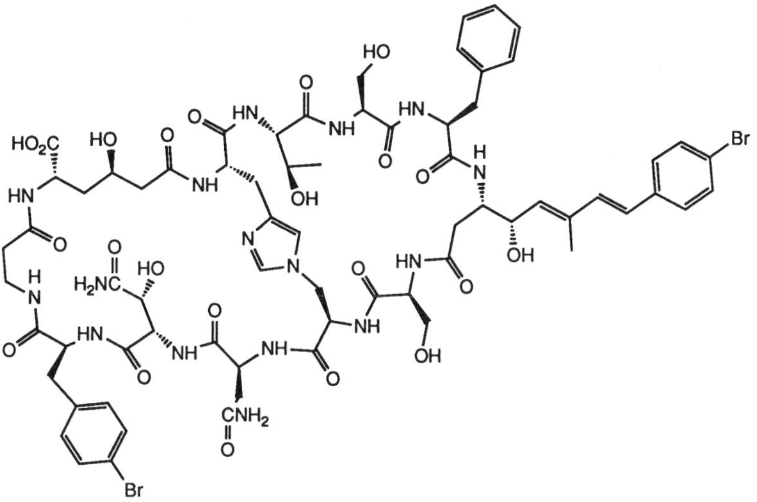

(431)

F (431) suggests that it is one member of a series of related compounds, this is not the case. The description F simply refers to a particular chromatographic fraction.

From a western Pacific sponge, *Hymeniacidon* sp., collected at Palau, Pettit *et al.* (1990a) isolated the cyclic octapeptide hymenistatin 1 (432), all amino acids therein having the *S* chirality. It showed both *in vitro* (ED_{50} 3.5 µg/ml) and *in vivo* activity (T/C 130) against P388 murine leukemia.

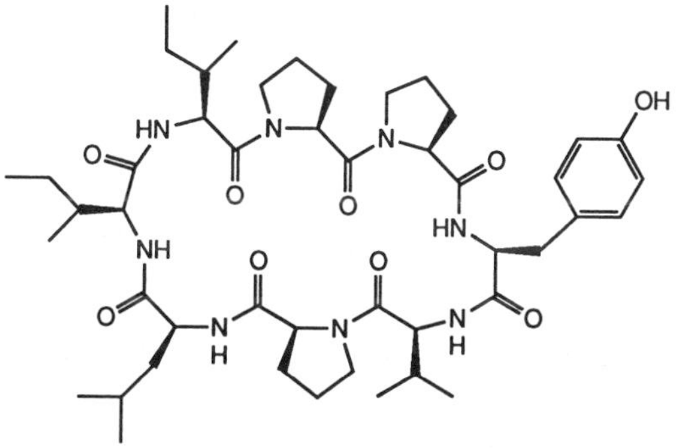

(432)

Gerwick *et al.* (1989) have provided preliminary information on hormotham-nin, a cyclic undecapeptide isolated from wild and cultured cells of the tropical marine cyanobacterium *Hormothamnion enteromorphoides*, collected in the Carib-bean. Cytotoxicity to several cell lines was established: IC_{50} for murine melanoma

B16-F10 was 0.13 μg/ml; for human lung SW1271 was 0.2 μg/ml; for human lung A529 was 0.16 μg/ml; and for human colon HCT-116 was 0.76 μg/ml. The peptide contains five uncommon or unknown amino acid residues and six standard residues: phenylalanine, leucine ($\times$ 2), isoleucine, glycine, hydroxyproline, and homoserine ($\times$ 2).

4.14. Nucleosides

The Fijian sponge *Jaspis johnstoni* yielded two cytotoxic nucleosides (Zabriskie and Ireland, 1989). These are 5-(methoxycarbonyl)tubercidin (**433**), previously reported as a synthetic material, and toyocamycin (**434**), reported

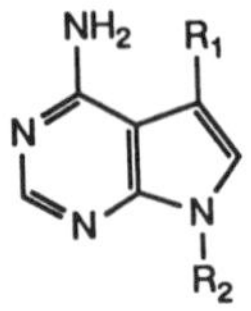

(**433**) R$_1$ = CO$_2$Me, R$_2$ = ribose
(**434**) R$_1$ = CN, R$_2$ = ribose

originally from an unidentified strain of *Streptomyces*. These nucleosides showed IC$_{50}$ values of 0.0026 and 0.27 μg/ml, respectively, against L1210. The 5-(methoxycarbonyl) tubercidin (**433**) had been reported earlier to have *in vivo* activity against L1210, increasing lifetimes by up to 39%. Since toyocamycin was originally isolated from a bacterium, it seems likely that bacteria are the ultimate source of the nucleosides isolated from the sponge.

4.15. Glycoproteins

A Japanese research group has reported cytotoxic glycoproteins from three different sea hares. Aplysianin A is a 350-kDa glycoprotein (9.8% neutral sugar) obtained from the albumin gland of *Aplysia kurodai* (Kamiya *et al.*, 1986), while aplysianin E is a 250-kDa glycoprotein (8% neutral sugar) isolated from the eggs of this sea hare (Yamazaki *et al.*, 1985; Kisugi *et al.*, 1987). Aplysianin A was reported to be cytotoxic to MM46 tumor cells (LD$_{50}$ 0.014 μg/ml), while aplysianin E was reported to have an ED$_{50}$ of 0.01–0.06 μg/ml against a variety of murine (MM46, MM48, L1210, EL4) and human (Raji, Molt-3, and K562 leukemia; PC-6 lung adenocarcinoma) tumor cell lines. Aplysianin E completely inhibits DNA, RNA, and protein synthesis by tumor cells within 2 hr. It is suggested that recognition of the sugar moiety is a key step to cytolysis induced by aplysianin E.

Dolabellanin A, a glycoprotein of mol. wt. 250 kDa which contains four subunits, was obtained from the albumin gland of the sea hare *Dolabella auricularia* (Yamazaki *et al.*, 1989c; Kisugi *et al.*, 1989b). It lyses tumor cells and is active toward a variety of tumor cells, with IC_{50} of 0.001–0.018 µg/ml. Tumor necrosis factor (TNF)-resistant cells were also lysed by dolabellanin A. It completely inhibits DNA and RNA synthesis in 1 hr. Dolabellanin A also prolonged the survival of mice bearing MM46 ascitic tumors. From the body fluid of *D. auricularia*, Kisugi *et al.* (1989a) have purified an antineoplastic glycoprotein designated dolabellanin C.

Kamiya *et al.* (1988) have also reported the isolation of antineoplastic glycoproteins from the genital mass and eggs of the sea hare *Aplysia juliana*. The compounds are reported to be similar to the aplysianins isolated from *A. kurodai*.

Three glycoproteins designated solnins A, B, and C were isolated from the red alga *Solieria robusta* (Hori *et al.*, 1988). They are monomeric proteins which showed mitogenic activity for mouse splenic lymphocytes and inhibited the growth *in vitro* of L1210 cells (IC_{50} 12, 15, and 18 µg/ml, respectively) and mouse FM3A tumor cells (IC_{50} 6,4, and 8 µg/ml, respectively). The authors note that the solnins stimulate mitosis and growth of normal cells involved in the immune system, while they inhibit the growth of the above tumor cells.

4.16. Proteins

Cytotoxic proteins have been separated from the purple fluid secreted by each of the three sea hares mentioned above. Aplysianin P, isolated from *Aplysia kurodai*, is a peptide of 60 kDa which lysed all tumor cells tested, but not normal cells, the half-maximal doses varying from 0.003 to 0.025 µg/ml (Yamazaki *et al.*, 1986, 1989a). Dolabellanin P is a polypeptide of 60 kDa obtained from the purple fluid of *Dolabella auricularia* (Yamazaki *et al.*, 1989b) which *non-specifically* lyses all cells tested at 0.05–0.02 µg/ml. An antineoplastic protein secretion has also been obtained from the sea hare *Aplysia juliana* (Kamiya *et al.*, 1989).

A cytostatic tumor growth inhibitory peptide has also been separated from the supernatant fraction of unfertilized homogenized ova from shad (*Alosa sapidissima*) (Sheid *et al.*, 1989).

5. POLYSACCHARIDES

A highly sulfated, agar-type polysaccharide which inhibited the transplantation of Ehrlich ascites carcinoma in mice was isolated from the red alga *Gracilaria dominguensis* collected in Cuba (Fernandez *et al.*, 1989). At a dose of 60 µg/kg it was reported that 9 of 10 mice were alive and tumor-free for over 30 days compared to a mean survival time of 20 days for untreated animals. Masuda *et al.* (1987) isolated an acidic polysaccharide, NRP-1, of mol. wt. 10,000 from the starfish *Asterias amurensis* Lutken and reported that it exhibits antitumor effects on three mouse tumors, Sarcoma 180, IMC carcinoma, and Meth-A fibrosoma. It also causes induction of cytolytic activity of peritoneal exudate cells and induction of interferon. Among the sugars comprising this polysaccharide are sialic and uronic acids.

6. SUMMARY

Clearly, many marine natural products that show some degree of cytotoxicity have been identified in the past 5½ years. The levels of inhibition of growth of various tumor cell lines *in vitro* range from 10^{-9} to ~30 µg/ml. For some compounds only toxicity to brine shrimp or inhibition of sea urchin egg development has been reported. Unfortunately, only a limited number of *in vivo* active compounds have been reported, but some of these show significant antitumor activity. However, it seems likely that the large majority of the compounds cited here have not been tested for *in vivo* activity due to either lack of a sufficient level of *in vitro* activity to warrant the *in vivo* testing, lack of routine access by many workers to *in vivo* testing, or, in many cases, lack of sufficient material.

Several major antitumor leads have emerged from work on marine organisms. Most prominent are didemnin B (**391**) and related compounds, bryostatin 1 (**100**) and relatives, the recently described ecteinascidins (**349–354**), and the dolastatins (**399–405**). One may hope that one or more clinical drugs will emerge from these leads. It is interesting to note that all of these major leads except for the ecteinascidins were known at the time of the earlier review by Munro *et al.* (1987).

Some of the metabolites initially discovered via cytotoxicity screening have subsequently been found to display other interesting activities. An example is okadaic acid (**55**), which has become an important biochemical tool because of its ability to selectively inhibit protein phosphatases I and IIa. Another is discodermolide (**56**), which displays immunosuppressive activity. Still another case in point is that of bryostatin 1 (**100**), for which a variety of interesting activities have been found (Suffness *et al.*, 1989; Pettit, 1991). These few examples illustrate the desirability of evaluating novel cytotoxins for a broad range of biological activities even though they may not display outstanding activities in *in vivo* antitumor tests. The fact that a compound is cytotoxic suggests a strong interaction with some biochemical receptor. Finding the specific target for the most powerful cytotoxins may lead to the discovery of new receptors or may yield more information about the structure and specificity of known receptor sites. If, as D. H. Williams *et al.* (1989) have proposed, "all natural products have evolved under pressure of natural selection to bind to specific receptors," then in-depth biochemical studies of these marine natural products is bound to result in new information.

It is interesting to compare the number of natural products reported from the different phyla of organisms. Of the 434 compounds described here, the approximate breakdown by source phylum is as follows: sponges, 193 compounds; ascidians (tunicates), 57; algae, 44; mollusks, 46; soft corals, 27; gorgonians, 20; dinoflagellates, 8; anemones, 8; echinoderms, 7; marine worms, 8; bryozoans, 5; bacteria, 3; hydroids, 3. Clearly, sponges have been a rich source of cytotoxins of widely varying structural types. However, some less prolific phyla have been the source of the most promising agents: didemnin B (**391**) and ecteinascidins (**349–354**) from tunicates, bryostatin 1 (**100**) from a bryozoan, and the dolastatins (**399–405**) from a sea hare.

A list of the most active cytotoxins cited in this chapter is given in Table V. The criterion for inclusion in the list was an ED_{50} of <0.005 µg/ml in a cell culture assay. Fifty-nine of the 434 compounds cited in this chapter meet this criterion. The compounds are grouped according to general structural type to give a simple overview of structure–activity relations. Macrolides, depsipeptides and polyether-type compounds account for approximately two-thirds of the most active metabolites. While most of the "very active" cytotoxins are medium to large molecules, relatively simple small molecules are also found in this group (group 6 plus **302**).

Any mention of source organisms brings up the question of the possible role of microorganisms in producing or influencing the production of metabolites isolated from macroscopic organisms. Dinoflagellates are a confirmed source of okadaic acid (**55**), which was originally extracted from sponges. A number of dinoflagellate macrolides (e.g., **58–63**) resemble other macrolides from tunicates and sponges. This suggests that dinoflagellates may be the ultimate source of many of these compounds. A few marine bacterial products are cited here, but much more work needs to be done to establish whether marine bacteria make a substantial contribution to host products. Obviously products of bacterial origin would be very desirable as anticancer drugs since they could be produced by

Table V. Most Active Cytotoxins ($ED_{50} < 10^{-3}$ µg/ml)
Organized According to General Class Type

1. Macrolides (18) **59**, **65**, **67**, **80–82**, **83–85**, **94**, **103–110**
2. Depsipeptides (13) **391**, **393**, **394**, **396**, **398**, **399–401**, **405**, **422–425**
3. Polyethers (11) **54**, **55**, **86–88**, **91**, **92**, **226**, **288–290**
4. Alkaloids (6) **302**, **322**, **323**, **349–351**
5. Complex polyketides ($\geqslant$C32) (4) **292–295**
6. Miscellaneous small metabolites (3) **43**, **194**, **433**
7. Steroid type (4) **266–269**

fermentation, thus avoiding the ecological problems inherent in harvesting large quantities of macroscopic organisms which may not be abundant. However, discovery of agents from microorganisms will be much slower than from macroscopic ones. Intensive efforts will be needed to culture and screen adequate numbers of microorganisms for cytotoxic activities. Substantial research is also needed to establish culture conditions under which to grow many marine bacteria, especially those that grow in association with host organisms.

Of the cytotoxic compounds reviewed here, few would be very water-soluble. This raises the question of whether indeed few such compounds are being produced or whether the extraction/processing schemes of most investigators is such that the water-solubles are overlooked.

Most of the compounds reviewed here appear to have been detected using whole-cell screens. Few, if any, appear to have been detected by mechanism-of-action assays. This may change in the future as more investigators and companies utilize the more rapid, semiautomated enzyme assays for screening. The debate regarding the merits of using specific mechanism-of-action screens versus screening against suitable cell lines continues and Suffness *et al.* (1989) have outlined the relative merits of the two approaches. The mechanism-based screens will find only those compounds whose mechanism is related to the screen, and hence ideally one should screen against *all* the enzymes and receptor systems important in the disease process. This would require a very large number of tests. The whole-cell assay is generally blind to mechanism of action and is likely to detect many compounds that are just general toxins. However, use of a whole-cell system guarantees that any actives so discovered can cross the cell membrane and if metabolic activation is needed to generate the active species, it has a good chance to occur. Furthermore, since numerous enzymes/receptors can be interfered with in the cell, fortuitous discovery of novel mechanisms may be made. Thus, while it appears that more emphasis will be given to mechanism-of-action screens in the future, whole-cell assays are also likely to remain important as screening tools.

Currently, an extensive human tumor cell line panel is being used at the National Cancer Institute in an effort to achieve "disease specific" or "tissue specific" drug discovery (Boyd *et al.*, 1988; Suffness, 1987). Panels of 6–10 human cancer cell lines of each of a variety of cancer types such as lung, colon, ovarian, renal carcinoma, etc., have been established to detect compounds that are selectively inhibitory to a given type of cancer. Such compounds can then be evaluated *in vivo* in athymic mice using tumors derived from cell lines already shown to be sensitive *in vitro*.

Marine organisms have clearly been established as a source of novel compounds with a variety of biological activities, including cytotoxic activity and *in vivo* antitumor activity. At least four leads are in clinical trials or appear to be destined for such testing. These leads and the promise of uncovering new biochemical tools that may be important in studying disease mechanisms provide a continuing impetus for further exploration of marine organisms as sources of potential anticancer drugs.

ACKNOWLEDGMENTS We thank the investigators who kindly sent us preprints of articles and so helped us be as current as possible in literature coverage. We thank Jeffrey Stewart for his generous help in proofreading and revising structural drawings and also thank him, Aeri Park, and Dr. Ali Abbas for proofreading.

REFERENCES

Abbruzzese, J., Ajani, J., Blackburn, R., Faintuch, J., Patt, Y., and Levin, B., 1988, Phase II study of didemnin-B in advanced colorectal cancer, *Proc. Am. Assoc. Cancer Res.* **29**:203.

Ahond, A., Zurita, M. B., Collin, M., Fizames, C., Laboute, P., Lavelle, F., Laurent, D., Poupat, C., Pusset, J., Pusset, M., Thoison, O., and Potier, P., 1988, Girolline, a new antitumoral compound extracted from the sponge, *Pseudaxinyssa cantharella* n. sp. (Axinellidae), *C. R. Acad. Sci. Paris* **307**:145–148.

Akee, R. K., Carroll, T. R., Yoshida, W. Y., Scheuer, P. J., Stout, T. J., and Clardy, J., 1990, Two imidazole alkaloids from a sponge, *J. Org. Chem.* **55**(6):1944–1946.

Alam, M., Sanduja, R., and Weinheimer, A. J., 1988, Isolation and structure of a cytotoxic epoxy sterol from the marine mollusk *Planaxis sulcatus*, *Steroids* **52**(1–2):45–50.

Alonso, R., and Brossi, A., 1988, Synthesis of hormothamnione, *Tetrahedron Lett.* **29**(7):735–738.

Andersson, L., Bohlin, L., Iorizzi, M., Riccio, R., Minale, L., and Moreno-Lopez, W., 1989, Biological activity of saponins and saponin-like compounds from starfish and brittle stars, *Toxicon* **27**(2):179–188.

Arabshahi, L., and Schmitz, F. J., 1987, Brominated tyrosine metabolites from an unidentified sponge, *J. Org. Chem.* **52**:3584.

Arabshahi, L., and Schmitz, F. J., 1988, Thiazole and imidazole metabolites from the ascidian *Aplydium pliciferum*, *Tetrahedron Lett.* **29**(10):1099–1102.

Asari, F., Kusumi, T., and Kakisawa, H., 1989, Turbinaric acid, a cytotoxic secosqualene carboxylic acid from the brown alga *Turbinaria ornata*, *J. Nat. Prod.* **52**(5):1167–1169.

Ayyanger, N. R., Khan, R. A., and Deshpande, V. H., 1988, Synthesis of hormothamnione, *Tetrahedron Lett.* **29**(19):2347–2348.

Bai, R., Pettit, G. R., and Hamel, E., 1990a, Dolastatin 10, a powerful cytostatic peptide derived from a marine animal. Inhibition of tubulin polymerization mediated through the vinca alkaloid binding domain, *Biochem. Pharmacol.* **39**(12):1941–1949.

Bai, R., Pettit, G. R., and Hamel, E., 1990b, Binding of dolastatin 10 to tubulin at a distinct site for peptide antimitotic agents near the exchangeable nucleotide and vinca alkaloid sites, *J. Biol. Chem.* **265**(28):17141–17149.

Banaigs, B., Jeanty, G., Francisco, C., Jouin, P., Poncet, J., Heitz, A., Cave, A., Prome, J. C., Wahl, M., and Lafargue, F., 1989, Didemnin B: Comparative study and conformational approach in solution, *Tetrahedron* **45**(1):181–190.

Barnekow, D. E., Cardellina, J. H., Zekder, A. S., and Martin, G. E., 1989, Novel cytotoxic and phytotoxic halogenated sesquiterpenes from the green alga *Neomeris annulata*, *J. Am. Chem. Soc.* **111**:3511–3517.

Barrow, C. J., Blunt, J. W., Munro, M. H. G., and Perry, N. B., 1988, Oxygenated furanosesterterpene tetronic acids from a sponge of the genus *Ircinia*, *J. Nat. Prod.* **51**:1294–1298.

Bartik, K., Braekman, J. C., Daloze, D., Stoller, C., Huysecom, J., Vandevyver, G., and Ottinger, R.,

1987, Topsentins, new toxic bis-indole alkaloids from the marine sponge *Topsentia genitrix*, *Can. J. Chem.* **65**:2118–2121.

Benslimane, A. F., Pouchus, Y. F., LeBoterff, J., Verbist, J. F., Roussakis, C., and Monniot, F., 1988, Cytotoxic and antibacterial substances of ascidian *Aplidium antillense*, *J. Nat. Prod.* **51**(3):582–583.

Bialojan, C., and Takai, A., 1988, Inhibitory effect of a marine-sponge toxin, okadaic acid, on protein phosphatases, *Biochem. J.* **256**:283–290.

Bloor, S. J., and Schmitz, F. J., 1987, A novel pentacyclic aromatic alkaloid from an ascidian, *J. Am. Chem. Soc.* **109**:6134–6136.

Blunt, J. W., Hartshorn, M. P., McLennan, T. J., Munro, M. H. G., Robinson, W. T., and Yorke, S. C., 1978, Thyrsiferol: A squalene-derived metabolite of *Laurencia thyrsifera*, *Tetrahedron Lett.* **1978**:69–72.

Blunt, J. W., Lake, R. J., and Munro, M. H. G., 1988, Antitumor Polycyclic Compounds from Marine *Ritterella sigillinoides* and Their Use and Preparation, WO 8800826 A1 11 February 1988.

Boyd, M. R., Shoemaker, R. H., McLemore, T. L., Johnston, M. R., Alley, M. C., Scudiero, D. A., Monks, A., Fine, D. L., Mayo, J. G., and Chabner, B. A., 1988, New drug developments, in: *Thoracic Oncology* (J. A. Roth, J. C. Ruckdescher, and T. H. Weisenburger, eds.), Saunders, New York, Chapter 51.

Braekman, J. C., Daloze, D., and Stoller, C., 1987a, Synthesis of topsentin-A, a bisindole alkaloid of the marine sponge *Topsentia genetrix*, *Bull. Soc. Chim. Belg.* **96**(10):809–812.

Braekman, J. C., Daloze, D., Moussiaux, B., and Riccio, R., 1987b, Jaspamide from the marine sponge *Jaspis johnstoni*, *J. Nat. Prod.* **50**(5):994–995.

Bruno, I., Minale, L., and Riccio, R., 1990, Starfish saponins, part 43. Structures of two new sulfated steroidal fucofuranosides (imbricatosides A and B) and six new polyhydroxysteroids from the starfish *Dermasterias imbricata*, *J. Nat. Prod.* **53**(2):366–374.

Burres, N. S., and Clement, J. J., 1989, Antitumor activity and mechanism of action of the novel marine natural products mycalamide-A and -B and onnamide, *Cancer Res.* **49**(11):2935–2940.

Burres, N. S., Sazesh, S., Gunawardana, G. P., and Clement, J. J., 1989, Antitumor activity and nucleic acid binding properties of dercitin, a new acridine alkaloid isolated from a marine *Dercitus* species sponge, *Cancer Res.* **49**(19):5267–5274.

Canonico, P. G., Pannier, W. L., Huggins, J. W., and Rinehart, Jr., K. L., 1982, Inhibition of RNA viruses *in vitro* and in Rift Valley fever-infected mice by didemnins A and B, *Antimicrob. Chemother.* **22**:696.

Cardellina, J. H., and Barnekow, D. E., 1988, Oxidised nakafuran 8 sesquiterpenes from the sponge *Dysidea etheria*. Structure, stereochemistry and biological activity, *J. Org. Chem.* **53**:882–884.

Cardellina, J. H., and Moore, R. E., 1979, The structures of pukeleimides A, B, D, E, F, and G, *Tetrahedron Lett.* **1979**:2007–2010.

Cardellina, J. H., Marner, F. J., and Moore, R. E., 1979, Malyngamide A, a novel chlorinated metabolite of the marine cyanophyte *Lyngbya majuscula*, *J. Am. Chem. Soc.* **101**(1):240–242.

Carmely, S., Roll, M., Loya, Y., and Kashman, Y., 1989a, The structure of eryloside A, a new antitumor and antifungal 4-methylated steroidal glycoside from the sponge *Erylus lendenfeldi*, *J. Nat. Prod.* **52**(1):167–170.

Carmely, S., Ilan, M., and Kashman, Y., 1989b, 2-Amino imidazole alkaloids from the marine sponge *Leucetta chagosensis*, *Tetrahedron* **45**(7):2193–2200.

Carroll, A. R., and Scheuer, P. J., 1990, Kuanoniamines A, B, C, and D: Pentacyclic alkaloids from a tunicate and its prosobranch mollusc predator *Chelynotus semperi*, *J. Org. Chem.* **55**:4426–4431.

Carter, D. C., Moore, R, E., Mynderse, J. S., Niemczura, W. P., and Todd, J. S., 1984, Structure of majusculamide C, a cyclic depsipeptide from *Lyngbya majuscula*, *J. Org. Chem.* **49**:236–241.

Chan, W. R., Tinto, W. F., Manchand, P. S., and Todaro, L., 1987, Stereostructures of geodiamolides A and B, novel cyclodepsipeptides from the marine sponge *Geordia* sp., *J. Org. Chem.* **52**(14):3091–3093.

Charyulu, G. A., McKee, T. C., and Ireland, C. M., 1989, Diplamine, a cytotoxic polyaromatic alkaloid from the tunicate *Diplosoma* sp., *Tetrahedron Lett.* **30**(32):4201–4202.

Cheng, J., Kobayashi, J., Nakamura, H., Ohizumi, Y., Hirata, Y., and Sasaki, T., 1988a, Penasterol, a novel antileukemic sterol from the Okinawan marine sponge *Penares* sp., *J. Chem. Soc. Perkin Trans. 1* **1988**(8):2403–2406.

Cheng, J., Ohizumi, Y., Walchli, M. R., Nakamura, H., Hirata, Y., Sasaki, T., and Kobayashi, J., 1988b, Prianosins B, C, and D, novel sulfur-containing alkaloids with potent antineoplastic activity from the Okinawan marine sponge *Prianos melanos*, *J. Org. Chem.* **53**(19):4621–4624.

Chun, H. G., Davies, B., Hoth, D., Suffness, M., Plowman, J., Flora, K., Grieshaber, C., and Leyland-Jones, B., 1986, Didemnin B, *Invest. New Drugs* **4**:279–284.

Cimino, G., DeGiulio, A., DeRosa, S., and DiMarzo, V., 1990, Minor bioactive polyacetylenes from *Petrosia Ficiformis*, *J. Nat. Prod.* **53**(2):345–353.

Cohen, P., Holmes, C. F. B., and Tsukitani, Y., 1990, Okadaic acid: A new probe for the study of cellular regulation, *Trends Biochem. Sci.* **15**:98–102.

Copp, B. R., Blunt, J. W., Munro, M. H. G., and Pannell, L. K., 1989, A biologically active 1,2,3-trithiane derivative from the New Zealand ascidian *Aplidium* sp. D, *Tetrahedron Lett.* **30**(28): 3703–3706.

Corley, D. G., Herb, R., Moore, R. E., Scheuer, P. J., and Paul, V. J., 1988a, Laulimalides. New potent cytotoxic macrolides from a marine sponge and a nudibranch predator, *J. Org. Chem.* **53**(15):3644–3646.

Corley, D. G., Moore, R. E., and Paul, V. J., 1988b, Patellazole B: A novel cytotoxic thiazole-containing macrolide from the marine tunicate *Lissoclinum patella*, *J. Am. Chem. Soc.* **110**(23):7920–7924.

Coval, S. J., Cross, S., Bernardinelli, G., and Jefford, C. W., 1988, Brianthein V, a new cytotoxic and antiviral diterpene isolated from *Briareum asbestinum*, *J. Nat. Prod.* **51**(5):981–984.

Crampton, S. L., Adams, E. G., Kuentzel, S. L., Li, L. H., Badiner, G., and Bhuyan, B. K., 1984, Biochemical and cellular effects of didemnins A and B, *Cancer Res.* **44**:1796–1801.

Crews, P., Manes, L. V., and Boehler, M., 1986, Jasplakinolide, a cyclodepsipeptide from the marine sponge, *Jaspis* sp., *Tetrahedron Lett.* **27**(25):2797–2800.

Crispino, A., DeGiulio, A., DeRosa, S., and Strazzullo, G., 1989, A new bioactive derivative of avarol from the marine sponge *Dysidea avara*, *J. Nat. Prod.* **52**(3):646–648.

D'Ambrosio, M., Guerriero, A., Debitus, C., Ribes, O., RicherdeForges, B., and Pietra, F., 1989, Corallistin A, a second example of a free porphyrin from a living organism. Isolation from the demosponge *Corallistes* sp. of the Coral Sea and inhibition of abnormal cells, *Helv. Chim. Acta* **72**(7):1451–1454.

D'Auria, M. V., Riccio, R., Minale, L., LaBarre, S., and Pusset, J., 1987, Novel marine steroid sulfates from Pacific ophiuroids, *J. Org. Chem.* **52**:3949.

DeGiulio, A., DeRosa, S., DiVincenzo, G., and Zavodnik, N., 1989, Terpenoids from the North Adriatic sponge *Spongia officinalis*, *J. Nat. Prod.* **52**(6):1258–1262.

Degnan, B. M., Hawkins, C. J., Lavin, M. F., McCaffrey, E. J., Parry, D. L., VandenBrenk, A. L., and Watters, D. J., 1989a, New cyclic peptides with cytotoxic activity from the ascidian *Lissoclinum patella*, *J. Med. Chem.* **32**(6):1349–1354.

Degnan, B. M., Hawkins, C. J., Lavin, M. F., McCaffrey, E. J., Parry, D. L., and Watters, D. J., 1989b, Novel cytotoxic compounds from the ascidian *Lissoclinum bistratum*, *J. Med. Chem.* **32**(6):1354–1359.

DeGuzman, F. S., and Schmitz, F. J., 1989, Chemistry of 2-bromoleptoclinidinone, Structure revision, *Tetrahedron Lett.* **39**(9):1069–1070.

Dell'Aquila, M. L., Nguyen, H. T., Herald, C. L., Pettit, G. R., and Blumberg, P. M., 1987, Inhibition by bryostatin 1 of the phorbol ester-induced blockage of differentiation in hexamethylene bisacetamide-treated Friend erythroleukemia cells, *Cancer Res.* **47**(22):6006–6009.

Dell'Aquila, M. L., Herald, C. L., Kamano, Y., Pettit, G. R., and Blumberg, P. M., 1988, Differential effects of bryostatins and phorbol esters on arachidonic acid metabolite release and epidermal growth factor binding in C3H 10T1/2 cells, *Cancer Res.* **48**:3702–3708.

DeRosa, S., DeStefano, S., and Zavodnik, N., 1988, Cacospongionolide. A new antitumoral sesterterpene, from the marine sponge *Cacospongia mollior*, *J. Org. Chem.* **53**(21):5020–5023.

Dilip de Silva, E., Andersen, R. J., and Allen, T. M., 1990, Geodiamolides C to F, new cytotoxic cyclodepsipeptides from the marine sponge *Pseudaxinyssa* sp., *Tetrahedron Lett.* **31**(4):489–492.

Drexler, H. G., Gignac, S. M., Jones, R. A., Scott, C. S., Pettit, G. R., and Hoffbrand, A. V., 1989, Bryostatin 1 induces differentiation of β-chronic lymphocytic leukemia cells, *Blood* **74**:1747–1757.

Drexler, H. G., Gignac, S. M., pettit, G. R., and Hoffbrand, A. V., 1990, Synergistic action of calcium ionophore A23187 and protein kinase C activator bryostatin 1 on human B cell activation and proliferation, *Eur. J. Immunol.* **20**(1):119–127.

Dubois, M. A., Higuchi, R., Komori, T., and Sasaki, T., 1988, Biologically active glycosides from asteroidea. XVI. Steroid oligoglycosides from the starfish *Asterina pectinifera* Muller et Troschel. 3. Structures of two new oligoglycoside sulfates, pectiniosides E and F, and biological activities of the six new pectiniosides, *Liebigs Ann. Chem.* **1988**(9):845–850.

Dumdei, E. J., Dillip de Silva, E., Andersen, R. J., Choudhary, M. I., and Clardy, J., 1989, Chromodorolide A, a rearranged diterpene with a new carbon skeleton from the Indian Ocean nudibranch, *Chromodoris cavae*, *J. Am. Chem. Soc.* **111**:2712–2713.

Fathi-Afshar, R., and Allen, T. M., 1988, Biologically active metabolites from *Agelas mauritiana*, *Can. J. Chem.* **66**:45–50.

Fathi-Afshar, R., Allen, T. M., Krueger, C. A., Cook, D. A., Clanachan, A. S., Vriend, R., Baer, H. P., and Cass, C. E., 1989, Some pharmacological activities of novel adenine-related compounds isolated from a marine sponge *Agelas mauritiana*, *Can. J. Physiol. Pharmacol.* **67**(4):276–281.

Faulkner, D. J., 1991, Marine natural products, *Nat. Prod. Rep.* **8**:97–147.

Fedoreev, S. A., Prokof'eva, N. G., Denisenko, V. A., and Rebachuk, N. M., 1988, Cytotoxic activity of aaptamines derived from *Suberitidae* sponges, *Khim. Farm. Zh.* **22**(8):943–946.

Fenical, W., and Paul, V. J., 1984, Antimicrobial and cytotoxic terpenoids from tropical green algae of the family *Udoteaceae*, *Hydrobiologia* **116/117**:135–140.

Fernandez, L. E., Valiente, O. G., Mainardi, V., Bello, J. L., Velez, H., and Rosado, A., 1989, Isolation and characterization of an antitumor active agar-type polysaccharide of *Gracilaria dominguensis*, *Carbohydr. Res.* **190**:77–83.

Francisco, C., Banaigs, B., Teste, J., and Cave, A., 1986, Mediterraneols: A novel biologically active class of rearranged diterpenoid metabolites from *Cystoseira mediterranea* (Pheophyta), *J. Org. Chem.* **51**:1115–1120.

Fusetani, N., Yasukawa, K., Matsunaga, S., and Hashimoto, K., 1985, Bioactive marine metabolites XII. Moritoside, an inhibitor of the development of starfish embryo, from the gorgonian *Euplexaura* sp., *Tetrahedron Lett.* **26**(52):6449–6452.

Fusetani, N., Asano, M., Matsunaga, S., and Hashimoto, K., 1987a, Acalycixeniolides, novel norditerpenes which inhibit cell division of fertilised star fish eggs, from the gorgonian *Acalycigorgia inermis*, *Tetrahedron Lett.* **28**(47):5837–5840.

Fusetani, N., Yasukawa, K., Matsunaga, S., and Hashimoto, K., 1987b, Dimorphosides A and B, novel steroid glycosides from the gorgonian *Anthoplexaura dimorpha*, *Tetrahedron Lett.* **28**(11):1187–1190.

Fusetani, N., Shiragaki, T., Matsunaga, S., and Hashimoto, K., 1987c, Bioactive marine metabolites XX. Petrosynol and petrosynone, antimicrobial C30 polyacetylenes from the marine sponge *Petrosia* sp.: Determination of the absolute configuration, *Tetrahedron Lett.* **28**(37):4313–4314.

Fusetani, N., Asano, M., Matsunaga, S., and Hashimoto, K., 1989a, Bioactive marine metabolites. Acalycixeniolides, novel norditerpenes with allene functionality from two gorgonians of the genus *Acalycigorgia, Tetrahedron* **45**(6):1647–1652.

Fusetani, N., Yasumuro, K., Matsunaga, S., and Hashimoto, K., 1989b, Mycalolides A–C, hybrid macrolides of ulapualides and halichondramide, from a sponge of the genus *Mycale, Tetrahedron Lett.* **30**(21):2809–2812.

Fusetani, N, Yasumuro, K., Matsunaga, S., and Hirota, H., 1989c, Haliclamines A and B, cytotoxic macrocyclic alkaloids from a sponge of the genus *Haliclona, Tetrahedron Lett.* **30**(49):6891–6894.

Fusetani, N., Yasumuro, K., Matsunaga, S., Hirota, H., Kawai, H., and Natori, T., 1989d, New cytotoxic compounds from marine sponges, *Tennen Yuki Kagobutsu Toronkai Koen Yoshishu* **31st**:340–347.

Fusetani, N., Nagata, H., Hirota, H., and Tsuyuki, T., 1989e. Astrogorgiadiol and astrogorgin, inhibitors of cell division in fertilized starfish eggs, from a gorgonian *Astrogorgia sp., Tetrahedron Lett.* **30**(50):7079–7082.

Fusetani, N., Yasumuro, K., Kawai, H., Natori, T., Brinen, L., and Clardy, J., 1990, Kalihinene and isokalihinol B, Cytotoxic diterpene isonitriles from the marine sponge *Acanthella klethra, Tetrahedron Lett.* **31**(25):3599–3602.

Garcia, M. O., and Rodriguez, A. D., 1990, Palominin, a novel furanosesterterpene from a Caribbean sponge *Ircinia sp., Tetrahedron* **46**(4):1119–1124.

Geran, R. I., Greenberg, N. H., McDonald, M. M., Schumacher, A. M., and Abbott, B. J., 1972, Protocols for screening chemical agents and natural products against animal tumors and other biological systems (third edition), *Cancer Chemother. Rep.* **3**:1–103.

Gerwick, W. H., 1989, 6-Desmethoxyhormothamnione, a new cytotoxic styrylchromone from the marine cryptophyte *Chrysophaeum taylori, J. Nat. Prod.* **52**(2):252–256.

Gerwick, W. H., Lopez, A., Van Duyne, G. D., Clardy, J., Ortiz, W., and Baez, A., 1986, Hormothamnione, a novel cytotoxic styrylchromone from the marine cyanophyte *Hormothamnion enteromorphoides* Grunow, *Tetrahedron Lett.* **27**(18):1979–1982.

Gerwick, W. H., Mrozek, C., Moghaddam, M. F., and Agarwal, S. K., 1989, Novel cytotoxic peptides from the tropical marine cyanobacterium *Hormothamnion enteromorphoides*. 1. Discovery, isolation and initial chemical and biological characterization of the hormothamnins from wild and cultured material, *Experientia* **45**(2):115–121.

Geschwendt, M., Kittstein, W., and Marks, F., 1989, The immunosuppressant FK-506, like cyclosporins and didemnin B, inhibits calmodulin-dependent phosphorylation of the elongation factor 2 *in vitro* and biological effects of the phorbol ester TPA on mouse skin *in vivo, Immunobiology* **179**:1–7.

Goo, Y. M., 1985, Antimicrobial and antineoplastic tyrosine metabolites from a marine sponge, *Aplysina fistularis, Arch. Pharmacol. Res.* **8**(1):21–30.

Grieco, P. A., and Perez-Medrano, A., 1988, Total synthesis of the mixed peptide-polypropionate based cyclodepsipeptide (+)-geodiamolide-B, *Tetrahedron Lett.* **29**(34):4225–4228.

Grieco, P. A., Hon, Y. A., and Perez-Medrano, A., 1988, A convergent, enantiospecific total synthesis of the novel cyclodepsipeptide (+)-jasplakinolide (jaspamide), *J. Am. Chem. Soc.* **110**:1630–1631.

Guella, G., Mancini, I., Chiasera, G., and Pietra, F., 1989, Sphinxolide, a 26-membered antitumoral macrolide isolated from an unidentified Pacific nudibranch, *Helv. Chim. Acta.* **72**(2):237–246.

Gunasekera, S. P., and Faircloth, G. T., 1990, New acetylenic alcohols from the sponge *Cribrochalina vasculum, J. Org. Chem.* **55**(25):6223–6225.

Gunasekera, S., Gunasekera, M., Gunawardana, G. P., McCarthy, P., and Burres, N., 1990a, Two new bioactive cyclic peroxides from the marine sponge *Plakortis angulospiculatis, J. Nat. Prod.* **53**(3):669–674.

Gunasekera, S. P., Gunasekera, M., Longley, R. E., and Schulte, A. K., 1990b, Discodermolide: A new bioactive polyhydroxylated lactone from the marine sponge *Discodermia dissoluta*, *J. Org. Chem.* **55**:4912–4915.

Gunawardana, G. P., Khomoto, S., Gunasekera, S. P., McConnell, O. J., and Koehn, F. E., 1988, Dercitin, a new biologically active acridine alkaloid from a deep-water marine sponge, *Dercitus* species, *J. Am. Chem. Soc.* **110**:4856–4858.

Gunawardana, G. P., Kohmoto, S., and Burres, N. S., 1989, New cytotoxic acridine alkaloids from two deep water marine sponges of the family Pachastrellidae, *Tetrahedron Lett.* **30**(33):4359–4362.

Gunawardana, G. P., Koehn, F. E., Lee, A. Y., Clardy, J., He, H., and Faulkner, D. J., 1992, Pyridoacridine alkaloids from deep-water marine sponges of the Family Pachastrellidae: structure revision of dercitin and related compounds and correlation with the kuanoiamines, *J. Org. Chem.* **57**:1523–1526.

Gustafson, K., Roman, M., and Fenical, W., 1989, The macrolactins, a novel class of antiviral and cytotoxic macrolides from a deep-sea marine bacterium, *J. Am. Chem. Soc.* **111**(19):7519–7524.

Guyot, M., Durgeat, M., and Morel, E., 1986, Ficulinic acid A and B, two novel cytotoxic straight-chain acids from the sponge *Ficulina ficus*, *J. Nat. Prod.* **49**(2):307–309.

Guyot, M., Davoust, D., and Morel, E., 1987, Isodidemnine-1, a cytotoxic cyclodepsipeptide isolated from a tunicate, *Trididemnum cyanophorum* (Didemnidae), *C. R. Acad. Sci. 2* **305**(8):681–686.

Hamada, Y., Shibata, M., and Shioiri, T., 1985a, New methods and reagents in organic synthesis. 58. A synthesis of patellamide A, a cytotoxic cyclic peptide from a tunicate. Revision of its proposed structure, *Tetrahedron Lett.* **26**(52):6501–6504.

Hamada, Y., Shibata, M., and Shioiri, T., 1985b, New methods and reagents in organic synthesis. 56. Total syntheses of patellamides B and C, cytotoxic cyclic peptides from a tunicate. 2. Their real structures have been determined by their syntheses, *Tetrahedron Lett.* **26**(42):5159–5162.

Hamada, Y., Kato, S., and Shioiri, T., 1985c, New methods and reagents in organic synthesis. 51. A synthesis of ascidiacyclamide, a cytotoxic cyclic peptide from ascidian—Determination of its absolute configuration, *Tetrahedron Lett.* **26**(27):3223–3226.

Hamada, Y., Kondo, Y., Shibata, M., and Shioiri, T., 1988, Efficient total synthesis of didemnins A and B, *J. Am. Chem. Soc.* **111**:669–673.

Hamada, Y., Hayashi, K., and Shioiri, T., 1991, New methods and reagents for organic synthesis. 95. Efficient stereoselective synthesis of dolastatin 10, an antineoplastic peptide from a sea hare, *Tetrahedron Lett.* **32**(7):931–934.

Harada, N., Sugioka, T., Ando, Y., Uda, H., and Kuriki, T., 1988, Total synthesis of (+)-halenaquinol and (+)-halenaquinone. Experimental proof of their absolute stereostructures theoretically determined, *J. Am. Chem. Soc.* **110**(25):8483–8487.

Hashimoto, M., Kan, T., Nozaki, K., Yanagiya, M., Shirahama, H., and Matsumoto, T., 1988, Total synthesis of (+)-thyrsiferol and (+)-venustratriol, *Tetrahedron Lett.* **29**(10):1143–1144.

Hashimoto, T., Fuseya, N., Takahashi, K., and Nohara, C., 1988, Isolation of Spermine Derivatives as Anticancer Agents, JP 63307849 A2 15 December 1988 Showa.

Hawkins, C. J., Lavin, M. F., Marshall, K. A., VandenBrenk, A. L., and Watters, D. J., 1990, Structure–activity relationships of the lissoclinamides: Cytotoxic cyclic peptides from the ascidian *Lissoclinum patella*, *J. Med. Chem.* **33**(6):1634–1638.

Hennings, H., Blumberg, P. M., Pettit, G. R., Herald, C. L., Shores, R., and Yuspa, S. H., 1987, Bryostatin 1, an activator of protein kinase C, inhibits tumor promotion by phorbol esters in SENCAR mouse skin, *Carcinogenesis* (London) **8**(9):1343–1346.

Herb, R., Carroll, A. R., Yoshida, W. Y., Scheuer, P. J., and Paul, V. J., 1990, Polyalkylated cyclopent-indoles: Cytotoxic fish antifeedants from a sponge, *Axinella* sp., *Tetrahedron* **46**(8):3089–3092.

Higa, T., Okuda, R. K., Severns, R. M., Scheuer, P. J., He, C. H., Xu, C., and Clardy, J., 1987, Unprecedented constituents of a new species of acorn worm, *Tetrahedron* **43**(6):1063–1070.

Higuchi, R., Noguchi, Y., Komori, T., and Sasaki, T., 1988, Biologically active glycosides from asteroidea. XVIII. Proton-NMR spectroscopy and biological activities of polyhydroxylated steroids from the starfish *Asterina pectinifera* Mueller et Troschel, *Liebigs Ann. Chem.* **1988** (12):1185–1189.

Hirata, Y., and Uemura, D., 1986, Halichondrins—Antitumor polyether macrolides from a marine sponge, *Pure Appl. Chem.* **58**(5):701–710.

Hirota, H., Takayama, S., Miyashiro, S., Ozaki, Y., and Ikegami, S., 1990a, Structure of a novel steroidal saponin, pachastrelloside A, obtained from a marine sponge of the genus *Pachastrella*, *Tetrahedron Lett.* **31**(23):3321–3324.

Hirota, H., Matsunaga, S., and Fusetani, N., 1990b, Bioactive marine metabolites. Part 32. Stellettamide A, an antifungal alkaloid from a marine sponge of the genus *Stelletta*, *Tetrahedron Lett.* **31**(29):4163–4164.

Hirsch, S., Miroz, A., McCarthy, P., and Kashman, Y., 1989, Etzionin, a new antifungal metabolite from a Red Sea tunicate, *Tetrahedron Lett.* **30**(32):4291–4294.

Hoffmann, H. M. R., and Rabe, J., 1985, Synthesis and biological activity of α-methylene-γ-butyrolactones, *Angew. Chem. Int. Ed. Engl.* **24**:94–110.

Hokama, S., Tanaka, J., Higa, T., Fusetani, N., Asano, M., Matsunaga, S., and Hashimoto, J., 1988, Bioactive marine metabolites. Ginamellene, a new norditerpene with allene functionality from four gorgonians of the genus *Acalycigorgia*, *Chem. Lett.* **1988**:855.

Holmes, C. F. B., Luu, H. A., Carrier, F., and Schmitz, F. J., 1990, Inhibition of protein phosphatases-1 and -2A with acanthifolicin. Comparison with diarrhetic shellfish toxins and identification of a region on okadaic acid important for phosphatase inhibition, *FEBS Lett.* **270**:216.

Holzapfel, C. W., Van Zyl, W. J., and Roos, M., 1990, The synthesis and conformation in solution of cyclo [L-Pro-L-Leu-L-(Gln)Thz-(Gly)Thz-L-Val] (dolastatin3), *Tetrahedron* **46**(2):649–660.

Honda, A., Yamamoto, Y., Mori, Y., Yamada, Y., and Kikuchi, H., 1985, Antileukemic effect of coral-prostanoids clavulones from the stolonifer *Clavularia viridis* on human myeloid leukemia (HL-60) cells, *Biochem. Biophys. Res. Commun.* **130**(2):515–523.

Honda, A., Mori, Y., Iguchi, K., and Yamada, Y., 1987, Antiproliferative and cytotoxic effects of newly discovered halogenated coral prostanoids from the Japanese stolonifer *Clavularia viridis* on human myeloid leukemia cells in culture, *Mol. Pharmacol.* **32**(4):530–535.

Honda, A., Mori, Y., Iguchi, K., and Yamada, Y., 1988a, Structure requirements for antiproliferative and cytotoxic activities of marine coral prostanoids from the Japanese stolonifer *Clavularia viridis* against human myeloid leukemia cells in culture, *Prostaglandins* **36**(5):621–630.

Honda, A., Mori, Y., Yamada, Y., Nakaike, S., Hayashi, H., and Otomo, S., 1988b, Prolonged survival time of sarcoma 180-bearing mice treated with lipid microspheres-entrapped antitumor marine coral prostanoids, *Res. Commun. Chem. Pathol. Pharmacol.* **61**(3):413–416.

Hori, K., Ikegami, S., Miyazawa, K., and Ito, K., 1988, Mitogenic and antineoplastic isoagglutinins from the red alga *Solieria robusta*, *Phytochemistry* **27**(7):2063–2067.

Hossain, M. B., Van der Helm, D., Antel, J., Sheldrick, G. M., Sanduja, S. K., and Weinheimer, A. J., 1988, Crystal and molecular structure of didemnin B, an antiviral and cytotoxic depsipeptide, *Proc. Natl. Acad. Sci. USA* **85**:4118–4122.

Ichiba, T., Sakai, R., Kohmoto, S., Saucy, G., and Higa, T., 1988, New manzamine alkaloids from a sponge of the genus *Xestospongia*, *Tetrahedron Lett.* **29**(25):3083–3086.

Iguchi, K., Kaneta, S., Mori, K., Yamada, Y., Honda, A., and Mori, Y, 1985, Marine natural products. Part XII. Chlorovulones, new halogenated marine prostanoids with an antitumor activity from the stolonifer *Clavularia viridis* Quoy and Gaimard, *Tetrahedron Lett.* **26**(47):5787–5790.

Iguchi, K., Kaneta, S., Mori, K., Yamada, Y., Honda, A., and Mori, Y., 1986, Bromovulone I and iodovulone I, unprecedented brominated and iodinated marine prostanoids with antitumor activity isolated from the Japanese stolonifer *Clavularia viridis* Quoy and Gaimard, *J. Chem. Soc. Chem. Commun.* **1986**(12):981–982.

Iguchi, K., Kaneta, S., Mori, K., and Yamada, Y., 1987, A new marine epoxy prostanoid with an antiproliferative activity from the stolonifer *Clavularia viridis* Quoy and Gaimard, *Chem. Pharm. Bull.* **35**(10):4375–4376.

Inman, W., and Crews, P., 1989, Novel marine sponge derived amino acids. 8 Conformational analysis of jasplakinolide, *J. Am. Chem. Soc.* **111**:2822–2829.

Inman, W. D., O'Neill-Johnson, M., and Crews, P., 1990, Novel marine sponge alkaloids. 1. Plakinidine A and B, anthelmintic active alkaloids from a *Plakortis* sponge, *J. Am. Chem. Soc.* **112**(1):1–4.

Ireland, C. M., and Scheuer, P. J., 1980, Ulicyclamide and ulithiacyclamide, two new small peptides from a marine tunicate, *J. Am. Chem. Soc.* **102**:5688–5691.

Ishibashi, M., Ohizumi, Y., Sasaki, T., Nakamura, H., Hirata, Y., and Kobayashi, J., 1987a, Pseudodistomins A and B, novel antineoplastic piperidine alkaloids with calmodulin antagonistic activity from the Okinawan tunicate *Pseudodistoma kanoko*, *J. Org. Chem.* **52**(3):450–453.

Ishibashi, M., Ohizumi, Y., Hamashima, M., Nakamura, H., Hirata, Y., Sasaki, T., and Kobayashi, J., 1987b, Amphidinolide B, a novel macrolide with potent antineoplastic activity from the marine dinoflagellate *Amphidinium* sp., *J. Chem. Soc. Chem. Commun.* **1987**(14):1127–1129.

Ishibashi, M., Ohizumi, Y., Cheng, J. F., Nakamura, H., Hirata, Y., Sasaki, T., and Kobayashi, J., 1988, Metachromins A and B, novel antineoplastic sesquiterpenoids from the Okinawan sponge *Hippospongia* cf. *metachromia*, *J. Org. Chem.* **53**(12):2855–2858.

Ishida, T., Inoue, M., Hamada, Y., Kato, S., and Shioiri, T., 1987, X-ray crystal structure of ascidiacylamide, a cytotoxic cyclic peptide from ascidian, *J. Chem. Soc. Chem. Commun.* **1987**(5): 370–371.

Ishida, T., Tanaka, M., Nabae, M., Inoue, M., Kato, S., Hamada, Y., and Shioiri, T., 1988, Solution and solid-state conformations of ascidiacyclamide, a cytotoxic cyclic peptide from ascidian, *J. Org. Chem.* **53**(1):107–112.

Ishida, T., Ohishi, H., Inoue, M., Kamigauchi, M., Sugiura, M., Takao, N., Kato, S., Hamada, Y., and Shioiri, T., 1989, Conformational properties of ulithiacyclamide, a strongly cytotoxic cyclic peptide from a marine tunicate, determined by ¹H nuclear magnetic resonance and energy minimization calculations, *J. Org. Chem.* **54**(22):5337–5343.

Ishida, T., In, Y., Doi, M., Inoue, M., Hamada, Y., and Shioiri, T., 1992, Molecular conformation of ascidiacyclamide, a cytotoxic cyclic peptide from ascidian: X-ray analyses of its free form and solvate crystals, *Biopolymers* **32**:131–143.

Ishitsuka, M. O., Kusumi, T. and Kakisawa, J., 1988, Antitumor xenicane and norxenicane lactones from the brown alga *Dictyota dichotoma*, *J. Org. Chem.* **53**(21):5010–5013.

Isobe, M., Ichikawa, Y., and Goto, T., 1986, Synthetic studies towards marine toxic polyethers [5]. The total synthesis of okadaic acid, *Tetrahedron Lett.* **27**(8):963–966.

Jouin, P., Poncet, J., Dufour, M. N., Pantaloni, A., and Castro, B., 1989, Synthesis of the cyclodepsipeptide nordidemnin B, a cytotoxic minor product isolated from the sea tunicate *Trididemnum cyanophorum*, *J. Org. Chem.* **54**(3):617–627.

Kakou, Y., Crews, P., and Bakus, G. J., 1987, Dendrolasin and lactrunculin a form the Fijian sponge *Spongia microfijiensis* and an associated nudibranch *Chromodoris lochi*, *J. Nat. Prod.* **50**(3): 482–484.

Kamano, Y., Kizu, H., Pettit, G. R., Herald, C. L., Tuinman, A. A., and Bontems, R. L., 1987, Structure of the marine animal antineoplastic constituent dolastatin 11 from *Dolabella auricularia*, *Tennen Yuki Kagobutsu Toronkai Koen Yoshishu* **29**:295–300.

Kamano, Y, Inoue, M., Pettit, G. R., Dufresne, C., Herald, D. L., and Christie, N. D., 1988, Chemistry of the powerful cell growth inhibitors cephalostatins, isolated from the south African marine worm *Cephalodiscus gilchristi*, *Tennen Yuki Kagobutsu Toronkai Koen Yoshishu* **30**:220–227.

Kamano, Y., Kizu, H., Pettit, G. R., Dufresne, C., Herald, D. L., Herald, C. L., Schmidt, J. M., and Cerny, R. L., 1989, Structure of the antineoplastic cyclic depsipeptide dolastatin 13 and dehydrodolastatin 13, isolated from the Indian Ocean sea hare *Dolabella auricularia*, *Tennen Yuki Kagobutsu Toronkai Koen Yoshishu* **31st**:641–647.

Kamiya, H., Muramoto, K., and Yamazaki, M., 1986, Aplysianin-A, an antibacterial and antineoplastic glycoprotein in the albumen gland of a sea hare, *Aplysia kurodai*, *Experientia* **42**(9):1065–1067.

Kamiya, H., Muramoto, K., Goto, R., Yamazaki, M., 1988, Studies on bioactive marine metabolites—VII. Characterization of the antibacterial and antineoplastic glycoproteins in a sea hare *Aplysia juliana*, *Nippon Suisan Gakkaishi* **54**(5):733–737.

Kamiya, H., Muramoto, K., Goto, R., Sakai, M., Endo, Y., and Yamazaki, M., 1989, Purification and characterization of an antibacterial and antineoplastic protein secretion of a sea hare, *Aplysia juliana*, *Toxicon* **27**(12):1269–1277.

Kashman, Y., Groweiss, A., and Shmueli, U., 1980, Latrunculin, a new 2-thiazolidinone macrolide from the marine sponge *Latrunculia magnifica*, *Tetrahedron Lett.* **21**:3629–3632.

Kashman, Y., Hirsh, S., Ohtani, I., Kusumi, T., and Kakisawa, H., 1989, Ptilomycalin A: A novel polycyclic guanidine alkaloid of marine origin, *J. Am. Chem. Soc.* **111**:8925–8926.

Kato, Y., Fusetani, N., Matsunaga, S., Hashimoto, K., Fujita, S., and Furuya, T., 1986a, Bioactive marine metabolites. Part 16. Calyculin A: A novel antitumor metabolite from the marine sponge *Discodermia calyx*, *J. Am. Chem. Soc.* **108**(10):2780–2781.

Kato, Y., Fusetani, N., Matsunaga, S., Hashimoto, K., Fujita, S., Furuya, T., and Koseki, K., 1986b, Structures of calyculins, novel antitumor substances from the marine sponge *Discodermia calyx*, *Tennen Yuki Kagobutsu Toronkai Koen Yoshishu* **28th**:168–175.

Kato, Y., Fusetani, N., Matsunaga, S., and Hashimoto, K., 1986c, Okinonellins A and B, two novel furanosesterterpenes which inhibit cell division of fertilized starfish eggs, from the marine sponge *Spongionella* sp., *Experientia* **42**(11–12):1299–1300.

Kato, Y., Fusetani, N., Matsunaga, S., Hashimoto, K., Sakai, R., Higa, T., and Kashman, Y., 1987a, Bioactive marine metabolites. Part XXIII. Antitumor macrodiolides isolated from a marine sponge *Theonella* sp.: Structure revision of misakinolide A, *Tetrahedron Lett.* **28**(49):6225–6228.

Kato, Y., Fusetani, N., Matsunaga, S., Hashimoto, K., Sakai, R., Higa, T., and Kashman, Y., 1987b, Structures of antitumor macrodiolides from a marine sponge *Theonella* sp., *Tennen Yuki Kagobutsu Toronkai Koen Yoshishu* **29**:301–308.

Kato, Y., Fusetani, N., Matsunaga, S., and Hashimoto, K., 1988a, Calyculins, potent antitumor metabolites from the marine sponge *Discodermia calyx*: Biological activities, *Drugs Exp. Clin. Res.* **14**(12):723–728.

Kato, Y., Fusetani, N., Matsunaga, S., Hashimoto, K., and Koseki, K., 1988b, Bioactive marine metabolites. 24. Isolation and structure elucidation of calyculins B, C, and D, novel antitumor metabolites, from the marine sponge *Discodermia calyx*, *J. Org. Chem.* **53**(17):3930–3932.

Kelly, T. R., and Maguire, M. P., 1985, A synthesis of aaptamine, *Tetrahedron* **41**(15):3033–3036.

Kernan, M. R., and Faulkner, D. J., 1987, Halichondramide, an antifungal macrolide from the sponge *Halichondria* sp., *Tetrahedron Lett.* **28**(25):2809–2812.

Kisugi, J., Kamiya, H., and Yamazaki, M., 1987, Purification and characterisation of aplysianin E, an antitumor factor from sea hare eggs, *Cancer Res.* **47**(21):5649–5653.

Kisugi, J., Kamiya, H., and Yamazaki, M., 1989a, Purification of dolabellanin-C an antineoplastic glycoprotein in the body fluid of a sea hare, *Dolabella auricularia*, *Dev. Comp. Immunol.* **13**(1):3–8.

Kisugi, J., Yamazaki, M., Ishii, Y., Tansho, S., Muramoto, K., and Kamiya, H., 1989b, Purification of a novel cytolytic protein from albumin gland of the sea hare *Dolabella auricularia*, *Chem. Pharm. Bull.* **37**(10):2773–2776.

Kitagawa, I., Kobayashi, M., Cui, Z., Kiyota, Y., and Ohnishi, M., 1986a, Marine natural products, XV. Chemical constituents of an Okinawan soft coral of *Xenia* sp. (Xeniidae), *Chem. Pharm. Bull.* **34**(11):4950–4596.

Kitagawa, I., Kobayashi, M., Lee, N. K., Shibuya, H., Kawata, Y., and Sakiyama, F., 1986b, Structure of theonellapeptolide 1D, a new bioactive peptide from an Okinawan marine sponge, *Theonella* sp. (Theonelliae), *Chem. Pharm. Bull.* **34**(6):2664–2667.

Kitagawa, I., Cui, Z., Son, B. W., Kobayashi, M., and Kyogoku, Y., 1987a, Marine natural products. XVII. Nephtheoxydiol, a new cytotoxic hydroperoxygermacrane sesquiterpene, and related sesquiterpenoids from an Okinawan soft coral of *Nephthea* sp. (Nephtheidae), *Chem. Pharm. Bull.* **35**(1):124–135.

Kitagawa, I., Lee, N. K., Kobayashi, M., and Shibuya, H., 1987b, Structure of theonellapeptolide Ie, a new tridecapeptide lactone from an Okinawan marine sponge, *Theonella* sp., *Chem. Pharm. Bull.* **35**:2129–2132.

Kitigawa, I., Kobayashi, M., Okamoto, Y., Yoshikawa, M., and Hamamoto, Y., 1987c, Structures of sarasinosides A_1, B_1, and C_1; New norlanostanetriterpenoid oligoglycosides from the Palauan marine sponge *Asteropus sarasinosum*, *Chem. Pharm. Bull.* **35**:5036–5039.

Kitagawa, I., Kobayashi, M., Lee, N. K., Oyama, Y., and Kyogoku, Y., 1989, Marine natural products XX. Bioactive scalarane-type bishomosesterterpenes from the Okinawan marine sponge *Phyllospongia foliascens*, *Chem. Pharm. Bull.* **37**(8):2078–2082.

Kitagawa, I., Kobayashi, M., Katori, T., Yamashita, M., Tanaka, J., Doi, M., and Ishida, T., 1990, Absolute stereostructure of swinholide A, a potent cytotoxic macrolide from the Okinawan marine sponge *Theonella swinhoei*, *J. Am. Chem. Soc.* **112**(9):3710–3712.

Kobayashi, J., Ishibashi, M., Nakamura, H., Ohizumi, Y., Yamasu, T., Sasaki, T., and Hirata, Y., 1986, Amphidinolide A, a novel antineoplastic macrolide from the marine dinoflagellate *Amphidinium* sp., *Tetrahedron Lett.* **27**(47):5755–5758.

Kobayashi, J., Cheng, J. F., Ishibashi, M., Nakamura, H., Ohizumi, Y. I., Hirata, Y., Sasaki, T., Lu, H., and Clardy, J., 1987, Prianosin A, a novel antileukemic alkaloid from the Okinawan marine sponge *Prianos melanos*, *Tetrahedron Lett.* **28**(43):4939–4942.

Kobayashi, J., Cheng, J. F., Nakamura, H., Ohizumi, Y., Hirata, Y., Sasaki, T., Ohta, T., and Nozoe, S., 1988a, Ascididemnin, a novel pentacyclic aromatic alkaloid with potent antileukemic activity from the Okinawan tunicate *Didemnum* sp., *Tetrahedron Lett.* **29**:1177–1180.

Kobayashi, J., Cheng, J., Ohta, T., Nakamura, H., Nozoe, S., Hirata, Y., Ohizumi, Y., and Sasaki, T., 1988b, Iejimalides A and B, novel 24-membered macrolides with potent antileukemic activity from the Okinawan tunicate *Eudistoma* c.f. *rigida*, *J. Org. Chem.* **53**(26):6147–6150.

Kobayashi, J., Cheng, J., Walchli, M. R., Nakamura, H., Hirata, Y., Sasaki, T., and Ohizumi, Y., 1988c, Cystodytins A, B, and C, novel tetracyclic aromatic alkaloids with potent antineoplastic activity from the Okinawan tunicate *Cystodytes dellechiajei*, *J. Org. Chem.* **53**(8):1800–1804.

Kobayashi, J., Ishibashi, M., Nakamura, H., Hirata, Y., Yamasu, T., Sasaki, T., and Ohizumi, Y., 1988d, Symbioramide, a novel Ca^{2+}-ATPase activator from the cultured dinoflagellate *Symbiodinium* sp., *Experientia* **44**(9):800–802.

Kobayashi, J., Ishibashi, M., Walchli, M., Nakamura, H., Hirata, Y., and Sasaki, T., 1988e, Amphidinolide C: The first 25-membered macrocyclic lactone with potent antineoplastic activity from the cultured dinoflagellate *Amphidinium* sp., *J. Am. Chem. Soc.* **100**:490–494.

Kobayashi, J., Ishibashi, M., Nakamura, H., Ohizumi, Y., Yamasu, T., Hirata, Y., Sasaki, T., Ohta, T., and Nozoe, S., 1989a, Cytotoxic macrolides from a cultured marine dinoflagellate of the genus *Amphidinium*, *J. Nat. Prod.* **52**(5):1036–1041.

Kobayashi, J., Murayama, T., Ohizumi, Y., Ohta, T., Nozoe, S., and Sasaki, T., 1989b, Metachromin C, a new cytotoxic sesquiterpenoid from the Okinawan marine sponge *Hippospongia metachromia*, *J. Nat. Prod.* **52**(5):1173–1176.

Kobayashi, J., Murayama, T., Ohizumi, Y., Sasaki, T., Ohta, T., and Nozoe, S., 1989c, Theonelladins A .apprx. D, novel antineoplastic pyridine alkaloids from the Okinawan marine sponge *Theonella swinhoei*, *Tetrahedron Lett.* **30**(36):4833–4836.

Kobayashi, J., Ishibashi, M., Murayama, M., Takamatsu, M., Iwamura, Y., Ohizumi, Y., and Sasaki, T., 1990a, Amphidinolide E: A novel antileukemic 19-membered macrolide from the cultured symbiotic dinoflagellate *Amphidinium* sp., *J. Org. Chem.* **55**:3421–3423.

Kobayashi, J., Cheng, J. F., Ohta, T., Nozoe, S., Ohizumi, Y., and Sasaki, T., 1990b, Eudistomins B, C, and D, novel antileukemic alkaloids from the Okinawan marine tunicate *Eudistoma glaucus*, *J. Org. Chem.* **55**:3666–3670.

Kobayashi, M., Son, B. W., Kyogoku, Y., and Kitagawa, I., 1986, Kericembrenolides A, B, C, D, and E, five new cytotoxic cembrenolides from the Okinawan soft coral *Clavularia koellikeri*, *Chem. Pharm. Bull.* **34**(5):2306–2309.

Kobayashi, M., Tanaka, J., Katori, T., Matsuura, M., and Kitagawa, I., 1989, Structure of swinholide A, a potent cytotoxic macrolide from the Okinawan marine sponge *Theonella swinhoei*, *Tetrahedron Lett.* **30**(22):2963–2966.

Kobayashi, M., Tanaka, J., Katori, T., and Kitagawa, I., 1990a, Marine natural products, XXIII. Three new cytotoxic dimeric macrolides, swinholides B and C and isoswinholide A, congeners of swinholide A, from the Okinawan marine sponge *Theonella swinhoei*, *Chem. Phar. Bull.* **38**(11):2960–2966.

Kobayashi, M., Tanaka, J., Katori, T., Matsuura, M., Yamashita, M., and Kitagawa, I., 1990b, Marine natural products XXII. The absolute stereostructure of swinholide A, a potent cytotoxic dimeric macrolide from the Okinawan marine sponge *Theonella swinhoei*, *Chem. Pharm. Bull.* **38**(9):2409–2418.

Kohda, K., Ohth, Y., Kawazoe, Y., Kato, T., Suzumura, Y., Hamada, Y., and Shioiri, T., 1989, Ulicyclamide is cytotoxic against L1210 cells *in vitro* and inhibits both DNA and RNA syntheses, *Biochem. Pharm.* **38**(24):4500–4502.

Kohmoto, S., McConnell, O. J., Wright, A., Koehn, F., Thompson, W., Lui, M., and Snader, K. M., 1987a, Puupehenone, a cytotoxic metabolite from a deep water marine sponge, *Strongylophora hartmani*, *J. Nat. Prod.* **50**(2):336.

Kohmoto, S., McConnell, O. J., Wright, A., and Cross, S., 1987b, Isospongiadiol, a cytotoxic and antiviral diterpene from a Caribbean deep water marine sponge, *Spongia* sp., *Chem. Lett.* **1987**(9):1687–1690.

Kohmoto, S., Kashman, Y., McConnell, O. J., Rinehart, K. L., Wright, A., and Koehn, F., 1988, Dragmacidin, a new cytotoxic bis(indole) alkaloid from a deep water marine sponge, *Dragmacidon* sp., *J. Org. Chem.* **53**(13):3116–3118.

Kondracki, M. L., and Guyot, M., 1987, Smenospongine: A cytotoxic and antimicrobial aminoquinone isolated from *Smenospongia* sp., *Tetrahedron Lett.* **28**(47):5815–5818.

Kondracki, M. L., and Guyot, M., 1989a, Biologically active quinone and hydroquinone sesquiterpenoids from the sponge *Smenospongia* sp., *Tetrahedron* **45**(7):1995–2004.

Kondracki, M. L., and Guyot, M., 1989b, Biologically active quinone and hydroquinone sesquiterpenoids from the sponge *Smenospongia* sp. [Erratum to document cited in CA **111**(9):75012z]. *Tetrahedron* **45**(24):7641.

Kondracki, M. L., Davoust, D., and Guyot, M., 1989a, Smenospondiol, a biologically active hydroquinone from the sponge *Smenospongia* sp., *J. Chem. Res. Synop.* **1989**(3):74–75.

Kondracki, M. L., Davoust, D., and Guyot, M., 1989b, Smenospondiol, a biologically active hydroquinone from the sponge *Smenospongia* sp. [Erratum to document cited in CA**111**(5): 36836b], *J. Chem. Res. Synop.* **1989**(12):400.

Kraft, A. S., William, F., Pettit, G. R., and Lilly, M. B., 1989, Varied differentiation responses of human leukemias to bryostatin 1, *Cancer Res.* **49**(5):1287–1293.

Kreuter, M. H., Bernd, A., Holzmann, H., Muller-Klieser, W., Maidhof, A., Weibmann, N., Kljajie, A., Batel, R, Schroder, H. C., and Muller, W. E. G., 1989, Cytostatic activity of aeroplysinin-1 against lymphoma and epithelioma cells, *Z. Naturforsch.* **44**(c):680–688.

Ksebati, M. B., Schmitz, F. J., and Gunasekera, S. P., 1988, Pouosides A–E, novel triterpene galactosides from a marine sponge, *Asteropus* sp., *J. Org. Chem.* **53**(17):3917–3921.

Ksebati, M. B., Schmitz, F. J., and Gunasekera, S. P., 1989, Pouosides A–E, novel, triterpene galactosides from a marine sponge, *Asteropus* sp. [Erratum to document cited in CA**109**(11): 89977p], *J. Org. Chem.* **54**(8):2026.

Kusumi, T., Uchida, H., Inouye, Y., Isahitsuka, M., Yamamoto, H., and Kakisawa, H., 1987, Novel cytotoxic monoterpenes having a halogenated tetrahydropyran from *Aplysia kurodai*, *J. Org. Chem.* **52**:4597–4600.

Kusumi, T., Ohtani, I., Inouye, Y., and Kakisawa, H., 1988a, Absolute configurations of cytotoxic marine cembranolides; consideration of Mosher's method, *Tetrahedron Lett.* **29**(37):4731–4734.

Kusumi, T., Uchida, H., Ishitsuka, M. O., Yamamoto, H., and Kakisawa, H., 1988b, Alcyonin, a new cladiellane diterpene from the soft coral *Sinularia flexibilis*, *Chem. Lett.* **1988**(6):1077–1078.

Kusumi, T., Igari, M., Ishitsuka, M. O., Ichikawa, A., Itezono, Y., Nakayama, N., and Kakisawa, H., 1990, A novel chlorinated biscembranoid from the marine soft coral *Sarcophyton glaucum*, *J. Org. Chem.* **55**(26):6286–6289.

Laatsch, H., and Pudleiner, H., 1989, Marine bacteria. I. Synthesis of pentabromopseudilin, a cytotoxic phenylpyrrole from *Alteromonas luteoviolaceus*, *Liebigs Ann. Chem.* **1989**(9):863–881.

Legrue, S. J., Sheu, T.-L., Carson, D. D., Laidlaw, J. L., and Sanduja, S. K., 1988, Inhibition of T-lymphocyte proliferation by the cyclic polypeptide didemnin B: No inhibition of lymphokine stimulation, *Lymphokine Res.* **7**:21–29.

Li, L. H., Timmins, L. G., Wallace, T. L., Krueger, W. C., Prairie, M. D., and Im, W. B., 1984, Mechanism of action of didemnin B, a depsipeptide from the sea, *Cancer Lett.* **23**:279–288.

Li, W.-R., Ewing, W. R., Harris, B. D., and Joullié, M. M., 1990, Total synthesis and structural investigations of didemnins A, B, and C, *J. Am. Chem. Soc.* **112**:7659–7672.

Lin, Y.-Y., Risk, M., Ray, S. M., Van Engen, D., Clardy, J., Golik, J., James, J. C., and Nakanishi, K., 1981, Isolation and structure of brevetoxin B from the "red tide" dinoflagellate *Ptychodiscus brevis* (*Gymnodinium breve*), *J. Am. Chem. Soc.* **103**:6773–6775.

Lindquist, N., Fenical, W., Sesin, D. F., Ireland, C. M., VanDuyne, G. D., Forsyth, C. J., and Clardy, J., 1988, Isolation and structure determination of the didemnenones, novel cytotoxic metabolites from tunicates, *J. Am. Chem. Soc.* **110**(4):1308–1309.

Lindquist, N., Fenical, W., Van Duyne, G. D., and Clardy, J., 1991, Isolation and structure determination of diazonamides A and B, unusual cytotoxic metabolites from the marine ascidian *Diazona chinensis*, *J. Am. Chem. Soc.* **113**:2303–2304.

Masuda, K., Funayama, S., Komiyama, K., Umezawa, I., and Ito, K., 1987, Antitumor acidic polysaccharide NRP-1 isolated from starfish; *Asterias amurensis* Lutken, *Kitasato Arch. Exp. Med.* **60**(3):95–103.

Matsunaga, S., Fusetani, N., Hashimoto, K., Koseki, K., and Noma, M., 1986, Kabiramide C, a novel antifungal macrolide from nudibranch eggmasses, *J. Am. Chem. Soc.* **108**:847.

Matsunaga, S., Fusetani, N., Hashimoto, K., and Walchli, M., 1989a, Theonellamide F. A novel antifungal bicyclic peptide from a marine sponge *Theonella* sp., *J. Am. Chem. Soc.* **111**(7):2582–2588.

Matsunaga, S., Fusetani, N., Hashimoto, K., Koseki, K., Noma, M., Noguchi, H., and Sankawa, U., 1989b, Bioactive marine metabolites. 25. Further kabiramides and halichondramides, cytotoxic macrolides embracing trisoxazole, from the *Hexabranchus* egg masses, *J. Org. Chem.* **54**(6):1360–1363.

May, W. S., Sharkis, S. J., Esa, A. H., Gebbia, V., Kraft, A. S., Pettit, G. R., and Sensenbrenner, L. L., 1987, Antineoplastic bryostatins are multipotential stimulators of human hematopoietic progenitor cells, *Proc. Natl. Acad. Sci. USA* **84**(23):8483–8487.

McKee, T. C., and Ireland, C. M., 1987, Cytotoxic and antimicrobial alkaloids from the Fijian sponge *Xestospongia caycedoi*, *J. Nat. Prod.* **50**(4):754–756.

Miyamoto, T., Higuchi, R., Marubayashi, N., and Komori, T., 1988, Studies on the constituents of marine opisthobranchia, IV. Two new polyhalogenated monoterpenes from the sea hare *Aplysia kurodai*, *Liebigs Ann. Chem.* **1988**:1191–1193.

Miyamoto, T., Togawa, K., Higuchi, R., Komori, T., and Sasaki, T., 1990, Constituents of Holothuroidea. II. Six newly identified biologically active triterpenoid glycoside sulfates from the sea cucumber *Cucumaria echinata*, *Liebigs Ann. Chem.* **1990**(5):453–460.

Molinski, T. F., and Ireland, C. M., 1988, Dysidazirine, a cytotoxic azacyclopropene from the marine sponge *Dysidea fragilis*, *J. Org. Chem.* **53**(9):2103–2105.

Molinski, T. F., and Ireland, C. M., 1989, Varamines A and B, new cytotoxic thioalkaloids from *Lissoclinum vareau*, *J. Org. Chem.* **54**(17):4256–4259.

Montgomery, D. W., and Zukoski, C. F., 1985, Didemnin B: A new immunosuppressive cyclic peptide with potent activity *in vitro* and *in vivo*, *Transplantation* **49**:49.

Moore, R. E., and Entzeroth, M., 1988, Majusculamide D and dehydroxymajusculamide D, two cytotoxins from *Lyngbya majuscula*, *Phytochemistry* **27**(10):3101–3103.

Moquin-Pattey, C., and Guyot, M., 1989, Grossularine-1 and grossularine-2, cytotoxic .alpha.-carbolines from the tunicate *Dendrodoa grossularia*, *Tetrahedron* **45**(11):3445–3450.

Mori, K., and Takeuchi, T., 1988, Synthesis of punaglandin 4 by means of enzymatic resolution of a key chlorocyclopropene derivative, *Tetrahedron* **44**(2):333–342.

Mori, K., Iguchi, K., Yamada, N., Yamada, Y., and Inouye, Y., 1987a, Stolonidiol, a new marine diterpenoid with a strong cytotoxic activity from the Japanese soft coral, *Tetrahedron Lett.* **28**(46):5673–5676.

Mori, K., Iguchi, K., Yamada, N., Yamada, Y., and Inouye, Y., 1987b, Isolation and structures of stolonidiol and its cogeners, new bioactive marine diterpenoids from the Okinawan soft coral *Clavularia* sp., *Tennen Yuki Kagobutsu Toronkai Koen Yoshishu* **29**:287–294.

Mori, K., Iguchi, K., Yamada, N., Yamada, Y., and Inouye, Y., 1988, Bioactive marine diterpenoids from Japanese soft coral of *Clavularia* sp., *Chem. Pharm. Bull.* **36**(8):2840–2852.

Moriarty, R. M., Roll, D. M., Ku, Y., Nelson, C., and Ireland, C. M., 1987, A revised structure for the marine bromoindole derivative citorellamine, *Tetrahedron Lett.* **28**(7):749–752.

Morris, S., and Andersen, R. J., 1989, Brominated bis(indole) alkaloids from the marine sponge *Hexadella* sp., *Tetrahedron* **46**(3):715–720.

Motzer, R., Scher, H., Bajorin, D., Sternberg, C., and Bosl, G. J., 1990, Phase II trial of didemnin B in patients with advanced renal cell carcinoma, *Invest. New Drugs* **8**:391–392.

Mueller, W. E. G., Zahn, R. K., Gasic, M. J., Dogovic, N., Maidhof, A., Becker, C., Diehl-Seifert, B., and Eich, E., 1985, Avarol, a cytostatically active compound from the marine sponge *Dysidea avara*, *Comp. Biochem. Physiol. C: Comp. Pharmacol. Toxicol.* **80**C(1):47–52.

Munro, M. H. G., Luibrand, R. T., and Blunt, J. W., 1987, The search for antiviral and anticancer compounds from marine organisms, in: *Bioorganic Marine Chemistry*, Vol. 1 (P. J. Scheuer, ed.), Springer-Verlag, New York, pp. 93–176.

Murakami, M., Oshima, Y., and Yasumoto, T., 1982, Identification of okadaic acid as a toxic component of marine dinoflagellate *Prorocentrum lima*, *Bull. Jpn. Soc. Sci. Fish.* **48**:69.

Nagaoka, H., Iguchi, K., Miyakoshi, T., Yamada, N., and Yamada, Y., 1986, Determination of absolute configuration of chlorovulones by CD measurement and by enantioselective synthesis of (−) chlorovulone II. *Tetrahedron Lett.* **27**:223–226.

Nakamura, H., Kobayashi, J., Nakamura, Y., Ohixumi, Y., Kondo, T., and Hirata, Y., 1986, Theonellamine B, a novel peptidal Na, K-ATPase inhibitor from an Okinawan marine sponge of the genus *Theonella*, *Tetrahedron Lett.* **17**(36):4319–4322.

Nakamura, H., Deng, S., Kobayashi, J., Ohizumi, Y., Tomotake, Y., Matsuzaki, T., and Hirata, Y., 1987a, Keramine A and B. Novel antimicrobial alkaloids from the Okinawan marine sponge *Pellina* sp., *Tetrahedron Lett.* **28**(6):621–624.

Nakamura, H., Kobayashi, J., Ohizumi, Y., and Hirata, Y., 1987b, Aaptamines, novel benzo[de][1,6]-naphthyridines from the Okinawan marine sponge *Aaptos aaptos*, *J. Chem. Soc. Perkin Trans. I* **1**:173–176.

Natori, T., Kawai, H., and Fusetani, N., 1990, Tubiporein, a novel diterpene from a Japanese soft coral *Tubipora* sp., *Tetrahedron Lett.* **31**(5):689–690.

Nishiwaki, S., Fujiki, H., Suganuma, M., Furuya-Suguri, H., Matsushima, R., Iida, Y., Ojika, M., Yamada, K., Uemura, D., Yasumoto, T., Schmitz, F. J., and Sugimura, T., 1990, Structure–

activity relationship within a series of okadaic acid derivatives, *Carcinogenesis* **11**(10):1837–1890.

O'Brien, E. T., Asai, D. J., Groweiss, A., Lipshutz, B. H., Fenical, W., Jacobs, R. S., and Wilson, L., 1986, Mechanism of action of the marine natural product stypoldione: Evidence for the reaction with sulfhydryl groups, *J. Med. Chem.* **29**:1851–1855.

Pathirana, C., and Andersen, R. J., 1986, Imbricatine, an unusual benzyltetrahydro isoquinoline alkaloid from the starfish *Dermasterias imbricata*, *J. Am. Chem. Soc.* **108**:8288–8289.

Pathirana, C., Andersen, R. J., and Wright, J. L. C., 1989, Hydrallmanol A, an interesting diphenyl-*p*-methane derivative of mixed biogenetic origin from the hydroid *Hydrallmania falcata*, *Tetrahedron Lett.* **30**(12):1487–1490.

Pathirana, C., Andersen, R. J., and Wright, J. L. C., 1990, Abietinarins A and B, cytotoxic metabolites of the marine hydroid *Abietinaria sp.*, *Can J. Chem.* **68**:394–396.

Patil, A. D., 1989, Novel Dioxolane Compositions from Halichondriidae Marine Sponges and Their Preparation and Use As Antitumor, Antibacterial, and Antifungal Agents, WO 87 04,708 [*Chem. Abstr.* **109**:17027f].

Paul, V. J., and Fenical, W., 1985, Diterpenoid metabolites from Pacific marine algae of the order Caulerpales (Chlorophyta), *Phytochemistry* **24**(10):2239–2243.

Perry, N. B., Blunt, J. W., McCombs, J. D., and Munro, M. H. G., 1986, Discorhabdin C, a highly cytotoxic pigment from a sponge of the genus *Latrunculia*, *J. Org. Chem.* **51**(26):5476–5478.

Perry, N. B., Blunt, J. W., Munro, M. H. G., and Pannell, L. K., 1988a, Mycalamide A, an antiviral compound from a New Zealand sponge of the genus *Mycale*, *J. Am. Chem. Soc.* **110**(14):4850–4851.

Perry, N. B., Blunt, J. W., and Munro, M. H. G., 1988b, Cytotoxic pigments from New Zealand sponges of the genus *Latrunculia*: Discorhabdins A, B, and C, *Tetrahedron* **44**(6):1727–1734.

Perry, N. B., Blunt, J. W., Munro, M. H. G., Higa, T., and Sakai, R., 1988c, Discorhabdin D, an antitumor alkaloid from the sponges *Latrunculia brevis* and *Piranos sp.*, *J. Org. Chem.* **53**(17):4127–4128.

Perry, N. B., Blunt, J. W., Munro, M. H. G., and Thompson, A. M., 1990, Antiviral and antitumor agents from a New Zealand sponge, *Mycale* sp. 2. Structures and solution conformations of mycalamides A and B, *J. Org. Chem.* **55**(1):223–227.

Pettit, G. R., 1991, The bryostatins, in: *Progress in the Chemistry of Organic Natural Products*, Vol. 57 (W. Herz, G. W. Kirby, W. Steglich and Ch. Tamm, eds.), Springer-Verlag, Berlin, pp. 153–195.

Pettit, G. R., Kamano, Y., Fujii, Y., Herald, C. L., Inoue, M., Brown, P., Gust, D., Kitahara, K., Schmidt, J. M., Doubek, D. L., and Michel, C., 1981, Marine animal biosynthetic constituents for cancer chemotherapy, *J. Nat. Prod.* **44**:482–488.

Pettit, G. R., Kamano, Y., Brown, P., Gust, O., Inoue, M., and Herald, C. Y., 1982, Structure of the cyclic peptide dolostatin 3 from *Dolabella auricularia*, *J. Am. Chem. Soc.* **104**:905–907.

Pettit, G. R., Kamano, Y., Aoyagi, R., Herald, C. L., Doubek, D. L., Schmidt, J. M., and Rudloe, J. J., 1985, Antineoplastic agents. 100. The marine bryozoan *Amathia convoluta*, *Tetrahedron* **41**(6):985–994.

Pettit, G. R., Kamano, Y., Herald, C. L., Schmidt, J. M., and Zubrod, C. G., 1986a, Relationship of *Bugula neritina* (Bryozoa) antineoplastic constituents to the yellow sponge *Lissodendoryx isodictyalis*, *Pure Appl. Chem.* **58**(3):415–421.

Pettit, G. R., Leet, J. E., Herald, C. L., Kamano, Y., and Doubek, D. L, 1986b, Antineoplastic agents, 116. An evaluation of the marine ascidian *Aplidium Californicum*, *J. Nat. Prod.* **49**(2):231–235.

Pettit, G. R., Kamano, Y., Holzapfel, C. W., Van Zyl, W. J., Tuinman, A. A., Herald, C. L., Baczynskyj, L., and Schmidt, J. M., 1987a, Antineoplastic agents. 150. The structure and synthesis of dolastatin 3, *J. Am. Chem. Soc.* **109**(24):7581–7582.

Pettit, G. R., Kamano, Y., Herald, C. L., Tuinman, A. A., Boettner, F. E., Kizu, H., Schmidt, J. M., Baczynskyj, L., Tomer, K. B., and Bontems, R. J., 1987b, The isolation and structure of a remarkable marine animal antineoplastic constituent, dolastatin 10, *J. Am. Chem. Soc.* **109**(22): 6883–6885.

Pettit, G. R., Inoue, M., Kamano, Y., Dufresne, C., Christie, N., Niven, M. L., and Herald, D. L., 1988a, Isolation and structure of the hemichordate cell growth inhibitors cephalostatins 2, 3, and 4, *J. Chem. Soc. Chem. Commun.* **1988**(13):865–867.

Pettit, G. R., Inoue, M., Kamano, Y., Dufresne, C., Christie, N., Niven, M. L., and Herald, D. L., 1988b, Isolation and structure of the hemichordate cell growth inhibitors cephalostatins 2, 3, and 4 [Erratum to document cited in CA**109**(13):107868k], *J. Chem. Soc. Chem. Commun.* **1988** (21):1440.

Pettit, G. R., Inoue, M., Kamano, Y., Herald, D. L., Arm, C., Dufresne, C., Christie, N. D., Schmidt, J. M., Doubek, D. L., and Krupa, T. S., 1988c, Antineoplastic agents. 147. Isolation and structure of the powerful cell growth inhibitor cephalostatin 1, *J. Am. Chem. Soc.* **110**(6):2006–2007.

Pettit, G. R., Kamano, Y., Dufresne, C., Cerny, R. L., Herald, C. L., and Schmidt, J. M., 1989a, Isolation and structure of the cytostatic linear depsipeptide dolastatin 15, *J. Org. Chem.* **54**(26): 6005–6006.

Pettit, G. R., Kamano, Y., Herald, C. L., Dufresne, C., Cerny, R. L., Herald, D. L., Schmidt, J. M., and Kizu, H., 1989b, Antineoplastic agent. 174. Isolation and structure of the cytostatic depsi-peptide dolastatin 13 from the sea hare *Dolabella auricularia*, *J. Am. Chem. Soc.* **111**(13):5015–5017.

Pettit, G. R., Kamano, Y., Kizu, H., Dufresne, C., Herald, C. L., Bontems, R. J., Schmidt, J. M., Boettner, F. E., and Nieman, R. A., 1989c, Antineoplastic agents. 173. Isolation and structure of the cell growth inhibitory depsipeptides dolastatins 11 and 12, *Heterocycles* **28**(2):553–558.

Pettit, G. R., Singh, S. B., Hogan, F., Lloyd-Williams, P., Herald, D. L., Burkett, D. D., and Clewlow, P. J., 1989d, Antineoplastic agents. Part 189. The absolute configuration and synthesis of natural (−)-dolastatin 10, *J. Am. Chem. Soc.* **111**(14):5463–5465.

Pettit, G. R., Kamano, Y., Dufresne, C., Inoue, M., Christie, N., Schmidt, J. M., and Doubek, D. L., 1989e, Antineoplastic agents. 165. Isolation and structure of the unusual Indian Ocean *Cephalo-discus gilchristi* components, cephalostatins 5 and 6, *Can. J. Chem.* **67**(10):1509–1513.

Pettit, G. R., Clewlow, P. J., Dufresne, C., Doubek, D. L., Cerny, R. L., and Rutzler, K., 1990a, Antineoplastic agents. 193. Isolation and structure of the cyclic peptide hymenistatin 1, *Can. J. Chem.* **68**(5):708–711.

Pettit, G. R., Kamano, Y., Herald, C. L., Dufresne, C., Bates, R. B., Schmidt, J. M., Cerny, R. L., and Kizu, H., 1990b, Antineoplastic agents. 190. Isolation and structure of the cyclodepsipeptide dolastatin 14, *J. Org. Chem.* **55**(10):2989–2990.

Pettit, G. R., Gao, F., Sengupta, D., Coll, J. C., Herald, C. L., Doubek, D. L., Schmidt, J. M., Van Camp, J. R., Rudloe, J. J., and Nieman, R. A., 1991a, Isolation and structure of bryostatins 14 and 15, *Tetrahedron* **47**(22):3601–3610.

Pettit, G. R., Herald, D. L., Gao, F., Sengupta, D., and Herald, C. L., 1991b, Antineoplastic agents. 200. Absolute configuration of the bryostatins, *J. Org. Chem.* **56**:1337–1340.

Pordesimo, E. O., and Schmitz, F. J., 1990, New bastadins from the sponge *Ianthella basta*, *J. Org. Chem.* **55**(15):4704–4709.

Prasad, A. V. K., and Shimizu, Y., 1989, The structure of hemibrevetoxin-B: A new type of toxin in the Gulf of Mexico red tide organism, *J. Am. Chem. Soc.* **111**:6476–6477.

Quinoa, E., and Crews, P., 1987a, Phenolic constituents of *Psammaplysilla*, *Tetrahedron Lett.* **28**(28): 3229–3232.

Quinoa, E., and Crews, P., 1987b, Niphatynes, methoxylamine pyridines from the marine sponge *Niphates* sp., *Tetrahedron Lett.* **28**(22):2467–2468.

Quinoa, E., Adamczeski, M., Crews, P., and Bakus, G. J., 1986a, Bengamides, heterocyclic anthelminthics from a Jaspidae marine sponge, *J. Org. Chem.* **51**:4494–4497.

Quinoa, E., Kho, E., Manes, L. V., Crews, P., and Bakus, G., 1986b, Heterocycles from the marine sponge *Xestospongia* sp., *J. Org. Chem.* **51**(22):4260–4264.

Quinoa, E., Kakou, Y., and Crews, P., 1988, Fijianolides, polyketide heterocycles from a marine sponge, *J. Org. Chem.* **53**(15):3642–3644.

Raub, M. F., Cardellina, J. H., Choudhary, I., Li, C., Clardy, J., and Alley, M. C., 1991, Clavipictins A and B, cytotoxic quinolizidines from the tunicate *Clavelina picta*, *J. Am. Chem. Soc.* **113**:3178–3180.

Rinehart, K. L., 1988, Didemnin and its biological properties, in: *Peptides, Chemistry and Biology* (Proceedings of the Tenth American Peptide Symposium) (G. R. Marshall, ed.), pp. 626–631.

Rinehart, K. L., 1989, Novel Anti-viral and Cytotoxic Agent. British Patent Application #8922026.3, September 29, 1989.

Rinehart, Jr., K. L., Gloer, J. B., Cook, Jr., J. C., Mizsak, S. A., and Scahill, T. A., 1981a, Structures of the didemnins, antiviral and cytotoxic depsipeptides from a Caribbean tunicate, *J. Am. Chem. Soc.* **103**:1857.

Rinehart, Jr., K. L., Gloer, J. B., Hughes, Jr., R. G., Riens, H. E., McGovren, J. P., Swynenberg, E. B., Stringfellow, D. A., Kuentzel, S. L., and Li, L. H., 1981b, Didemnins: Antiviral and antitumor depsipeptides from a Caribbean tunicate, *Science* **212**:933.

Rinehart, Jr., K. L., Gloer, J. B., Wilson, G. R., Hughes, Jr., R. G., Li, L. H., Renis, H. E., and McGovren, J. P., 1983, Antiviral and antitumor compounds from tunicates, *Fed. Proc. Fed. Am. Soc. Exp. Biol.* **48**:87.

Rinehart, K. L., Kishore, V., Nagarajan, S., Lake, R. J., Gloer, J. B., Bozich, F. A., Li, K., Maleczka, R. E., Todsen, W. L., Munro, M. H. G., Sullins, D. W., and Sakai, R., 1987, Total synthesis of didemnins A, B, and C, *J. Am. Chem. Soc.* **109**:6846–6848.

Rinehart, K. L., Holt, T. G., Fregeau, N. L., Stroh, J. G., Keifer, P. A., Sun, F., Ki, L. H., and Martin, D. G., 1990a, Antitumor agents from the Caribbean tunicate *Ecteinascidia turbinata*, *J. Org. Chem.* **55**(15):4512–4515.

Rinehart, Jr., K. L., Sakai, R., and Stroh, J. G., 1990b, Cytotoxic Cyclic Depsipeptides from the Tunicate *Trididemnum solidum*, U.S. Patent 4,948,791, August 14, 1990.

Rinehart, K. L., Holt, T. G., Fregeau, N. L., Keifer, P. A., Wilson, G. R., Perun, Jr., T. J., Sakai, R., Thompson, A. G., Stroh, J. G., Shield, L. S., Seigler, D. S., Li, D. H., Martin, D. G., and Gäde, G., 1990c, Bioactive compounds from aquatic and terrestrial sources, *J. Nat. Prod.* **53**(4):771–792.

Roesener, J. A., and Scheuer, P. J., 1986, Ulapualide A and B, extraordinary antitumor macrolides from nudibranch eggmasses, *J. Am. Chem. Soc.* **108**(4):846–847.

Roll, D. M., and Ireland, C. M., 1985, Citorellamine, a new bromoindole derivative from *Polycitorella mariae*, *Tetrahedron Lett.* **26**(36):4303–4306.

Roll, D. M., Ireland, C. M., Lu, H. S. M., and Clardy, J., 1988, Fascaplysin, an unusual antimicrobial pigment from the marine sponge *Fascaplysinopsis* sp., *J. Org. Chem.* **53**(14):3276–3278.

Rudi, A., and Kashman, Y., 1990, Three new norrisolide related rearranged spongians, *Tetrahedron* **46**(11):4019–4022.

Sakai, R., Higa, T., and Kashman, Y., 1986a, Misakinolide-A, an antitumor macrolide from the marine sponge *Theonella* sp., *Chem. Lett.* **1986**(9):1499–1502.

Sakai, R., Higa, T., Jefford, C. W., and Bernardinelli, G., 1986b, Manzamine A, a novel antitumor alkaloid from a sponge, *J. Am. Chem. Soc.* **108**(20):6404–6405.

Sakai, R., Higa, T., Jefford, C. W., and Bernardinelli, G., 1986c, The absolute configurations and biogenesis of some halogenated chamigrenes from the sea hare *Aplysia dactylomela*, *Helv, Chim. Acta* **69**(1):91–105.

Sakai, R., Kohmoto, S., Higa, T., Jefford, C. W., and Bernardinelli, G., 1987, Manzamines B and C, two novel alkaloids from the sponge *Haliclona* sp., *Tetrahedron Lett.* **28**(45):5493–5496.

Sakemi, S., and Higa, T., 1987, 2,3-Dihydrolinderazulene, a new bioactive azulene pigment from the Gorgonian *Acalycigorgia* sp., *Experientia* **43**(6):624–625.

Sakemi, S., Higa, T., Anthoni, U., and Christophersen, C., 1987, Antitumor cyclic peroxides from the sponge *Plakortis lita*, *Tetrahedron* **43**(1):263–268.

Sakemi, S., Ichiba, T., Kohmoto, S., Saucy, G., and Higa, T., 1988, Isolation and structure elucidation of onnamide A, a new bioactive metabolite of a marine sponge, *Theonella* sp., *J. Am. Chem. Soc.* **110**(14)4851–4853.

Schmidt, U., and Greisser, H., 1986, Synthesis and structure determination of patellamide B, *Tetrahedron Lett.* **27**(2):163–166.

Schmitz, F. J., and Bloor, S. J., 1988, Xesto- and halenaquinone derivatives from a sponge, *Adocia* sp., from Truk lagoon, *J. Org. Chem.* **53**(17):3922–3925.

Schmitz, F. J., and Yasumoto, T., 1991, The 1990 United States–Japan Seminar on Bioorganic Marine Chemistry, Meeting Report, *J. Nat. Prod.* **54**:1469–1490.

Schmitz, F. J., Hollenbeak, K. H., and Campbell, D. C., 1978, Marine natural products: Halitoxin, toxic complex of several marine sponges of the genus *Haliclona*, *J. Org. Chem.* **43**:3916.

Schmitz, F. J., Prasad, R. S., Yalamanchili, G., Hossain, M. B., van der Helm, D., and Schmidt, P., 1981, Acanthifolicin, a new episulfide-containing polyether carboxylic acid from extracts of the marine sponge *Pandaros acanthifolium*, *J. Am. Chem. Soc.* **103**:2467–2469.

Schmitz, F. J., Michaud, D. P., and Schmidt, P. G., 1982, Marine natural products: Parguerol, deoxyparguerol, and isoparguerol. New brominated diterpenes with modified pimarane skeletons from the sea hare *Aplysia dactylomela*, *J. Am. Chem. Soc.* **104**:6415.

Schmitz, F. J., Agarwal, S. K., Gunasekera, S. P., Schmidt, P. G., and Shoolery, J. N., 1983, Amphimedine, new aromatic alkaloid from a Pacific sponge, *Amphimedon* sp. carbon connectivity determination from natural abundance $^{13}C–^{13}C$ coupling constants, *J. Am. Chem. Soc.* **105**:4835.

Schmitz, F. J., Ksebati, M. B., Gunasekera, S. P., and Agarwal, S., 1988, Sarasinoside A: A saponin containing amino sugars isolated from a sponge, *J. Org. Chem.* **53**(25):5941–5947.

Schmitz, F. J., Ksebati, M. B., Chang, J. S., Wang, J. L., Hossain, M. B., Van der Helm, D., Engel, M. H., Serban, A., and Silfer J. A., 1989, Cyclic peptides from the ascidian *Lissoclinum patella*: Conformational analysis of patellamide D by x-ray analysis and molecular modeling, *J. Org. Chem.* **54**(14):3463–3472.

Schreiber, S. L., 1991, Chemistry and biology of the immunophilins and their immunosuppressive ligands, *Science* **251**:283–287.

Sesin, D. F., Gaskell, S. J., and Ireland, C. M., 1986, The chemistry of *Lissoclinum patella*, *Bull. Soc. Chim. Belg.* **95**:583–867.

Sharma, P., and Alam, M., 1988, Sclerophytins A and B. Isolation and structures of novel cytotoxic diterpenes from the marine coral *Sclerophytum capitalis*, *J. Chem. Soc. Perkin Trans. 1* **1988**(9):2537–2540.

Sheid, B., Prat, J. C., and Gaetjens, E., 1989, A tumor growth inhibitory factor and a tumor growth promoting factor isolated from unfertilized ova of shad (*Alosa sapidissima*), *Biochem. Biophys. Res. Commun.* **159**(2):713–719.

Shioiri, T., Hamada, Y., Kato, S., Shibata, M., Kondo, Y., Nakagawa, H., and Kohda, K., 1987, Cytotoxic activity of cyclic peptides of marine origin and their derivatives: Importance of oxazoline functions, *Biochem. Pharm.* **36**(23):4181–4185.

Sigel, M. M., Wilhelm, L. L. Lichter, W., Dudeck, L. E., Gargus, J. L., and Lucas, L. H., 1970, in: *Food-Drugs from the Sea, Proceedings, 1969* (H. W. Youngken, ed.), Marine Technology Society, Washington, D. C., pp. 281–295.

Snader, K. M., and Higa, T., 1987, Antitumor Chamigrene Derivatives and Their Methods of Use, WO 87 01,279 A2 12 March 1987 [*Chem. Abstra.* **107**:205175k].

Still, I. W. J., and Shi, Y., 1987, A stereoselective synthesis of clavularin A from 2-cycloheptenone, *Tetrahedron Lett.* **28**(22):2489–2490.

Stone, R. M., Sariban, E., Pettit, G. R., and Kufe, D. W., 1988, Bryostatin 1 activates protein kinase C and induces monocytic differentiation of HL-60 cells, *Blood* **72**(1):208–213.

Suffness, M., 1987, New approaches to the discovery of antitumor agents, in: *Biologically Active Natural Products (Annual Proceedings of the Phytochemical Society of Europe*, K. Hostettmann and P. J. Lea, Eds.), Clarendon Press, Oxford, pp. 85–104.

Suffness, M., and Pezzuto, J. M., 1991, Assays related to cancer drug discovery, in: *Methods in Plant Biochemistry*, Vol. 6. *Biological Techniques*, Academic Press, London, pp. 71–131.

Suffness, M., Newman, D. J., and Snader, K., 1989, Discovery and development of antineoplastic agents from natural sources, in: *Bioorganic Marine Chemistry* (P. J. Scheuer, ed.), Springer-Verlag, New York, pp. 131–168.

Suganuma, M., Fujiki, H., Suguri, H., Yoshizawa, S., Hirota, M., Nakayasu, M., Ojika, M., Wakamatsu, K., Yamada, K., and Sugimura, T., 1988, Okadaic acid: An additional non-phorbol-12-tetradecanoate-13-acetate-type tumor promoter, *Proc. Natl. Acad. Sci. USA* **85:**1768–1771.

Sun, H. H., Sakemi, S., Burres, N., and McCarthy, P., 1990, Isobatzellines A, B, C, and D. Cytotoxic and antifungal pyrroloquinoline alkaloids from the marine sponge *Batzella* sp., *J. Am. Chem. Soc.* **55:**4964–4966.

Suzuki, M., Morita, Y., Yanagisawa, A., Baker, B. J., Scheuer, P. J., and Noyori, R., 1988, Synthesis and structural revision of (7E)- and (7Z)-punaglandin 4, *J. Org. Chem.* **53:**286–295.

Suzuki, T., Suzuki, M., Furusaki, A., Matsumoto, T., Kato, A., Imanaka, Y., and Kurosawa, E., 1985, Constituents of marine plants. 62. Teurilene and thyrsiferyl 23-acetate, meso and remarkably cytotoxic compounds from the marine red alga *Laurencia obtusa* (Hudson) Lamouroux, *Tetrahedron Lett.* **26**(10)**:**1329–1332.

Suzuki, T., Takeda, S., Suzuki, M., and Kurosawa, E., 1987a, The constituents of the marine red alga *Laurencia obtusa* (Hudson) Lamouroux, *Tennen Yuki Kagobutsu Toronkai Koen Yoshishu* **29:**576–583.

Suzuki, T., Takeda, S., Suzuki, M., Kurosawa, E., Kato, A., and Imanaka, Y., 1987b, Constituents of marine plants. Part 67. Cytotoxic squalene-derived polyethers from the marine red alga *Laurencia obtusa* (Hudson) Lamouroux, *Chem. Lett.* **1987**(2)**:**361–364.

Suzuki, T., Takeda, S., Hayama, N., Tanaka, I., and Komiyama, K., 1989, The structure of a brominated diterpene from the marine red alga *Laurentia obtusa* (Hudson) Lamouroux, *Chem. Lett.* **1989:**969–970.

Tachibana, K., Scheuer, P. J., Tsukitani, Y., Kikuchi, H., Van Engen, D., Clardy, J., Gopichand, Y., and Schmitz, F. J., 1981, Okadaic acid, a cytotoxic polyether from two marine sponges of the genus *Halichondria*, *J. Am. Chem. Soc.* **103:**2469.

Takeda, S., Matsumoto, T., Komiyama, K., Kurosawa, E., and Suzuki, T., 1990a, Constituents of marine plants. Part 76. A new cytotoxic diterpene from the marine alga *Laurencia obtusa* (Hudson) Lamouroux, *Chem. Lett.* **1990**(2)**:**277–280.

Takeda, S., Kurosawa, E., Kumiyama, K., and Suzuki, T., 1990b, The structures of cytotoxic diterpenes containing bromine from the marine red alga *Laurencia obtusa* (Hudson) Lamouroux, *Bull. Chem. Soc. Jpn.* **63:**3066–3072.

Tanaka, J., Higa, T., Kobayashi, M., and Kitagawa, I., 1990, Marine natural products, XXIV. The absolute stereostructure of misakinolide A, a potent cytotoxic dimeric macrolide from an Okinawan marine sponge *Theonella* sp., *Chem. Pharm. Bull.* **38**(11)**:**2967–2970.

Tapiolas, D. M., Roman, M., Fenical, W., Stout, T. J., and Clardy, J., 1991, Octalactins A and B, cytotoxic 8-membered ring lactones from the marine bacterium, *Streptomyces* sp., *J. Am. Chem. Soc.* **113:**4682–4683.

Torigoe, K., Murata, M., Yasumoto, T., and Iwashita, T., 1988, Prorocentrolide, a toxic nitrogenous macrocycle from the marine dinoflagelate, *Prorocentrum lima*, *J. Am. Chem. Soc.* **110:**7876–7877.

Toth, S. I., 1990, Advancements in Natural Products Chemistry, Ph.D. Dissertation, University of Sydney, Sydney, Australia.

Trenn, G., Pettit, G. R., Takayama, H., Hu-Li, J., and Sitkovsky, M. V., 1988, Immunomodulating properties of a novel series of protein kinase C activators. The bryostatins, *J. Immunol.* **140**(2):433–439.

Tsujii, S., and Rinehart, K. L., 1988, Topsentin, bromotopsentin, and dihydrodeoxytopsentin: New antiviral and antitumor bis(indolyl) imidazoles from Caribbean deep sea sponges of the Family Halichondriidae. Structural and synthetic studies, *J. Org. Chem.* **53**(23):5446–5453.

Uchio, Y., Eguchi, S., Kuramoto, J., Nakayama, M., and Hase, T, 1985, Denticulatolide, an ichthyotoxic peroxide-containing cambranolide from the soft coral *Lobophytum denticulatum*, *Tetrahedron Lett.* **26**(37):4487–4490.

Weinheimer, A. J., and Spraggins, R. L., 1969, The occurrence of two new prostaglandin derivatives (15-*epi*PGA$_2$ and its acetate, methyl ester) in the gorgonian *Plexaura homomalla*, *Tetrahedron Lett.* **59**:5185–5188.

Wender, P. A., Cribbs, C. M., Koehler, K. F., Sharkey, N. A., Herald, C. L., Kamano, Y., Pettit, G. R., and Blumberg, P. M., 1988, Modeling of the bryostatins to the phorbol ester pharmacophore on protein kinase C, *Proc. Natl. Acad. Sci. USA*, **85**(19):7197–7201.

West, R. R., and Cardellina, J. H., 1988, Isolation and identification of eight new polyhydroxylated sterols from the sponge *Dysidea etheria*, *J. Org. Chem.* **53**(12):2782–2787.

West, R. R., Mayne, C. L., Ireland, C. M., Brinen, L. S., and Clardy, J., 1990, Plakinidines: Cytotoxic alkaloid pigments from the Fijian sponge *Plakortis* sp., *Tetrahedron Lett.* **31**(23):3271–3274.

White, J. D., and Amedio, J. C., 1989, Total synthesis of geodiamolide A, a novel cyclodepsipeptide of marine origin, *J. Org. Chem.* **54**(4):736–738.

Williams, D. E., Moore, R. E., and Paul, V. J., 1989, The structure of ulithiacyclamide B. Antitumor evaluation of cyclic peptides and macrolides from *Lissoclinum patella*, *J. Nat. Prod.* **52**(4): 732–739.

Williams, D. H., Stone, M. J., Hauck, P. R., and Rahman, S. K., 1989, Why are secondary metabolites (natural products) biosynthesized? *J. Nat. Prod.* **52**(6):1189–1208.

Wright, A. E., and Thompson, W. C., 1987, Antitumor Compositions Containing Imidazolylpyrrolo-azepines, WO 8707274 A2 3 December 1987 [*Chem. Abstr.* **110**(17):147852c].

Wright, A. E., McConnell, O. J., Kohmoto, S., Lui, M. S., Thompson, W., and Snader, K. M., 1987a, Duryne, a new cytotoxic agent from the marine sponge *Cribrochalina dura*, *Tetrahedron Lett.* **28**(13):1377–1380.

Wright, A. E., Pomponi, S. A., McConnell, O. J., Kohmoto, S., and McCarthy, P. J., 1987b, (+)-Curcuphenol and (+)-curcudiol, sesquiterpene phenols from shallow and deep water collections of the marine sponge *Didiscus flavus*, *J. Nat. Prod.* **50**(5):976–978.

Wright, A. E., Burres, N. S., and Schulte, G. K., 1989, Cytotoxic cembranoids from the gorgonian *Pseudopterogorgia bipinnata*, *Tetrahedron Lett.* **30**(27):3491–3494.

Wright, A. E., Forleo, D. A., Gunawardana, G. P., Gunasekera, S. P., Koehn, F. E., and McConnell, O. J., 1990, Antitumor tetrahydroisoquinoline alkaloids from the colonial ascidian *Ecteinascidia turbinata*, *J. Org. Chem.* **55**(15):4508–4512.

Yamazaki, M., Kisugi, J., Kimura, K., Kamiya, H., and Mizuno, D., 1985, Purification of antineoplastic factor from eggs of a sea hare, *FEBS Lett.* **185**(2):295–298.

Yamazaki, M., Kimura, K., Kisugi, J., and Kamiya, H., 1986, Purification of a cytolytic factor from purple fluid of a sea hare, *FEBS Lett.* **198**(1):25–28.

Yamazaki, M., Kimura, K., Kisugi, J., Muramoto, K., and Kamiya, H., 1989a, Isolation and characterization of a novel cytolytic factor in purple fluid of the sea hare *Dolabella auricularia*, *Cancer Res.* **49**:3834–3838.

Yamazaki, M., Tansho, S., Kisugi, J., Muramoto, K., and Kamiya, H., 1989b, Purification and characterization of a cytolytic protein from purple fluid of the sea hare *Dolabella auricularia*, Chem. Pharm. Bull. **37**(8):2179–2182.

Yamazaki, M., Kisugi, J., and Kamiya, H., 1989c, Biopolymers from marine invertebrates. XI.

Characterization of an antineoplastic glycoprotein, dolabellanin A, from the albumen gland of a sea hare, *Dolabella auricularia, Chem. Pharm. Bull.* **37**(12):3343–3346.

Zabriskie, T. M., and Ireland, C. M., 1989, The isolation and structure of modified bioactive nucleosides from *Jaspis johnstoni, J. Nat. Prod.* **52**(6):1353–1356.

Zabriskie, T. M., Klocke, J. A., Ireland, C. M., Marcus, A. H., Molinski, T. F., Faulkner, J. D., Xu, C., and Clardy, J. C., 1986, Jaspamide, a modified peptide from a *jaspis* sponge with insecticidal and antifungal activity, *J. Am. Chem. Soc.* **108**:3123–3124.

Zabriskie, T. M., Mayne, C. L., and Ireland, C. M., 1988, Patellazole C: A novel cytotoxic macrolide from *Lissoclinum patella, J. Am. Chem. Soc.* **110**(23):7919–7920.

Zheng, G. C., Hatano, M., Ishitsuka, M. O., Kusumi, T., and Kakisawa, H., 1990a, Novel seco- and seconorsesquiterpenes having a cylopropane ring from the Okinawan actinia *Anthopleura pacifica* Uchida, *Tetrahedron Lett.* **31**(18):2617–2618.

Zheng, G. C., Ichikawa, A., Ishitsuka, M. O., Kusumi, T., Yamamoto, H., and Kakisawa, H., 1990b, Cytotoxic hydroperoxylepidozenes from the actinia *Anthopleura pacifica* uchida, *J. Org. Chem.* **55**:3677–3679.

REVIEW ARTICLES ON CYTOTOXIC/ANTITUMOR COMPOUNDS FROM MARINE ORGANISMS

Fusetani, N., 1989, Antitumor substances from marine invertebrates, *Bio. Ind.* **6**(7):572–579.

Higa, T., 1987, Bioactive substances produced by marine organisms in the coral reef of Okinawa Island, *Kagaku Zokan* (Kyoto) **1987**(111):155–164.

Kamano, Y., Herald, C. L., Leet, J. E., and Pettit, G. R., 1985, Marine antineoplastic substances: Chemistry of the bryostatins, *Tennen Yuki Kagobutsu Toronkai Koen Yoshishu* **27th**:383–388.

Kitagawa, I., and Kobayashi, M., 1989, Antitumor natural products isolated from marine organisms, *Gan to Kagaku Ryoho* **16**(1):1–8.

Kobayashi, J., 1985, Bioactive substances in ascidians, *Kagaku to Seibutsu* **23**(2):119–123.

Krebs, H. C., 1986, Recent developments in the field of marine natural products with emphasis on biologically active compounds, *Fortschr. Chem. Org. Naturst.* **49**:151.

Munro, M. H. G., Blunt, J. W., Barns, G., Battershill, C. N., Lake, R. J., and Perry N. B., 1989, Biological activity in New Zealand marine organisms, *Pure Appl. Chem.* **61**(3):529–534.

Son, B. W., 1990, Bioactive marine natural products, *Kor. J. Pharmacogn.* **21**(1):1–48.

Tang, W., 1985, Antineoplastic substances from marine organisms, *Zhongcaoyao* **16**(2):83–89.

Uemura, D., 1987, Antitumor polyethermacrolides produced by sponges, *Kagaku Zokan* (Kyoto) **1987**(111):145–153.

8

Antiviral Substances

Kenneth L. Rinehart, Lois S. Shield, and Martha Cohen-Parsons

1. INTRODUCTION

Even today, antiviral drugs are a rarity (Becker, 1976, 1983; Rothschild *et al.*, 1978; R. T. Walker *et al.*, 1979; Collier and Oxford, 1980; De Clercq and Walker, 1984; Mandell *et al.*, 1985; Rinehart, 1992): acyclovir is a notable success in reducing the severity of genital *herpes* infections, and newer analogues are under development; azidothymidine (AZT) is widely employed to extend the lifetime of AIDS sufferers, while other compounds with anti-AIDS potential are under investigation; vidarabine is approved for the treatment of idoxuridine- and acyclovir-resistant infections; ribavirin is an intranasal inhalant effective against respiratory syncitial virus (Stephen *et al.*, 1980); amantidine has been used for many years in treating some forms of influenza (Davies *et al.*, 1964). Even fewer antiviral agents were available in the 1970s when we began our systematic surveys designed to assess the bioactivity of marine organisms.

As the need to cope with human viruses becomes more pressing, interest in antiviral agents continues to mount (Munro *et al.*, 1987). The antiviral potential of extracts from red algae and clams was noted in a 1964 conference on antiviral substances sponsored by the New York Academy of Sciences and the National Institute of Allergy and Infectious Diseases (NIAID) (Hermann, 1965). Today,

Kenneth L. Rinehart, Lois S. Shield, and Martha Cohen-Parsons • Department of Chemistry, University of Illinois, Urbana, Illinois 61801.

Marine Biotechnology, Volume 1: Pharmaceutical and Bioactive Natural Products, edited by David H. Attaway and Oskar R. Zaborsky. Plenum Press, New York, 1993.

both the National Cancer Institute (NCI) and NIAID are involved in screening for anti-AIDS drugs, some of which are derived from marine sources (Kolberg, 1991). Moreover, research groups around the world regularly test marine organisms for antiviral activity, including cold- and deep-water species (Higa, 1986; Cross and Lewis, 1987; Munro *et al.*, 1989).

2. MARINE-DERIVED ANTIVIRAL PROGRAM AT THE UNIVERSITY OF ILLINOIS

The marine natural products chemistry program at the University of Illinois in Urbana was launched in earnest in 1974 with an 8-week expedition on board the *R/V Alpha Helix*, during which over 800 species of marine plants and animals found in Baja California waters were surveyed for antimicrobial activity (Hager *et al.*, 1976; Rinehart *et al.*, 1976; Shaw *et al.*, 1976). Subsequently some of the extracts were tested for antiviral activity with positive results. Antiviral activity and cytotoxicity were associated most frequently with tunicates (ascidians, sea squirts; phylum Chordata, subphylum Tunicata or Urochordata). Extracts of certain *Polyandrocarpa* and *Aplidium* species, for example, were modestly antiviral and cytotoxic; they yielded the polyandrocarpidines (Cheng and Rinehart, 1978; Rinehart *et al.*, 1983a) and aplidiasphingosine (Carter and Rinehart, 1978b), respectively. Although these compounds were of interest primarily for their antimicrobial and cytotoxic potentials, we were sufficiently impressed with the prospects for antiviral agents from marine sources that our screening protocol was expanded during a 5-week Caribbean expedition in 1978 to include a shipboard antiviral/cytotoxicity assay using herpes simplex virus type 1 (HSV-1) grown in monkey kidney (CV-1) cells (Rinehart *et al.*, 1981a). Such testing (Rinehart, 1988a) remains vital to an ongoing search for antiviral, antitumor, antimicrobial, and immunomodulatory metabolites in our marine collection of over 3000 samples from Florida, Texas, Maine, Washington, Alaska, Central America, the Bahamas, and the western Mediterranean (Rinehart, 1989; Rinehart *et al.*, 1990a,b).

3. ANTIVIRAL ASSAYS

Dependable antiviral assays must be the heart of any search for antiviral compounds in deciding which organisms to investigate, in following the isolation of the pure compounds, and in measuring the potency of the isolated compounds. Such assays may, of course, be *in vitro* or *in vivo*, but very few marine-derived compounds have been tested *in vivo*.

In our laboratory we have employed HSV-1, a DNA virus, as our primary *in vitro* screen. This is a plaque assay in which CV-1 cells are infected with HSV-1, which proceeds to grow plaques or aggregates (Schroeder *et al.*, 1981) in cells. Reduction in plaque formation is deemed a positive assay. Similar plaque assays have been employed in determining antiviral activity against *Vesicular stomatitis* virus, an RNA virus, in CV-1 or baby hamster kidney (BHK) cells. Another RNA virus extensively employed for *in vitro* assays is the A59 corona virus in NCTC 1469 cells (Cross and Lewis, 1987). A variety of DNA and RNA viruses have been employed for secondary antiviral assays (Canonico *et al.*, 1982; Rinehart *et al.*, 1983b). For prediction of AIDS inhibition both the HIV virus itself and another retrovirus, a Visna (sheep) virus, have been employed (Frank *et al.*, 1987).

In an interesting spin-off, the antiviral assays also provide an assessment of cytotoxicity against normal (CV-1, BHK) cells and thus serve as a measure of potential for using the compounds as cytotoxic agents.

In the subsequent discussion we shall arbitrarily divide the antiviral compounds into those regarded as very active (IC_{50} < 1 μg/ml or well), active (IC_{50} 1–10 μg/ml or well), and modestly active (IC_{50} > 10 μg/ml or well) in *in vitro* assays. Unfortunately, many reports of "antiviral" activity do not include any quantitative measurements. Compounds lacking quantitation are, again arbitrarily, assigned to the modestly active group. In addition, only isolated compounds are included in the present review. Observations carried out on crude extracts, while providing useful stimuli, have often proved irreproducible and are omitted.

Following successful *in vitro* assays, *in vivo* studies must be carried out to measure the efficacy of an antiviral agent. Our most antiviral compounds, the didemnins and eudistomins, were tested successfully in a topical mouse vaginal herpes infection (HSV-2), which records prevention of death of the mice (Rinehart *et al.*, 1983b). Other *in vivo* herpes assays include rabbit eye herpes and herpes encephalitis assays. The latter is a difficult hurdle, since the drug must pass through the blood–brain barrier to be effective. Still other *in vivo* assays used with marine natural products include the A59 corona virus (Munro *et al.*, 1989) and Rift Valley fever virus (Canonico *et al.*, 1982). We note again that very few marine natural products have been tested *in vivo*, in part at least due to cost.

4. VERY ACTIVE ANTIVIRAL AGENTS

From the time of our earliest observations of marine antiviral activity (see above), we have become increasingly aware of the antiviral potential of substances derived from tunicates (Rinehart and Shield, 1983). That new emphasis was rewarded by the detection of strong antiviral activity in several species collected during the 1978 expedition (Rinehart *et al.*, 1981a). Our subsequent discovery of the didemnins and eudistomins was the beginning of a wide-ranging and continuing effort by chemists in this laboratory and others around the world, as well as by virologists, cancer researchers, taxonomists, and algologists. Structures have been assigned and confirmed by syntheses, *in vitro* and *in vivo* evaluations have been carried out, and analogues are currently being prepared for determining the relationship between structure and activity. At the same time, we continue to search for new, more potent members of the chemical families. The current status of our two furthest advanced projects—didemnins and eudistomins—will be reviewed first, with emphasis on their antiviral aspects, followed by other strongly antiviral compounds.

4.1. Didemnins

The didemnins (Rinehart *et al.*, 1987a; Rinehart, 1988b, and references therein) are a family of cyclic depsipeptides isolated from a *Trididemnum* species (family Didemnidae) that is found most often as a gray-green, flat, pancake-like coating on rocks or coral at depths to 120 feet. We first collected *Trididemnum* samples during our 1978 expedition in the waters off Belize, Honduras, Mexico, Colombia, and Panama. Independently, the bioactivity of such extracts attracted the attention of the NCI (Chun *et al.*, 1986). The structure elucidation of didemnins A, B, and C employed several types of mass spectrometry (Rinehart *et al.*, 1981b) and provided an early success with fast atom bombardment mass spectrometry when that then new technique became available. Modifications made in the course of our synthetic efforts and confirmed in other laboratories resulted in structures **1–9**, including two novel elements–2S,4S-hydroxyisovalerylpropionic acid (Hip) and (3S,4R,5S)-isostatine (Ist).

In shipboard testing, *Trididemnum* extracts inhibited HSV-1, with underlying cytotoxicity to the CV-1 cells. Both antitumor and antiviral activities were detected for the first didemnins isolated (Rinehart *et al.*, 1981a,c, 1983b; Canonico *et al.*, 1982), as well as for later members of the family, and, indeed, attention was soon focused on the anticancer potential of didemnin B and the ensuing clinical trials (Chun *et al.*, 1986). The didemnin family now includes many antitumor compounds from *T. solidum* (Rinehart *et al.*, 1990c; Sakai, 1991), including the even more potent dehydrodidemnin B from a Mediterranean tunicate, *Aplidium albicans* (Rinehart, 1990, 1992). Phase II clinical trials of didemnin B are expected to extend into 1993, with evaluation to follow. It was noted in early studies that **1** and **2** showed quantitatively altered bioactivities although they differed only in their side chain, suggesting that chemical modifications might lead to improved therapeutic potential.

After shipboard testing against HSV-1, the several samples of *Trididemnum* collected in 1978 were tested by Renis at the Upjohn Company, Kalamazoo, Michigan, against a battery of RNA (Coxsackie A21 virus, COE: equine rhino

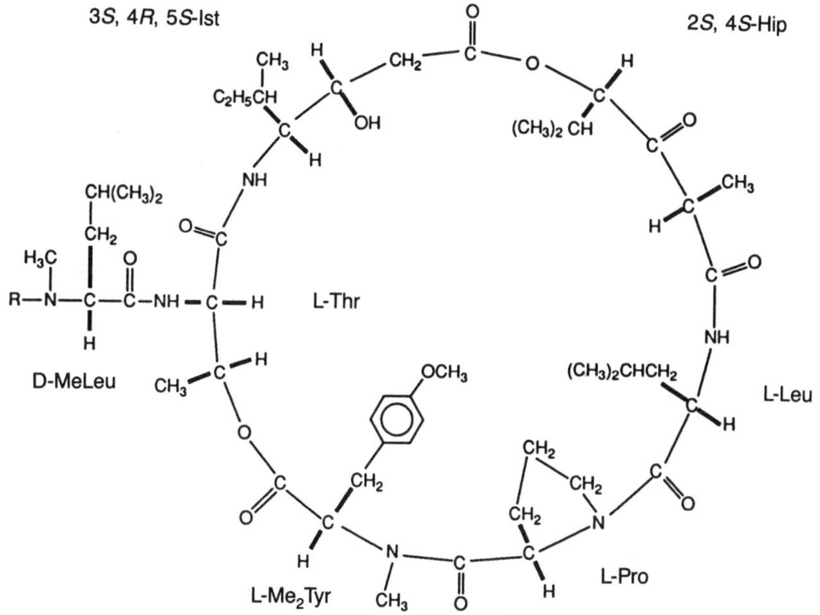

3S, 4R, 5S-Ist

2S, 4S-Hip

L-Thr

D-MeLeu

L-Leu

L-Me₂Tyr

L-Pro

(**1**) Didemnin A: R = H

(**2**) Didemnin B: R = CH₃CHOHC—N——C —
      L-Lac

     L-Pro

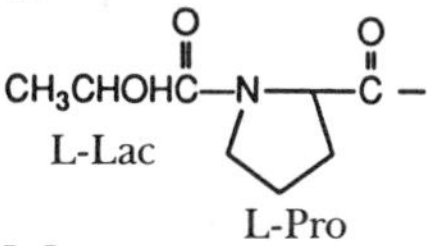

(**3**) Didemnin C: R = L-Lac
(**4**) Didemnin D: R = L-pGlu—(L-Gln)₃—L-Lac—L-Pro—
(**5**) Didemnin E: R = L-pGlu—(L-Gln)₂—L-Lac—L-Pro—
(**6**) Didemnin G: R = CHO
(**7**) Didemnin X: R = Hydec—(L-Gln)₃—L-Lac—L-Pro—
(**8**) Didemnin Y: R = Hydec—(L-Gln)₄—L-Lac—L-Pro—
      Hydec = n-C₇H₁₅

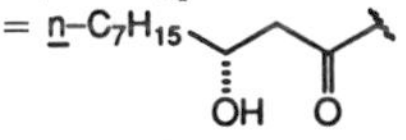

(**9**) Dehydrodidemnin B: R = CH₃C-C-N——C —
      Pyruv

     L-Pro

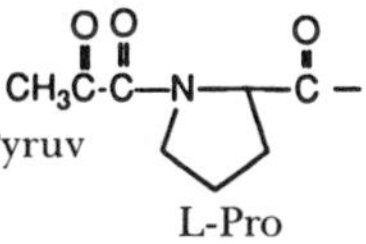

virus, ER; influenza virus, PR8; parainfluenza-3 virus, HA-1) and DNA (HSV-1; HSV-2; vaccinia, vacc) viruses. Later, individual didemnins (A, the major component; B, less abundant; and C, a trace component) from the initial extract were tested against the seven viruses. *In vitro*, didemnin B was 10–100 times as active as didemnin A. Didemnin B caused a >3 log reduction at 0.5 μg/ml in the growth of HSV-1 and HSV-2, while didemnin A caused a 1–2 log reduction at 5.0 μg/ml.

Topical application of didemnins A (at 1 mg/ml) and B (at 0.2 mg/ml) to mice inoculated intravaginally with HSV-2 produced improved survival rates and decreased virus titers. Neither didemnin A nor B was active against lethal Semliki Forest virus infections, and skin irritation was high for both compounds.

Didemnins A and B were tested by Canonico *et al.* (1982) against several virulent human pathogens for which neither treatment nor prevention is available— Rift Valley fever (RVF, Zagazig 501), Venezuelan equine encephalomyelitis (VEE, Trinidad donkey), yellow fever (YF, Asibi), and Pichinde arenavirus (PIC, AN3739), to give ID_{50} values of 0.04, 0.08, 0.08, and 0.22 μg/ml, respectively, for didemnin B and 1.37, 0.43, 0.4, and 2.9 μg/ml for didemnin A. Didemnin B was also effective in limiting the mortality of RVF-infected mice so that most lived for the duration of the study. Didemnin B is toxic, however; slightly higher doses proved lethal to mice. It is possible that, despite their low antiviral therapeutic indices, the didemnins could be modified or used in combination with other antiviral agents to combat such difficult disease.

4.2. Eudistomins

Extracts of *Eudistoma olivaceum* (family Polycitoridae), a shallow-water tunicate first collected during the Alpha Helix Caribbean Expedition in 1978, gave the strongest antiviral results in shipboard HSV-1 testing. The tunicate has since been recollected in Mexico, Belize, and Florida, usually by snorkeling or wading among mangrove roots, and 17 members of the eudistomin family have been isolated in this laboratory from toluene and chloroform extraction of the tunicate (J. Kobayashi *et al.*, 1984; Rinehart *et al.*, 1984, 1986, 1987b). The eudistomins we isolated showed a range of antiviral and antimicrobial activity and were characterized as a family of β-carbolines of four types—unsubstituted (**10–13**), pyrrolyl-substituted (**19,20**), pyrrolinyl-substituted (**21–25**), and tetrahydro-β-carbolines containing an oxathiazepine ring (**14–18**), a unique condensed ring system. Eudistomins G, H, I, and P were isolated and partially described simultaneously by the Cardellina group in Montana; they refined the separation process for those eudistomins and subsequently identified eudistomins R (**26**), S (**27**), and T (**28**) (Kinzer and Cardellina, 1987). Eudistomins D, E, H, and I have also been isolated from the Okinawan tunicate *Eudistoma glaucus*, together with related compounds such as eudistomidin B (**29**) (J. Kobayashi *et al.*, 1990).

From the compound ascidian *Ritterella sigillinoides*, Munro and Blunt and co-workers (Blunt *et al.*, 1987; Lake *et al.*, 1988a,b) isolated the trifluoroacetate salt of **17**. In the course of that work they revised the stereochemistry of the N–O bond of **17** (and **14**, **15**, and **18**), as confirmed later by x-ray analysis of the *p*-bromobenzoate derivative.

The *in vitro* antiviral potency of the eudistomins ranges from 5 to 500 ng/disk and follows the trend C, E, K, and L (oxathiazepino-tetrahydro-β-carbolines)

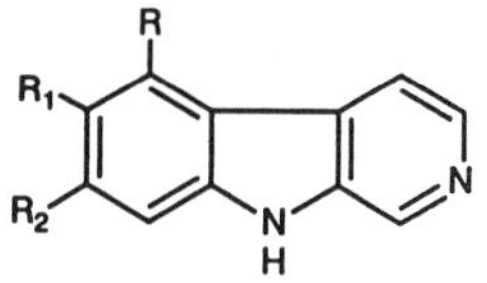

(10) Eudistomin D: R = Br, R_1 = OH,
    R_2 = H
(11) Eudistomin J: R = H, R_1 = OH,
    R_2 = Br
(12) Eudistomin N: R = R_2 = H, R_1 = Br
(13) Eudistomin O: R = R_1 = H, R_2 = Br

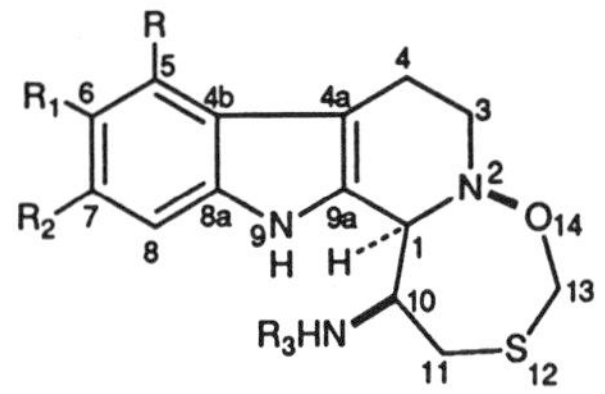

(14) Eudistomin C: R = R_3 = H,
    R_1 = OH, R_2 = Br
(15) Eudistomin E: R = Br, R_1 = OH,
    R_2 = R_3 = H
(16) Eudistomin F: R = H, R_1 = OH,
    R_2 = Br, R_3 = $C_2H_3O_2$
(17) Eudistomin K: R = R_1 = R_3 = H,
    R_2 = Br
(18) Eudistomin L: R = R_2 = R_3 = H,
    R_1 = Br

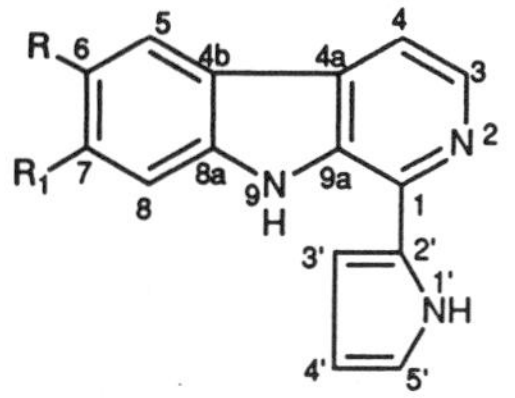

(19) Eudistomin A: R = OH, R_1 = Br
(20) Eudistomin M: R = OH, R_1 = H

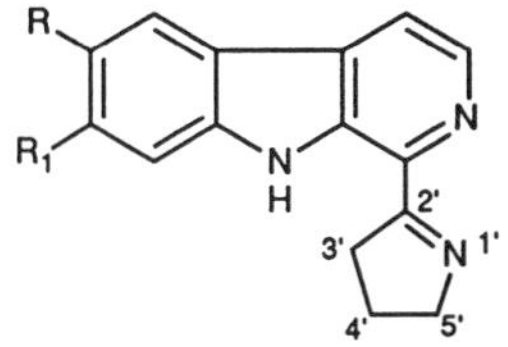

(21) Eudistomin G: R = H, R_1 = Br
(22) Eudistomin H: R = Br, R_1 = H
(23) Eudistomin I: R = R_1 = H
(24) Eudistomin P: R = OH, R_1 = Br
(25) Eudistomin Q: R = OH, R_1 = H

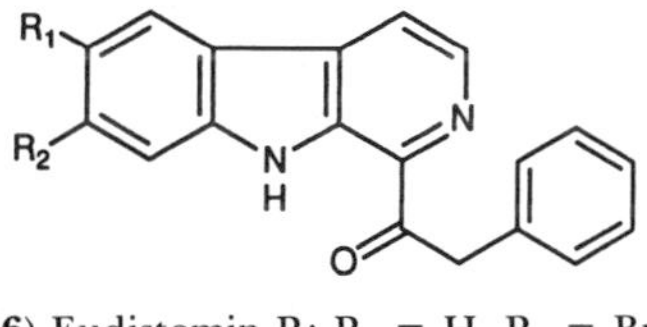

(26) Eudistomin R: R_1 = H, R_2 = Br
(27) Eudistomin S: R_1 = Br, R_2 = H
(28) Eudistomin T: R_1 = R_2 = H

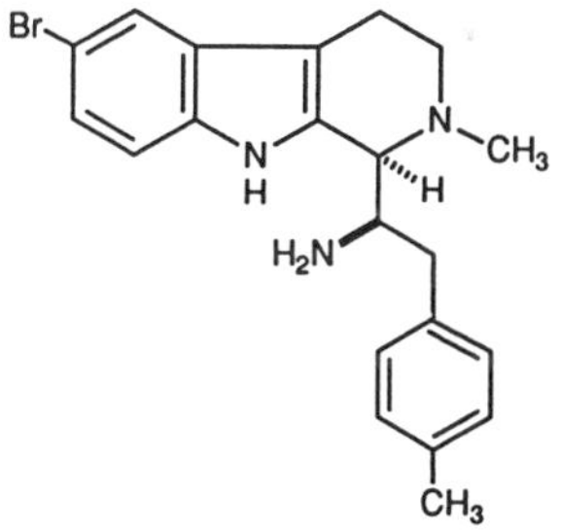

(29) Eudistomidin B

$\gg$ H and P (1-pyrrolinyl-substituted) = D and N (+0) (1-unsubstituted) > A (1-pyrrolyl-substituted, no inhibition). The influence of Br and/or OH substituents on the β-carboline benzenoid ring follows the order E (5-Br, 6-OH) = C (6-OH, 7-Br) > L (6-Br) = K (7-Br), and P (6-OH, 7-Br) = H (6-Br) > G (7-Br) = Q (6-OH) = I (no substitution). Despite the challenge presented by the oxathiazepine ring, synthetic efforts in several laboratories around the world (Kirkup *et al.*, 1989;

Nakagawa *et al.*, 1989; Still and Strautmanis, 1989; Hermkens *et al.*, 1990) have resulted in successful schemes for obtaining the most active eudistomins in amounts which should be sufficient for extensive *in vivo* testing.

4.3. Mycalamides and Onnamide A

Perry *et al.* (1988) and Munro *et al.* (1988) reported the isolation and structure determination of mycalamide A (**30**) from a New Zealand sponge, a *Mycale* sp. (phylum Porifera). They found that a material consisting of 2% mycalamide A was effective against A59 coronavirus *in vivo* in mice at 0.2 μg/kg per day with 100% survival after 14 days. When pure mycalamide A was obtained,

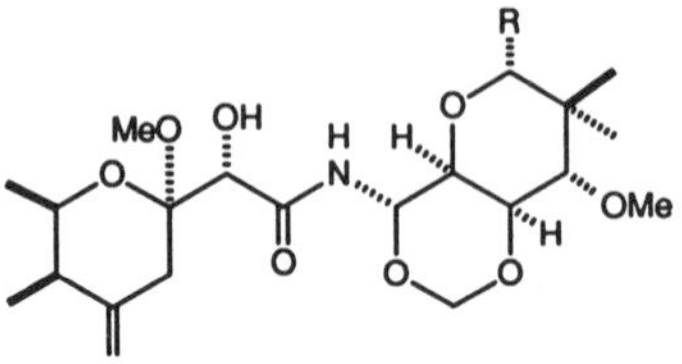

(**30**) Mycalamide A: R =

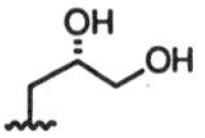

(**31**) Mycalamide B: R =

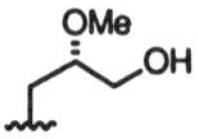

(**32**) Onnamide A: R =

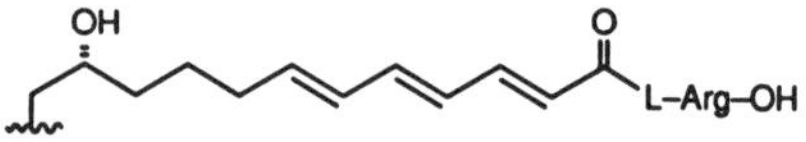

it inhibited HSV-1 or polio virus type I at 5 ng/disk. The structure was assigned based on MS and NMR data, including HETCOR, COSY, long-range HETCOR, and difference NOE experiments, and by comparison with the known compound pederin, isolated from a terrestrial beetle. A related compound, onnamide A, was isolated from a Japanese sponge at about the same time (see the following).

Blunt *et al.* (1989) and Perry *et al.* (1990) reported further work on the mycalamides, including the antiviral and antitumor mycalamide B (**31**). Mycalamide B had greater antiviral activity and cytotoxicity than mycalamide A; *in vitro* antiviral testing showed a minimum dose of 1–2 ng/disk for B and 3.5–5.0 ng/disk for A. Neither has been tested *in vivo*.

Onnamide A (**32**) was extracted from a *Theonella* sp. (phylum Porifera) collected off the coast of Okinawa (Sakemi *et al.*, 1988). Both the extract and the isolated compound were reported to have "potent activity" *in vitro* against HSV-1, VSV, and A59 coronavirus. Onnamide A was isolated in a procedure utilizing CCC and was assigned a structure based on UV, MS, and NMR data, including COSY, HETCOSY, and NOE difference experiments, and by analogy to pederin and mycalamide A (Sakemi *et al.*, 1988). Although details concerning the antiviral activity of **32** have not been published, its structural similarity to **30** and **31** suggests that the compound is probably quite active. In keeping with the antiviral, antitumor, and antifungal activity, Higa *et al.* (1989) are pursuing the use of onnamide A and its derivatives as agricultural and medical fungicides and virucides.

Mycalamides A and B, pederin, and onnamide A are protein synthesis inhibitors with commonalities in structure. Although the active solution conformation for this group of compounds is unknown, a correlation between substructure and bioactivity has been suggested (Perry *et al.*, 1990). However, due to the high variability in antiviral potency between these related compounds, it will probably be necessary to investigate a wide range of derivatives before any such correlation can be established (Perry *et al.*, 1990).

4.4. Avarol and Avarone

Avarol (**33**) and avarone (**34**) were recently reported to inhibit human immunodeficiency virus at doses of 0.1–1 μg/ml *in vitro* and thus are of potential use in treatment of AIDS (Sarin *et al.*, 1987). Extracted from the sponge *Disidea avara* (phylum Porifera), the compounds were identified by IR and NMR spectra

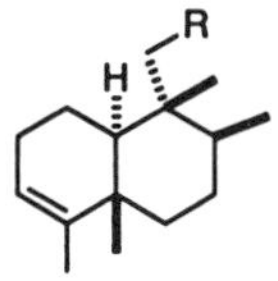

(**33**) Avarol: R = 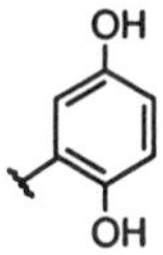   (**34**) Avarone: R =

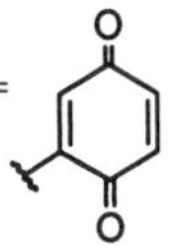

as sesquiterpenes attached to a quinone or hydroquinone unit and are related to a number of other compounds, including aureol, zonarol, chromazonarol, panicein, kamalonen, spongiaquinone, and ilimaquinone (Minale *et al.*, 1974). Avarol and avarone are of particular interest in the development of clinical application because of their high therapeutic indices and ability to cross the blood–brain barrier (Sarin *et al.*, 1987).

4.5. Ptilomycalin A and Crambescidins

Kashman *et al.* (1989a) reported the isolation of ptilomycalin A (**35**) from the Caribbean sponge *Ptilocaulis spiculifer* and a Red Sea sponge, a *Hemimycale* sp. (phylum Porifera). Activity against HSV was observed at a concentration of 0.2 μg/ml (Kashman *et al.*, 1989a). In addition to the high antiviral activity, the compound exhibited antitumor and antifungal activity. Its polycyclic guanidine structure was assigned based on UV, IR, MS, and NMR spectroscopy, including

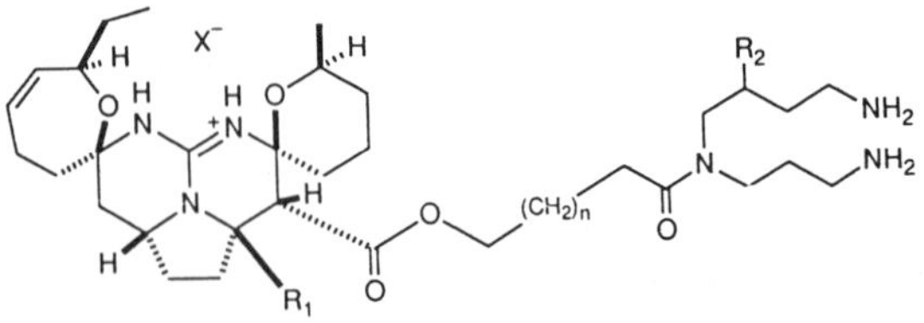

(35) Ptilomycalin A: $R_1 = R_2 = H$, n = 13
(36) Crambescidin 816: $R_1 = R_2 = OH$, n = 13
(37) Crambescidin 830: $R_1 = R_2 = OH$, n = 14
(38) Crambescidin 844: $R_1 = R_2 = OH$, n = 15
(39) Crambescidin 800: $R_1 = H$, $R_2 = OH$, n = 13

COLOC, NOESY, ROESY, HMBC, and HOHAHA data. The discovery of **35** revealed a new class of alkaloids linked to spermidine via an ω-hydroxy acid.

Very recently (Jares-Erijman *et al.*, 1991) a series of compounds related to **35** was isolated from the Mediterranean sponge *Crambe crambe* (phylum Porifera). The structures of these new compounds, the crambescidins (**36–39**), were assigned based on FABMS/MS, HRFABMS, and a series of NMR studies including HMBC. Compounds **36–39** differ from **35** by the presence of a hydroxyspermidine unit and from one another in the chain length of the long-chain hydroxy acid and in the presence or absence of a hydroxyl group in the guanidine-containing heterocyclic system. All of the crambescidins show activity against HSV-1 at 1.25 μg/ml and exhibit 98% inhibition of L1210 cell growth at 0.1 μg/ml.

4.6. Hennoxazoles

Hennoxazoles A–D (**40–43**) were isolated from a sponge, a *Polyfibrospongia* sp. (phylum Porifera), collected on the island of Miyako in Okinawa (Ichiba *et al.*, 1991). The names derive from the presence of two oxazole units in the molecules, which otherwise appear to be formed from a polyketide-amino acid biosynthetic pathway. In addition to displaying analgesic activity, hennoxazole A, the major component (0.01% of wet weight), showed strong activity against HSV-1 (IC_{50} = 0.6 μg/ml).

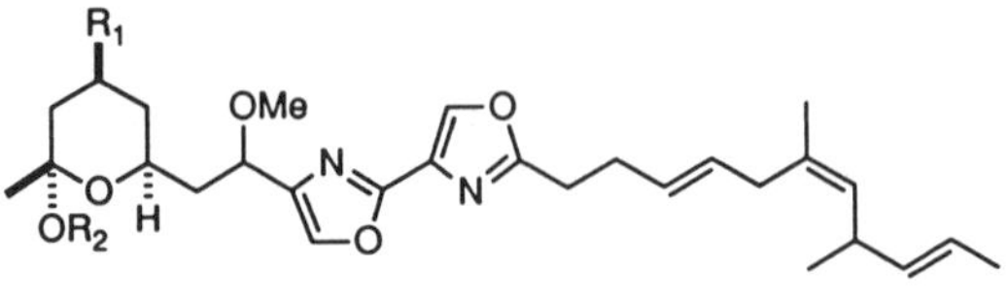

(40) Hennoxazole A: $R_1 = OH$, $R_2 = CH_3$
(41) Hennoxazole B: $R_1 = OH$, $R_2 = CH_2CH_3$
(42) Hennoxazole C: $R_1 = OH$, $R_2 = CH_2CH_2CH_2CH_3$
(43) Hennoxazole D: $R_1 = H$, $R_2 = CH_3$

4.7. Thyrsiferol and Related Triterpenes

Blunt *et al.* (1978) isolated thyrsiferol (**44**) from the red alga *Laurencia thyrsifera* (phylum Rhodophyta) collected in New Zealand. Although no biological activity was observed at that time, its structure was assigned as a squalene-derived triterpene tetracyclic ether. Years later, Gonzalez *et al.* (1984) isolated a

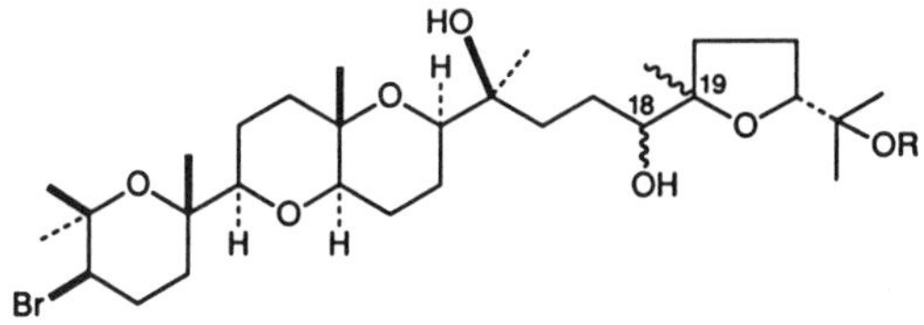

(**44**) Thyrsiferol: 18*S*, 19*R*; R = H
(**45**) Thyrsiferol acetate: 18*S*, 19*R*; R = Ac
(**46**) Venustatriol: 18*R*, 19*S*; R = H

number of inactive compounds from *L. pinnatifida*, some of which were terpenoids related to thyrsiferol (especially dehydrothyrsiferol and thyrsiferol mono-acetate). By contrast, Suzuki *et al.* (1985) reported that a crude extract of *L. obtusa* was strongly cytotoxic to P388 cells (ED_{50} = 0.18 μg/ml), and they isolated thyrsiferol, thyrsiferol acetate, and teurilene from the extract. Structures for these compounds were assigned based on NMR, IR, and x-ray data.

In subsequent studies, thyrsiferol-23 acetate (**45**) and a related compound, venustatriol (**46**), proved to be strongly antiviral when isolated from an extract of an Okinawan sample of *L. venusta* which showed "significant activity" against VSV and HSV-1 (Sakemi *et al.*, 1986). Based on MS, NMR, and x-ray data, venustatriol was found to be a stereoisomer of thyrsiferol. Still later, all three compounds that had been isolated by Sakemi *et al.* (1986), viz. thyrsiferol, thyrsiferol-23 acetate, and venustatriol, showed *in vitro* activity against VSV and HSV-1 (Higa *et al.*, 1988a), with efficacies reported for all three compounds at levels of 0.1–0.5 μg/well (Rinehart, 1992). Some accompanying cytotoxicity has also been observed as well as slight activity against A59 corona virus without concurrent cytotoxicity (Rinehart, 1992). Suzuki *et al.* (1987) isolated additional compounds of this type from *L. obtusa* collected in Japan. Included were five new compounds with some cytotoxicity against P388 *in vitro*—15(28)-anhydrothyrsiferol diacetate, 15-anhydrothyrsiferol diacetate; magireols A, B, and C. Their structures were established on the basis of IR, NMR, and HR mass spectra compared to known compounds of this type as well as chemical derivatization. Antiviral activity has not yet been reported for the additional compounds.

The cytotoxicity and antiviral activity reported for thyrsiferol-related compounds stimulated interest in their synthesis. As a result, thyrsiferol was partially synthesized by Hashimoto *et al.* (1987) and total syntheses of (+)-thyrsiferol (Hashimoto *et al.*, 1988) and (+)-venustatriol (Hashimoto *et al.*, 1988; Corey and Ha, 1988) were subsequently achieved.

4.8. Solenolides and Briantheins

Groweiss *et al.* (1988) reported the isolation of solenolides A–F (**47–52**) from a *Solenopodium* sp. (phylum Coelenterata), a newly identified Indopacific gorgonian collected at Palau. The diterpenoid lactone structures of **47–52** were assigned from NMR data, including NOESY and NOE difference experiments; UV data; and chemical derivatization. These diterpenoid compounds represent a

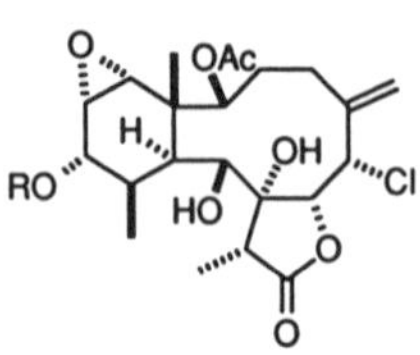

(**47**) Solenolide A: R = C$_5$H$_{11}$CO   (**49**) Solenolide C: R = H   (**51**) Solenolide E
(**48**) Solenolide B: R = Ac   (**50**) Solenolide D: R = Ac

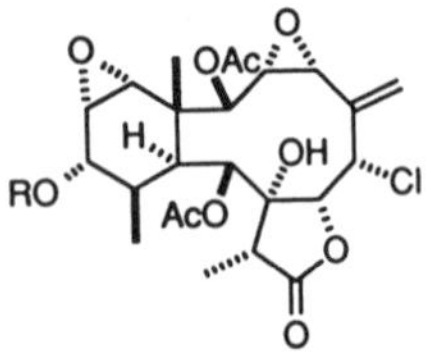

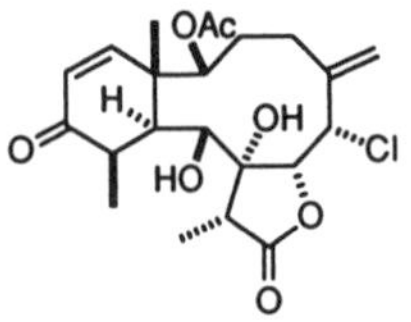

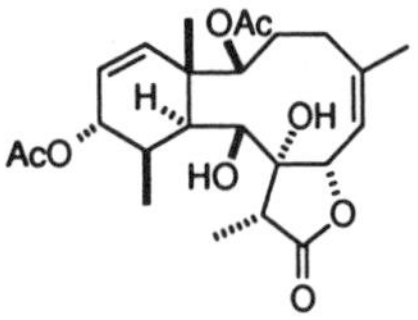

(**52**) Solenolide F 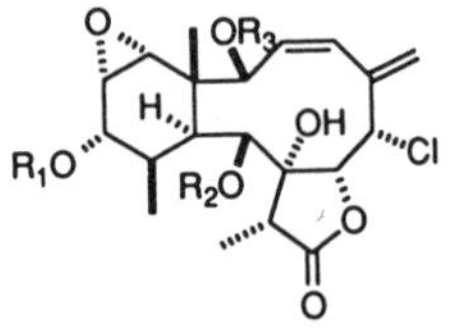   (**53**) Brianthein V: R$_1$ = R$_3$ = COCH$_2$CH$_2$CH$_3$, R$_2$ = Ac
                           (**54**) Brianthein X: R$_1$ = R$_2$ = Ac, R$_3$ = H
                           (**55**) Brianthein Y: R$_1$ = R$_2$ = Ac, R$_3$ = COCH$_2$CH$_2$CH$_3$
                           (**56**) Brianthein Z: R$_1$ = R$_2$ = R$_3$ = Ac

variation on the class of briarein marine products previously found to have biomedical potential. Three of the five new solenolide compounds exhibited antiviral activity, the most notable of which were the inhibitions of rhinovirus by solenolides A (IC$_{50}$ = 0.39 μg/ml) and E (IC$_{50}$ = 12.5 μg/ml). Additional findings included activities against HSV (solenolides A and E), polio III (solenolide A), Ann Arbor (solenolides A, D, and E), Maryland (solenolide A), and Semliki Forest viruses (solenolide D). Solenolides A, D, E, and F also exhibited anti-inflammatory activity. The solenolides are closely related to the briantheins (see below), which have also been reported to be antiviral.

The briarein and asbestinin series of diterpenes consists of highly oxidized compounds isolated from gorgonian coral (*Briareum asbestinum* and *B. poly-anthes*; phylum Coelenterata) found in Caribbean and Bermudan waters (Stierle *et al.*, 1980; Grode *et al.*, 1983a). These compounds are very closely related chemically to the solenolides but are much less active as antiviral agents. Among the briareins, briantheins V, Y, and Z inhibited A59 mouse corona virus *in vitro* at 50, 400, and 80 μg/ml, respectively (Coval *et al.*, 1988). Brianthein Z, first isolated by Grode *et al*. (1983a), inhibited HSV-1 at 80 μg/ml. Briantheins Z and V displayed *in vitro* cytotoxicity toward P388 (Coval *et al*., 1988), and brianthein Y has been noted for its insecticidal potential (Grode *et al*., 1983b). In light of certain taxonomic discrepancies (for example, *B. polyanthes* had also been known as *Ammothea polyanthes*, *Erythropodium polyanthes*, and *B. asbestinum*), compounds V, Y, and Z were of interest as potential chemotaxonomic markers. The brianthein structures were assigned by x-ray analysis of V (**53**) (Coval *et al*., 1988) and X–Z (**54–56**) (Grode *et al*., 1983b).

4.9. Spongiadiol and Related Compounds

Kazlauskas *et al*. (1979) isolated eight tetracyclic furanoditerpenes from an Australian sponge of the genus *Spongia* (phylum Porifera) collected from the Great Barrier Reef. The compounds were originally given one of three trivial classifications, spongiadiol [3α,19-dihydroxyspongia-13(16),14-dien-2-one], spongiatriol [3α,17,19-trihydroxyspongia-13(16),14-dien-2-one], and epispongiadiol [3β,19-dihydroxyspongia-13(16),14-dien-2-one]. Structures were determined on the basis of NMR, x-ray analysis of one of the compounds, and CD data. Earlier, degraded C-21 terpenes of this general form had been isolated from a *Spongia* sp. (Cimino *et al*., 1974; Kazlauskas *et al*., 1976), as had the diterpene spongi-12-en-16-one (Kazlauskas *et al*., 1976), but no bioactivities were reported. Later, Kohmoto *et al*. (1987) reported the isolation from a deep-water Caribbean *Spongia* sp. of spongiadiol (**57**), epispongiadiol (**58**), and the new isospongiadiol [2α,19-dihydroxyspongia-13(16),14-dien-3-one] (**59**). In their study, spongiadiol was

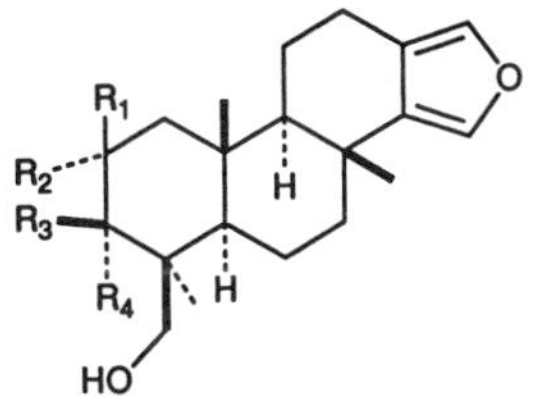

(**57**) Spongiadiol: $R_1 + R_2 = O$, $R_3 = H$, $R_4 = OH$
(**58**) Epispongiadiol: $R_1 + R_2 = O$, $R_3 = OH$, $R_4 = H$
(**59**) Isospongiadiol: $R_1 = H$, $R_2 = OH$, $R_3 + R_4 = O$

isolated as 0.13% of the frozen weight, epispongiadiol as 0.87%, and isospongiadiol as 0.2% in an isolation process that involved CCC, and structures were assigned based on IR, MS, and NMR data, including COSY and NOE experiments. Kohmoto *et al*. (1987) reported both antiviral activity and cytotoxicity for all three spongiols. *In vitro* assays against HSV-1 revealed a spectrum of activities ranging from the very active spongiadiol ($IC_{50} = 0.25$ μg/ml) to the modestly active epispongiadiol ($IC_{50} = 12.5$ μg/ml), with isospongiadiol exhibiting intermediate activity (IC_{50} = 2.0 μg/ml). In additional reports of the antitumor and antiviral activities of these three furanoditerpenoids, spongiadiol and isospongiadiol gave 100% inhibition of HSV-1 plaque formation at 20 and 0.5 μg/(6-mm disk), and epispongiadiol gave partial inhibition at 12.5 μg/disk (Kohmoto *et al.*, 1988).

4.10. Ara-A

A family of potent antiviral and antitumor compounds including two presently in clinical use as antiviral or antitumor agents (i.e., ara-A, 9-β-D-arabinofuranosyladenine, **60**; ara-C, 1-β-D-arabinosylcytosine, **61**) is related to the arabinosides isolated in the early 1950s from the marine sponge *Cryptotethia crypta* (Bergmann and Feeney, 1950, 1951). Bergmann first collected *C. crypta* in

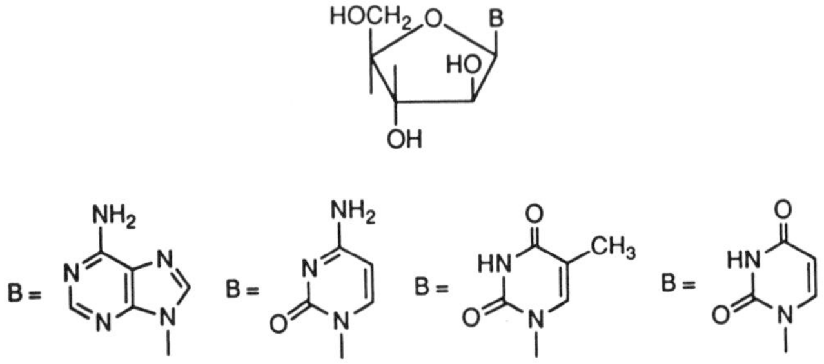

(60) ara-A   (61) ara-C   (62) ara-T   (63) ara-U

1945, and in the next few years he reported the presence of spongothymidine (ara-T, 1-β-D-arabinofuranosylthymidine, **62**), spongouridine (ara-U, 1-β-D-arabinofuranosyluracil, **63**), and spongosine (1-β-D-ribofuranosyl-2-methoxyadenine) [reviewed by Cohen (1966)]. Cimino *et al.* (1984) identified ara-U, as well as ara-A and the 3'-*O*-acetyl derivative of ara-A, in the 1-butanol extract of the gorgonian *Eunicella cavolini* (phylum Coelenterata) on the basis of UV, IR, NMR, and MS data, and by comparison with authentic samples. This was the first discovery of ara-A in a natural marine source, although it had been synthesized as one of many bioactive variations on the naturally occurring spongouridine.

Early *in vitro* studies showed the antiviral activity of the arabinosides to vary depending upon whether the challenge was against HSV-1 or -2. Using rabbit kidney and human skin fibroblast cultures, De Clercq *et al.* (1977) reported MICs (minimum inhibitory concentration) as low as 0.02 and 1 μg/ml for ara-C and ara-A, respectively, against HSV-1; and 200 and 10 μg/ml, respectively, against HSV-2. Significant *in vitro* activity has also been observed for a number of xylofuranonucleosides against three DNA viruses (HSV-1, HSV-2, vaccinia) and one RNA virus (rhinovirus-9) (Gosselin *et al.*, 1986).

5. ACTIVE ANTIVIRAL AGENTS

5.1. Dercitin

A fused pentacyclic aromatic alkaloid, dercitin (**64**), was isolated from a *Dercitus* sp. (sponge; phylum Porifera) by Gunawardana *et al.* (1988). It was observed to be a violet pigment having antitumor, antiviral, and immunomodulatory activity *in vitro* and antitumor activity *in vivo*. Dercitin was obtained in a yield of 0.69% of the wet weight of the sponge, and its structure was assigned as

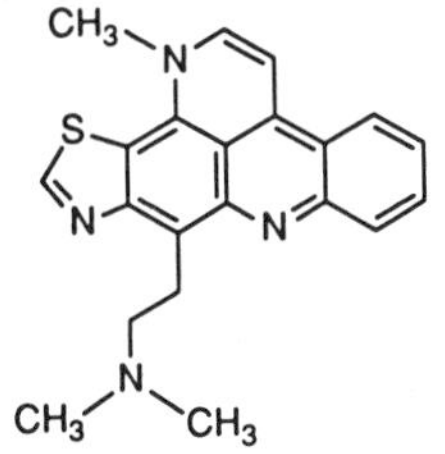

(64)

N,*N*,1-trimethyl-1*H*-pyrido[4,3,2-*mn*]thiazolo[5,4-b]acridine-9-ethanamine on the basis of UV, MS, and NMR data, including COSY, NOE, HETCOSY, COLOC, and INADEQUATE experiments. The presence of the fused thiazole unit was thought to be unique to this pentacyclic aromatic alkaloid. Its cytotoxicity and antiviral activity were reported as 10, +++ at 5 μg/well against HSV-1 and 0, +++ at 1 μg/well against A59 murine corona virus. Further study showed that the antitumor activity of dercitin was associated with its intercalation into nucleic acids (Burres *et al.*, 1989). Other bioactive compounds isolated from sponges of the family Pachastrellidae have been found to contain the same pyrido[4,3,2-*mn*]-acridine skeleton as **64** (Gunawardana *et al.*, 1989).

5.2. Indolocarbazole

Knübel *et al.* (1990) extracted a bioactive blue-green alga, *Nostoc sphaericum* (phylum Cyanophyta), from an Oahu mud sample. From the cultured Hawaiian alga they isolated indolo[2,3A]carbazole compounds and found the major component, 6-cyano-5-methoxy-12-methylindolo[2,3A]carbazole (**65**), to

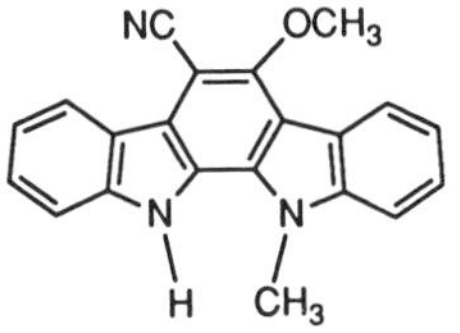

(65)

be responsible for most of the antiviral activity and cytotoxicity. The virus titer in mink lung cells infected with HSV-2 was reduced 95% at ca. 1 μg/ml, but some virus remained before the cytotoxic MIC was reached at 100 μg/ml. Similar activity was observed for the 12-demethyl analogue. The major compound was obtained in a 0.22% dry weight yield, and its structure was assigned on the basis of UV, MS, and NMR, including COSY, HMQC, HMBC, and NOE experiments. Although this *Nostoc* species is not, strictly speaking, a marine blue-green alga, other cyanobacteria are found in the ocean.

5.3. Topsentins

The bioactivity (P388, HSV-1) of the genus *Spongosorites* (phylum Porifera) was found in our laboratory to be associated with the bis(indolyl)imidazoles shown here–topsentin (**66**), bromotopsentin (**67**), and isotopsentin (**68**) (Tsujii *et al.*, 1988; Gunasekera *et al.*, 1989), whose structures were assigned based on HREIMS and NMR. Synthetic work undertaken to confirm the structure assignments and to study structure–activity relationships afforded the family of com-

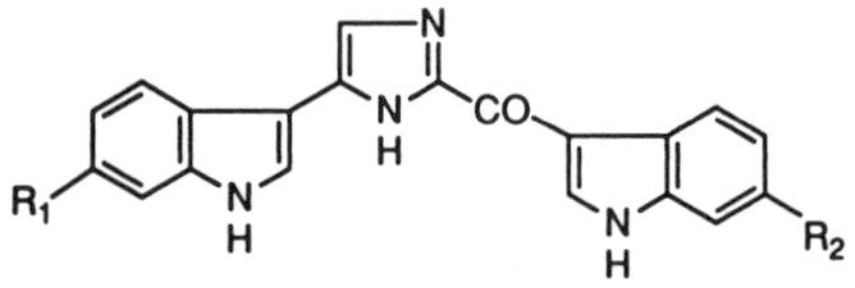

(**66**) Topsentin: R_1 = H, R_2 = OH
(**67**) Bromotopsentin: R_1 = Br, R_2 = OH
(**68**) Isotopsentin: R_1 = OH, R_2 = H
(**69**) Hydroxytopsentin: R_1 = R_2 = OH
(**70**) Deoxytopsentin: R_1 = R_2 = H

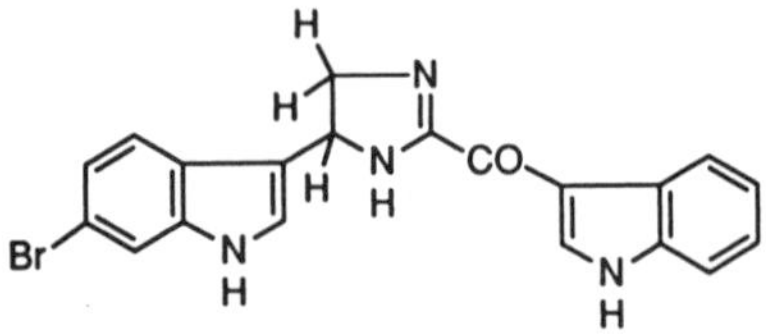

(**71**) 4,5-dihydro-6″-deoxybromotopsentin

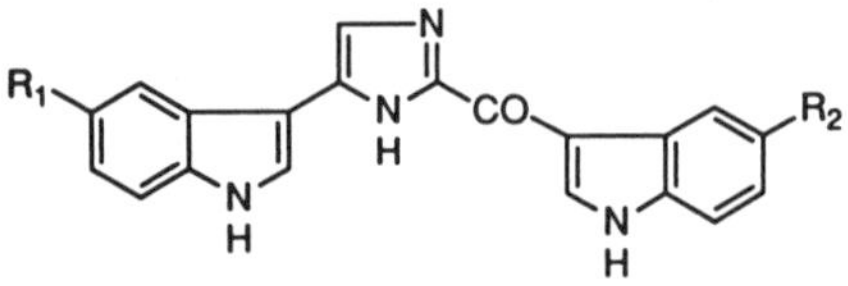

(**72**) Neotopsentin: R_1 = H, R_2 = OH
(**73**) Neoisotopsentin: R_1 = OH, R_2 = H
(**74**) Neohydroxytopsentin: R_1 = R_2 = OH

pounds **69–74**. The most active antiviral compound, topsentin, inhibited A59 corona virus at 2 μg/disk and HSV-1 at 50 μg/disk in tissue culture assays.

5.4. Variabilin

Faulkner (1973) reported the isolation of several furanosesterterpenes from a sponge indigenous to New Zealand and belonging to the genus *Ircinia* (phylum Porifera). Among them was variabilin (**75**), an antimicrobial agent (vs. *Staphylococcus aureus*) accounting for 0.2% of the dry weight of the sponge. Its structure

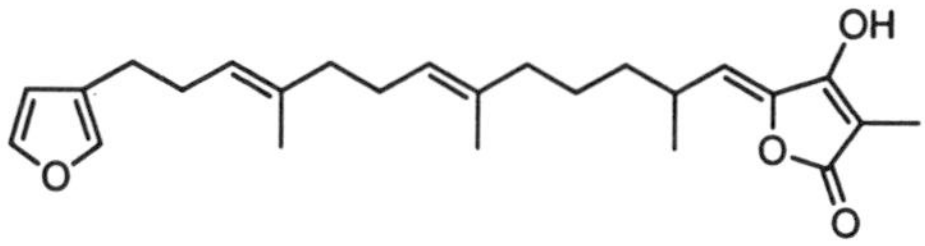

(75)

was assigned on the basis of UV, IR, and NMR data and comparisons with other tetronic acids reported earlier, such as ircinins 1 and 2 and fasciculatin. Due to its ubiquitous presence in the genus *Ircinia*, variabilin served as a valuable taxonomic marker and was later used as a chemotaxonomic marker to facilitate the study of sponges of the order Dictyoceratida (Perry *et al.*, 1987). Variabilin proved to be a major component in extracts of six *Ircinia*, three *Psammocinia*, and one *Sarcotragus* samples.

Four new furanosesterterpene tetronic acids were identified by Barrow *et al.* (1988b). In the course of that work, crude *Ircinia* extracts displayed *in vitro* antiviral activity against VSV-1 and polio virus type I in BSC (green monkey kidney) host cells. Some cytotoxicity at 2 μg/disk was also observed. Although variabilin purportedly showed varying antiviral behavior, an additional study found the compound to be cytotoxic but not antiviral (Barrow *et al.*, 1988a). Nevertheless, the stereochemistry of variabilin, the major bioactive component in a *Sarcotragus* sample, was completed and three new terpenes of the same general type were reported. Later, Barrow *et al.* (1989) studied the decomposition products of variabilin and obtained some of its bioactive yet stable analogues. The derivatives were more stable in the presence of light and air than was variabilin, but there remained the problem of any useful antiviral effect (either *in vitro* or *in vivo*) being overshadowed by cytotoxicity. In the group of compounds examined, the activity observed at 2–20 μg/disk depended not on the presence of a furan or a tetronic acid unit, but on the presence of such terminal groups as hydroxyl or carboxyl.

5.5. Reiswigins

Kashman *et al.* (1987) reported the isolation of reiswigins A (**76**) and B (**77**), bioactive terpenes from the sponge *Epipolasis reiswigi* (phylum Porifera). Their

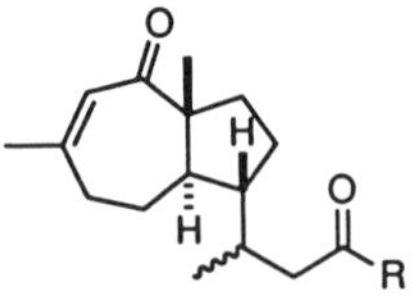

(**76**) Reiswigin A: R = CH$_2$CH(CH$_3$)$_2$
(**77**) Reiswigin B: R = –CH=C(CH$_3$)$_2$

structures were assigned on the basis of UV, IR, MS, and NMR data, including COSY, INADEQUATE, NOESY, and NOE experiments. Both compounds inhibited HSV-1 completely at 2 µg and A59 virus partially at 20 µg (++), and reiswigin A completely inhibited VSV at 2 µg without accompanying cytotoxicity. Antiviral activity was reported for a series of six related diterpenes, including **76** and **77** (Kashman *et al.*, 1989b).

5.6. Prostaglandins

Activity against both RNA and DNA viruses has been recorded for a number of prostaglandins (Santoro *et al.*, 1980; Ankel *et al.*, 1985), some of which occur in the marine environment among the soft corals. This observation of antiviral activity has been extended to the more unusual clavulone II (**78**), a prostanoid isolated from the soft coral *Clavularia viridis* (phylum Cnidaria) and identified

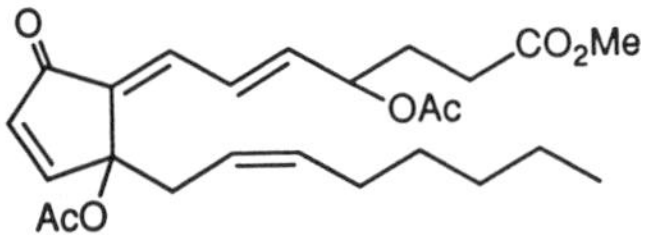

(**78**) Clavulone II 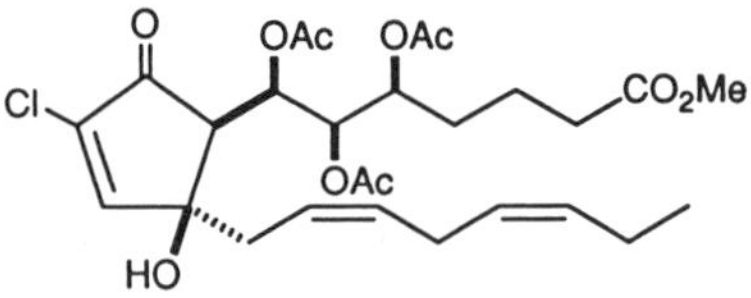   (**79**) Punaglandin-1

by UV, IR, and NMR data (Kikuchi *et al.*, 1982; M. Kobayashi *et al.*, 1982). Clavulone II was found to be the most active prostanoid in tests conducted against VSV (IC$_{50}$ ca. 2 µg/ml) and encephalomyocarditis (EMC) (Bader *et al.*, 1991).

Punaglandins (halogenated eicosanoids, e.g., **79**) are unusual prostaglandins obtained from the octocoral *Telesto riisei*. Although the original descriptions of the natural products did not report antiviral activity, subsequent patent applications (Noyori *et al.*, 1987a,b) indicated that some punaglandin derivatives were antiviral agents.

5.7. Macrolactin A

Gustafson *et al.* (1989) reported the isolation of macrolactins A–F from the culture broth of a deep-sea bacterium that could not be classified taxonomically.

Structures were established on the basis of UV, IR, MS, and NMR data, including COSY, HETCOR, and COLOC experiments. The compounds were found to be 24-membered ring lactones and their glucose β-pyranoside analogues and included the open-chain macrolactinic and isomacrolactinic acids. Macrolactin A (**80**) showed some activity against *Bacillus subtilis* and *S. aureus* as well as B16-F10 murine melanoma cells *in vitro*. Against HSV-1 (strain LL) and HSV-2 (strain

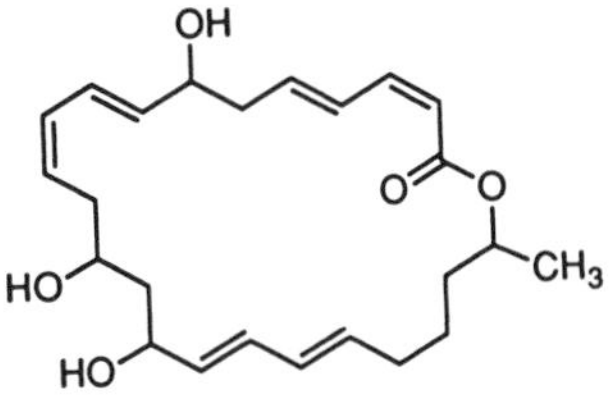

(**80**)

G), the IC_{50} was 5.0 and 8.3 μg/ml, respectively. Although no cytotoxicity data were provided, Gustafson *et al.* (1989) indicated that the potential therapeutic index fell in the range 10–100. In ongoing tests conducted by the NCI, a concentration of 10 μg/ml of macrolactin A gave maximum protection against human HIV replication (Gustafson *et al.*, 1989).

6. MODESTLY ACTIVE ANTIVIRAL AGENTS

6.1. Misakinolide A and Bistheonellides

Misakinolide A was first isolated from a *Theonella* sp. (phylum Porifera) collected in Okinawa (Sakai *et al.*, 1986). *In vitro* antiviral and antifungal activities were reported. On the basis of MS and NMR data, including COSY, Sakai *et al.* (1986) assigned a 20-membered macrolide structure similar to that of swinholide A, a known antifungal compound isolated from a sponge of the same genus (Carmely and Kashman, 1985). Comparisons with swinholide A led to the assignment of a monomeric structure as found in the scytophycins from the blue-green alga *Scytonema pseudohofmani*. Upon further study, Kato *et al.* (1987) revised the structure of misakinolide A from a monomeric to a dimeric macrolide and concluded that the dimeric structure was identical to that of bistheonellide A (**81**), newly isolated from a *Theonella* sp. The structure of bistheonellide B (**82**), a related compound, was also assigned. These are the first reports of dimeric macrolides having a 40-membered ring (Kato *et al.*, 1987). The dimeric structure determination included FABMS data and was confirmed by chemical degradation. Kato *et al.* (1987) also reported that bistheonellides A and B inhibited starfish (*Asterina pectinifera*) embryo development, a finding suggestive of *in vivo*

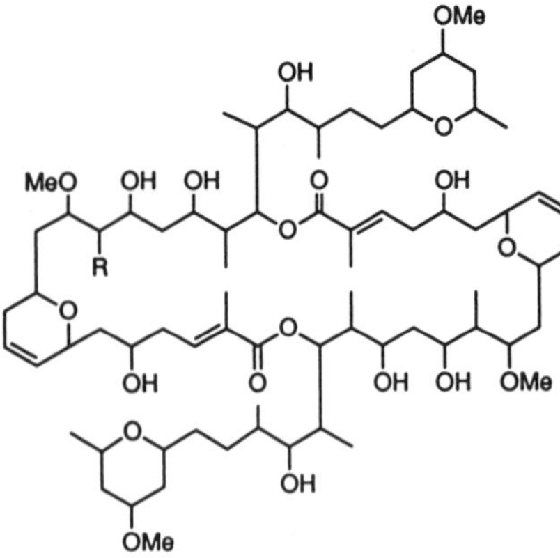

(81) Bistheonellide A: R = Me
(82) Bistheonellide B: R = H

cytotoxicity. In addition, Higa *et al.* (1988b) recorded antitumor, antiviral, and antifungal activity for misakinolide A (bistheonellide A), citing activity against HSV-1 and VSV in CV-1 cells at 8 μg/0.5 ml.

6.2. Sceptrins and Ageliferins

Extracts of the Caribbean sponge *Agelas conifera* (phylum Porifera) yielded the diacetate salts of the series of bromopyrroles shown here (**83–87, 88–90**)

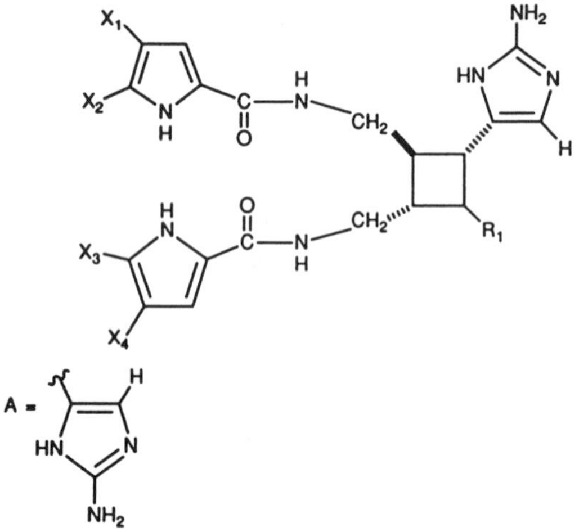

		X_1	X_2	X_3	X_4	R_1
(83)	Sceptrin:	Br	H	H	H	A
(84)	Debromosceptrin:	Br	H	H	Br	A
(85)	Dibromosceptrin	Br	Br	Br	Br	A

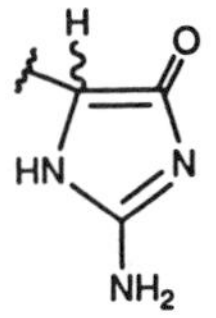

		X_1	X_2	X_3	X_4	R_1
(86)	Debromooxysceptrin:	H	H	H	Br	B
(87)	Oxysceptrin:	Br	H	H	Br	B

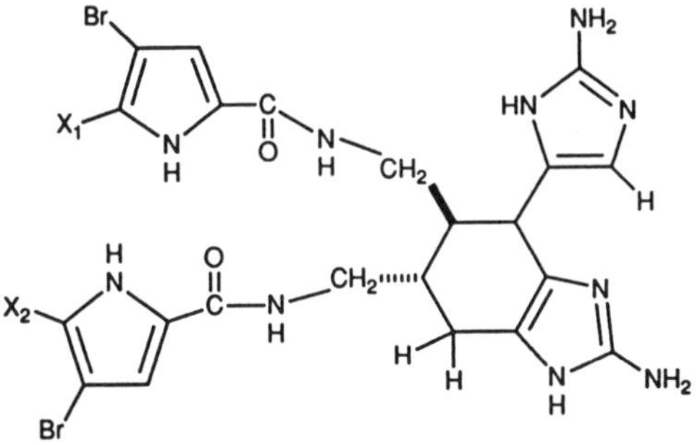

(88) Ageliferin: $X_1 = X_2 = H$
(89) Bromoageliferin: $X_1 = Br, X_2 = H$
(90) Dibromoageliferin: $X_1 = X_2 = Br$

(Rinehart, 1988c, Keifer *et al.*, 1991). Based on spectroscopic comparisons to the known sceptrin (R. P. Walker *et al.*, 1981), as well as on FABMS and NMR data, the structures assigned included the oxysceptrins and ageliferins. The latter compounds have sceptrinlike formulas with less symmetrical structures. Compounds of the sceptrin and ageliferin groups are active against HSV-1 at 20 μg/disk and VSV at 100 μg/disk, while the oxysceptrins are less active (Keifer *et al.*, 1991).

6.3. Halitunal

Halitunal (91), a diterpene isolated from the green alga *Halimeda tuna* (phylum Chlorophyta) by Koehn *et al.* (1991), was collected near Chub Point in the Bahamas, and constituted 0.01% of the wet weight of the alga. The molecular formula was assigned mainly from NMR spectroscopic measurements and required extensive use of HMBC correlations. Halitunal showed ca. 50% inhibition of viral replication of A59 murine corona virus in NCTC 1469 mouse liver cells at a dose of 20 μg per test well.

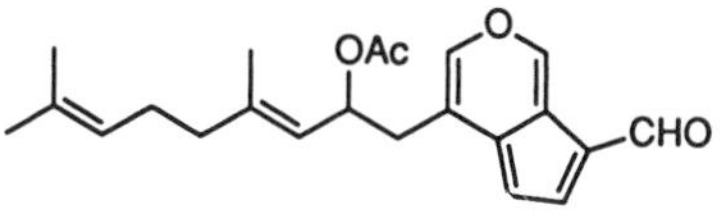

(91)

6.4. Sesquiterpenoid Isocyanide

Wright *et al.* (1988) have reported the antitumor, antiviral, and antifungal activities of a sesquiterpenoid isocyanide (**92**) isolated from the marine sponge *Bubaris* (phylum Porifera). At 20 μg/0.5 ml, the A59 coronavirus in mouse liver cells was partially inhibited, indicating that the sesquiterpenoid compound is only weakly virucidal.

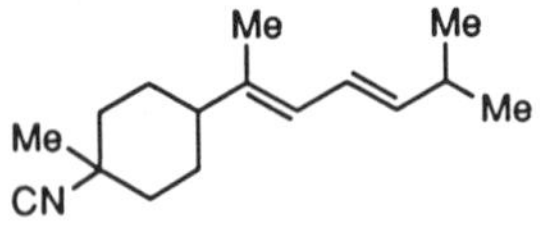

(92)

6.5. Acarnidines and Polyandrocarpidines

Acarnidines 1a–1c (**93–95**) were isolated from *Acarnus erithacus* (de Laubenfels), a sponge (phylum Porifera), and were among the antiviral substances identified from our collections in the Gulf of California (Carter and Rinehart, 1978a). The homospermidine skeleton common to these three guanidino compounds was assigned based on GC/MS data, and the compounds were distinguished from one another by their fatty acid constituents. In addition to

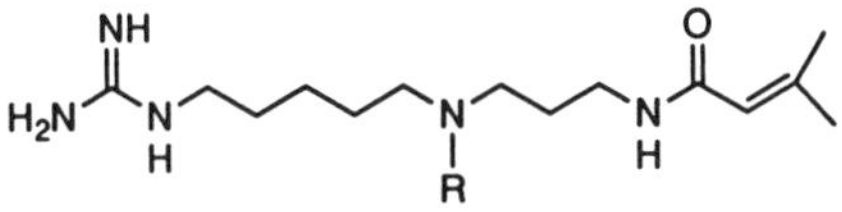

(**93**) Acarnidine 1a: R = CO(CH$_2$)$_{10}$CH$_3$
(**94**) Acarnidine 1b: R = CO(CH$_2$)$_3$CH=CH(CH$_2$)$_5$CH$_3$ (Z)
(**95**) Acarnidine 1c: R = COC$_{13}$H$_{21}$

some antibacterial activity, we observed activity against HSV-1 at 100 μg/disk. However, Munro *et al.* (1987) reported a lack of activity against "a range of DNA and RNA viruses" despite observations of cytotoxicity and antibacterial activity.

We obtained a mixture of the homologues, polyandrocarpidines I and II, from an extract of a *Polyandrocarpa* sp. (tunicate, phylum Chordata) collected in Baja California (Cheng and Rinehart, 1978). The mixture displayed antibacterial activity and was cytotoxic to CV-1 cells at 200 μg/well. We also observed slight antiviral activity against HSV-1. From studies utilizing NMR data, Carté and Faulkner (1982) found that each homologue was a mixture of γ-methylene-γ-lactam isomers (**96, 97; 98, 99**) in varying proportions and the structure assignment was confirmed by synthesis of derivatives (Rinehart *et al.*, 1983a).

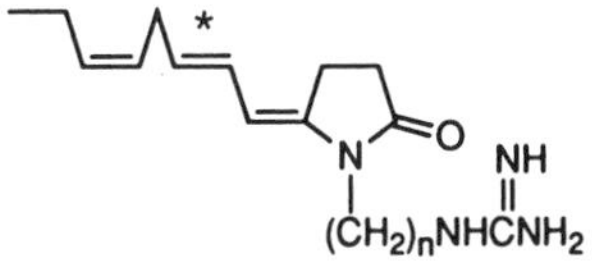

(**96**) Polyandrocarpidine A: n = 5; * *cis* isomer
(**97**) Polyandrocarpidine B: n = 5; * *trans* isomer
(**98**) Polyandrocarpidine C: n = 4; * *cis* isomer
(**99**) Polyandrocarpidine D: n = 4; * *trans* isomer

6.6. Tubastrine

Sakai and Higa (1987) isolated tubastrine (**100**), a guanidino styrene compound obtained from the Okinawan coral *Tubastrea aurea* (phylum Coelenterata). For the extract they reported mild activity against HSV-1 and VSV. The structure was assigned as β-(aminoiminomethyl)-amino-3,4-dihydroxystyrene on the basis of spectroscopic data and chemical derivatization. An additional report claimed that tubastrine completely inhibits VSV and HSV-1 in CV-1 cells at 200 μg/0.5 ml (Higa and Sakai, 1988).

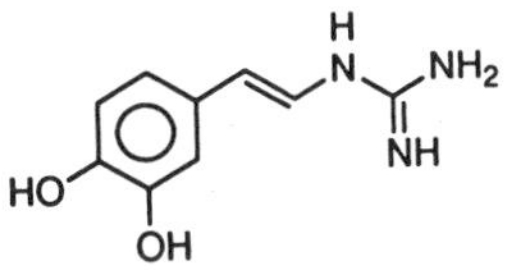

(**100**)

6.7. Saponins

The steroidal glycoside saponins obtained from a variety of starfish exhibit a wide array of activities of biological importance. Shimizu (1971) provided an early example of saponin antiviral activity when he discovered that an extract of the common Atlantic starfish *Asterias forbesi* (phylum Echinodermata) had activity against influenza virus in chick embryos. The purified active components were also obtained from *Acanthaster planci* and *Asterias pectinifera* and found to be polyhydroxylated steroidal glycosides, i.e., asterosaponins (Shimizu, 1971). More recently, Andersson *et al.* (1989) assayed 18 compounds derived from nine starfish and two brittle-stars and identified two polyhydroxylated steroidal glycoside saponins (crossasterosides B and D, **101** and **102**) showing more than a 25% reduction of SHV-1 (Suid herpes virus) plaque formation in porcine kidney-15 cells. The two compounds also showed moderate or weak cytotoxicity and activity against *S. aureus*.

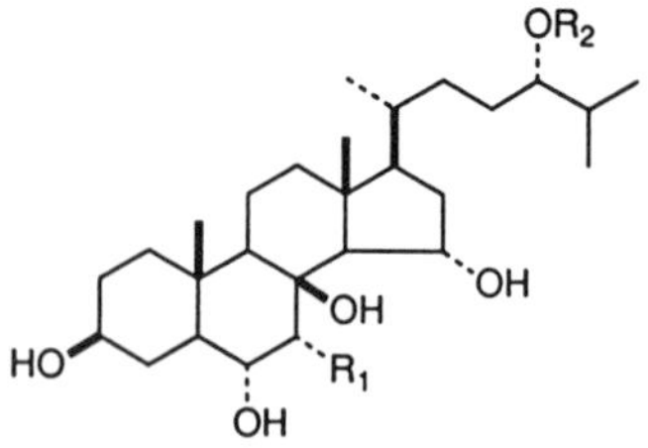

(**101**) Crossasteroside B: R_1 = H, R_2 = 4-O-Me-Xyl$\xrightarrow{1\rightarrow2}$3-O-Me-Xyl-
(**102**) Crossasteroside D: R_1 = OH, R_2 = Xyl$\xrightarrow{1\rightarrow2}$3-O-Me-Xyl-

6.8. BDS-1

Citing unpublished data, Driscoll *et al.* (1989a) reported the presence of antiviral activity in the antihypertensive protein, BDS-I (**103**). They isolated BDS-I from the sea anemone *Anemonia sulcata* (phylum Coelenterata) and determined its three-dimensional solution structure using NMR techniques, including

```
1    5    10    15    20    25    30    35    40
                                        L
A A P C F C S G K P G R G D L W I F R G T C P G G Y G Y T S N C Y K W P N I C C Y P H
```

(**103**)

NOESY, DQF-COSY, HOHAHA, and E-COSY (Driscoll *et al.*, 1989b). BDS-I showed neither the cardiotoxicity nor the neurotoxicity usually associated with sea anemone peptides, but at an unreported concentration it protected mouse liver cells "completely" from the mouse hepatitis virus strain MSV-A59. BDS-I is an approximately 1:1 mixture of the isoproteins (Leu[18])- and (Phe[18])-BDS-I.

6.9. Aplidiasphingosine

Our collection of tunicates from the Gulf of California included an *Aplidium* species (phylum Chordata) from which we isolated aplidiasphingosine (**104**) (Carter and Rinehart, 1978b). The terpenoid structure, identified as a derivative of sphingosine, was assigned on the basis of IR and NMR data. Broad-spectrum antibacterial activity as well as antifungal, antitumor, and modest antiviral activity (HSV-1) were observed.

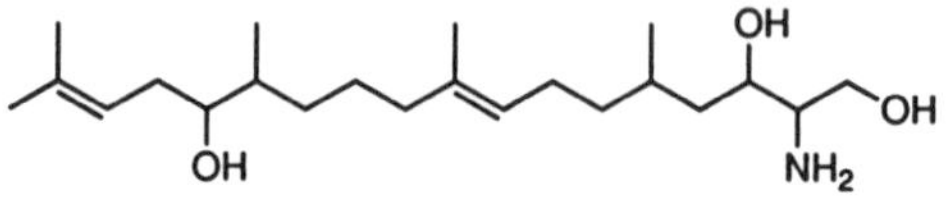

(**104**)

6.10. Cyclohexadienone

Extraction of the red alga *Desmia hornemanni* (phylum Rhodophyta) yielded a series of octodene-type halogenated cyclohexadienones (monoterpenes) identified by IR, MS, and NMR spectral data (Higa, 1985; Higa *et al.*, 1985; Snader and Higa, 1986a). Along with some of their derivatives, the compounds were tested against L1210, HSV-1, and VSV; of these the acetate shown (**105**) was reported to have "potent antiviral activity" against HSV-1 and VSV.

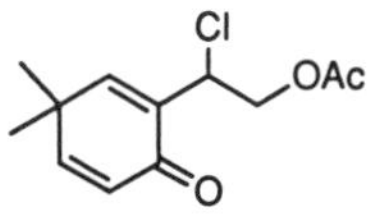

(**105**)

6.11. Reticulatines

Reticulatines A and B (**106** and **107**) were isolated from the Fijian sponge *Fascaplysinopsis reticulata* (phylum Porifera) and were found to be closely related to the known fascaplysins, isolated from the same species (Jiménez *et al.*, 1991). Positive and negative ion FABMS played an important role in assigning the structures of these β-carbolinium salts. The cationic structure was an unusual feature for the molecules, and both **106** and **107** were said to show "potency" in antiviral assays, although no data were provided.

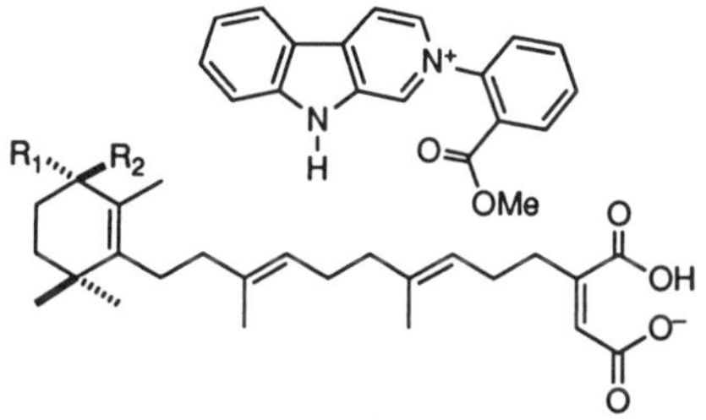

(**106**) Reticulatine A: $R_1 + R_2 = O$
(**107**) Reticulatine B: $R_1 = R_2 = H$

6.12. Chamigrene Derivatives

Snader and Higa (1986b) obtained chamigrene derivatives (e.g., **108**) from the sea hare *Aplysia dactylomela* (phylum Mollusca). Although no data were provided, *in vitro* HSV-1 and VSV inhibitions were claimed.

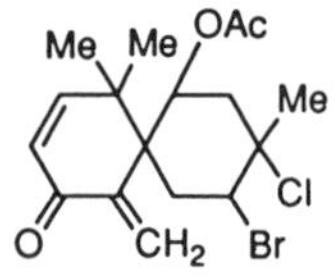

(108)

6.13. Polysaccharides

Carrageenan is a cell-wall polysaccharide constructed from galactose with varying amounts of sulfate substituents and is isolated in large quantity from red algae (phylum Rhodophyta). Samples collected in Senegal, including *Hypnea musciformis, Anatheca montagnei, Agardhiella tenera,* and *Euchema cottonii,* inhibited the activity of yellow fever virus by up to 25.8% (Ferrer-Di Martino *et al.*, 1985). A carrageenan sample obtained commercially (Sigma Chem. Co.) by Gonzalez *et al.* (1987) inhibited HSV-1 cell growth in HeLa cells without becoming cytotoxic when concentrations were maintained as high as 200 μg/ml. Their studies indicated that the time course of HSV-1 infection is a critical factor in determining the success of carrageenan treatment. Neushul (1991) observed activity against HIV in water-soluble substances extracted from *Schizymenia california* and reported that a component of the extract, carrageenan, inhibited reverse transcriptase. The use of marine-derived polysaccharides in the treatment of retroviruses had previously been proposed (Muto *et al.*, 1988).

7. CONCLUSIONS

From the foregoing discussion of antiviral substances found in marine extracts, two general observations stand out. First, antiviral activity is by no means limited to any one class of chemical compounds any more than it is to any one phylum of marine species. Peptides, heterocycles, and terpenes all contribute compounds with confirmed antiviral activity. From this it follows that a number of different mechanisms of action will be found for this disparate collection of compounds. Very little is known about these mechanisms, but studies of modes of action of the compounds should provide an active area of investigation in years to come.

The second generality is less optimistic. Only in very few cases have any of the compounds discussed above been tested *in vivo*, an obvious prerequisite for any attempt to introduce an antiviral agent into the clinic. Moreover, where *in vivo* activity has been measured, the specificities and margins of safety have been relatively narrow. Thus, toxicity seems likely to be a serious problem with most marine-derived drugs as with most other antiviral agents. The one clinically useful compound at present, ara-A, originally resulted from a structure–activity relationship study of arabinosyl nucleosides but was subsequently found in nature.

Although one can envision cases where these antiviral compounds could be used in life-threatening situations, in the main we are still a long way from introducing any marine natural products as marketable antiviral agents. A potential area for introduction of a marine-derived drug would be in treatment of AIDS and perhaps efforts should be increased in this direction.

REFERENCES

Andersson, L., Bohlin, L., Iorizzi, M., Riccio, R., Minale, L., and Moreno-López, W., 1989, Biological activity of saponins and saponin-like compounds from starfish and brittle-stars, *Toxicon* 27:179–188.

Ankel, H., Mittnacht, S., and Jacobsen, H., 1985, Antiviral activity of prostaglandin A on encephalo-myocarditis virus-infected cells: A unique effect unrelated to interferon, *J. Gen. Virol.* 66:2355–2364.

Bader, T., Yamada, Y., and Ankel, H., 1991, Antiviral activity of the prostanoid clavulone II against vesicular stomatitis virus, *Antiviral Res.* 16:341–355.

Barrow, C. J., Blunt, J. W., Munro, M. H. G., and Perry, N. B., 1988a, Variabilin and related compounds from a sponge of the genus *Sarcotragus*, *J. Nat. Prod.* 51:275–281.

Barrow, C. J., Blunt, J. W., Munro, M. H. G., and Perry, N. B., 1988b, Oxygenated furanosesterterpene tetronic acids from a sponge of the genus *Ircinia*, *J. Nat. Prod.* 51:1294–1298.

Barrow, C. J., Blunt, J. W., and Munro, M. H. G., 1989, Autooxidation studies on the marine sesterterpene tetronic acid, variabilin, *J. Nat. Prod.* 52:346–359.

Becker, Y., 1976, *Antiviral Drugs, Mode of Action and Chemotherapy of Viral Infections of Man*, (Monographs in Virology, Volume 11), S. Karger, Basel.

Becker, Y., 1983, *Molecular Virology, Molecular and Medical Aspects of Disease-Causing Viruses of Man and Animals*, M. Nijhoff, The Hague.

Bergmann, W., and Feeney, R. J., 1950, The isolation of a new thymine pentoside from sponges, *J. Am. Chem. Soc.* 72:2809–2810.

Bergmann, W., and Feeney, R. J., 1951, Contributions to the study of marine products. XXXII. The nucleosides of sponges. I, *J. Org. Chem.* 16:981–987.

Blunt, J. W., Hartshorn, M. P., McLennan, T. J., Munro, M. H. G., Robinson, W. T., and Yorke, S. C., 1978, Thyrsiferol: A squalene-derived metabolite of *Laurencia thyrsifera*, *Tetrahedron Lett.* **1978**: 69–72.

Blunt, J. W., Lake, R. J., Munro, M. H. G., and Toyokuni, T., 1987, The stereochemistry of eudistomins C, K, E, F and L, *Tetrahedron Lett.* 28:1825–1826.

Blunt, J. W., Munro, M. H. G., Perry, N. B., and Thompson, A. M., 1989, Preparation of Mycalamides and Their Derivatives As Antitumor and Antiviral Agents, U.S. Patent No. 4,868,204, September 19, 1989 [*Chem. Abstr.* 113:114949y (1990)].

Burres, N. S., Sazesh, S., Gunawardana, G. P., and Clement J. J., 1989, Antitumor activity and nucleic acid binding properties of dercitin, a new acridine alkaloid isolated from a marine *Dercitus* species sponge, *Cancer Res.* 49:5267–5274.

Canonico, P. G., Pannier, W. L., Huggins, J. W., and Rinehart, Jr., K. L., 1982, Inhibition of RNA viruses *in vitro* and in Rift Valley fever-infected mice by didemnins A and B, *Antimicrob. Agents Chemother.* 22:696–697.

Carmely, S., and Kashman, Y., 1985, Structure of swinholide-A, a new macrolide from the marine sponge *Theonella swinhoei*, *Tetrahedron Lett.* **26:**511–514.

Carté, B., and Faulkner, D. J., 1982, Revised structures for the polyandrocarpidines, *Tetrahedron Lett.* **23:**3863–3866.

Carter, G. T., and Rinehart, Jr., K. L., 1978a, Acarnidines, novel antiviral and antimicrobial compounds from the sponge *Acarnus erithacus* (de Laubenfels), *J. Am. Chem. Soc.* **100:**4302–4304.

Carter, G. T., and Rinehart, Jr., K. L., 1978b, Aplidiasphingosine, an antimicrobial and antitumor terpenoid from an *Aplidium* sp. (marine tunicate), *J. Am. Chem. Soc.* **100:**7441–7442.

Cheng, M. T., and Rinehart, Jr., K. L., 1978, Polyandrocarpidines: Antimicrobial and cytotoxic agents from a marine tunicate (*Polyandrocarpa* sp.) from the Gulf of California, *J. Am. Chem. Soc.* **100:** 7409–7411.

Chun, H. G., Davies, B., Hoth, D., Suffness, M., Plowman, J., Flora, K., Grieshaber, C., and Leyland-Jones, B., 1986, Didemnin B. The first marine compound entering clinical trials as an antineoplastic agent, *Invest. New Drugs* **4:**279–284.

Cimino, G., De Stefano, S., and Minale, L., 1974, Oxidized furanoterpenes from the sponge *Spongia officinalis*, *Experientia* **30:**18–20.

Cimino, G., De Rosa, S., and De Stefano, S., 1984, Antiviral agents from a gorgonian, *Eunicella cavolini*, *Experientia* **40:**339–340.

Cohen, S. S., 1966, Introduction to the biochemistry of D-arabinosyl nucleosides, in: *Progress in Nucleic Acid Research and Molecular Biology*, Vol. 5 (J. N. Davidson and W. E. Cohn, eds.), Academic Press, New York, pp. 1–88.

Collier, L. H., and Oxford, J. (eds), 1980, *Developments in Antiviral Therapy*, Academic Press, London.

Corey, E. J., and Ha, D.-C., 1988, Total synthesis of venustatriol, *Tetrahedron Lett.* **29:**3171–3174.

Coval, S. J., Cross, S., Bernardinelli, G., and Jefford, C. W., 1988, Brianthein V, a new cytotoxic and antiviral diterpene isolated from *Briareum asbestinum*, *J. Nat. Prod.* **51:**981–984.

Cross, S. S., and Lewis, T. W., 1987, Development of rapid assay for screening compounds for antiviral activity against RNA viruses, in: *Advances in Experimental Medicine and Biology*, Vol. 28 (Proceedings of the Third International Coronaviruses Symposium, September 14–18, 1986, Asilomar, California), pp. 275–276.

Davies, W. L., Grunert, R. R., Haff, R. F., McGahen, J. W., Neumayer, E. M., Paulshock, M., Watts, J. C., Wood, T. R., Hermann, E. C., and Hoffmann, C. E., 1964, Antiviral activity of 1-adamantanamine (Amantadine), *Science* **144:**862–863.

De Clercq, E., and Walker, R. T., (Eds.), 1984, *Targets for the Design of Antiviral Agents* (NATO Advanced Study Institute on Targets for the Design of Antiviral Agents, 1983, Les Arcs, France), Plenum Press, New York.

De Clercq, E., Krajewska, E., Descamps, J., and Torrence, P. F., 1977, Anti-herpes activity of deoxythymidine analogues: Specific dependence on virus-induced deoxythymidine kinase, *Mol. Pharmacol.* **13:**980–984.

Driscoll, P. C., Clore, G. M., Beress, L., and Gronenborn, A. M., 1989a, A proton nuclear magnetic resonance study of the antihypertensive and antiviral protein BDS-I from the sea anemone *Anemonia sulcata*: Sequential and stereospecific resonance assignment and secondary structure, *Biochemistry* **28:**2178–2187.

Driscoll, P. C., Gronenborn, A. M., Beress, L., and Clore, G. M., 1989b, Determination of the three-dimensional solution structure of the antihypertensive and antiviral protein BDS-I from the sea anemone *Anemonia sulcata*: A study using nuclear magnetic resonance and hybrid distance geometry-dynamical simulated annealing, *Biochemistry* **28:**2188–2198.

Faulkner, D. J., 1973, Variabilin, an antibiotic from the sponge, *Ircinia variabilis*, *Tetrahedron Lett.* **1973:**3821–3822.

Ferrer-Di Martino, M., Ba, D., Kornprobst, J.-M., Combaut, G., and Digoutte, J.-P., 1985, Caracterisation chimique et activité virostatique *in vitro* vis à vis du virus de la fievre jaune de quelques carraghenanes extraits d'algues rouges senegalaises, in: *Vth IUPAC*, Paris, Abstract PA-44.

Frank, K. B., McKernan, P. A., Smith, R. A., and Smee, D. F., 1987, Visna virus as an *in vitro* model for human immunodeficiency virus and inhibition by ribavarin, phosphonoformate, and $2',3'$-dideoxynucleosides, *Antimicrob. Agents Chemother.* **31**:1369–1374.

Gonzalez, A. G., Arteaga, J. M., Fernandez, J. J., Martin, J. D., Norte, M., and Ruano, J. Z., 1984, Terpenoids of the red alga *Laurencia pinnatifida*, *Tetrahedron* **40**:2751–2755.

Gonzalez, M. E., Alarcón, B., and Carrasco, L., 1987, Polysaccharides as antiviral agents: Antiviral activity of carrageenan, *Antimicrob. Agents Chemother.* **31**:1388–1393.

Gosselin, G., Bergogne, M.-C., de Rudder, J., De Clercq, E., and Imbach, J.-L., 1986, Systematic synthesis and biological evaluation of α- and β-xylofuranosyl nucleosides of the five naturally occurring bases in nucleic acids and related analogues, *J. Med. Chem.* **29**:203–213.

Grode, S. H., James, T. R., and Cardellina, II, J. H., 1983a, Brianthein Z, a new polyfunctional diterpene from the gorgonian *Briareum polyanthes*, *Tetrahedron Lett.* **24**:691–694.

Grode, S. H., James, Jr., T. R., Cardellina II, J. H., and Onan, K. D., 1983b, Molecular structures of the briantheins, new insecticidal diterpenes from *Briareum polyanthes*, *J. Org. Chem.* **48**:5203–5207.

Groweiss, A., Look, S. A., and Fenical, W., 1988, Solenolides, new antiinflammatory and antiviral diterpenoids from a marine octocoral of the genus *Solenopodium*, *J. Org. Chem.* **53**:2401–2406.

Gunasekera, S. P., Cross, S. S., Kashman, Y., Lui, M. S., Rinehart, K. L., and Tsujii, S., 1989, Topsentin Compounds Effective Against Viruses and Certain Tumors, U.S. Patent No. 4,866,084, September 12, 1989 [*Chem. Abstr.* **112**:185775d].

Gunawardana, G. P., Kohmoto, S., Gunasekera, S. P., McConnell, O. J., and Koehn, F. E., 1988, Dercitin, a new biologically active acridine alkaloid from a deep water marine sponge, *Dercitus* sp., *J. Am. Chem. Soc.* **110**:4856–4858.

Gunawardana, G. P., Kohmoto, S., and Burres, N. S., 1989, New cytotoxic acridine alkaloids from two deep water marine sponges of the family *Pachastrellidae*, *Tetrahedron Lett.* **30**:4359–4362.

Gustafson, K., Roman, M., and Fenical, W., 1989, The macrolactins, a novel class of antiviral and cytotoxic macrolides from a deep-sea marine bacterium, *J. Am. Chem. Soc.* **111**:7519–7524.

Hager, L. P., White, R. H., Hollenberg, P. F., Doubek, D. L., Brusca, R. C., and Guerrero, R., 1976, A survey of organic halogens in marine organisms, in: *Food-Drugs from the Sea Proceedings 1974* (H. H. Webber and G. D. Ruggieri, eds.), Marine Technology Society, Washington, D.C., pp. 421–428.

Hashimoto, M., Kan, T., Yanagiya, M., Shirahama, H., and Matsumoto, T., 1987, Synthesis of A-B-C-ring segment of thyrsiferol construction of a strained tetrahydropyran ring existent as a boat form, *Tetrahedron Lett.* **28**:5665–5668.

Hashimoto, M., Kan, T., Nozaki, K., Yanagiya, M., Shirahama, H., and Matsumoto, T., 1988, Total syntheses of (+)-thyrsiferol and (+)-venustatriol, *Tetrahedron Lett.* **29**:1143–1144.

Hermann, Jr., E. C., 1965, Antiviral substances, *Ann. N. Y. Acad. Sci.* **130**:1–482.

Hermkens, P. H. H., v. Maarseveen, J. H., Ottenheijm, H. C. J., Kruse, C. G., and Scheeren, H. W., 1990, Intramolecular Pictet–Spengler reaction of *N*-alkoxytryptamines. 3. Stereoselective synthesis of (−)-debromoeudistomin L and (−)-*O*-methyldebromoeudistomin E and their stereoisomers, *J. Org. Chem.* **55**:3998–4006.

Higa, T., 1985, 2-(1-Chloro-2-hydroxyethyl)-4,4-dimethylcyclohexa-2,5-dienone: A precursor of 4,5-dimethylbenzo[*b*]furan from the red alga *Desmia hornemanni*, *Tetrahedron Lett.* **26**:2335–2336.

Higa, T., 1986, Biological activities of marine organisms from Okinawa, in: *Japan–U.S. Seminar on Bio-organic Marine Chemistry, Okinawa, June 30–July 5, 1986*, Abstract V-4, p. 22.

Higa, T., and Sakai, R., 1988, Antiviral Guanidine Derivative Compositions and Their Methods of Use, PCT International Application WO 88 00,181, January 14, 1988; U.S. Patent Application 879,079, June 26, 1986 [*Chem. Abstr.* **109**:104790t].

Higa, T., Sakai, R., Snader, K. M., Cross, S. S., and Theiss, W., 1985, Antitumor and antiviral cyclohexadienones from the red alga *Desmia hornemanni*, *Vth IUPAC*, Paris, Abstract C:22.

Higa, T., Sakemi, S., and Cross, S. S., 1988a, Antiviral Organic Triterpene Compositions and Derivatives, and Their Manufacture from Red Alga, PCT International Application WO 88 00,194, January 14, 1988; U.S. Patent Application 879,092, June 26, 1986 [*Chem. Abstr.* **109:** 21643w].

Higa, T., Sakai, R., and Lui, M. S., 1988b, Antibiotic and Antitumor Misakinolide Compositions and Their Derivatives, PCT International Application WO 88 00,195, January 14, 1988 [*Chem. Abstr.* **111:**17702p].

Higa, T., Sakemi, S., and Cross, S. S., 1989, Isolation of Onnamide A Derivatives as New Antiviral, Antitumor and Antifungal Agents, European Patent Application EP 299,713, January 18, 1989; U.S. Patent Application 74,977, July 17, 1987 [*Chem. Abstr.* **111:**167390z].

Ichiba, T., Yoshiba, W. Y., and Scheuer, P. J., 1991, Hennoxazoles: Bioactive bisoxazoles from a marine sponge, *J. Am. Chem. Soc.* **113:**3173–3174.

Jares-Erijman, E. A., Sakai, R., and Rinehart, K. L., 1991, Crambescidins, new antiviral and cytotoxic compounds from the sponge *Crambe crambe J. Org. Chem.* **56:**5712–5715.

Jiménez, C., Quiñoá, E., and Crews, P., 1991, Novel marine sponge alkaloids. 3. β-carbolinium salts from *Fascaplysinopsis reticulata*, *Tetrahedron Lett.* **32:**1843–1846.

Kashman, Y., Hirsch, S., Koehn, F., and Cross, S., 1987, Reiswigins A and B, novel antiviral diterpenes from a deepwater sponge, *Tetrahedron Lett.* **28:**5461–5464.

Kashman, Y., Hirsch, S., McConnell, O. J., Ohtani, I., Kusumi, T., and Kakisawa, H., 1989a, Ptilomycalin A: A novel polycyclic guanidine alkaloid of marine origin, *J. Am. Chem. Soc.* **111:** 8925–8926.

Kashman, Y., Hirsch, S., Cross, S. S., and Koehn, F., 1989b, Antiviral Compositions Derived from Marine Sponge *Epipolasis reiswigi* and Their Methods of Use, European Patent Application EP 306,282, March 8, 1989; U.S. Patent Application 91,078, August 31, 1987 [*Chem. Abstr.* **111:** 140473s].

Kato, Y., Fusetani, N., Matsunaga, S., Hashimoto, K., Sakai, R., Higa, T., and Kashman, Y., 1987, Antitumor macrodiolides isolated from a marine sponge *Theonella* sp.: Structure revision of misakinolide A, *Tetrahedron Lett.* **28:**6225–6228.

Kazlauskas, R., Murphy, P. T., Quinn, R. J., and Wells, R. J., 1976, Tetradehydrofurospongin-1, a new C-21 furanoterpene from a sponge, *Tetrahedron Lett.* **16:**1331–1332.

Kazlauskas, R., Murphy, P. T., Wells, R. J., Noack, K., Oberhänsli, W. E., and Schönholzer, P., 1979, A new series of diterpenes from Australian *Spongia* species, *Aust. J. Chem.* **32:**867–880.

Keifer, P. A., Schwartz, R. E., Koker, M. E. S., Hughes, Jr., R. G., Rittschof, D., and Rinehart, K. L., 1991, Bioactive bromopyrrole metabolites from the Caribbean sponge *Agelas conifera*, *J. Org. Chem.* **56:**2965–2975.

Kikuchi, H., Tsukitani, Y., Iguchi, K., and Yamada, Y., 1982, Clavulones, new type of prostanoids from the stolonifer *Clavularia viridis* Quoy and Gaimard, *Tetrahedron Lett.* **23:**5171–5174.

Kinzer, K. F., and Cardellina II, J. H., 1987, Three new β-carbolines from the Bermudian tunicate *Eudistoma olivaceum*, *Tetrahedron Lett.* **28:**925–926.

Kirkup, M. P., Shankar, B. B., McCombie, S., Ganguly, A. K., and McPhail, A. T., 1989, A concise route to the oxathiazepine containing eudistomin skeleton and some carba-analogs, *Tetrahedron Lett.* **30:**6809–6812.

Knübel, G., Larsen, L. K., Moore, R. E., Levine, I. A., and Patterson, G. M. L., 1990, Cytotoxic, antiviral indolocarbazoles from a blue-green alga belonging to the Nostocaceae, *J. Antibiot.* **43:** 1236–1239.

Kobayashi, J., Harbour, G. C., Gilmore, J., and Rinehart, Jr., K. L., 1984, Eudistomins A, D, G, H, I, J, M, N, O, P, and Q, bromo-, hydroxy-, pyrrolyl-, and 1-pyrrolinyl-β-carbolines from the antiviral Caribbean tunicate *Eudistoma olivaceum*, *J. Am. Chem. Soc.* **106:**1526–1528.

Kobayashi, J., Cheng, J., Ohta, T., Nozoe, S., Ohizumi, Y., and Sasaki, T., 1990, Eudistomidins B, C, and D: Novel antileukemic alkaloids from the Okinawan marine tunicate *Eudistoma glaucus*, *J. Org. Chem.* **55**:3666–3670.

Kobayashi, M., Yasuzawa, T., Yoshihara, M., Akutsu, H., Kyogoku, Y., and Kitagawa, I., 1982, Four new prostanoids: Claviridenone-A, -B, -C, and -D, *Tetrahedron Lett.* **23**:5331–5334.

Koehn, F. E., Gunasekera, S. P., Neal, D. N., and Cross, S. S., 1991, Halitunal, an unusual diterpene aldehyde from the marine alga *Halimeda tuna*, *Tetrahedron Lett.* **32**:169–172.

Kohmoto, S., McConnell, O. J., Wright, A., and Cross, S., 1987, Isospongiadiol, a cytotoxic and antiviral diterpene from a Caribbean deep water marine sponge, *Spongia* sp., *Chem. Lett.* **1987**: 1687-1690.

Kohmoto, S., McConnell, O. J., and Cross, S. S., 1988, Antitumor and Antiviral Furanoditerpenoids from a Marine Sponge, European Patent Application EP 285,301, October 5, 1988; U.S. Patent Application 30,727, March 25, 1987 [*Chem. Abstr.* **111**:50424x].

Kolberg, R., 1991, Critics call for a smarter way to screen for drugs, *J. NIH Res.* **3**:25–26.

Lake, R. J., Brennan, M. M., Blunt, J. W., Munro, M. H. G., and Pannell, L. K., 1988a, Eudistomin K sulfoxide–An antiviral sulfoxide from the New Zealand ascidian *Ritterella sigillinoides*, *Tetrahedron Lett.* **29**:2255–2256.

Lake, R. J., McCombs, J. D., Blunt, J. W., Munro, M. H. G., and Robinson, W. T., 1988b, Eudistomin K: Crystal structure and absolute stereochemistry, *Tetrahedron Lett.* **29**:4971–4972.

Mandell, G. L., Douglas, Jr., R. G., and Bennett, J. E. (eds.), 1985, *Antiinfective Therapy*, Wiley, New York.

Minale, L., Riccio, R., and Sodano, G., 1974, Avarol, a novel sesquiterpenoid hydroquinone with a rearranged drimane skeleton from the sponge *Disidea avara*, *Tetrahedron Lett.* **1974**:3401–3404.

Munro, M. H. G., Luibrand, R. T., and Blunt, J. W., 1987, The search for antiviral and anticancer compounds from marine organisms, in: *Bioorganic Marine Chemistry*, Vol. 1 (P. J. Scheuer, ed.), Springer-Verlag, Berlin, pp. 93–176.

Munro, M. H. G., Perry, N. B., and Blunt, J. W., 1988, Isolation and Testing of the *Mycale* Metabolite Mycalamide As a Virucide and Neoplasm Inhibitor, European Patent Application EP 289,203, November 2, 1988; U.S. Patent Application 43,700, April 29, 1987 [*Chem. Abstr.* **110**:88610x].

Munro, M. H. G., Blunt, J. W., Barns, G., Battershill, C. N., Lake, R. J., and Perry, N. B., 1989, Biological activity in New Zealand marine organisms, *Pure Appl. Chem.* **61**:529–534.

Muto, S., Nimura, K., Oohara, M., Oguchi, Y., Matsunaga, K., Hirose, K., Kakuchi, J., Sugita, N., and Furusho, T., 1988, Polysaccharides from Marine Algae and Antiviral Drugs Containing the Same As Active Ingredient, European Patent Application EP 295,956, December 21, 1988; Japanese Patent Application 87/152,086, June 18, 1987 [*Chem. Abstr.* **111**:54116w].

Nakagawa, M., Liu, J.-J., and Hino, T., 1989, Total synthesis of (−)-eudistomin L and (−)-debromoeudistomin L, *J. Am. Chem. Soc.* **111**:2721–2722.

Neushul, M., 1991, Antiviral carbohydrates from marine red algae, in: *Bioactive Compounds from Marine Organisms*, (M.-F. Thompson, R. Sarojini, and R. Nagabhushanam, eds.), Oxford & IBH Publishing Co. Pvt. Ltd., New Delhi, India, pp. 275–281.

Noyori, R., Suzuki, M., and Kurozumi, S., 1987a, Preparation of Punaglandin Derivatives, Japanese Kokai Patent No. 62,059,258, March 14, 1987 [*Chem. Abstr.* **107**:39505w].

Noyori, R., Suzuki, M., Morita, Y., and Yanagisawa, A., 1987b, Preparation of Punaglandin Derivatives, Japanese Kokai Patent No. 62,207,254, September 11, 1987 [*Chem. Abstr.* **108**: 221488r].

Perry, N. B., Battershill, C. N., Blunt, J. W., Fenwick, G. D., Munro, M. H. G., and Bergquist, P. R., 1987, Occurrence of variabilin in New Zealand sponges of the order Dictyoceratida, *Biochem. Syst. Ecol.* **15**:373–376.

Perry, N. B., Blunt, J. W., Munro, M. H. G., and Pannell, L. K., 1988, Mycalamide A, an antiviral compound from a New Zealand sponge of the genus *Mycale*, *J. Am. Chem. Soc.* **110**:4850–4851.

Perry, N. B., Blunt, J. W., Munro, M. H. G., and Thompson, A. M., 1990, Antiviral and antitumor agents from a New Zealand Sponge, *Mycale* sp. 2. Structures and solution conformations of mycalamides A and B, *J. Org. Chem.* **55**:223–227.

Rinehart, K. L., 1988a, Screening to detect biological activity, in: *Biomedical Importance of Marine Organisms* (Memoirs of the California Academy of Sciences Number 13; D. G. Fautin, ed.), California Academy of Sciences, San Francisco, pp. 13–22.

Rinehart, K. L., 1988b, Didemnin and its biological properties, in: *Peptides, Chemistry and Biology* (Proceedings of the Tenth American Peptide Symposium; G. R. Marshall, ed.), ESCOM, Leiden, pp. 626–631.

Rinehart, K. L., 1988c, Bioactive metabolites from the Caribbean Sponge *Agelas coniferin*, U.S. Patent 4,737,510, April 12, 1988 [*Chem. Abstr.* **109**:216002u].

Rinehart, K. L., 1989, Biologically active marine natural products, *Pure Appl. Chem.* **61**:525–528.

Rinehart, K. L., 1990, Novel Anti-Viral and Cytotoxic Agents, U.S. Patent Application P-82,663, December 13, 1990; British Patent Application 8922026.3, September 29, 1989.

Rinehart, K. L., 1992, Antiviral agents from novel marine and terrestrial sources, in: *Innovations in Antiviral Development and the Detection of Virus Infections* (L. R. Walsh, T. M. Block, R. L. Crowell, and D. L. Jungkind, eds.), Plenum Press, New York, pp. 41–60.

Rinehart, Jr., K. L., and Shield, L. S., 1983, In search of tunicates: Source of an antitumor compound, *Aquasphere J. N. Engl. Aquarium* **17**:8–13.

Rinehart, Jr., K. L., Johnson, R. D., Paul, I. C., McMillan, J. A., Siuda, J. F., and Krejcarek, G. E., 1976, Identification of compounds in selected marine organisms by gas chromatography-mass spectrometry, field desorption mass spectrometry, and other physical methods, in: *Food-Drugs from the Sea Conference Proceedings 1974* (H .H. Webber and G. D. Ruggieri, eds.), Marine Technology Society, Washington, D.C., pp. 434–442.

Rinehart, Jr., K. L., Shaw, P. D., Shield, L. S., Gloer, J. B., Harbour, G. C., Koker, M. E. S., Samain, D., Schwartz, R. E., Tymiak, A. A., Weller, D. L., Carter, G. T., Munro, M. H. G., Hughes, Jr., R. G., Renis, H. E., Swynenberg, E. B., Stringfellow, D. A., Vavra, J. J., Coats, J. H., Zurenko, G. E., Kuentzel, S. L., Li, L. H., Bakus, G. J., Brusca, R. C., Craft, L. L., Young, D. N., and Connor, J. L., 1981a, Marine natural products as sources of antiviral, antimicrobial, and antineoplastic agents, *Pure Appl. Chem.* **53**:795–817.

Rinehart, Jr., K. L., Gloer, J. B., Cook, Jr., J. C., Mizsak, S. A., and Scahill, T. A., 1981b, Structures of the didemnins, antiviral and cytotoxic depsipeptides from a Caribbean tunicate, *J. Am. Chem. Soc.* **103**:1857–1859.

Rinehart, Jr., K. L., Gloer, J. B., Hughes, Jr., R. G., Renis, H. E., McGovren, J. P., Swynenberg, E. B., Stringfellow, D. A., Kuentzel, S. L., and Li, L. H., 1981c, *Science* **22**:933–935.

Rinehart, Jr., K. L., Harbour, G. C., Graves, M. D., and Cheng, M. T., 1983a, Synthesis of hexahydropolyandrocarpidine (a revised structure), *Tetrahedron Lett.* **1983**:1593–1596.

Rinehart, Jr., K. L., Gloer, J. B., Wilson, G. R., Hughes, Jr., R. G., Li, L. H., Renis, H. E., McGovren, J. P., 1983b, Antiviral and antitumor compounds from tunicates, *Fed. Proc.* **42**: 87–90.

Rinehart, Jr., K. L., Kobayashi, J., Harbour, G. C., Hughes, Jr., R. G,. Mizsak, S. A., and Scahill, T. A., 1984, Eudistomins C, E, K, and L, potent antiviral compounds containing a novel oxathiazepine ring from the Caribbean tunicate *Eudistoma olivaceum*, *J. Am. Chem. Soc.* **106**:1524–1526.

Rinehart, Jr., K. L., Harbour, G. C., and Kobayashi, J., 1986, Antiviral Eudistomins from a Marine Tunicate, U.S. Patent No. 4,631,149, December 23, 1986; European Patent Application EP 133,000, February 13, 1985 [*Chem. Abstr.* **102**:226023w].

Rinehart, K. L., Kishore, V., Nagarajan, S., Lake, R. J., Gloer, J. B., Bozich, F. A., Li, K.-M., Maleczka, Jr., R. E., Todsen, W. L., Munro, M. H. G., Sullins, D. W., and Sakai, R., 1987a, Total synthesis of didemnins A, B, and C, *J. Am. Chem. Soc.* **109**:6846–6848.

Rinehart, Jr., K. L., Kobayashi, J., Harbour, G. C., Gilmore, J., Mascal, M., Holt, T. G., Shield, L. S., and Lafargue, F., 1987b, Eudistomins A-Q, β-carbolines from the antiviral Caribbean tunicate *Edistoma olivaceum*, *J. Am. Chem. Soc.* **109:**3378–3387.

Rinehart, K. L., Holt, T. G., Fregeau, N. L., Keifer, P. A., Wilson, G. R., Perun, Jr., T. J., Sakai, R., Thompson, A. G., Stroh, J. G., Shield, L. S., Seigler, D. S., Li, L. H., Martin, D. G., Grimmelikhuijzen, C. J. P., and Gäde, G., 1990a, Bioactive compounds from aquatic and terrestrial sources, *J. Nat. Prod.* **53:**771–792.

Rinehart, K. L., Sakai, R., Holt, T. G., Fregeau, N. L., Perun, Jr., T. J., Seigler, D. S., Wilson, G. R., and Shield, L. S., 1990b, Biologically active natural products, *Pure Appl. Chem.* **62:**1277–1280.

Rinehart, K. L., Sakai, R., Stroh, J. G., 1990c, Novel Cytotoxic Cyclic Depsipeptides from the Tunicate *Trididemnum solidum*, U.S. Patent No. 4,948,791, August 14, 1990 [*Chem. Abstr.* **114:** 214413h].

Rothschild, H., Allison, Jr., F., and Howe, C., 1978, *Human Diseases Caused by Viruses. Recent Developments*, Oxford University Press, Oxford.

Sakai, R., 1991, Biologically Active Compounds from Tunicates and a Sponge. Ph.D. Thesis, University of Illinois, Urbana, Illinois.

Sakai, R., and Higa, T., 1987, Tubastrine, a new guanidinostyrene from the coral *Tubastrea aurea*, *Chem. Lett.* **1987:**127–128.

Sakai, R., Higa, T., and Kashman, Y., 1986, Misakinolide-A, an antitumor macrolide from the marine sponge *Theonella* sp., *Chem. Lett.* **1986:**1499–1502.

Sakemi, S., Higa, T., Jefford, C. W., and Bernardinelli, G., 1986, Venustatriol, a new, anti-viral, triterpene tetracyclic ether from *Laurencia venusta*, *Tetrahedron Lett.* **27:**4287–4290.

Sakemi, S., Ichiba, T., Kohmoto, S., Saucy, G., and Higa, T., 1988, Isolation and structure elucidation of onnamide A, a new bioactive metabolite of a marine sponge, *Theonella* sp., *J. Am. Chem. Soc.* **110:**4851–4853.

Santoro, M. G., Benedetto, A., Carruba, G., Garaci, E., and Jaffe, B. M., 1980, Prostaglandin A compounds as antiviral agents, *Science* **209:**1032–1034.

Sarin, P. S., Sun, D., Thornton, A., and Müller, W. E. G., 1987, Inhibition of replication of the etiologic agent of acquired immune deficiency syndrome (human T-lymphotropic retrovirus/lymphadenopathy-associated virus) by avarol and avarone, *J. Natl. Cancer Inst.* **78:**663–666.

Schroeder, A. C., Hughes, Jr., R. G., and Block, A., 1981, Synthesis and biological effects of acyclic pyrimidine nucleoside analogues, *J. Med. Chem.* **24:**1078–1083.

Shaw, P. D., McClure, W. O., Van Blaricom, G., Sims, J., Fenical, W., and Rude, J., 1976, Antimicrobial activities from marine organisms, in: *Food-Drugs from the Sea 1974* (H. H. Webber and G. D. Ruggieri, eds.), Marine Technology Society, Washington, D.C., pp. 429–433.

Shimizu, Y., 1971, Antiviral substances in starfish, *Experientia* **27:**1188–1189.

Snader, K. M., and Higa, T., 1986a, Antiviral and Antitumor Cyclohexadienone Compositions, PCT International Application WO 86 03,738, July 3, 1986; U.S. Patent Application 682,278, December 17, 1984 [*Chem. Abstr.* **105:**150026p].

Snader, K. M., and Higa, T., 1986b, Antiviral Chamigrene Derivative, PCT International Application WO 86 03,739, July 3, 1986; U.S. Patent Application 682,896, December 18, 1984 [*Chem. Abstr.* **106:**12959q].

Stephen, E. L., Jones, D. E., Peters, C. J., Eddy, G. A., Loizeaux, P. S., and Jahrling, P. B., 1980, Ribavirin treatment of toga-, arena- and bunyavirus infections in subhuman primates and other laboratory animal species, in: *Ribavirin: A Broad Spectrum Antiviral Agent* (R. A. Smith and W. Kirkpatrick, eds.), Academic Press, New York, pp. 169–183.

Stierle, D. B., Carté, B., Faulkner, D. J., Tagle, B., and Clardy, J., 1980, The asbestinins, a novel class of diterpenes from the gorgonian *Briareum asbestinum*, *J. Am. Chem. Soc.* **102:**5088–5092.

Still, I. W. J., and Strautmanis, J. R., 1989, Synthesis of N(10)-acetyleudistomin L, *Tetrahedron Lett.* **30:**1041–1044.

Suzuki, T., Suzuki, M., Furusaki, A., Matsumoto, T., Kato, A., Imanaka, Y., and Kurosawa, E., 1985, Teurilene and thyrsiferyl 23-acetate, *meso* and remarkably cytotoxic compounds from the marine red alga *Laurencia obtusa* (Hudson) Lamouroux, *Tetrahedron Lett.* **26:**1329–1332.

Suzuki, T., Takeda, S., Suzuki, M., Kurosawa, E., Kato, A., and Imanaka, Y., 1987, Cytotoxic squalene-derived polyethers from the marine red alga *Laurencia obtusa* (Hudson) Lamouroux, *Chem. Lett.* **1987:**361–364.

Tsujii, S., Rinehart, K. L., Gunasekera, S. P., Kashman, Y., Cross, S. S., Lui, M. S., Pomponi, S. A., and Diaz, M. C., 1988, Topsentin, bromotopsentin, and dihydrodeoxybromotopsentin: Antiviral and antitumor bis(indolyl)imidazoles from Caribbean deep-sea sponges of the family Halichondriidae. Structural and synthetic studies, *J. Org. Chem.* **53:**5446–5453.

Walker, R. P., Faulkner, D. J., Van Engen, D., and Clardy, J., 1981, Sceptrin, an antimicrobial agent from the sponge *Agelas sceptrum*, *J. Am. Chem. Soc.* **103:**6772–6773.

Walker, R. T., De Clercq, E., and Eckstein, F. (eds.), 1979, *Nucleoside Analogues. Chemistry, Biology, and Medical Applications* (NATO Advanced Study Institute on Nucleoside Analogues, 1979, Urbino), Plenum Press, New York.

Wright, A. E., McCarthy, P., Cross, S. S., Rake, J. B., and McConnell, O. J., 1988, Sesquiterpenoid Isocyanide Purification from a Marine Sponge and Its Use As a Neoplasm Inhibitor, Virucide, and Fungicide, European Patent Application EP 285,302, October 5, 1988; U.S. Patent Application 32,289, March 30, 1987 [*Chem. Abstr.* **111:**50414u].

9

The Search for Antiparasitic Agents from Marine Animals

Phil Crews and Lisa M. Hunter

1. INTRODUCTION

Infectious diseases caused by parasites occur in people (and animals) through-out the world. It is well known that improperly prepared pork can cause trichi-nosis, an intestinal infection by the helminth *Trichinella spiralis*. Hookworm disease is widespread throughout nearly all tropical and subtropical countries and is an infection of the small intestine by the helminth *Necator americanus*. Campers often see warning signs that water in a nearby stream could be contaminated with the protozoan *Giardia lamblia*. Travelers to the deep tropics (as well as the natives) are advised to use prophylaxis to protect against malarial infections from *Plasmodium* parasites. These examples vividly illustrate the ubiquity of parasite diseases, which, in most cases, can be controlled by chemotherapy (Gutteridge, 1989b). There are well-known examples of clinically active antiparasitic agents derived from plant and fermentation sources (Hart *et al.*, 1989; James and Gilles, 1985). Just beginning to emerge are marine natural products with activity against helminth and protozoal organisms. These will be the subject of our review.

The largest number of chemically diverse secondary marine metabolites

Phil Crews and Lisa M. Hunter • Department of Chemistry and Biochemistry, and Institute of Marine Sciences, University of California, Santa Cruz, California 95064.

Marine Biotechnology, Volume 1: Pharmaceutical and Bioactive Natural Products, edited by David H. Attaway and Oskar R. Zaborsky. Plenum Press, New York, 1993.

emanate from seaweeds, sponges, and coelenterates (Faulkner, 1991, and references therein). Alternatively, tunicates, bryozoans, and echinoderms have also yielded novel metabolites (Ireland *et al.*, 1988). Both shallow- and deep-water plants and invertebrates (Reed and Pomponi, 1989) as well as cultured organisms (Moore *et al.*, 1988) are important sources of compounds with potency against important disease targets such as cancer (Munro *et al.*, 1987), viruses (Munro *et al.*, 1987), microbes (Faulkner, 1978), or fungi (Fusetani, 1988), whereas few such metabolites have been evaluated against parasites (indeed, no review has appeared on this subject).

A quick look at the highlights of past work on marine bioactive substances provides an interesting perspective. The first examples of a bioactivity-directed study of marine organisms appeared in the 1950s and included Nigrelli's (1959) work on sponges and Burkholder and Burkholder's (1958) studies on corals. The current literature shows that marine antimicrobial agents are still being extensively isolated (Jiménez and Crews, 1991). Poisonous marine animals and plants represent another obvious source of bioactive substances; recent announcements of the labyrinthine structures of palytoxin (Moore, 1985) and of ciguatoxin (Murata *et al.*, 1990) culminate work begun in the early 1960s. Enduring research efforts begun in the 1970s have revealed important anticancer agent leads such as didemnin B (Rinehart *et al.*, 1981, 1987; Li *et al.*, 1990), bryostatin 1 (Pettit *et al.*, 1984; Suffness *et al.*, 1989), and dolastatin 10 (Pettit *et al.*, 1987), which are in, or about to enter, advanced stages of clinical development. A number of interesting anti-inflammatory substances were identified during the 1980s. Heading this list is a sesterterpenoid, manoalide, whose inhibition of phospholipase A_2 renders it an important tool for the study of both inflammation mechanisms and the role of Ca^{2+} as a secondary messenger (Mayer and Jacobs, 1988). Kainic acid (**1**) is the singular example of a clinically active antiparasitic agent based on a marine natural product. It is listed in the *Merck Index* as both an antiparasitic agent against nematodes and a useful tool in neurobiology. Kainic acid and related compounds were first isolated during the 1950s from the red alga *Digenea simplex*. A single dose of kainic acid at 5–10 mg/adult affords up to a 70% reduction of parasitic intestinal worms, while the related lactone **2** is *in vitro* inactive (Baslow, 1977).

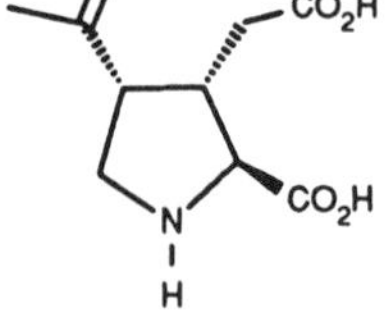

(1) 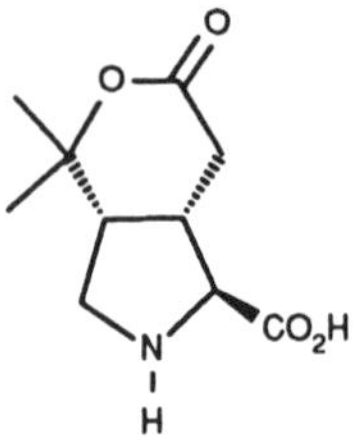   (2)

Until the middle 1980s there were no other *in vitro* active antiparasitic marine natural products described in the literature.

In 1984 our group at the University of California at Santa Cruz (UCSC) established a collaboration with Dr. Tom Matthews and his co-workers at the Institute of Antiviral and Antimicrobial Chemotherapy of Syntex Research, Inc., Palo Alto, California. Our aim was to use antiparasitic primary screens to identify the most promising among hundreds of sponge extracts that were either in hand or that would be obtained from Indo-Pacific organisms. A number of antiparasitic-active compounds were subsequently purified, many possessed novel chemotypes, and some of their properties have been described in both the patent and primary chemical literature. In this review we will combine published literature with unpublished results so that the diversity of active (and inactive) structures can be appreciated. Furthermore, it is our desire to stimulate future research on antiparasitics from marine organisms by indicating the extent of knowledge gaps and by citing case examples which show that *in vivo* active antiparasitic natural products can be discovered.

1.1. Terminology and Scope of Parasitic Diseases

Most commonly, the appellation "parasitic disease" denotes invasion by worms (helminths) or single-celled organisms (protozoans). The helminths are responsible for diseases such as roundworm, hookworm, and tapeworm. The various types of helminth parasitic diseases, grouped by causative organism, are summarized in Table I. River blindness (onchocerciasis) represents a notable example of an ancient but still problematic helminth disease. Protozoans cause a wide range of diseases and the most important of these are listed in Table II. Malaria and giardiasis represent protozoal diseases that are a problem beyond the Third World.

The actual number of people affected by parasitic diseases is difficult to assess. Individuals may suffer anything from severe to mild symptoms, or may be asymptomatic. Nonetheless, the minimum cases which are in the literature are staggering. For example, an estimated 800 million people are infected with hookworm; some 18 million suffer river blindness and more than 80 million more are threatened; there are at least 100 million cases of malaria worldwide; and the World Health Organization estimates that 43% of the world population lives in malarious areas.

1.2. General Approaches to Combatting Parasitic Diseases

There are several important strategies used to combat helminth or protozoal infections. Prevention of infection constitutes a major tactic. Effective preventa-

Table I. *Prevalent Helminth Parasitic Diseases*

Disease	Causative parasite	Chemotherapeutic of choice[a]
Roundworms (nematodes) intestinal		
Roundworm	*Ascaris lumbricoides*	Mebendazole
Whipworm	*Trichuris trichiura*	Mebendazole
Pinworm	*Enterobius vermicularis*	Mebendazole
Strongyloidiasis	*Strongloides stercoralis*	Thiabendazole
Roundworm (nematodes) tissue		
Filariases	*Wuchereria bancrofti*	Diethylcarbamizine
	Grugia (W.) *malayi*	
	Mansonella ozzardi	
	Loa loa	
River blindness	*Onchocerca volvulus*	Ivermectin
Guinea worm	*Dracunculus medinensis*	None satisfactory[b]
Trichinosis	*Trichinella sprialis*	Thiabendazole[c]
Visceral larva migrans	*Toxocara* sp.	Diethylcarbamizine[c] or thiabendazole
Angiostronglyliasis	*Angiostrongylus* sp.	None satisfactory[d]
Anisakiasis	*Anasakis* sp.	None[d]
Flatworms		
Tapeworm (cestodes)	*Taenia saginata* (beef)	Niclosamide or praziquantel[c]
	Taenia solium pork)	
	Diphyllobothrium latum (fish)	
Hydatid disease	*Echinocuccus* sp.	None satisfactory[d]
Dwarf tapeworm	*Hymenolepis nana*	Praziquantel[c]
Flukes (trematodes)		
Schistosomiasis	*Schistosoma* sp.	Praziquantel
Intestinal fluke	*Fasciola buski*	Praziquantel
Chinese liver fluke	*Clonorchis sinensis*	Praziquantel
Lung fluke	*Paragonimus westermani*	Praziquantel
Sheep liver fluke	*Fasciola hepatica*	Bithional

[a]Medical letter (1988).
[b]Cook (1990).
[c]Drug is still in investigative stage for treatment of humans.
[d]Cook (1991).

tive measures include the use of repellents to rebuff parasite-carrying organisms (e.g., mosquito lotion), proper cooking of foods which may harbor parasites, and appropriate hygiene practices. The development of vaccines represents a desirable way to protect populations in areas where parasitic vectors are present in large numbers. Unfortunately, this approach has been unsuccessful to date in part because fundamental immunological information is still being collected (Agabian and Cerami, 1990; Gutteridge, 1989a). Alternatively, chemotherapy has been a long-standing effective instrument for battling parasitic infections; in many cases insights gained from the structures of natural products are of key importance. For

Table II. Prevalent Protozoal Parasitic Diseases

Disease	Causitive parasite	Chemotherapeutic of choice[a]
Sleeping sickness	*Trypanosoma cruzi*	Nifurimox
Chagas' disease	*Trypanosoma gambiense*	Suramin
	Trypanosoma rhodesiense	Late disease: melarsopol
Kala-azar, espundia	*Leishmania* spp.	Stibogluconate sodium
Trichomoniasis	*Trichomonas vaginalis*	Metronidazole
Giardiasis	*Giardia lamblia*	Quinacrine HCl
Amoebiasis	*Entamoeba histolyica*	Iodoquinol, metronidazole
Amoebic meningoencephalitis	*Naegleria fowleri*	Amphotericin B[b]
Toxoplasmosis	*Toxoplasma gondii*	Pyrimethamine, trisulfapyrimidines
Malaria	*Plasmodium* spp.	Mefloquine, chloroquine, artemisinin[b]
Babesiosis	*Babesia* spp.	Clindamycin,[b] quinine
Pneumocystis pneumonia	*Pneumocystis carinii*	Trimethoprimsulfamethoxazole
Balantidiasis	*Balatidium coli*	Tetracycline[b]

[a]Medical letter (1988).
[b]Drug is still in investigative stage for treatment of humans.

example, the use of chloroquine, whose structure is parallel to that of quinine from Cinchona bark, prevents the development of blood parasitemia (malaria prophylaxis). The simple synthetic compound O-11, an analog of myristic acid, exhibits selective toxicity to African trypanosomas (sleeping sickness) because it disrupts the ability of the parasite to uptake critically needed myristates from its host (Doering *et al.*, 1991).

The disruption of mechanisms that are molecular target-oriented rather than the use of toxicity-oriented compounds represents another antiparasite chemotherapy approach. In this context, verapamil (Martin *et al.*, 1987) is under study as a potential means to reverse chloroquine resistance. It is believed that this Ca channel blocker might act against a multidrug resistance gene which is amplified in some *Plasmodium falciparum* (malaria) and has been successfully cloned (Foote *et al.*, 1990). Agents such as suramin, used to treat African Chagas' disease, may be potent because of action on a specific molecular target. Suramin blocks the cell surface binding of a number of important growth factors, such as PDGF (platelet-derived growth factor) or TGF-β (transforming growth factor) (Powis, 1991). The anthelmintic ivermectin is an agent which operates by yet a different mechanism. Its potency against nematodes occurs by disruption of the GABA-mediated transmission of nerve signals (Campbell, 1985).

Although therapeutic treatment of parasitic diseases has made considerable progress in the last few decades, there are many unsolved problems. The drugs used for treatment often have adverse side effects, and in some cases complete eradication of the parasite is not guaranteed (Cook, 1991). In other words, new chemotherapeutic tools are needed.

2. SCREENING METHODS FOR ANTHELMINTICS

2.1. The Disease Targets

The parasitic worms responsible for helminth diseases belong to two major groups: roundworms (nematodes) and flatworms (platyhelminths), as summarized in Table I. The flatworms can be further divided into tapeworms (cestodes) and flukes (trematodes), as shown in Table III. Infection occurs by transmission of eggs or larvae, which develop into adult worms within their host. The adult form reproduces and releases its eggs or larvae into the environment.

2.2. Bioassay Strategies

Biological screens used by different investigators for evaluating potential anthelmintic agents often employ a variety of targets. However, the general approach is usually the same. Initial evaluations involve a primary *in vitro* screen with two or more helminth targets. Compounds which are active are next advanced to a secondary animal screen which often uses two or more parasite targets. Compounds active in primary and secondary screens are then progressed to more advanced testing to consider issues of toxicity, pharmacokinetics, and bio-availability. It would also be of importance to obtain positive results in *in vitro* tests which employ the actual disease organisms. Our discovery program in collaboration with Syntex utilized the parasite models denoted by a # in Table III.

Some major obstacles arise in the approach outlined above. One important factor is that parasitic helminths (and protozoans) are complex organisms with highly specialized life cycles, which means that *in vitro* studies have limited applicability to clinical use. Alternatively, studies of the isolated parasite provide valuable information about fundamental biochemical properties. A difficulty of *in vitro* studies is establishing proper culture conditions.

Table III. **Examples of Parasites Used for Evaluation of Anthelmintic Activity**[a]

Disease	*In vitro* assay	*In vivo* assay (animal model)
Roundworms (nematodes)		
Intestinal	*#Nippostrongylus brasiliensis*	*#Nematospiroides dubius* (mouse)
Systemic	*Brugia malayi*	*Litosomoides* (rat)
	Trichinella spirals	*Trichinella spiralis* (mouse)
Flatworms		
Tapeworms (cestodes)	*Hymenopis nana*	*#Hymenopis nana*
		Taenia taeniformis
Flukes (trematodes)	*Schistosoma* spp.	*Schistosoma* spp. (rodents, primates)
	Fasciola hepatica	*Fasciola hepatica* (rodents)

[a]The # denotes a model used in our discovery program in collaboration with Syntex.

3. ANTHELMINTIC-ACTIVE NATURAL PRODUCTS

Structures of a variety of anthelmintic agents will now be surveyed. The nonmarine natural products that stimulated development of clinically relevant anthelmintics will be considered first; this literature is not very extensive. A detailed survey of contemporary marine natural products will then be examined. To date, only the marine natural products isolated from UCSC have been systematically evaluated for anthelmintic properties. The process began by subjecting crude extracts to a primary screen against the fourth larval stage of a single roundworm (nematode), *Nippostrongylus brasiliensis* (Jenkins *et al.*, 1980). Each extract or compound was tested at a concentration of 50 μg/ml. Parameters used to determine activity were the ability of the larvae to molt to the adult (cast formation), viability, and motility. Figure 1 shows a diagram of the experimental method used for the *in vitro* evaluation. In order to be considered for followup, an extract had to show >60% activity against casts, and pure compounds that showed >95% activity against casts were classified as very active. Active pure compounds, or occasionally compound mixtures, were advanced to an *in vivo* assay summarized in Fig. 2. A mouse model was employed and a mixed helminth infection of *Nematospirodes dubius* (a roundworm) and *Hymenolepis nana* (tapeworm) was challenged with the test material presented in feed pellets at concentrations of 200–1000 ppm; oxfendazole was the positive activity control. Actives were those that reduced one or both worms by >60%.

3.1. Agents of Nonmarine Origin

Ivermectin (**3**) stands as the "magic bullet" among clinically active anthelmintics that are based on a natural product. It was introduced in 1983 (Campbell *et al.*, 1983) and is a semisynthetic derivative of avermectin B_1, produced by the

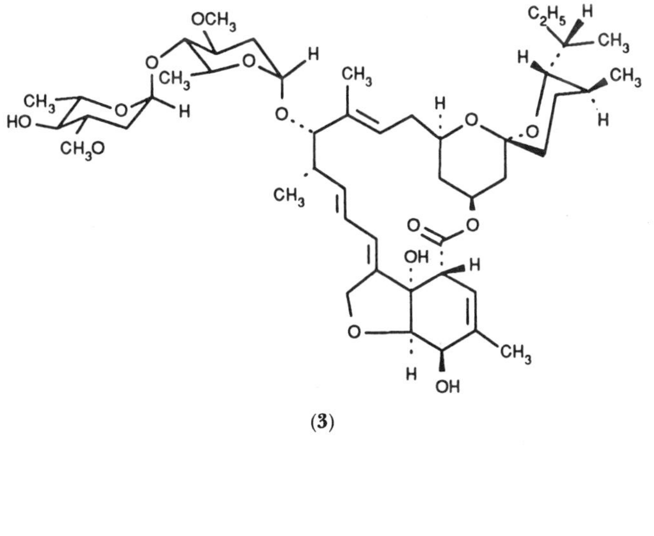

(**3**)

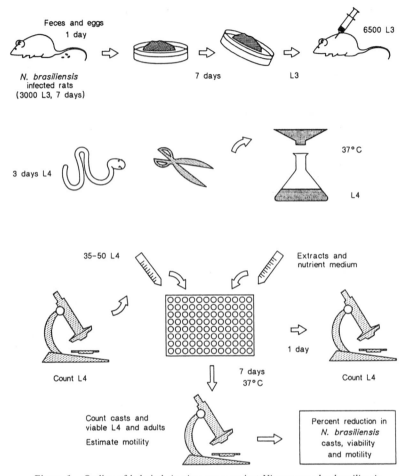

Figure 1. Outline of helminth *in vitro* assay against *Nippostrongylus brasiliensis*.

actinomycete *Streptomyces avermitilus*. Ivermectin is effective against intestinal nematodes and is currently in use as a broad-spectrum anthelmintic (Campbell, 1985; Freedman *et al.*, 1989). Of particular importance is its activity against river blindness (White *et al.*, 1986) and this important drug has been referred to as the most important chemotherapeutic for tropical medicine in this century (Eckholm, 1989).

Another important class of anthelmintics based on a natural product lead are those containing the benzimidazole ring system. This substructure is embedded in ribofuranosylbenzimidazole, which is a substructure of vitamin B_{12}. Since the

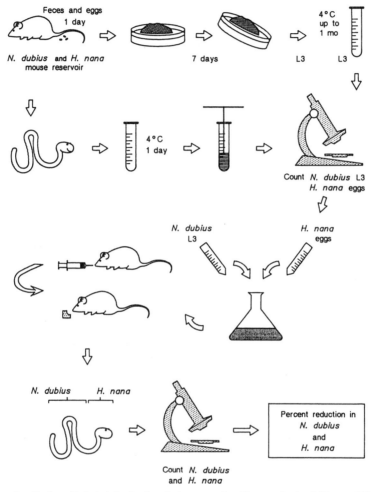

Figure 2. Outline of helminth *in vivo* (in mice) assay against *Nematospirodes dubious* and *Hymeno-lepis nana*.

discovery of the anthelmintic activity of benzimidazoles, thousands of benzimid-azole derivatives have been synthesized for anthelmintic screening (Townsend and Wise, 1990). From this tremendous effort fewer than 20 benzimidazoles have reached commercial use, and only three are currently in clinical use for humans: albendazole (**4**), flubendazole (**5**), and mebendazole (**6**) (Cook, 1990). Although there have been major advances in the clinical treatment of helmintic diseases, there are still no ideal broad-spectrum anthelmintics (Cook, 1991).

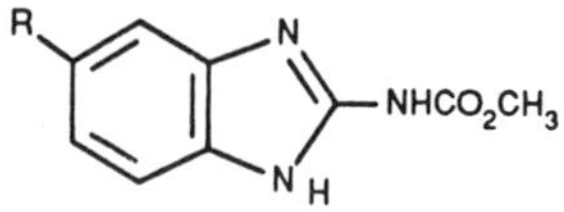

(4) R = CH$_3$CCH$_2$CH$_2$S–

(5) R = 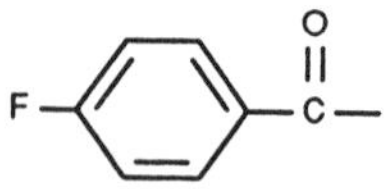 F—〈benzene ring〉—C(=O)—

(6) R = 〈benzene ring〉—C(=O)—

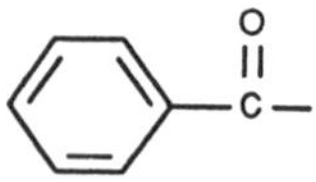

3.2. Marine Sponge-Derived Compounds

Our search of the literature revealed a variety of anthelmintic sponge-derived compounds. Interestingly, there are no publications outside of those from the UCSC group which report the structures of new or known sponge metabolites along with antiparasitic activity data.

From 1984 to 1990, extracts and pure compounds from 407 different sponge samples, as well as those of a small number of other marine invertebrates, were screened in the UCSC–Syntex collaborative research program. A variety of known and new compounds were examined and they are summarized in this section. *In vitro* activity was assessed in a primary screen against *N. brasiliensis* (Jenkins *et al.*, 1980). The actives were divided into those that were strongly active (>95% action at 50 μg/ml) or moderately active (50–95% action at 50 μg/ml) and Table IV shows a summary of the results. The extracts from sponges provided the most impressive activity and they served as a source of all of the subsequent chemistry reported below.

Pure compounds were then sought from the sponge extracts (as noted above) and prioritized according to their activity and the amount on hand. Those compounds with confirmed *in vitro* activity and that could be reisolated in greater than 50 mg amounts were then tested for *in vitro* activity against a mixed helminth infection in mice. Reductions in intestinal worm burden of the nematode *Nematospirodes dubius* and the cestode *Hymenolepis nana* were measured following treatment with the test agent.

Our first implementation of the above procedure produced encouraging results. Very active extracts were pinpointed whose active constituents proved to be new chemotypes. The chronology of these early events is as follows. A first batch of Indo-Pacific sponge extracts was submitted to the Syntex screens during the fall of 1984. These consisted of 17 extracts of Tongan sponges (collected in July 1983) and 14 extracts of Fijian sponges (collected in July 1984); actives included

Table IV.   Marine Samples Evaluated for in Vitro Anthelmintic Activity[a]

	Number submitted	Moderate activity[b]	Strong activity[c]
Extracts			
Sponge	407	21	16
Tunicate	30	0	0
Byrozoan	3	0	0
Hydroids	9	0	0
Soft coral	3	2	0
Nudibranch	3	2	0
Zoanthid	2	0	0
Total	457	25	16
Pure compounds			
Sponge	85	17	19
Soft coral	1	0	0
Nudibranch	1	0	1
Zoanthid	2	0	1
Total	89	17	21

[a] Assay organism *Nippostrongylus brasiliensis*.
[b] 50–95% action at 50 μg/ml.
[c] >95% action at 50 μg/ml.

four of the former and five of the latter. The active constituent of one of the Tongan sponges proved to be (S)-1-tridecoxy-2,3-propanediol (Myers and Crews, 1983). The simplicity of this structure made synthetic preparation of additional material feasible. Subsequent *in vivo* evaluation against the mixed helminths in mice showed that these compounds were only slightly active. Paucity of extract from all of the Fijian sponges required that they be recollected. This was accomplished during the summer of 1985, providing 5.5 kg of *Jaspis* sp. (now identified as *Jaspis johnstoni*) and 1.4 kg of an undescribed Jaspidae sponge (now identified as *Jaspis* sp.). The cycle of bioassay-guided isolation eventually yielded the novel alkaloids jasplakinolide and the bengamides. These exciting leads generated great interest on the part of the Syntex collaborators and demonstrated, at a critical point, the value of considering marine natural products in helminth lead compound development efforts. The long list of *in vitro* active compounds that were subsequently discovered will be described below, grouped according to their biogenetic constitution.

3.2.1. Alkaloids

Jasplakinolide (7; see Scheme 1) (Crews *et al.*, 1986) (or jaspamide) (Zabriskie *et al.*, 1986), is a 19-member macrocyclic ketide-cylodepsipeptide constituent the orange Indo-Pacific *Jaspis johnstoni*. Its structure was simulta-

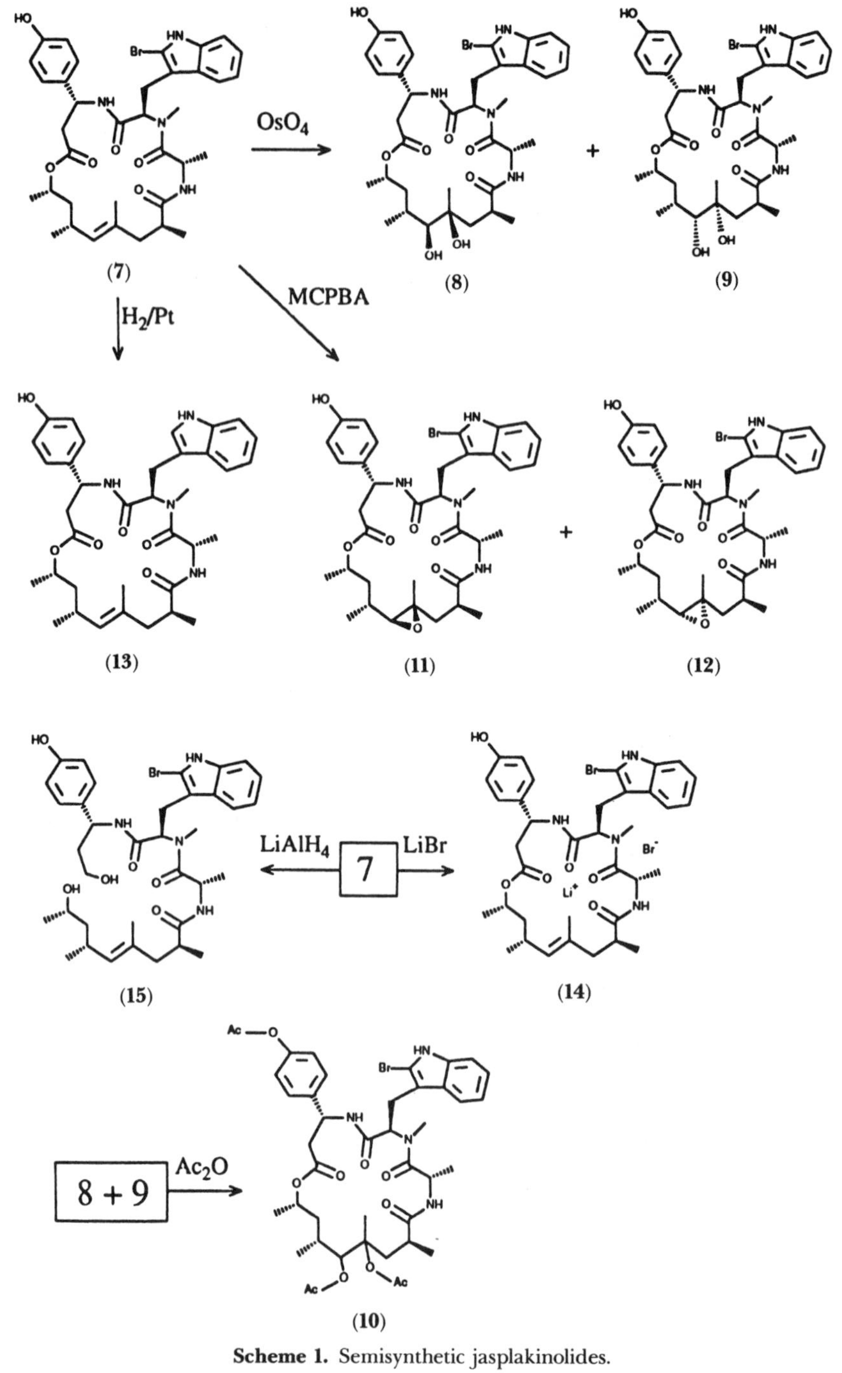

Scheme 1. Semisynthetic jasplakinolides.

neously reported by two groups, both engaged in the study of Fijian material. A subsequent examination of a Papua New Guinea collection of *J. johnstoni* has also yielded jasplakinolide (Braekman *et al.*, 1987b). We have extensively examined the conformational (Inman and Crews, 1989) and ionophoric properties (Inman *et al.*, 1989) of this macrolide and its lithium complex, and jasplakinolide has been the subject of successful total syntheses (Grieco *et al.*, 1988; Schmidt *et al.*, 1988). Pooling material from several collections made it possible for us to prepare several semisynthetic jasplakinolide derivatives (7–15 in Scheme 1). Most of these were examined further in the primary screen and all were either very active or moderately active, as shown in Fig. 3. Thus, the potency of jasplakinolide (7) is closely matched by the debromo derivative 13, the Li$^+$ complex 14, and the epoxides 11 and 12.

The bengamides (Quiñoá *et al.*, 1986) are a series of novel heterocyclic ketide amino acids which were isolated from *Jaspis* sp. Bengamides A (16) and B (17) were the first of this series to be isolated and a possible biosynthetic composition of the bengamide skeleton is shown in Fig. 4. Additional members of

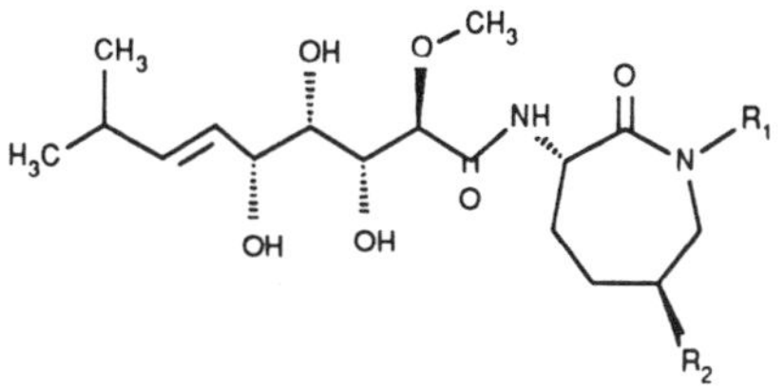

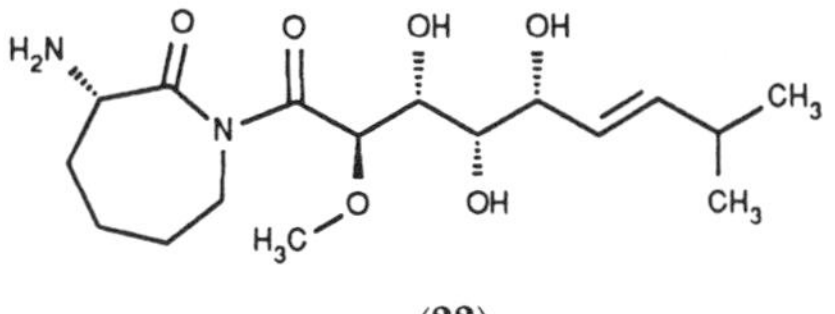

	R$_1$	R$_2$
(16)	H	-O$_2$C(CH$_2$)$_{12}$CH$_3$
(17)	CH$_3$	-O$_2$C(CH$_2$)$_{12}$CH$_3$
(18)	H	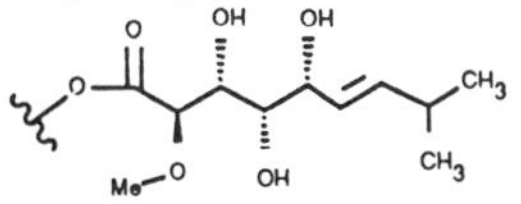
(19)	CH$_3$	
(20)	H	H
(21)	CH$_3$	H

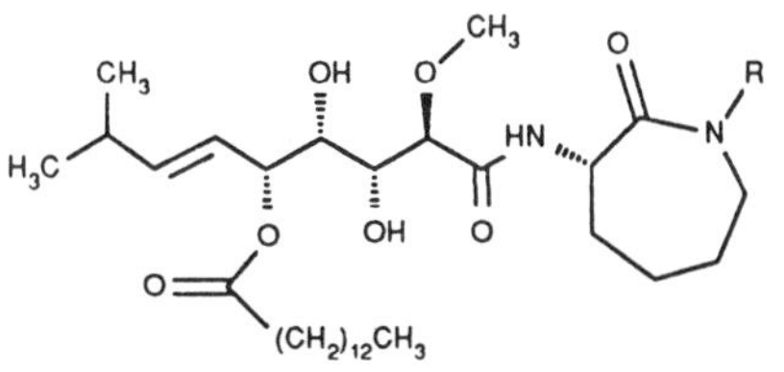

(22)

(23) R = H
(24) R = CH$_3$

this group that have been characterized include bengamides C–H (18–21, 23, 24) and isobengamide E (22) (Adamczeski *et al.*, 1989). Their absolute stereochemistry was recently proposed based on non-x-ray methods (Adamczeski *et al.*, 1990) and confirmed from total syntheses (Chida *et al.*, 1991; Broka and Ehrler, 1991).

The bengamides were isolated in varying ratios among sponge collections from Fiji or the Solomon Islands. Anthelmintic evaluation of these derivatives is summarized in Fig. 5. Bengamides A (16), B (17), G (23), and H (24) are all powerfully active *in vitro*, but none of these was tested *in vivo*. Fortunately, two other *in vitro* active Bengamides E (20) and F (21) were screened *in vivo*; bengamide E (20) was inactive, but bengamide F (21) showed potency against the cestode *H. nana* but not against the nematode *N. dubius*, as summarized in Fig. 6.

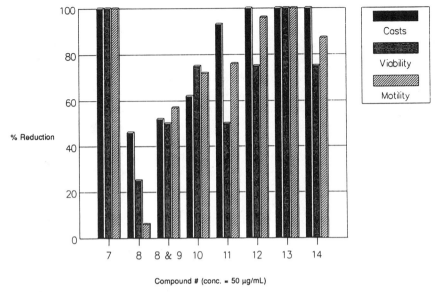

Figure 3. *In vitro* anthelmintic activity of jasplakinolides and semisynthetic derivatives.

Accompanying the bengamide metabolites discussed above were the bengazoles, which are also anthelmintic-active ketide-amino acids. A large recollection from the Benga Lagoon, Fiji, during the summer of 1986 of *Jaspis* sp., normally a source of bengamides A and B, provided only trace amounts of these metabolites. Surprisingly, its major components were bengazoles A (**25**) and B (**26**) (Adamczeski *et al.*, 1988). An analysis of their ketide-amino acid biogenetic origin is shown in Scheme 2. Bengazole A showed 100% activity (against casts, viability, and motility) in the *in vitro* screen against *N. brasiliensis*, while bengazole B

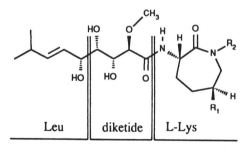

Figure 4. Biogenesis of the bengamides.

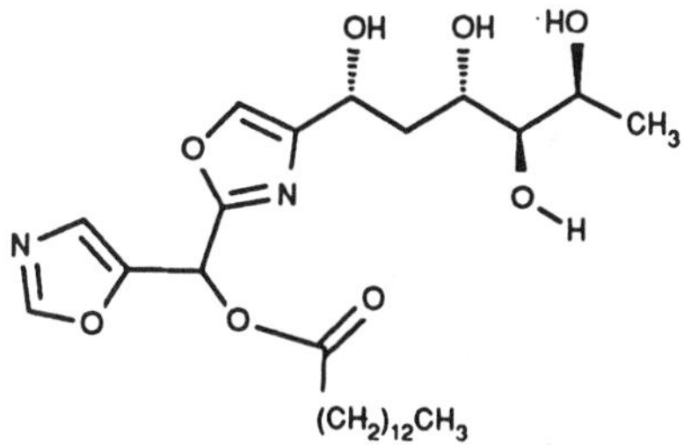

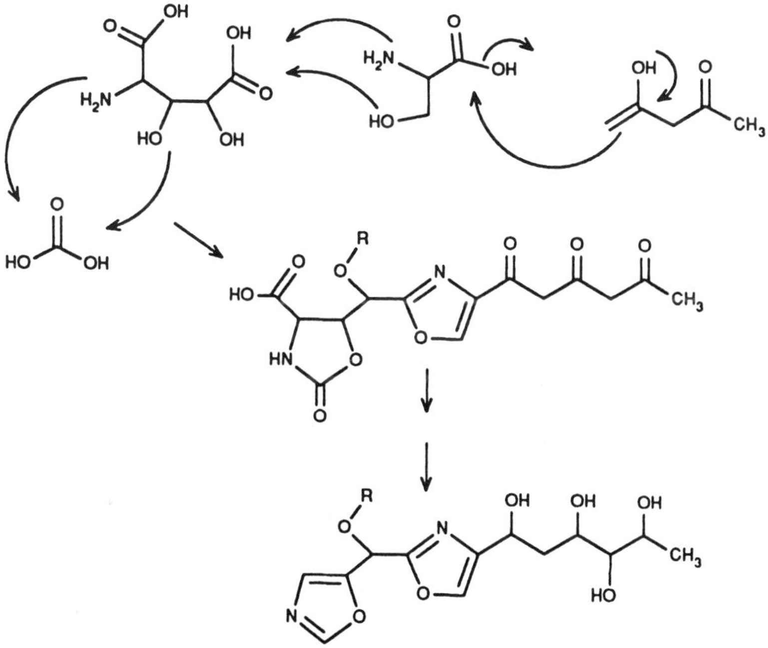

Scheme 2. Biogenesis of bengazoles.

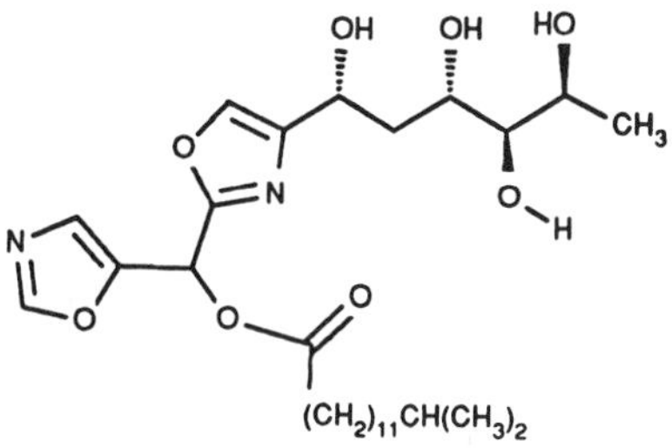

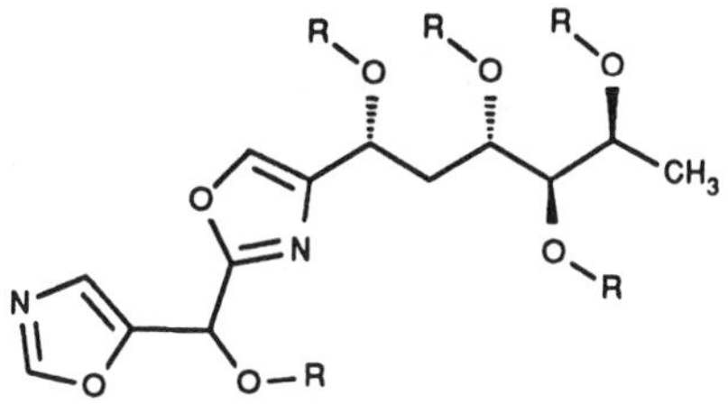

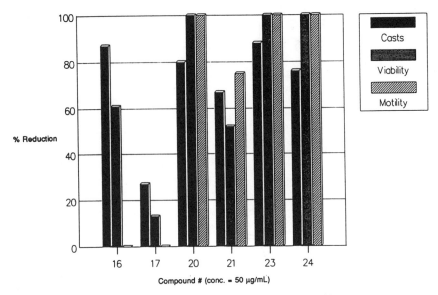

Figure 5. *In vitro* anthelmintic activity of bengamides.

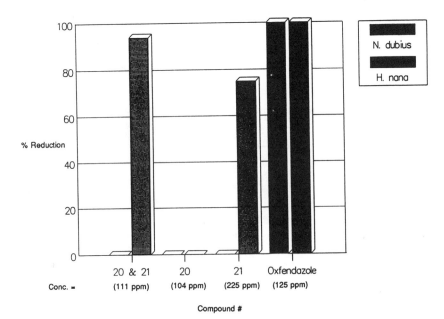

Figure 6. *In vivo* anthelmintic activity of bengamides.

decomposed before it could be submitted for assay (see below). Two bengazole A derivatives, the hydrolysis product **27** and the pentaacetate **28**, were not available in large enough quantities for testing. A large sample of bengazole A was submitted for *in vivo* assay; however, an *in vitro* retest indicated that the activity was lost. Analysis of the sample showed that bengazole A was present as a minor component and was accompanied by major fragmentation products **29** and **30** (see Scheme 3) whose formation could be rationalized by cycloaddition with singlet oxygen accompanied by cleavage during HPLC purification in methanol as diagrammed in Scheme 3. These latter two compounds, when individually screened, did not show activity.

Other anthelmintic-active alkaloids were isolated from a Fijian sponge of the family Spongiidae, originally identified as *Spongia mycofijiensis* (Kakou *et al.*, 1987). This sponge, currently identified as *Leiosella ?levis* (C. Diaz, P. Crews, and

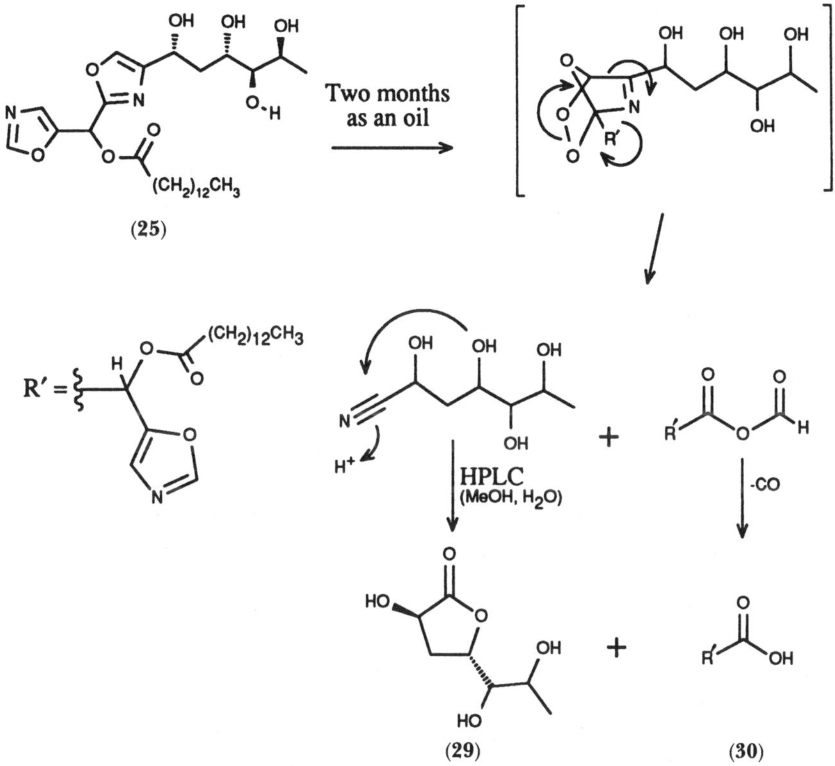

Scheme 3. Fragmentative decomposition of bengazole A.

R. W. H. van Soest unpublished), has a nudibranch associant, *Chromodoris lochi*, and is a source of the fijianolides (Quiñoá *et al*., 1988), macrolide polyketides, which are also isolated from the sponge nudibranch pair *Hyattella* sp.–*Chromodoris lochi* (Corely *et al*., 1988). Our Fijian collections of this sponge yielded latrunculin A (**31**) (Kashman *et al*., 1980), which showed excellent *in vitro* activity at 50 μg/ml

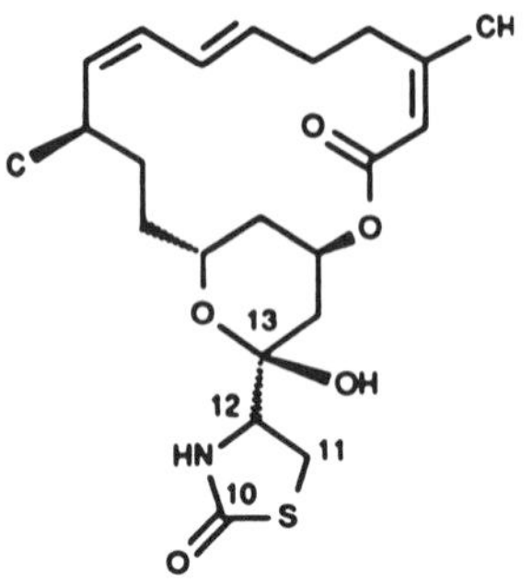

(31) 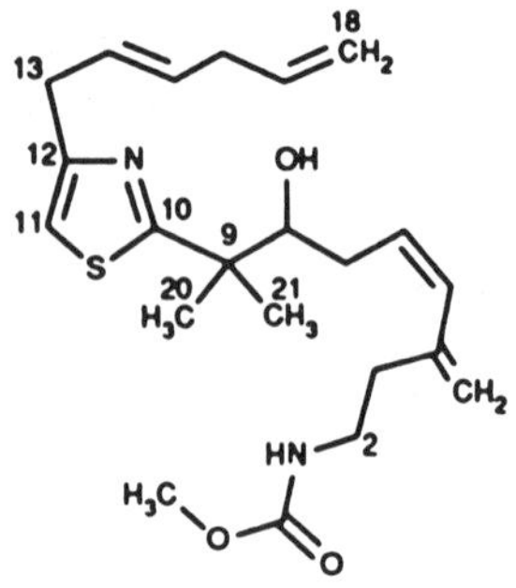   (32) 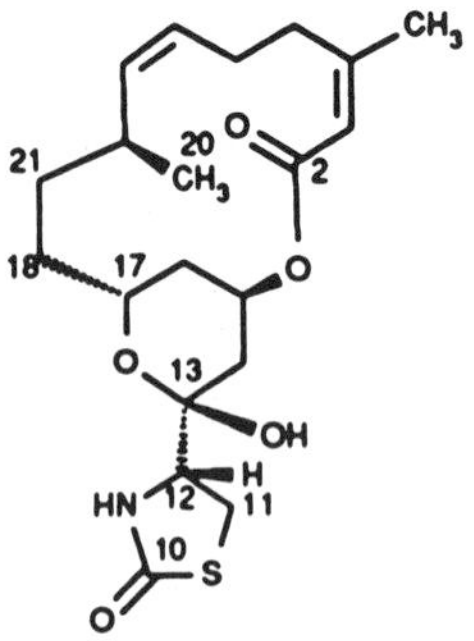   (33)

against *N. brasiliensis*. Vanuatuan collections of *L.* ?*levis* (Crews *et al*., 1988) yielded **31** accompanied by mycothiazole (**32**) and the latter compound was 100% active in the primary screen. The *in vivo* followup was complicated because both **31** and **32** were toxic to mice and no efficacy could be observed at subtoxic doses. We also obtained latrunculin B (**33**) from workup of the Red Sea sponge *Latrunculia magnifica* (Groweiss *et al*., 1983), but it proved to be inactive in the primary screen as shown in Fig. 7. This trio of compounds are related in that the S·C-11·C-12·N atoms in each are undoubtedly derived from cysteine while the remaining carbon skeleton is of polyketide origin. The atom numbers accompanying mycothiazole (**32**) and latrunculin B (**33**) summarize our proposal that their carbon skeletons have a parallel genesis (Crews *et al*., 1988).

The chemistry of sponges in the order Verongida is especially relevant because all its genera possess brominated tyrosine derivatives. For example, the two species *Psammaplysilla purpurea* and *P. crassa* (synonyms for *P. crassa* are *Ianthella ardis* and *Aiolochroia crassa*) are Verongida sponges of the family Aplysinellidae. The few natural products known from the *P. purpurea* include a simple yet highly modified tyrosine monomer, aeroplysinin I (Makarieva *et al*., 1981), and a more complex metabolite, ianthelline (Litaudon and Guyot, 1986). A similar small body of literature can be found for the *P. crassa* and includes 2-hydroxy-3,5-dibromo-4-methoxyphenylacetamide (Chang and Weinheimer, 1977), complex unsymmetrical tyrosine dimeric derivatives such as the psammaplysins (Roll *et al*., 1985), purealin (Nakamura *et al*., 1985) or the lipopurealins (Wu *et al*., 1986), and dimeric and tetrameric dipeptides such as psammaplin A (Quiñoá and Crews, 1987; Arabshahi and Schmitz, 1987; Rodriguez *et al*., 1987)

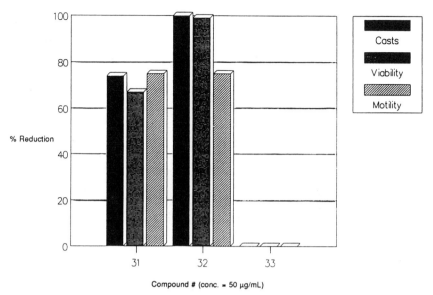

Figure 7. *In vitro* anthelmintic activity of cysteine-containing polyketides.

and bisaprasin (Rodriguez *et al.*, 1987). Tyrosine-derived constituents psamma-plins A–D (**34–37**) were isolated from *Psammaplysilla purpurea* collected from Tonga and Fiji (Quiñoá and Crews, 1987; Jiménez *et al.*, 1991a). Three of these, psammaplins A, C, and D, were evaluated in the primary screen, but only psammaplin D (**37**) was active at 50 μg/ml with 66% inhibition against casts, as

(**34**) X = -S-S
(**35**) X = -SCN
(**36**) X = -SO₂NH₂
(**37**) X = -S-S

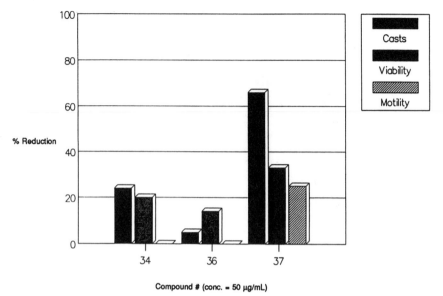

Compound # (conc. = 50 µg/mL)

Figure 8. *In vitro* anthelmintic activity of psammaplins.

shown in Fig. 8. The related sponge *Pseudoceratina ianthelliformis*, collected from Fiji, yielded the bromotyrosine derivatives **38** (Sharma *et al.*, 1968), **39** (Norte *et al.*, 1988), and **40** (Gopichand and Schmitz, 1979). These were all inactive in the primary anthelmintic screen.

The crude extract of a small collection of a nudibranch, *Notodoris gardineri*, exhibited positive activity in the primary screen. The active constituent was

(38)   (39)   (40)

expected to be a 2-aminoimidazole derivative. This bright yellow nudibranch accentuated with black markings is often observed perched on the yellow calcareous sponge *Leucetta* sp. 2-Aminoimidazoles have been reported from both of these organisms (Alvi *et al.*, in press). These alkaloids include naamidines, isonaamidines, naamines, and isomaamine from the Red Sea nudibranch–sponge pair *N. citrina–L. chagosensis* (Carmely *et al.*, 1989); kealiiquinone and pyronaamidine, from Indo-Pacific specimens of *Leucetta* sp. (Akee *et al.*, 1990); and an unusual ionophoric 2-aminoimidazole, clathridine, from the calcareous sponge *Clathrina clathrus* (Ciminiello *et al.*, 1989). Our bioassay-guided search for the active constituents of *N. gardineri* afforded dorimidazole A (**41**), accompanied

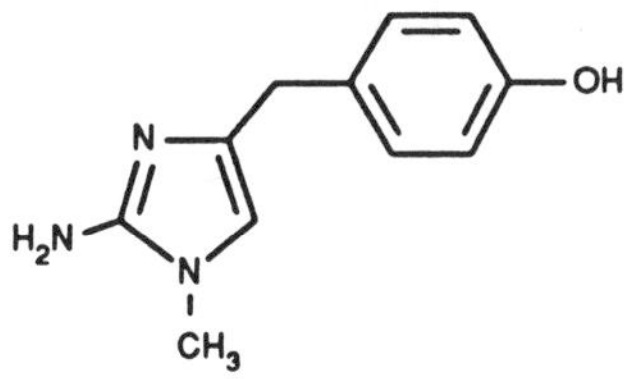

(41) 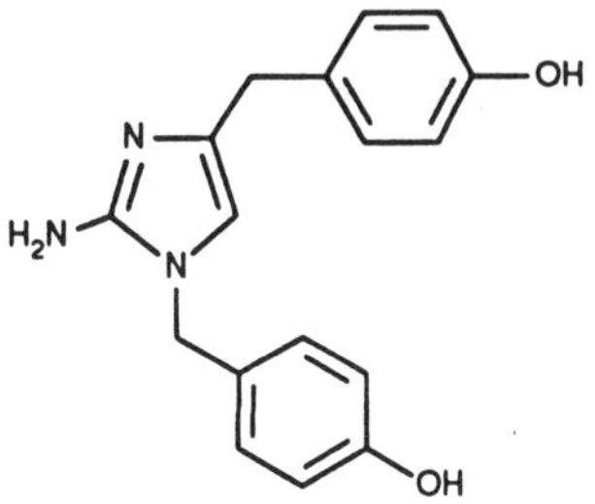   (42)

by a small amount of isonaamine A (**42**) (Alvi *et al.*, 1991a). The former was extremely active and at 50 μg/ml, showing 100% inhibition against casts, but this activity level was reduced to 12% at 5 μg/ml. Surprisingly, reassay of **41** obtained from a total synthesis and tested as the HBr salt revealed inactivity. Trace amounts of unknown 2-aminoimidazoles which appear to be present in the natural sample of **41** were proposed to explain this difference in bioactivity behavior. Dorimidazole A, discussed above, and 2-aminoimidazole, isolated from *Reniera cratera* (Cimino *et al.*, 1974b), are the simplest examples containing this moiety to be isolated from a nudibranch or sponge. Marine phytoplankton have been shown to degrade tyrosine to *p*-hydroxyphenyl pyruvate (**43**) (Barrow, 1983). Thus, guanidine

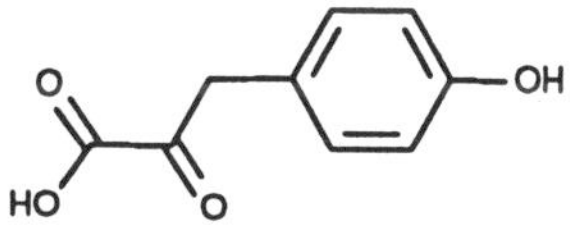

(43)

and **43** could condense to generate an intermediate which could then serve as a precursor to both **41** and **42**. Such a pathway could also be the basis for the genesis of some 40 other 2-aminoimidazole-containing compounds that have been isolated from marine sponges. Interestingly, many of these structures can be dissected into guanidine plus amino acid subunits; for example keramadine (**44**) (Nakamura *et al.*, 1984) appears to comprise guanidine + lysine + proline moieties.

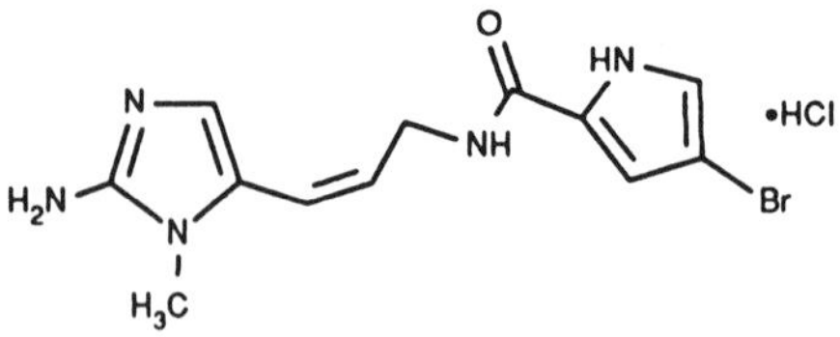

(44)

The massive red-brown sponge *Fascaplysinopsis reticulata* has been the subject of several studies which have reported sesterterpenes, alkaloids, and sesterterpene-alkaloid salts (Kazlauskas *et al.*, 1977; Roll *et al.*, 1988; Jiménez *et al.*, 1991a,b). Lead compounds from this sponge are dehydroluffariellolide diacid (**S1**), fascaplysin (**45**), fascaplysin A (**46**), and reticulatine A (**51**). Additional components include a sesterterpene-alkaloid salt **47** and alkaloids **48–50**.

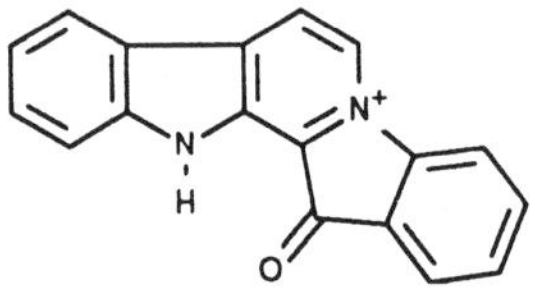

(45) Anion = Cl⁻
(46) Anion = S1⁻

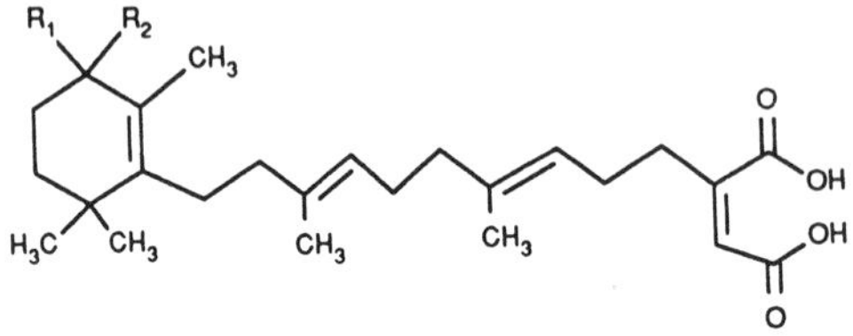

(S1) $R_1 = R_2 = H$
(S2) $R_1 = R_2 = O$

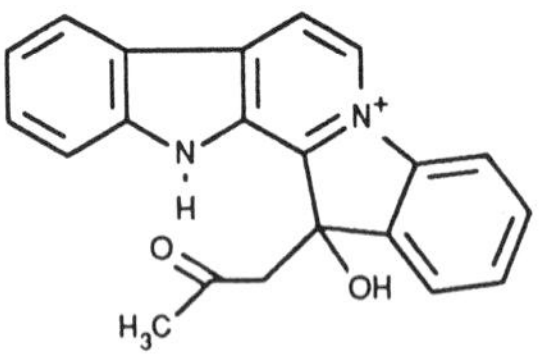

(47) Anion = S1⁻ 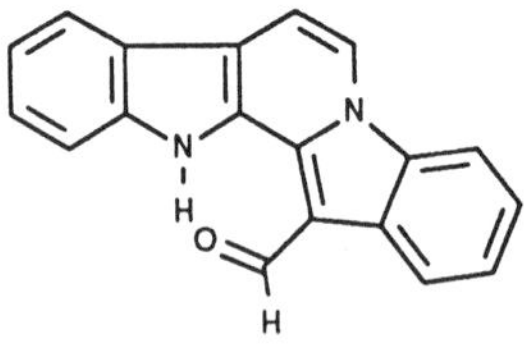   (48) 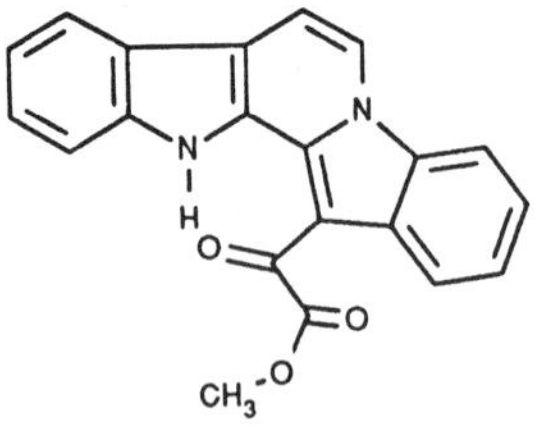   (49)

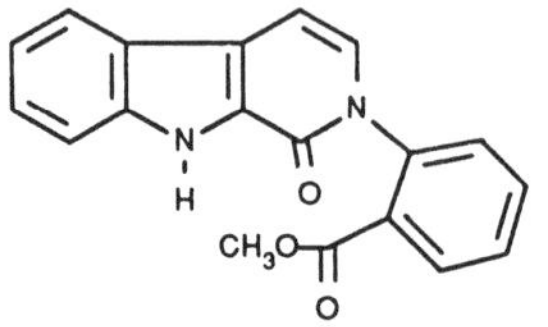

(50) 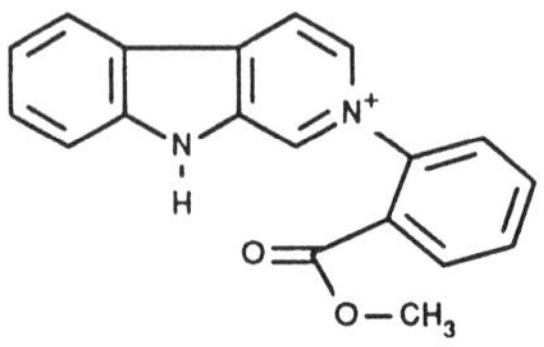   (51) Anion = S2⁻

Three alkaloids of this group (**47–49**) were examined in the primary screen and none was active. That these compounds are β-carbolines is intimated by the structures of secofascaplysin A (**50**) and reticulatine A (**51**). The exact relationship of these along with **45** and **47–49** to tryptophan is uncertain regarding the ability of this precursor to contribute one or both nitrogens to the final alkaloid skeletons.

Agelas sponges are a source of bromopyrrole derivatives (Keifer *et al.*, 1991) which are most likely derived, in part, from proline. Our investigation of the primary-screen-active extract of *Agelas mauritiana* (C. Jimenez and P. Crews, unpublished) yielded the known bromopyrrole dibromophakellin (**52**) (Sharma and Magdoff-Fairchild, 1977) and a new bromopyrrole, isomidpacamide (**53**). The former showed moderate activity, while the latter was completely inactive.

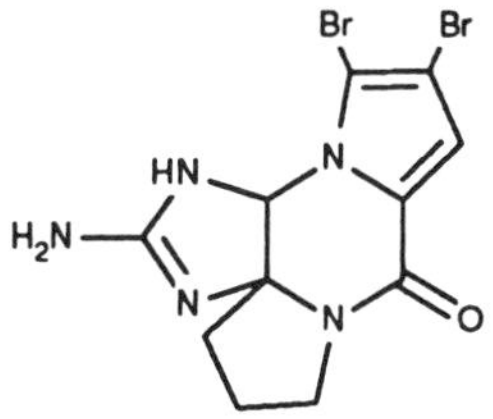

(52) 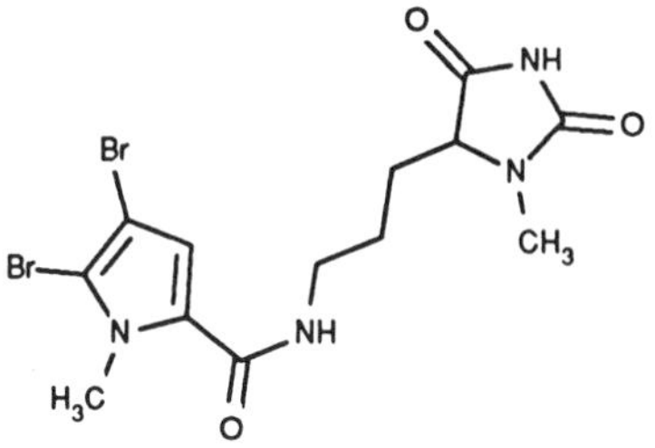   (53)

Several simple amino acid derivatives were evaluated in the helminth primary screen. Octopamine (**54**), which has been investigated in skeletal-muscle receptor-binding assays (Evans *et al.*, 1988), was isolated from *Fascaplysinopis reticulata* and found to be inactive in the primary screen. The sponge *Myrmekioderma styx* contained five simple amino acid derivatives (**55–58b**) which were all *in vitro* inactive. Three related metabolites, xestoaminols A–C (**59–61**), which are undoubtedly condensation products of unbranched C_{11} fatty acids and *L*-alanine,

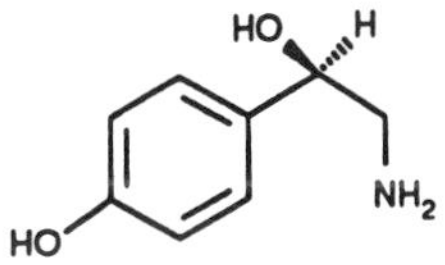

(54) 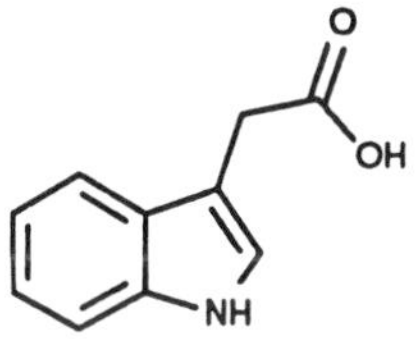   (55)

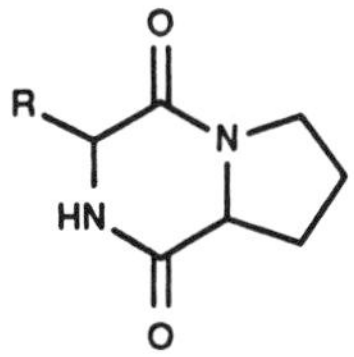

(56) R = CH_3
(57) R = $CH_2CH(CH_3)_2$
(58a) R = $CH(CH_3)_2$
(58b) R = $CH_2C_6H_5$

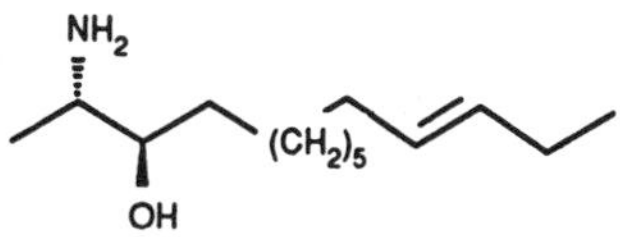

(59)

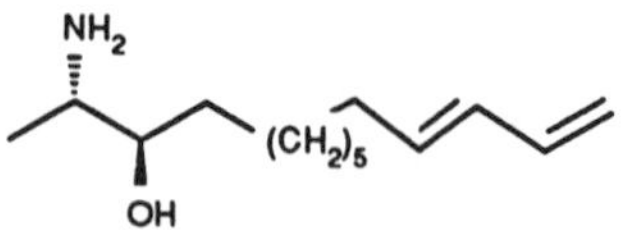

(60) 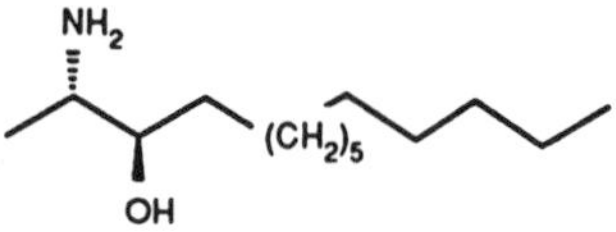   (61)

were isolated from *Xestospongia* sp. (Jiménez and Crews, 1990). All were tested, but only **59** was active at 50 μg/ml with 100% reduction of viability and motility and 90% reduction of casts. An *in vivo* followup at 570 ppm in the mixed mouse helminth model failed to reveal activity. Related amino alcohols were isolated from the sponge *Xestospongia* sp. (Gulavita, 1989); a subsequent synthesis corrected the absolute configuration of these compounds, as well as the absolute configuration of **59–61** (Mori, 1992).

   Alkaloids which do not have obvious amino acid subunits have been investigated for their anthelmintic properties. Two novel pentacyclic aromatic alkaloids, plakinidines A (**62**) and B (**63**), isolated from a *Plakortis* sp. collected from Vanuatu (Inman *et al.*, 1990), showed excellent initial *in vitro* activity as shown in Fig. 9. However, the followup dose–response relation showed inactivity at 5 μg/ml, as shown in Fig. 9 for **62** and **63**. A third member of this series, plakinidine C (**64**), has also been reported (West *et al.*, 1990). A known polycyclic alkaloid, aaptamine (**65**) was isolated from the bright yellow sponge *Aaptos aaptos* (Nakamura *et al.*, 1982), which is also a source of other related alkaloids (Nakamura *et al.*, 1987). Compound **65** was not active in the primary screen, as shown in Fig. 10. A tiny amount of a new nucleoside **66** was isolated from *Geodia neptuni* and it

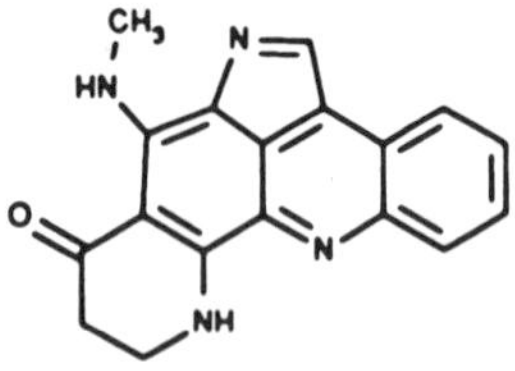

(62) 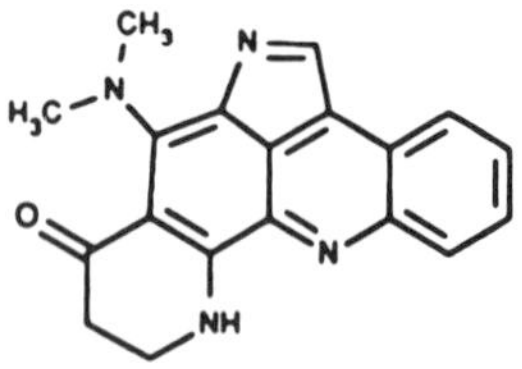   (63) 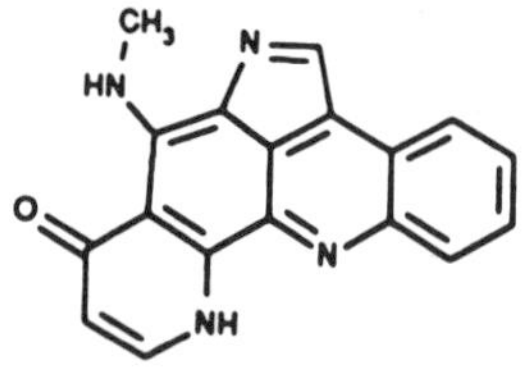   (64)

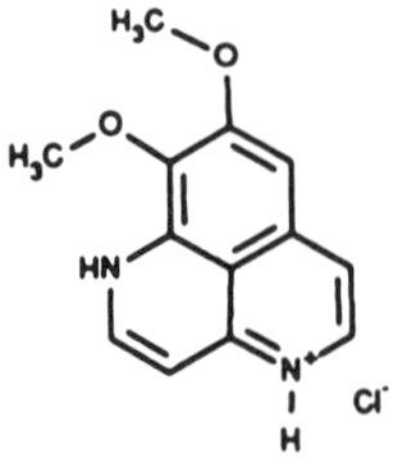

(65) 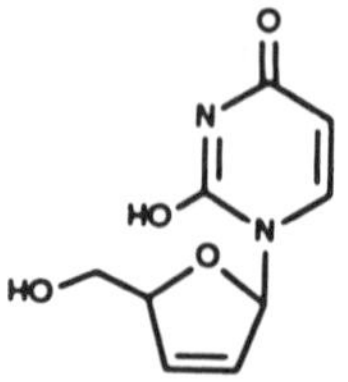   (66) 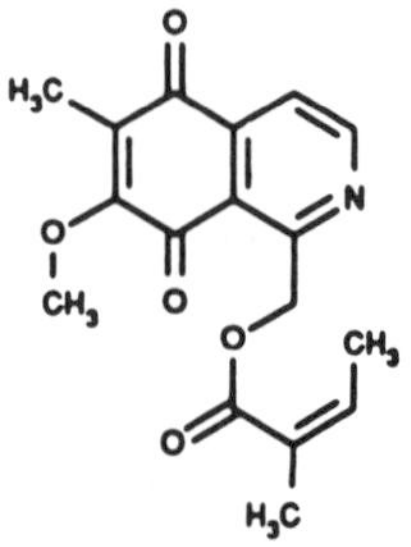   (67)

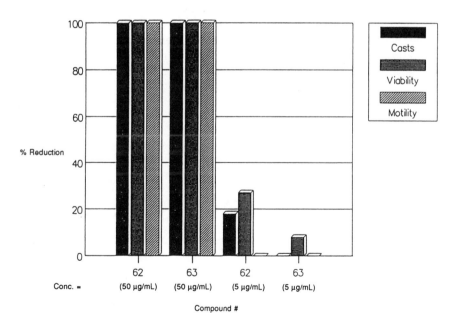

(68)   (69)   (70)

was moderately active, as shown in Fig. 10. We isolated (K. A. Alvi and P. Crews, unpublished) four inactive quinones, including renierone (**67**) (McIntyre *et al.*, 1979) and **68–70** from a *Haliclona* sp., but these compounds have all been previously reported from *Reniera* (Frincke and Faulkner, 1982).

3.2.2. Polyketides

Sponges are a source of a variety of cytotoxic polyketides comprising acetates, propionates, or a mixture of these building blocks (Munro *et al.*, 1987).

Figure 9. *In vitro* anthelmintic activity of plakinidines.

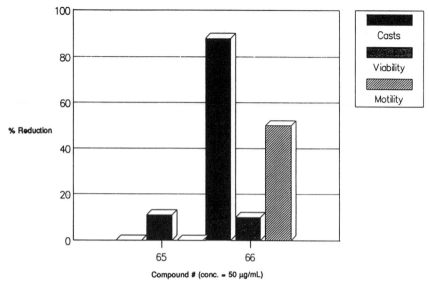

Figure 10. *In vitro* anthelmintic activity of miscellaneous alkaloids.

Very few members of this fascinating biosynthetic class were evaluated in the helminth primary screen and none was discovered to be active. The polycyclic ketide halenaquinone (**71**) (Roll *et al.*, 1983) was isolated from a Fijian collection of *Xestospongia ?carbonaria* (Lee *et al.*, 1992). A macrolide, fijianolide B (**72**) (Quiñoá *et al.*, 1988) [laulimalide (Corely *et al.*, 1988)], was isolated from Vanuatuan and Indonesian collections of *Leiosella ?levis*.

(71)   (72)

3.2.3. Benzenoids

No activity was found among any of the simple benzenoid compounds tested. This included phenyl acetic acid (**73**), isolated from *Callyspongia ?ramosa*, and

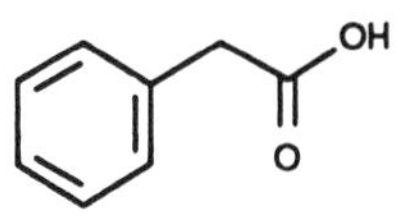

(73) 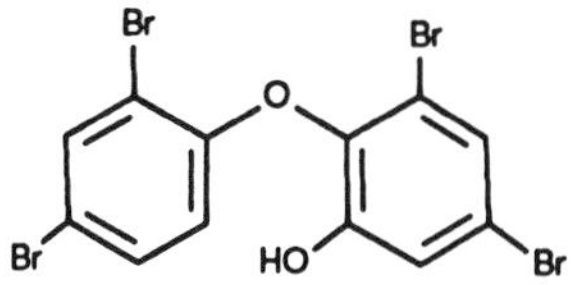   (74) 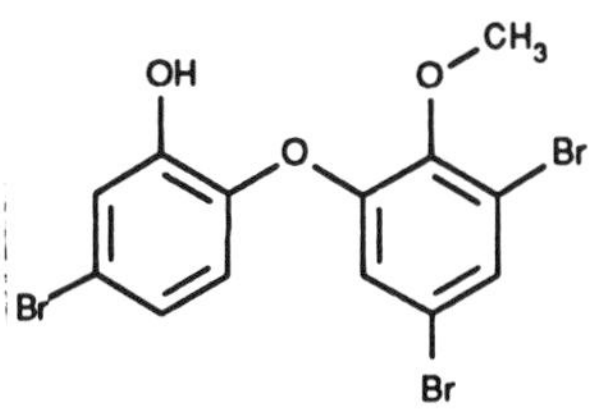   (75)

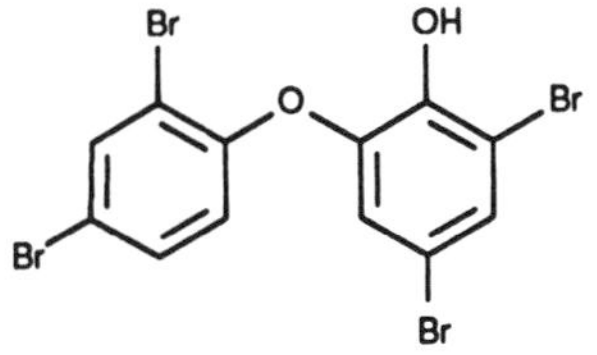

(76) 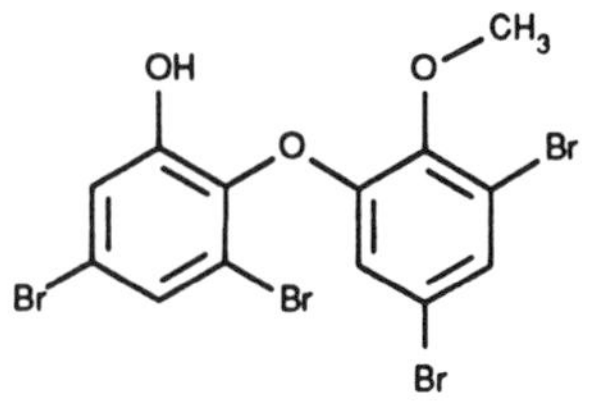   (77)

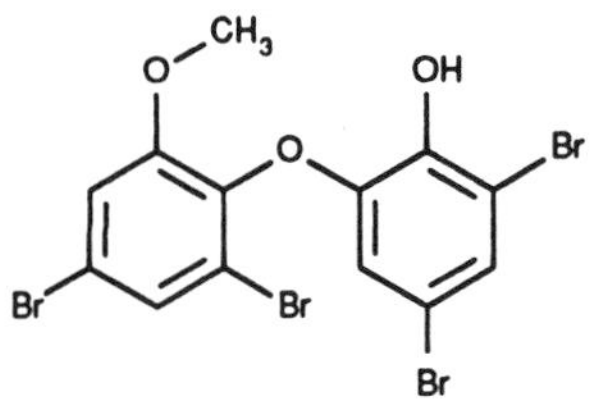

(78) 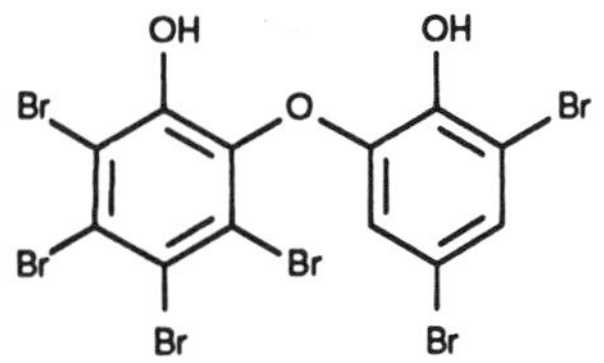   (79)

diphenyl bromophenol ether (74) (Capon *et al.*, 1981; Carté and Faulkner, 1981), from an unidentified Thorectidae sponge. Additional inactive diphenyl bromophenol ethers 75 (C. Jiménez and P. Crews, unpublished), 76 (Carté and Faulkner, 1981), 77 (Norton *et al.*, 1981), 78 (C. Jiménez and P. Crews, unpublished), and 79 (Norton *et al.*, 1981), were isolated from an unidentified Fijian Dictyoceratida sponge.

3.2.4. Sesquiterpenoids

Many Dictyoceratida sponges are sources of sesquiterpenes. A sponge of this order, *Dysidea herbacea* (family Dysideidae), was the subject of a bioassay-guided isolation project because the crude extract of a tan-colored specimen collected in Fiji (No. 86070) showed excellent anthelmintic activity at 50 µg/ml, as shown in Fig. 11. *Dysidea herbacea* is a source of tricyclic furanosesquiterpenes, which include the dysinins and the dysins (Horton *et al.*, 1990; Kazlauskas *et al.*, 1978). Five compounds of the former class (80–84) were isolated (Horton *et al.*, 1990) and the *in vitro* active constituents were identified as furodysinin (80) and a mixture of 82 and 84. Their activity properties are shown in Fig. 11. A larger amount of these samples was reisolated and subjected to an *in vivo* assay.

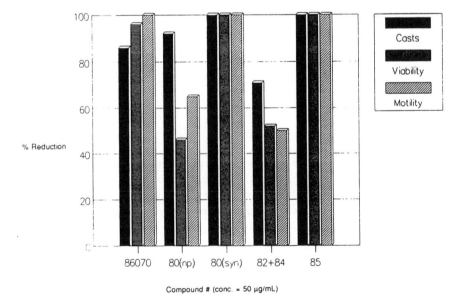

Figure 11. *In vitro* anthelmintic activity of dysinins.

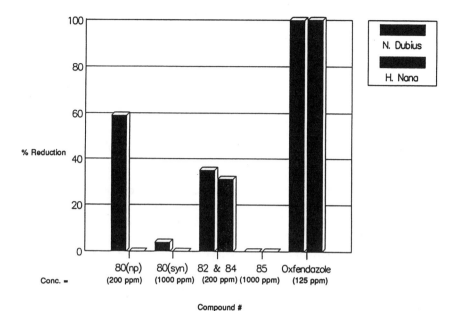

Figure 12. *In vivo* anthelmintic activity of dysinins.

Furodysinin (**80**) was partially active, while the mixture of **82** and **84** was inactive, as shown in Fig. 12. An evaluation of synthetic samples was conducted in parallel with the work on the natural products. A sample of (−)-furodysinin (**80**) was supplied from a total synthesis (Richou *et al.*, 1989); it was active *in vitro* (Fig. 11), but was inactive *in vivo* (Fig. 12). Disappointingly, as shown in Figs. 11 and 12, a parallel pattern of *in vitro* activity and *in vivo* inactivity was shown for the tricyclic model **85**, which was available from synthesis by Syntex chemists (P. Nelson, unpublished).

Several oxygen and nitrogen functionalized sesquiterpenes showed interesting *in vitro* activity, as displayed in Fig. 13. Heading this list is dendrolasin (**86**) (Vanderah and Schmitz, 1974), a sesquiterpene furan from *Leiosella ?levis* (Kakou *et al.*, 1987) which showed strong *in vitro* activity, but did not show any significant activity *in vivo*. Eventually, the positive *in vitro* activity was ascribed to impurities present in the natural product, as dendrolasin obtained from synthesis (P. Nelson,

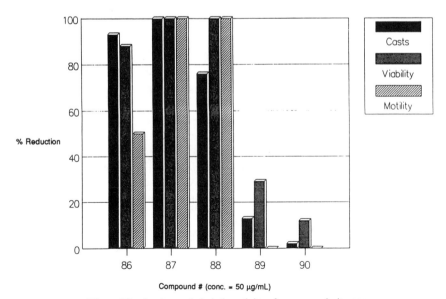

unpublished) showed very little *in vitro* activity. An undescribed sponge from Vanuatu was a source of the biosaboline amine **87** (Sullivan *et al.*, 1986; Kitagawa *et al.*, 1987), which was completely active *in vitro* but not *in vivo* (see Fig. 13). Two phenols (**88, 89**) (Wright *et al.*, 1987; Fusetani *et al.*, 1987) related to curcuphenol were isolated from *Didiscus havus* collected from Honduras (P. Kambhampati and P. Crews, unpublished); the former showed strong *in vitro* activity, while the latter was inactive. We have isolated a number of sesquiterpene quinones and quinols related to arenarone-ilimaquinone, as exemplified by **90** (Djura *et al.*, 1980) from *Dactylospongia elegans* (Rodriguez *et al.*, 1992), and none of these displays anthelmintic activity.

Figure 13. In vitro anthelmintic activity of oxygen and nitrogen.

3.2.5. *Isonitrile and Related Terpenoids*

Isonitrile, isothiocayanato, and related functionalized terpenes are characteristic metabolites of sponges belonging to the order Halichondrida. Several sesquiterpene examples of these compounds were isolated and tested for anthelmintic activity. Four amorphane sesquiterpenes (**91–94**) were isolated from the

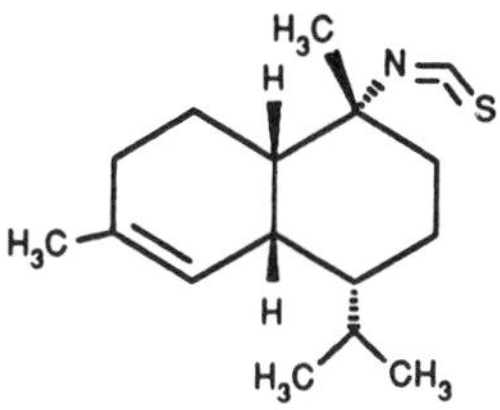

(91) 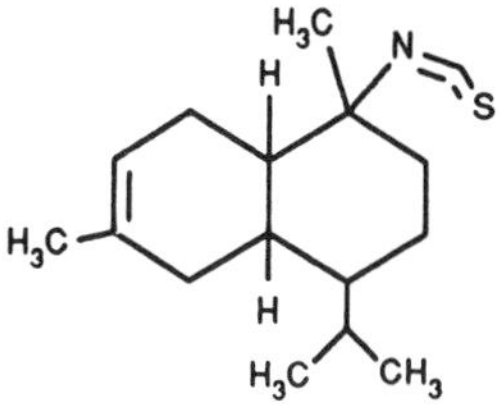   (92) 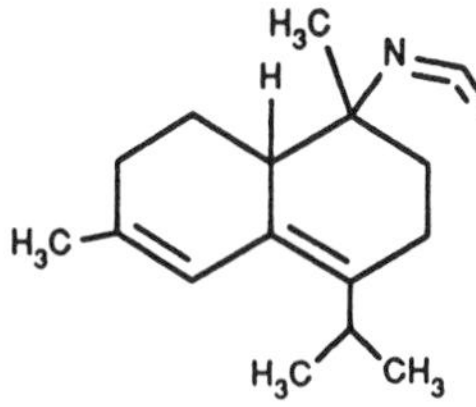   (93)

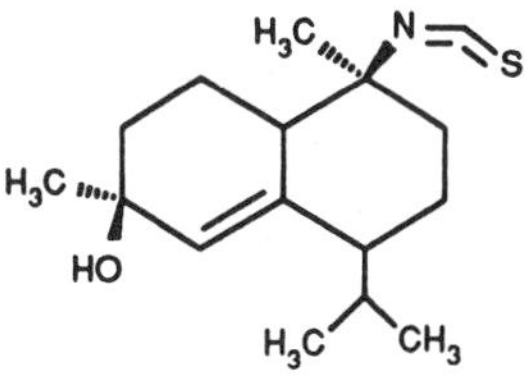

(94) 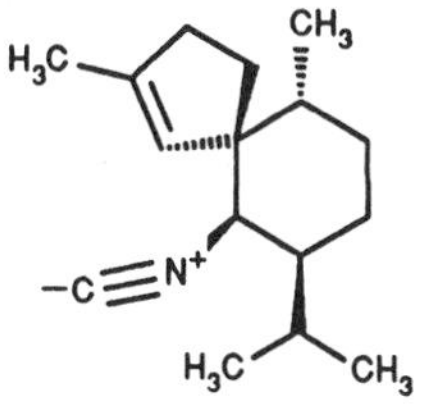   (95)

Fijian sponge *Axinyssa fenestratus* (Alvi *et al.*, 1991b); **91** was previously isolated from *Halichondria* sp. collected off of Hawaii (Burreson *et al.*, 1975). Three of these were extremely active *in vitro*, as shown in Fig. 14, but this compound was inactive below 50 μg/ml. Another sesquiterpene, axisonitrile-3 (**95**) (Braekman *et al.*, 1987a; Di Blassio *et al.*, 1976), was isolated as the single component of a *Topsentia* sp. from Thailand (Alvi *et al.*, 1991b). This compound showed superior activity *in vitro* at 50 μg/ml, but was not active *in vivo*.

The bright orange sponge *Acanthella cavernosa* produces a series of highly functionalized isonitrile and isothiocyanate diterpenes (Alvi *et al.*, 1991b; Chang

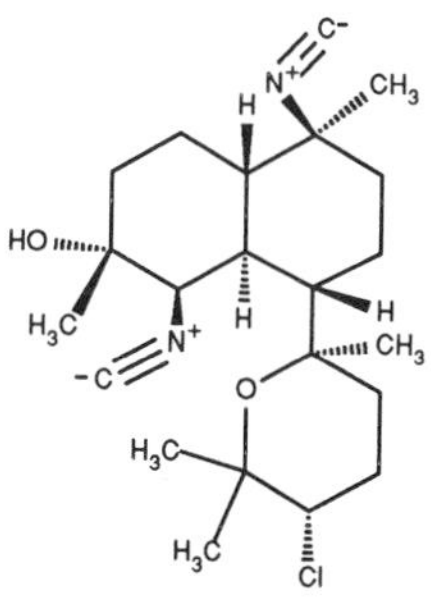

(96) 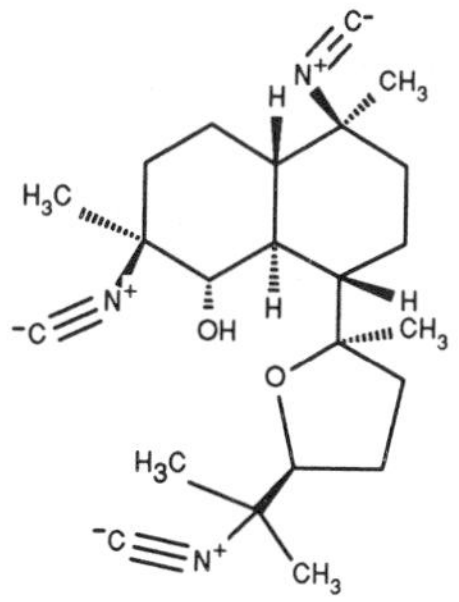   (97) 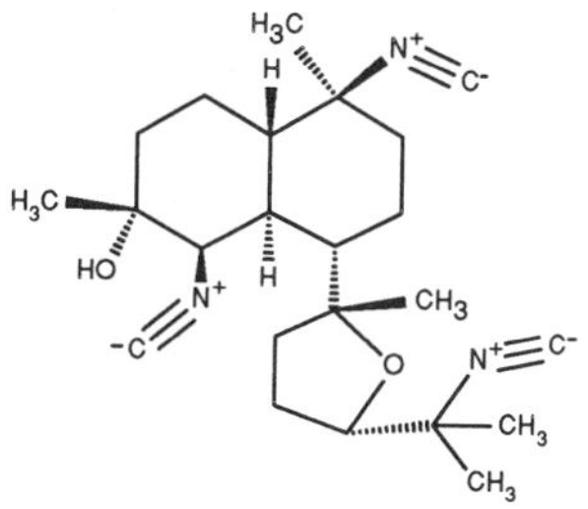   (98)

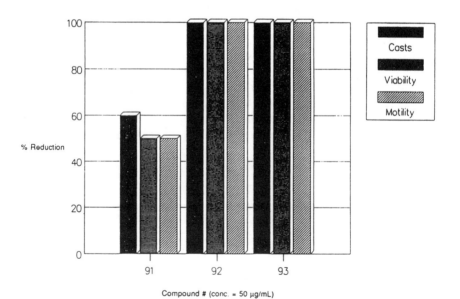

Figure 14. *In vitro* anthelmintic activity of isothiocyanate terpenoids.

et al., 1987; Omar *et al.*, 1988), including **96–102**. As summarized in Fig. 15, kalihinols Y (**101**) and J (**100**) showed potent *in vitro* activity. Kalihinols X (**99**), Z (**102**), and A (**96**) were moderately active. Kalihinol Y (**101**) was screened for *in vivo* activity, but did not show any reductions in the worm burden. *In vitro* tests have not been run on kalihinol F (**98**) nor have *in vivo* tests been run on Kalihinol J (**100**). Isokalihinol F (**97**), which has a slightly different isonitrile substitution pattern versus all other kalihinols (Omar *et al.*, 1988), did not show significant *in vitro* activity.

3.2.6. Diterterpenoids

A pair of strongylophorines (**103, 104**), members of a series of known meroditerpenes (Braekman *et al.*, 1978; Salva and Faulkner, 1990), were isolated from *Oceanapia* sp. collected in the Solomon Islands. Both were inactive in the primary *in vitro* screen.

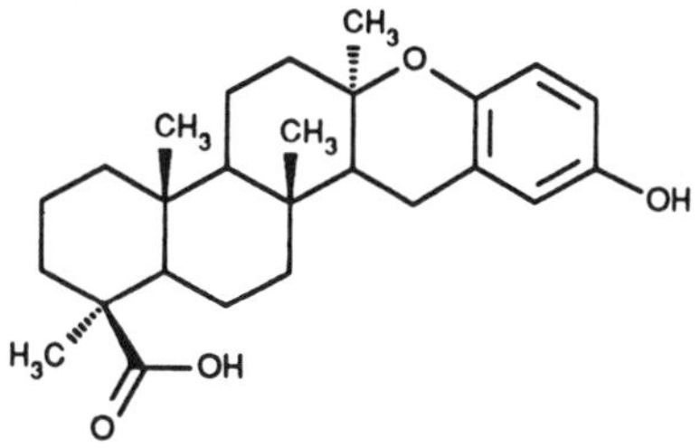

(103) 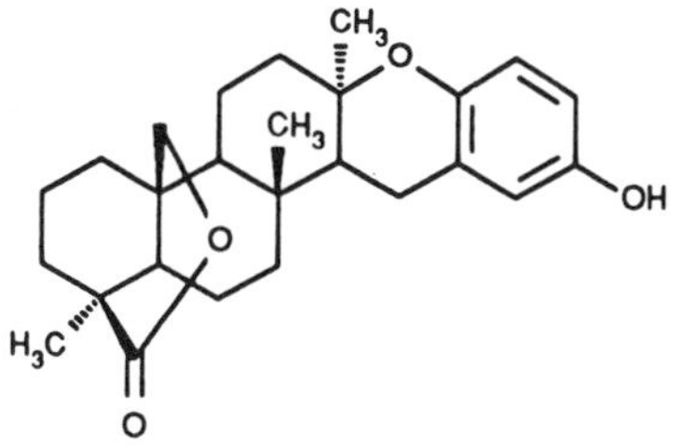   (104)

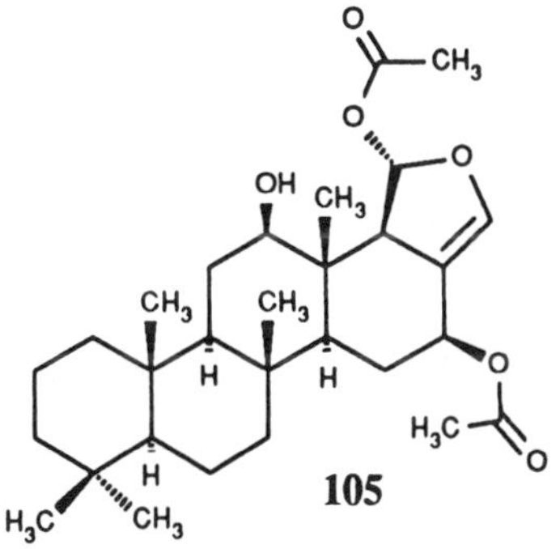

105

(105)

3.2.7. Sesterterpenoids

Sesterterpenes from the sponge *Hyrtios erecta* showed *in vitro* anthelmintic activity. Heteronemin (**105**) (Kazlauskas *et al.*, 1976; Kashman and Rudi, 1977), the major product of the sponge, was tested several times *in vitro* with varying

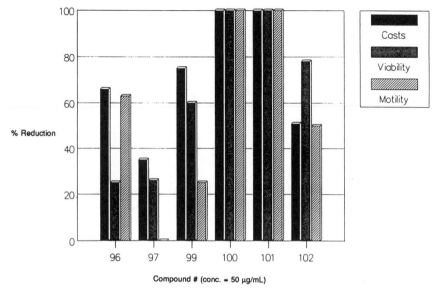

Figure 15. *In vitro* anthelmintic activity of kalihinols.

results, and showed activity a majority of times. Another compound, 12-epi-scalarin (**106**) (Crews and Bescansa, 1986; Cimino *et al.*, 1977), exhibited moderate *in vitro* activity, while inactive and closely related structures included 12-epi-scalaradial (**107**) (Cimino *et al.*, 1974a), 12-deacetyl-12-epi-scalaradial (**108**) (Crews and Bescansa, 1986), and scalarafuran (**109**) (Walker *et al.*, 1980).

(106)   (107)

3.3. Non-Sponge-Derived Compounds

The number of marine invertebrate compounds evaluated for anthelmintic activity from sources other than sponges is exceptionally small. A zoanthid

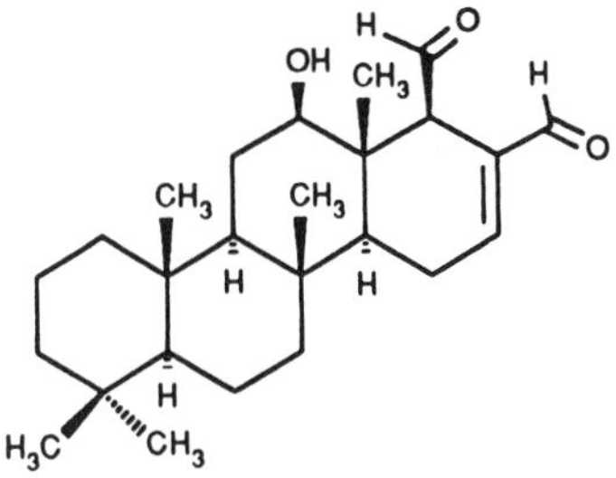

(108)

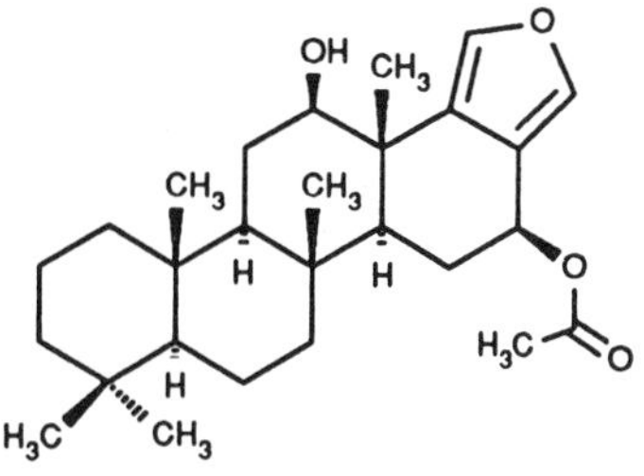

(109)

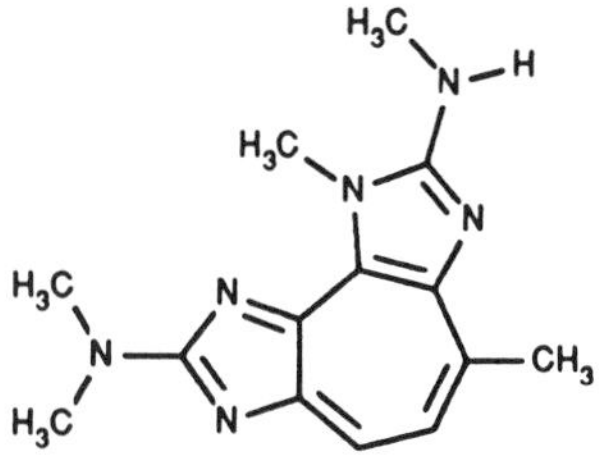

(110)

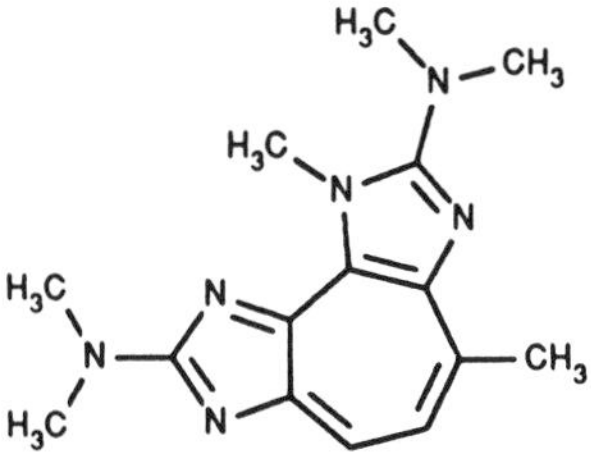

(111)

collected in Fiji yielded the zoanthoxanthins **110** (C. Jimenez and P. Crews, in press) and **111** (Schwart *et al.*, 1976), which were both inactive against *N. brasiliensis*.

4. SCREENING METHODS FOR ANTIPROTOZOAL AGENTS

4.1. The Disease Targets

The parasitic protozoans are single-celled organisms that are responsible for diseases such as malaria, sleeping sickness, and giardiasis. The most prevalent infectious diseases caused by parasitic protozoans are summarized in Table II. The protozoans can be transmitted to a host through contaminated water or via an infected insect carrier (vector). They create a wide range of medical conditions by invading the epithelia, tissue fluids (lymph or plasma), or the tissue cells of their host.

4.2. Bioassay Strategies

The diversity of protozoal organisms makes it difficult to engage in a broad-spectrum screening of natural products. Alternatively, a variety of excellent *in vitro* and *in vivo* antiprotozoal assays have been described in the literature. A feeling for these methods can be gained by considering the status of screens for

Trypanosoma (Scott and Matthews, 1987), *Giardia* (Edlind *et al.*, 1990; Keister, 1983), and *Plasmodium* (Mons and Sinden, 1990). A battery of primary, secondary, and *in vivo* assays against this trio would constitute coverage of the most important protozoal diseases of Table II.

5. ANTIPROTOZOAL-ACTIVE NATURAL PRODUCTS

The effective agents of choice as chemotherapeutics to treat protozoal parasites were summarized in Table II. Some entries in Table II, such as amphotericin B and tetracycline, are antimicrobial natural products. Others, such as mefloquine, are indirectly based on the structure of a nonmarine natural product. Very little published data can be found on the systematic evaluation of marine natural products for their antiprotozoal potential.

5.1. Known Agents of Nonmarine Origin

The natural products of nonmarine origin will always be important models for new antiprotozoal substances. One major reason is that the ethanobotanical literature contains a wealth of information on medicinal plants and their use against these diseases. Surprisingly, little evidence can be found in the literature of bioassay-guided studies to isolate antiprotozoal substances from these or related plants.

Malaria is the major tropical disease of humans, and despite decades of research, the need for both prophylactics and therapeutic agents is urgent. Of the four species of *Plasmodium* causing malaria, *P. falciparum* develops drug-resistant strains at an alarming rate. There is no drug available which will guarantee complete protection from *P. falciparum* (Peters, 1989). The most commonly used antimalarials are compounds related to quinine (**112**), a natural product isolated from *Cincona* bark. The resistance of *P. falciparum* to quinine and chloroquine (**113**), and the recent reports of resistance to mefloquine (**114**), which

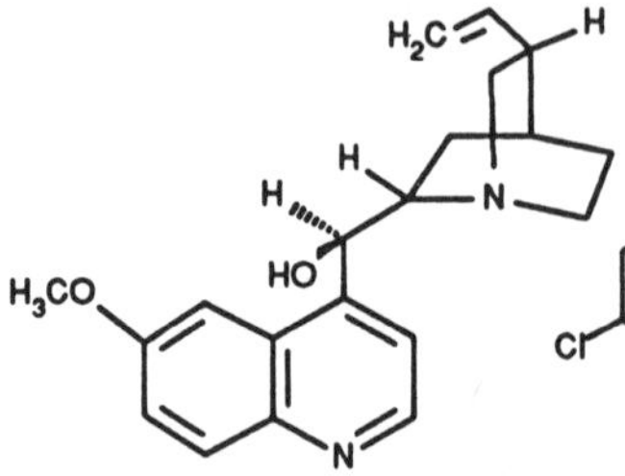

(112) 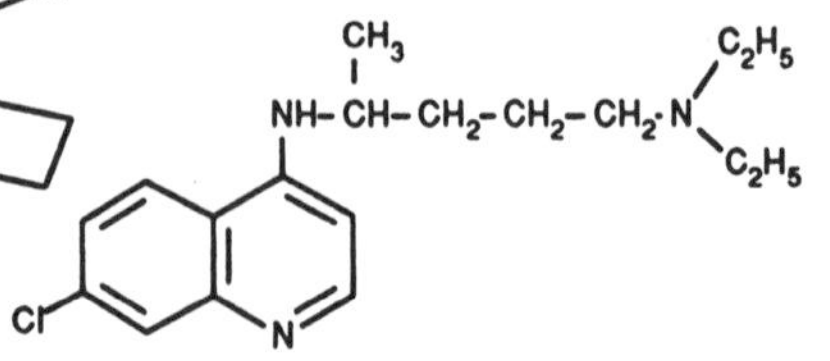   (113)

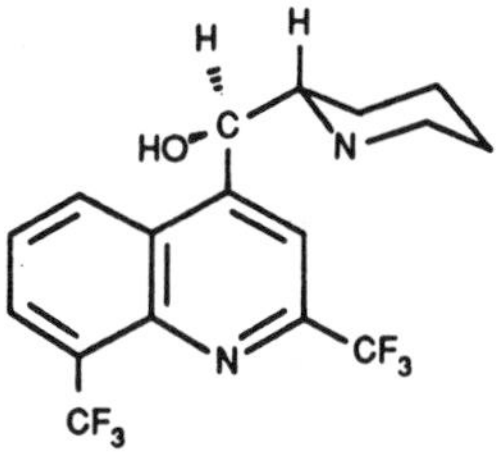

(114)

was introduced in 1984 (Peters, 1989), show the ease with which resistance is developed within this group of compounds. Some important new antimalarial agents in development are based on an ancient Chinese herbal malarial remedy. The lead compound qinghaosu (**115**), also known as artemisinin, is isolated from the plant *Artemesia annua*, and exhibits an IC_{50} of 1 ng/ml against both chloroquine-susceptible and chloroquine-resistant strains of *P. falciparum* (Kepler *et al.*, 1988)

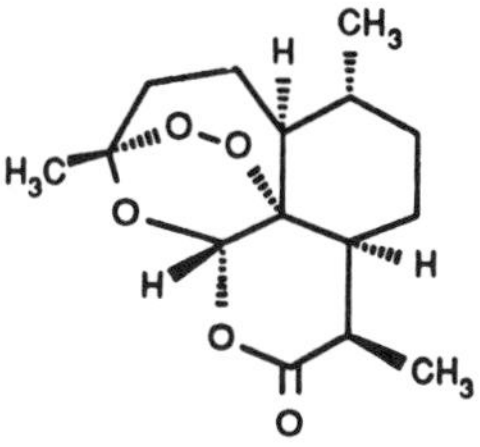

(115) 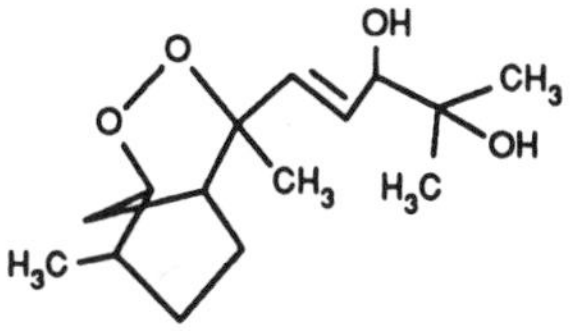   (116)

and is extremely potent against other *Plasmodium* strains, as summarized in Table V (König *et al.*, 1991). Qinghaosu exhibits some adverse side effects; consequently, synthetic derivatives are being advanced for clinical trials (Gutteridge,

Table V.   *In Vitro Antimalarial Activity Using Plasmodium falciparum[a]*

Compound	IC$_{50}$ (ng/ml) *Plasmodium falciparum* clone	
	West Africa (D-6)	Indochina (W-2)
Axisonitrile-3 (**95**)	32	900
Quinine (**112**)	19	66
Chloroquine (**113**)	3	35
	14[b]	16[b]
Mefloquine (**114**)	13	3
Qinghaosu (**115**)	5	2

[a]Data from König *et al.* (1991), except where noted.
[b]Data from Martin *et al.* (1987).

1989a). Yingzhaosu (**116**) is another traditional Chinese herbal medicine that has provided a lead for new antimalarials. A series of bicyclic peroxides in current development are patterned on its structure (Gutteridge, 1989a).

Other important antiparasitics based on natural products are as follows. Amphotericin B (**117**) can be used in the treatment of leishmaniases, which are a group of parasitic diseases with a wide geographic distribution caused by protozoans of the genus *Leishmania*. Amphotericin B, first isolated from *Streptomyces*

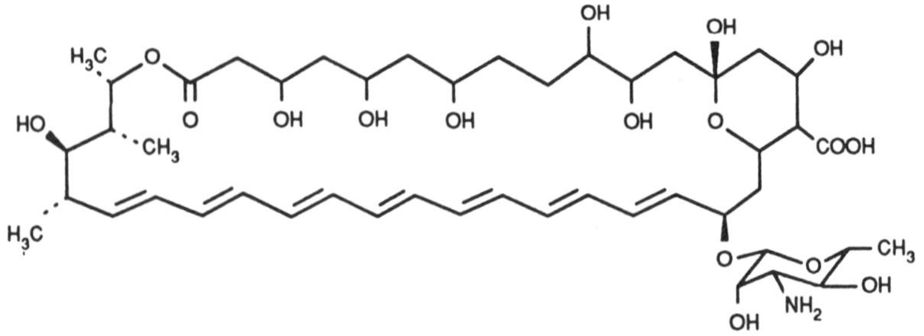

(**117**)

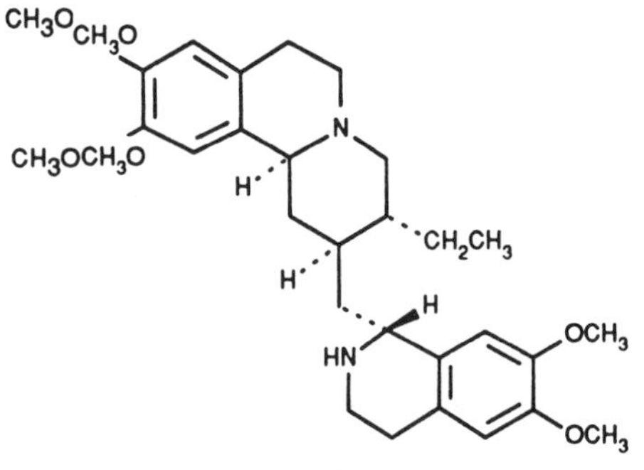

(**118**)

spp. (James and Gilles, 1985), is also under development for treatment of amoebic meningoencephalitis. Emetine (**118**) (Van Tamelen *et al.*, 1969), the first effective drug for treating amoebiasis, is an isoquinoline alkaloid isolated from roots of *Uragoga ipecacuanha*, a Brazilian shrub.

5.2. Marine-Derived Compounds

The list of marine natural products that have been evaluated for antiprotozoal properties is exceedingly short.

A series of six triterpenoid glycoside sulfates were isolated from the sea cucumber (Holothuroidea) *Cucumaria echinata*. One compound, cucumechino-

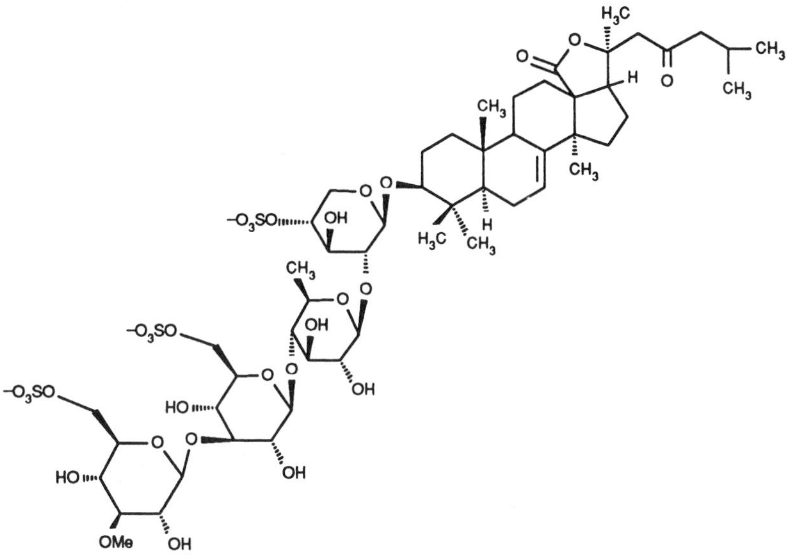

(119)

side F (119), was active, with a MIC (minimum growth inhibition concentration) of 10 μg/ml against a protozoan, *Trichomonas foetus* (Miyamoto *et al.*, 1990).

At one time a *Giardia* primary screen was used in the cooperative UCSC–Syntex program. A single batch of 24 extracts of sponges collected from Vanuatu was evaluated at a concentration of 50 μg/ml against *Giardia lamblia*. Three of these extracts were highly active and six were active. One of these latter extracts, 87014, was pursued in bioassay-guided isolation. Two compounds, plakinidines A (62) and B (63) (Inman *et al.*, 1990), were eventually isolated, but the *Giardia* assay had been discontinued before these could be reassayed. Alternatively, these compounds were observed to have excellent *in vitro* anthelmintic activity, as summarized in Fig. 9. Unfortunately, the deletion of *Giardia lamblia* as an assay target made it impossible to pursue the other promising active extracts.

A small number of marine algal and sponge natural products have been tested in antimalarial assays (König *et al.*, 1991). Only one such study has been reported and some 34 marine natural products were tested for *in vitro* sensitivities against *P. falicaparum*. Two strains were employed, including a chloroquine-sensitive West African clone (D-6) and a chloroquine-resistant Indochina clone (W-2). Axisonitrile-3 (95) proved to be active against strain D-6, but not against W-6, as shown by the data of Table V. No other marine natural products can be found in the literature with comparable activity against *Plasmodium*.

6. *FUTURE PROSPECTS*

A vast array of compounds of unprecedented structures and functionality has been isolated from marine invertebrate organisms during the last few decades. As we have shown above, only a fraction of these have been considered in testing programs using antiparasitic disease models. This troubling situation will undoubtedly persist in research carried out in the United States. Few of the major U.S. pharmaceutical companies invest in antiparasitic drug discovery programs. It is simply not cost-effective; the major market for such drugs are the people of developing countries, who usually cannot afford exotic medications (Vagelos, 1991). Likewise, very little funding to discover new antiparasitics is available from the U.S. National Institute of Allergy and Infectious Diseases (NIH AID) Institute. For example, in a recent conference on marine-derived pharmaceuticals, it was reported that among 500 research support awards within NIH AID, 43 were research projects dealing with natural products and none appeared to have antiparasitic agents as a discovery target (Delappe, 1988). The U.S. Army Medical Research and Development Command at Fort Detrick, Frederick, Maryland, does invest in antiparasitic agent research in connection with the Army Research Program on Antiparasitic Drugs, and malaria continues to be a prime disease target. Likewise, the Edna McConnell Clark Foundation in New York funds basic research on tropical parasite disease targets, especially against onchocerciasis (river blindness).

Collectively, the preceding discussions show that the surface has only been scratched in terms of using marine natural products as a tool to uncover antiparasitic drug leads. The clinically effective anthelmintic marine algal agent kainic acid, mentioned in the introduction, represents a successful early milestone. The current status of the search for antiparasitic marine natural products can be summed up by the 17 moderately and 21 strongly *in vitro* active anthelmintic compounds from sponges, a nudibranch, and a zoanthid, as shown in Table IV; the confirmed *in vivo* active anthelmintic compound bengamide F (**21**) discussed in Section 3.2.1; and the moderately *in vitro* antiprotozoal-active sea cucumber product cucumechinoside F (**119**) discussed in Section 5.2.

The results summarized above represent a modest but encouraging beginning. There are many opportunities for the future. An emphasis on mechanism-based assays against biochemical targets as exemplified by those discussed in Section 1.2 should be pursued. Another strategy for the future is implied by the intriguing recent discovery that levamasole (Van den Bossche and Janssen, 1969), an *in vivo* active anthelmintic (nematodes) assay standard, is also effective in the clinic against Dukes' C colon carcinoma (Laurie *et al.*, 1989; Moertel *et al.*, 1990). This suggests that the myriad of marine natural products, reviewed elsewhere in this book, which exhibit *in vitro* cytotoxicity to cancer cell lines or *in vivo* antitumor activity ought to be examined for their potential as antiparasitic-active substances.

ACKNOWLEDGMENTS Our research discussed in this review represents the efforts of many individuals. A number of important interactions and data came from the Syntex Research group of Dr. Tom Matthews and his staff, Virginia Scott, Elizabeth Fraser-Smith, and Stanford Bingham, Jr. Synthetic organic chemists at Syntex Research, Dr. Peter Nelson and Dr. David Loughhead, made important contributions. The efforts of our UCSC co-workers have, of course, been extremely important, and individuals participating in this research have included Dr. Barbara Myers (Sea Grant Trainee, SGT), Pedro Bescansa, Yao Kakou, Dr. Lawrence Manes (SGT), Dr. Wayne Inman (SGT), Dr. Madeline Adamczeski (SGT), Paul Horton (SGT), Dr. Emilio Quiñoá, Prof. Siraj Omar, Dr. Carlos Jiménez, Carolyn Albert, Dr. Tahsine Fanni, Dr. Khisal Alvi, Dr. Pullaiah Kambhampati, and Dr. Barbara Peters. Partial funding for the UCSC research was from NOAA, National Sea Grant College Program, Department of Commerce, University of California project Nos. R/MP-45, R/MP-41, and R/MP-33. Sea Grant Traineeships (SGT) are also gratefully appreciated. We are indebted to E. Fraser-Smith for providing the concept of Figs. 1 and 2. Thanks to Dr. R. Emrich, Dr. B. Peters, and Dr. D. Ball for their comments about this manuscript. We are grateful to A. Michels for her tireless work on the preparation of this manuscript.

REFERENCES

Adamczeski, M., Quiñoá, E., and Crews, P., 1988, Novel sponge derived amino acids 3. Unusual anthelminthic oxazoles from a marine sponge, *J. Am. Chem. Soc.* **110**:1598–1602.

Adamczeski, M., Quiñoá, E., and Crews, P., 1989, Novel sponge derived amino acids 5. Structures, stereochemistry and synthesis of several new heterocyclics, *J. Am. Chem. Soc.* **111**:647–654.

Adamczeski, M., Quiñoá, E., and Crews, P., 1990, Novel sponge derived amino acids 11. The entire absolute stereochemistry of the bengamides, *J. Org. Chem.* **55**:240–242.

Agabian, N., and Cerami, A. (eds.), 1990, *Parasites: Molecular Biology, Drug and Vaccine Design*, Wiley-Liss, New York.

Akee, R. H., Carroll, T. R., Yoshida, W. Y., Scheuer, P. J., Stout, T. J., and Clardy, J., 1990, Two imidazole alkaloids from a sponge, *J. Org. Chem.* **55**:1944–1946.

Alvi, K. A., Crews, P., and Loughhead, D. G., 1991a, Structures and total synthesis of 2-aminoimidazoles from a *Notodoris* nudibranch, *J. Nat. Prod.* **54**:1509–1515.

Alvi, K. A., Tenenbaum, L., and Crews, P., 1991b, Anthelmintic polyfunctional nitrogen containing terpenes from marine sponges, *J. Nat. Prod.* **54**:71–78.

Alvi, K. A., Peters, B. M., Hunter, L. M., and Crews, P., 1993, 2-Aminoimidazoles and their zinc complexes from Indo-Pacific *Leucetta* sponges and *Notodoris* nudibranchs, *Tetrahedron*, in press.

Arabshahi, L., and Schmitz, F. J., 1987, Brominated tyrosine metabolites from an unidentified sponge, *J. Org. Chem.* **52**:3584–3586.

Barrow, K. D., 1983, Biosynthesis of marine metabolites, in: *Marine Natural Products, Chemical and Biological Perspectives*, Vol. 5 (P. J. Scheuer, ed.), Academic Press, New York, pp. 60–61.

Baslow, M. H., 1977, *Marine Pharmacology*, Krieger, New York, pp. 69–70.

Braekman, J. C., Daloze, D., Hulot, G., Tursch, B., Declerq, J. P., Germain, G., and Van Meerssche, M., 1978, Chemical studies of marine invertebrates, XXXVII. Three novel meroditerpenoids from the sponge *Strongylophora durissima*, *Bull. Soc. Chim. Belg.* **87:**917–926.

Braekman, J. C., Daloze, D., Deneubourg, F., Huysecom, J., and Vandevyer, G., 1987a, 1-Isocyano-aromadendrane, a new isonitrile sesquiterpene from the sponge *Acanthella acuta*, *Bull. Soc. Chem. Belg.* **96:**539–543.

Braekman, J. C., Daloze, D., Moussiaux, B., and Riccio, R., 1987b, Jaspamide from the marine sponge *Jaspis johnstoni*, *J. Nat. Prod.* **50:**994–995.

Broka, C., and Ehrler, J., 1991, Enantioselective total synthesis of bengamides B and E, *Tetrahedron Lett.* **32:**5907–5910.

Burkholder, P. R., and Burkholder, L. M., 1958, Antimicrobial activity of horny corals, *Science* **127:** 1174–1175.

Burreson, B. J., Christophersen, C., and Scheuer, P. J., 1975, Co-occurrence of two terpenoid isocyanide–formamide pairs in a marine sponge (*Halichondria* sp.), *Tetrahedron* **31:**2015–2018.

Campbell, W. C., 1985, Ivermectin: An update, *Parasitol. Today* **1:**10–16.

Campbell, W. C., Fisher, M. H., Stapley, E. O., Alberg-Schonberg, G., and Jacob, T. A., 1983, Ivermectin: A potent new antiparasitic agent, *Science* **221:**823–828.

Capon, R., Ghisalberti, E. L., Jefferies, P. R., Skelton, B. W., and White, A. H., 1981, Structural studies of halogenated diphenyl ethers from a marine sponge, *J. Chem. Soc. Perkin Trans. I* **1981:** 2464–2467.

Carmely, S., Ilan, M., and Kashman, Y., 1989, 2-Amino imidazole alkaloids from the marine sponge *Leucetta chagosensis*, *Tetrahedron* **45:**2193–2200.

Carté, B., and Faulkner, D. J., 1981, Polybrominated diphenyl ethers from *Dysidea herbacea*, *Dysidea chlorea* and *Phyllospongia foliascens*, *Tetrahedron* **37:**2335–2339.

Chang, C. W. J., and Weinheimer, A. J., 1977, 2-Hydroxy-3,5-dibromo-4-methoxyphenylacetamide. A dibromotyrosine metabolite from *Psammaplysilla purea*, *Tetrahedron Lett.* **1977:**4005–4008.

Chang, C. W., Patra, A., Baker, J. A., and Scheuer, P. J., 1987, Kalihinols, multifunctional diterpenoid antibiotics from marine sponges *Acanthella* spp., *J. Am. Chem. Soc.* **109:**6119–6123.

Chida, N., Tobe, T., and Ogawa, S., 1991, Total synthesis of bengamide E, *Tetrahedron Lett.* **32:** 1063–1066.

Ciminiello, P., Fattorusso, E., Magno, S., and Mangoni, A., 1989, Clathridine and its zinc complex, novel metabolites from the marine sponge *Clathrina clathrus*, *Tetrahedron* **45:**3873–3878.

Cimino, G., De Stefano, S., and Minale, L., 1974a, Scalaradial, a third sesterterpene with the tetra-carbocyclic skeleton of scalarin, from the sponge *Cacospongia mollior*, *Experientia* **30:**846–847.

Cimino, G., De Stefano, S., and Minale, L., 1974b, Occurrence of hydroxyhydroquinone and 2-aminoimidazole in sponges, *Comp. Biochem. Physiol.* **47B:**895–897.

Cimino, G., De Stefano, S., Minale, L., and Trivellone, E., 1977, 12-Epi-scalarin and 12-epi-deoxyscalarin, sesterterpenes from the sponge *Spongia nitens*, *J. Chem. Soc. Perkin Trans. I* **1977:** 1587–1588.

Cook, G. C., 1990, Use of benzimidazole chemotherapy in human helminthiases: Indications and efficacy, *Parasitol. Today* **6:**133–136.

Cook, G. C., 1991, Anthelminthic agents: Some recent developments and their clinical application, *Postgrad. Med. J.* **67:**16–22.

Corely, D. G., Herb, R., Moore, R. E., Scheuer, P. J., and Paul, V. J., 1988, Lauliamalides: New potent cytotoxic macrolides from a marine sponge and nudibranch predator, *J. Org. Chem.* **53:**3644–3646.

Crews, P., and Bescansa, P., 1986, Sesterterpenes from a common marine sponge, *Hyrtios erecta*, *J. Nat. Prod.* **49:**1041–1052.

Crews, P., Manes, L. V., and Boehler, M., 1986, Novel sponge derived amino acids 1. Jasplakinolide, a cyclodepsipeptide from the marine sponge, *Jaspis* sp., *Tetrahedron Lett* **27**:2797–2800.

Crews, P., Kakou, Y., and Quiñoá, E., 1988, Novel sponge derived amino acids 4. Mycothiazole, a polyketide heterocycle from a marine sponge, *J. Am. Chem. Soc.* **110**:4365–4368.

Delappe, I. P., 1988, *Pharmaceuticals and the Sea* (C. W. Jefford, K. L. Rinehart, and L. S. Shield, eds.), Technomic, Basel, pp. 23–26.

Di Blassio, G., Fattorusso, E., Mango, S., Mayol, L., Pedone, C., Santacroce, C., and Sica, D., 1976, Axisonitrile-3, axisothiocyanate-3, and axamide-3, sesquiterpenes with a novel spiro[4,5]decane skeleton from the sponge *Axinella cannabina*, *Tetrahedron* **32**:473–478.

Djura, P., Stierle, D. B., Sullivan, B., and Faulkner, D. J., 1980, Some metabolites of the marine sponges *Smenospongia aurea* and *Smenospongia (= Polyfibrospongia) echina*, *J. Org. Chem.* **45**: 1435–1441.

Doering, T. L., Raper, J., Buxbaum, L. U., Adams, S. P., Gordon, J. I., Hart, G. W., and Englund, P. T., 1991, An analog of myristic acid with selective toxicity for African trypanosomas, *Science* **252**: 1852–1854.

Eckholm, E., 1989, *N. Y. Times Mag.* **1989**(January):20.

Edlind, T. D., Hang, T. L., and Chakraborty, P. R., 1990, Activity of the anthelmintic benzimidazoles against *Giardia lamblia in vitro*, *J. Infect. Dis.* **162**:1408–1411.

Evans, P. D., Thonor, C., and Midgley, J. M., 1988, Activities of octopamine and synephrine stereoisomers on octopaminergic receptor subtypes in locust skeletal muscle, *J. Pharm. Pharmacol.* **40**:855–861.

Faulkner, D. J., 1978, Antibiotics from marine organisms, in: *Topics in Antibiotic Chemistry*, Vol. 2 (P. G. Sammes, ed.), Ellis Harwood, pp. 9–58.

Faulkner, D. J., 1991, Marine natural products, *Nat. Prod. Rep.* **7**:97–147.

Foote, S. J., Thompson, J. K., Marshall, V., Cowan, A. F., Biggs, B. A., Brown, G. V., and Kemp, D. J., 1990, The miltidrug resistance gene of *P. falciparum*; Does it mediate chloroquine resistance? in: *Parasites: Molecular Biology, Drug and Vaccine Design* (N. Agabian and A. Cerami, eds.), Wiley-Liss, pp. 325–334.

Freedman, D. O., Zierdt, W. S., Lujan, A., and Nutman, T. B., 1989, The efficiency of ivermectin in the chemotherapy of gastrointestinal helminthiasis in humans, *J. Infect. Dis.* **159**:1151–1153.

Frincke, J. M., and Faulkner, D. J., 1982, Antimicrobial metabolites of the sponge *Reniera* sp., *J. Am. Chem. Soc.* **104**:265–269.

Fusetani, N., Antifungal substances from marine invertebrates, 1988, *Ann. N. Y. Acad. Sci.* **554**: 113–127.

Fusetani, N., Sugano, M., Matsunaga, S., Hashimoto, K., 1987, (+)-Curcuphenol and dehydrocurcuphenol, novel sesquiterpenes which inhibit H,K-ATPase, from a marine sponge *Epipolasis* sp., *Experientia* **43**:1234–1235.

Gopichand, Y., and Schmitz, F. J., 1979, Marine natural products: Fistularin-1 and -3 from the sponge *Aplysina fistularis* forma *fulva*, *Tetrahedron Lett.* **41**:3921–3924.

Grieco, P. A., Hon, Y. S., and Perez-Medrano, A., 1988, A convergent enentiospecific total synthesis of the novel cyclodepsipeptide (+)-jasplakinolide, (jaspamide), *J. Am. Chem. Soc.* **110**:1630–1631.

Groweiss, A., Shmueli, U., and Kashman, Y., 1983, Marine toxins of *Latrunculia magnifica*, *J. Org. Chem.* **48**:3512–3516.

Gulavita, N. K., and Scheuer, P. J., 1989, Isolation of the two epimers of aminotetradecadienol from a *Xestospongia* species, *J. Org. Chem.* **54**:366–369.

Gutteridge, W. E., 1989a, Antimalarial drugs currently in development, *J. R. Soc. Med.* **82**(Suppl. 17): 63–66.

Gutteridge, W. E., 1989b, Parasite vaccines versus anti-parasite drugs: Rivals or running mates, *Parasitology* **98**:S87–S97.

Hart, D., Langridge, A., Barlow, D., and Sutton, B., 1989, Antiparasitic drug design, *Parasitol. Today* **5**:114–120.

Horton, P., Inman, W. D., and Crews, P., 1990, Anthelmintic enantiomeric heterocycles from *Dysidea* marine sponges, *J. Nat. Prod.* **53**:143–151.

Inman, W. D., and Crews, P., 1989, Novel sponge derived amino acids 8. Conformational analysis of jasplakinolide, *J. Am. Chem. Soc.* **111**:2822–2829.

Inman, W. D., Crews, P., and McDowell, R., 1989, Novel sponge derived amino acids 9. Lithium complexation of jasplakinolide, *J. Org. Chem.* **54**:2523–2526.

Inman, W. D., O'Neill-Johnson, M., and Crews, P., 1990, Novel marine sponge alkaloids 1. Plakinidine a and b, anthelmintic active alkaloids from a *Plakortis* sponge, *J. Am. Chem. Soc.* **112**:1–4.

Ireland, C. M., Roll, D. M., Molinski, T. F., McKee, T. C,. Zabriskie, T. M., and Swersey, J. C., 1988, Uniqueness of the marine chemical environment: Categories of marine natural products from invertebrates, in: *Biomedical Importance of Marine Organisms* (D. G. Fautin, ed.), California Academy of Sciences, San Francisco, pp. 41–57.

James, D. M., and Gilles, H. M., 1985, *Human Antiparasitic Drugs: Pharmacology and Usage*, Wiley, New York.

Jenkins, D. C., Armitage, R., and Carrington, T. S., 1980, A new primary screen test for anthelmintics utilizing the parasitic stages of *Nippostrongylus brasiliensis*, *in vitro*, *Z. Parasitenkd.* **63**:261–269.

Jiménez, C., and Crews, P., 1990, Novel sponge derived amino acids 10. Xestoaminols from *Xestospongia* sp., *J. Nat. Prod.* **53**:978–982.

Jiménez, C., and Crews, P., 1991, Novel sponge-derived amino acids 13. Additional psammaplin derivatives from *Psammaplysilla purpurea*, *Tetrahedron* **47**:2097–2102.

Jiménez, C., Quiñoá, E., Adamczeski, M., Hunter, L. M., and Crews, P., 1991a, Novel sponge derived amino acids 12. Tryptophan type derivatives from *Fascaplysinopsis reticulata*, *J. Org. Chem.* **56**:3403–3410.

Jiménez, C., Quiñoá, E., and Crews, P., 1991b, Novel marine sponge alkaloids 3. β-Carbolinium salts from *Fascaplysinopsis reticulata*, *Tetrahedron Lett.* **47**:3585–3600.

Jiménez, C. and Crews, P., 1993, Carbon-13 NMR assignments and cytotoxicity assessment of zoanthoxanthin alkaloids from zoanthid corals, *J. Nat. Prod.*, in press.

Kakou, Y., Crews, P., and Bakus, G. J., 1987, Dendrolasin and latrunculin A from the Fijian sponge *Spongia mycofijiensis* (Bakus) and *Chromodoris lochi*, *J. Nat. Prod.* **50**:482–484.

Kashman, Y., and Rudi, A., 1977, The ^{13}C-NMR spectrum and stereochemistry of heteronemin, *Tetrahedron* **33**:2997–2998.

Kashman, Y., Groweiss, A., and Shmueli, U., 1980, Latrunculin, a new 2-thiazolidinone macrolide from the marine sponge *Latrunculia magnifica*, *Tetrahedron Lett.* **21**:3629–3632.

Kazlauskas, R., Murphy, P. T., Quinn, R. J., and Wells, R. J., 1976, Heteronemin a new scalarin type sesterterpene from the sponge *Heteronema erecta*, *Tetrahedron Lett.* **1976**:2631–2634.

Kazlauskas, R., Murphy, P. T., Quinn, R. J., and Wells, R. J., 1977, Aplysinopsin, a new tryptophan from a sponge, *Tetrahedron Lett.* **1977**:61–64.

Kazlauskas, R., Murphy, P. T., Wells, R. J., Daly, J. J., and Schonholzer, P., 1978, Two sesquiterpene furans with new carbocyclic ring systems and related thiol acetates from a species of the sponge genus *Dysidea*, *Tetrahedron Lett.* **49**:4951–4954.

Keifer, P. A., Schwartz, R. E., Koker, M. E. S., Hughes, R. G. Jr., Rittschof, D., and Rinehart, K. L., 1991, Bioactive bromopyrrole metabolites from the Caribbean sponge *Agelas conifera*, *J. Org. Chem.* **56**:2965–2975.

Keister, D. B., 1983, Axenic culture of *Giardia lamblia* in TYI-S-33 medium supplemented with bile, *Trans. R. Soc. Trop. Med. Hyg.* **77**:487–488.

Kepler, J. A., Philip, A., Lee, Y. W., Morey, M. C., and Carroll, F. I., 1988, 1,2,4-Trioxanes as potential antimalarial agents, *J. Med. Chem* **31**:713–716.

Kitagawa, I., Yoshioka, N., Kamba, C., Yoshikawa, M., and Hamamoto, Y., 1987, Four new bisabolene-type aminosesquiterpenes from an Okinawan marine sponge, *Theonella* sp. (Theonellidae), *Chem. Pharm. Bull.* **35**:928–931.

König, G. M., Wright, A. D., Sticher, O., Jurcic, K., Offermann, F., Redl, K., Wagner, H., Angerhofer, C. K., and Pezzuto, J. M., 1991, Evaluation of the cytotoxic, antimalarial and antiinflammatory activities of compounds isolated from marine organisms, in: *International Research Congress on Natural Products, 32nd Annual Meeting of the American Society of Pharmacognosy*, Abstract P-129.

Laurie, J. A., Moertel, C. G., Fleming, T. R., Wieand, H. S., Leigh, J. E., Rubin, J., McCormac, G. W., Gerstner, J. B., Krook, J. E., and Malliard, J., 1989, Surgical adjuvant therapy of large-bowel carcinoma: An evaluation of levamisole and the combination of levamisole and fluorouracil. A study of the north central cancer treatment group and the Mayo clinic, *J. Clin. Oncol.* **7**:1447–56.

Lee, R. H., Slate, D. L., Moretti, R., Alvi, K. A., and Crew, P., 1992, Marine sponge polyketide inhibitors of protein tyrosine kinase, *Biochem. Biophys. Res. Commun.* **184**:765–772.

Li, W. R., Ewing, W. R., Harris, B. D., and Joullié, M. M., 1990, Total synthesis and structural investigations of didemnins A, B, and C, *J. Am. Chem. Soc.* **112**:7659–7672.

Litaudon, M., and Guyot, M., 1986, Ianthelline, un nouveau derivé de la dibromo-3,5-tyrosine, isolé de l'éponge *Ianthella ardis* (Bahamas), *Tetrahedron Lett.* **27**:4455–4456.

Makarieva, T. N., Stonik, V. A., Alcolado, P., and Elyakov, Y. B., 1981, Comparative study of the halogenated tyrosine derivatives from demospongiae (Porifera), *Comp. Biochem. Physiol.* **68B**:481–484.

Martin, S. K., Odvola, A. M. J., and Milhous, W. K., 1987, Reversal of chloroquine resistance in *Plasmodium falciparum* by verapamil, *Science* **235**:899–901.

Mayer, A. M. S., and Jacobs, R. S., 1988, Manoalide: An antiinflammatory and analgesic marine natural product, in: *Biomedical Importance of Marine Organisms* (D. G. Fautin, ed.), California Academy of Sciences, San Francisco, pp. 125–132.

McIntyre, D. E., Faulkner, D. J., Van Engen, D., and Clardy, J., 1979, Renierone, an antimicrobial metabolite from a marine sponge, *Tetrahedron Lett.* **1979**:4163–4166.

Medical Letter, 1988, Drugs for Parasitic Infections, *Medical Letter on Drugs and Therapeutics*, Medical Letter, Inc., Vol. 30, No. 759 (12 February).

Miyamoto, T., Togawa, K., Higuchi, R., Komori, T., and Sasaki, T., 1990, Six newly identified biologically active triterpenoid glycoside sulfates from the sea cucumber *Cucumaria echinata*, *Liebigs Ann. Chem.* **1990**:453–460.

Moertel, C. G., Fleming, T. R., MacDonald, J. S., Haller, D. G., Laurie, J. A., Goodman, P. J., Ungerleider, J. S., Emerson, W. A., Tormey, D. C., and Glick, J. H., 1990, Levamisole and fluorouracil for adjuvant therapy of resected colon carcinoma, *N. Engl. J. Med.* **322**:352–358.

Mons, B., and Sinden, R. E., 1990, Laboratory models for research *in vivo* and *in vitro* on malaria parasites of mammals: Current status, *Parasitol. Today* **6**:3–7.

Moore, R. E., 1985, Structure of palytoxin, *Prog. Chem. Org. Nat. Prod.* **48**:81–202.

Moore, R. E., Patterson, G. M. L., and Carmichael, W. W., 1988, New pharmaceuticals from cultured blue-green algae, in: *Biomedical Importance of Marine Organisms* (D. G. Fautin, ed.), California Academy of Sciences, San Francisco, pp. 143–150.

Munro, M. H. G., Luibrand, R. T., and Blunt, J. W., 1987, The search for antiviral and anticancer compounds from marine organisms, *Bioorgan. Mar. Chem.* **1**:94–176.

Murata, M., Legrand, A. M., Ishibashi, Y., Fukui, M., and Yasumoto, T., 1990, Structures and configurations of ciguatoxin from the moray eel *Gymnothorax javanicus* and its likely precursor from the dinoflagellate *Gambierdiscus toxicus*, *J. Am. Chem. Soc.* **112**:4380–4386.

Myers, B. L., and Crews, P., 1983, Chiral ether glycerides from a marine sponge, *J. Org. Chem.* **48**:3583–3585.

Nakamura, H., Kobayashi, Y., Ohizumi, Y., and Hirata, Y., 1982, Isolation and structure of aaptamine, a novel heteroaromatic substance possessing α-blocking activity from the sea sponge *Aaptos aaptos*, *Tetrahedron Lett.* **23**:5555–5558.

Nakamura, H., Ohizumi, Y., Kobayashi, J., and Hirata, Y., 1984, Keramadine, a novel antagonist of serotonergic receptors isolated from the Okinawan sea sponge *Agelas* sp., *Tetrahedron Lett.* **25:** 2475–2479.

Nakamura, H., Wu, H., Kobayashi, J., Makamura, Y., and Ohizumi, Y., 1985, Puralin, a novel enzyme activator from the Okinawan marine sponge *Psammaplysilla purea*, *Tetrahedron Lett.* **26:**4517– 4520.

Nakamura, H., Kobayashi, J., and Ohizumi, Y., 1987, Aaptamines. Novel benzo[*de*][1,6]naph-thyridines from the Okinawan marine sponge *Aaptos aaptos*, *J. Chem Soc. Perkin Trans. I* **1987:** 173–176.

Nigrelli, R. F., Jakowska, S., and Calventi, I., 1959, Ectyonin, an antimicrobial agent from the sponge *Microciona prolifera*, *Zoologica* **44:**173–176.

Norte, M., Rodriguez, M. L., Fernandez, J. J., Eguren, L., and Estrada, D. M., 1988, Aplysinadiene and (R,R) 5 [3,5-dibromo-4-[(2-oxo-5-oxazolidinyl)] methoxyphenyl]-2-oxazolidinone, two novel metabolites from *Aplysina aerophoba*. Synthesis of aplysinadiene, *Tetrahedron* **44:**4973– 4980.

Norton, R. S., Croft, K. D., and Wells, R. J., 1981, Polybrominated oxydiphenol derivatives from the sponge *Dysidea herbacea*, *Tetrahedron* **37:**2341–2349.

Omar, S., Albert, C., Fanni, T., and Crews, P., 1988, Polyfunctional diterpene isonitriles from a marine sponge, *Acanthella cavernosa*, *J. Org. Chem.* **53:**5971–5972.

Peters, W., 1989, Changing pattern of antimalarial drug resistance, *J. R. Soc. Med.* **82:**14–17.

Pettit, G. R., Kamano, Y., Herald, C. L., and Tozawa, M., 1984, Structure of bryostatin 4. An important antineoplastic constituent of geographically diverse *Bugula neritina* (Bryozoa), *J. Am. Chem. Soc.* **106:**6768–6771.

Pettit, G. R., Kamano, Y., Herald, C. L., Tuinman, A. A., Boettner, F. E., Kiza, H., Schmidt, J. M., Baczynskyj, L., Tomer, K. B., and Bontems, R. J., 1987, The isolation and structure of a remark-able marine animal antineoplastic constituent: Dolastatin 10, *J. Am. Chem. Soc.* **109:**6883–6885.

Powis, G., 1991, *Trends Pharmacol. Sci.* **12:**188–194.

Quiñoá, E., and Crews, P., 1987, Novel sponge derived amino acids 6. Phenolic constituents of *Psammaplysilla*, *Tetrahedron Lett.* **28:**3229–3232.

Quiñoá, E., and Adamczeski, M., and Crews, P., 1986, Bengamides, heterocyclic anthelmintics from a Jaspidae marine sponge, *J. Org. Chem.* **51:**4494–4497.

Quiñoá, E., Kakou, Y., and Crews, P., 1988, Fijianolides, polyketide heterocycles from a marine sponge, *J. Org. Chem.* **53:**3642–3644.

Reed, J. K., and Pomponi, S. L., 1989, Biomedical research in the sea, a search for drugs and novel compounds, in: *Diving for Science, 1989; Proceedings of the American Academy of Underwater Sciences Ninth Annual Scientific Diving Symposium* (M. A. Lang and W. C. Jaap, eds.), American Academy of Underwater Sciences, Costa Mesa, California, pp. 273–287.

Richou, O., Vaillancourt, V., Faulkner, D. J., and Albizati, K. F., 1989, Synthesis and absolute configuration of (−)-Furodysinin. New transformations of camphor derivatives, *J. Org. Chem.* **54:**4729–4730.

Rinehart, K. L., Gloer, J. B., Cook, J. C., Mizsak, S. A., and Scahill, T. A., 1981, Structures of the didemnins, antiviral and cytotoxic depsipeptides from a Caribbean tunicate, *J. Am. Chem. Soc.* **103:**1857–1859.

Rinehart, K. L., Kishore, V., Nagarajan, S., Lake, R. J., Gloer, J. B., Bozich, F. A., Li, K. M., Maleczka, R. E., Todsen, W. L., Munro, M. H. G., Sullins, D. W., and Sakai, R., 1987, Total synthesis of didemnins A, B, and C, *J. Am. Chem. Soc.* **109:**6846–6848.

Rodriguez, A. D., Akee, R. K., and Scheuer, P. J., 1987, Two bromotyrosine-cysteine derived metabolites from a sponge, *Tetrahedron Lett.* **28:**4989–4992.

Rodriguez, J., Quiñoá, E., Riguera, R., Peters, B. M., Abrell, L. M., and Crews, P., 1992, The structures and stereochemistry of cytotoxic sesquiterpene quinones from *Dactylospongia ele-gans*, *Tetrahedron Lett.* **48:**6667–6680.

Roll, D. M., Scheuer, P. J., Matsumoto, G. K., and Clardy, J., 1983, Halenaquinone, a pentacyclic polyketide from a marine sponge, *J. Am. Chem. Soc.* **105**:6177–6178.

Roll, D. M., Chang, C. W. J., Scheuer, P. J., Gray, G. A., Shoolery, J. N., Matsumoto, G. K., Van Duyne, G. D., and Clardy, J., 1985, Structure of the psammaplysins, *J. Am. Chem. Soc.* **107**: 2916–2920.

Roll, D. M., Ireland, C. M., Lu, H. S. M., and Clardy, J., 1988, Fascaplysin, an unusual antimicrobial pigment from the marine sponge *Fascaplysinopsis* sp., *J. Org. Chem.* **53**:3276–3278.

Salva, J., and Faulkner, D. J., 1990, Metabolites of the sponge *Strongylophora durissima* from Maricaban Island, Philippines, *J. Org. Chem.* **55**:1941–1943.

Schmidt, U., Siegel, W., and Mundinger, K., 1988, Total synthesis of jaspamide (jasplakinolide) and geodiamolide A and B—1. Stereoselective synthesis of (2*S*,4*E*,6*R*,8*S*)-8-hydroxy-2,4,6-trimethyl-4-nonenoic acid, *Tetrahedron Lett.* **28**:1269–1270.

Schwart, R. E., Yunker, M. B., Scheuer, P. J., and Ottersen, T., 1976, Constituents of bathyal marine organisms: A new zoanthoxanthin from a coelenterate, *Tetrahedron Lett.* **26**:2235–2238.

Scott, V. R., and Matthews, T. R., 1987, The efficacy of an N-substituted imidazole, RS-49676, against a *Trypanosoma cruzi* infection in mice, *Am. J. Trop. Med. Hyg.* **37**:308–313.

Sharma, G. M., and Magdoff-Fairchild, B., 1977, Natural products of marine sponges 7. The constitution of weakly basic guanidine compounds, dibromophakellin and monobromophakellin, *J. Org. Chem.* **42**:4118–4124.

Sharma, G. M., Vig, B., Burkholder, P. R., 1968, in: *Drugs from the Sea* (H. D. Freedenthal, ed.), Marine Technology Society, Washington, D.C.

Suffness, M., Newman, D. J., and Snader, K., 1989, Discovery and development of antineoplastic agents from natural sources, *Bioorgan. Mar. Chem.* **3**:132–168.

Sullivan, B. W., Faulkner, D. J., Okamoto, K. T., Chen, M. H. M., and Clardy, J., 1986, (6*R*,7*S*)-7-Amino-7,8-dihydro-α-bisabolene, an antimicrobial metabolite from the marine sponge *Halichondria* sp., *J. Org. Chem.* **51**:5134–5136.

Townsend, L. B., and Wise, D. S., 1990, The synthesis and chemistry of certain anthelmintic benzimidazoles, *Parasitol. Today* **6**:107–133.

Vagelos, P. R., 1991, Are prescription drug prices high? *Science* **252**:1080–1084.

Van den Bossche, H., and Janssen, P. A. J., 1969, Biochemical mechanism of action of antinematodal drug tetramisole, *Biochem. Pharmacol.* **18**:35–39.

Vanderah, D. J., and Schmitz, F. J., 1974, Marine natural products: Isolation of dendrolasin from the sponge *Oligoceras hemorrhages*, *J. Nat. Prod.* **38**:271–272.

Van Tamelen, E. E., Placeway, C., Schiemenz, G. P., and Wright, I. G., 1969, Total synthesis of *d,l*-ajmalicine and emetine, *J. Am. Chem. Soc.* **91**:7359–7371.

Walker, R. P., Thompson, J. E., and Faulkner, D. J., 1980, Sesterterpenes from *Spongia idia*, *J. Org. Chem.* **45**:4976–4979.

West, R. R., Mayne, C. L., Ireland, C. M., Brinen, L. S., and Clardy, J., 1990, Plakinidines: Cytotoxic alkaloid pigments from the Fijian sponge *Plakortis* sp., *Tetrahedron Lett.* **23**:3271–3274.

White, A. T., Newland, H. S., Taylor, H. R., Erttmann, K. D., Williams, P. N., and Greene, B. M., 1986, Controlled trial and dose finding study of ivermectin for treatment of onchocerciasis, *Trop. Med. Parasitol.* **37**:96–97.

Wu, H., Nakamura, H., Kobayashi, J., and Hirata, Y., 1986, Lipopurealins, novel bromotyrosine derivatives with long chain acyl groups, from the marine sponge *Psammaplysilla purea*, *Experientia* **42**:855–856.

Wright, A. E., Pomponi, S. A., McConnell, O. J., Kohmoto, S., and McCarthy, P. J., 1987, (+)-Curcuphenol and (+)-curcudiol, sesquiterpene phenols from shallow and deep water collections of the marine sponge *Didiscus flavus*, *J. Nat. Prod.* **50**:976–978.

Zabriskie, T. M., Klocke, J. A., Ireland, C. M., Marcus, A. H., Molinski, T. F., Faulkner, D. J., Xu, C., and Clardy, J. C., 1986, Jaspamide, a modified peptide from a *Jaspis* sponge, with insecticidal and antifungal activity, *J. Am. Chem. Soc.* **108**:3123–3124.

10

Dinoflagellates as Sources of Bioactive Molecules

Yuzuru Shimizu

1. INTRODUCTION

Marine organisms, particularly marine invertebrates, have been shown to be a rich source of highly bioactive compounds. At the same time, there is growing appreciation that many of the fascinating molecules found in marine animals are actually derived from marine microorganisms through the food chain or symbiotic relationship. Since most of the marine animals which contain active principles are severely limited as collectable resources, more and more attention has been drawn to the culturable and dependable sources for potentially important medicinals. In fact, dinoflagellates were the first microorganisms recognized as the primary source of certain secondary metabolites, i.e., shellfish toxins, found in marine invertebrates. Recently, a number of other bioactive molecules such as polyethers and macrolides have been found to have their origins in dinoflagellates, and efforts have been made to achieve the cultural production of these potentially useful substances. This chapter is intended to provide an overview of the current status of

Yuzuru Shimizu • Department of Pharmacognosy and Environmental Sciences, College of Pharmacy, The University of Rhode Island, Kingston, Rhode Island 02881.

Marine Biotechnology, Volume 1: Pharmaceutical and Bioactive Natural Products, edited by David H. Attaway and Oskar R. Zaborsky. Plenum Press, New York, 1993.

the research on the exploration or development of biomedical agents from the dinoflagellates.

2. ORGANISMS

2.1. Classification of Dinoflagellates

Dinoflagellates are unicellular microalgae found in both marine and fresh-water environments. They are eukaryotes, but the histone contents of dinoflagellate nuclei are normally very low, and some people place the organisms somewhere between prokaryotes and eukaryotes, calling them mesokaryotes. While dinoflagellates are generally considered to be plants, many of them are actually devoid of chlorophylls and live by heterotrophy. In that respect, dinoflagellates also occupy a unique position between the plants and the animals. In fact, some of the genera are concurrently classified as animals. The taxonomy of dinoflagellates has been always controversial, and individual taxonomists have presented it in their own terms. It is far from defined, and changes in generic names are frequently made. Recent comprehensive treatises on dinoflagellate taxonomy are by Taylor (1987) and Doge (1984).

Photosynthetic or nonphotosynthetic, dinoflagellates can be divided into three categories: free-living, symbiotic, and parasitic (Table I). Of the free-living species, some live in close association with certain macroalgae and other organisms. This association may be very significant in considering the culturability of the organisms and the production of specific metabolites.

Table I. Some Common Dinoflagellate Genera and Their Ordinary Living and Nutritional Modes

Genum	Living mode	Major nutritional mode
Alexandrium	Free-living	Autotrophy
Amphidinium	Free-living, symbiotic	Autotrophy, heterotrophy, facultative heterotrophy
Dinophysis	Free-living	Heterotrophy?, facultative heterotrophy
Gymnodinium	Free-living	Autotrophy, heterotrophy, phagotrophy
Noctiluca	Free-living	Heterotrophy, phagotrophy
Oodinium	Parasitic	Heterotrophy
Ostreopsis	Free-living	Autotrophy
Polykritos	Free-living	Heterotrophy (phagotrophy)
Prorocentrum	Free-living	Autotrophy, heterotrophy
Zooxanthella (Symbiodinium)	Symbiotic, free-living	Facultative heterotrophy, autotrophy, heterotrophy

2.2. Culturing of Dinoflagellates

While many people are enthusiastically considering dinoflagellates as a promising source of medicinals, two crucial underlying issues which are not fully appreciated are (1) whether the particular organism can be cultured and (2) whether the organism can produce the desired compound in culture. Discouragingly, out of the probably 4000 known organisms, only a very limited number of species are successfully cultured (Guillard and Keller, 1984). This difficulty of growing dinoflagellates is largely due to their unique living modes. Many of them live in complex ecosystems—symbiosis, association, and other forms of biointeraction, where nutrient and metabolite transfers are prevalent. Many of them can grow only under a delicate balance of nutrients and physical conditions, such as the state of heavy metal ions, salinity, and temperature. Many organisms are speculated to require sexual reproduction, and there may be a problem of incomplete reproductive cycle in culture. The most recent and extensive review of this subject was given by Guillard and Keller (1984).

Based on their nutritional requirements, dinoflagellates can be classified into autotrophic, heterotrophic, and facultative heterotrophic organisms. However, the actual nutrient requirements of most species are difficult to determine. There is no clear-cut division between these categories, and both autotrophic and heterotrophic organisms can be found in the same genus. It is speculated that many organisms actually have a dual life, autotrophic and heterotrophic. Many of the autotrophic organisms engage in heterotrophy, depending on living conditions (Gaines and Elbrächter, 1987). *Thus, the fact that a certain organism can be easily cultured in the laboratory does not necessarily mean that it is under conditions close to those of its natural habitat.* This possible variance in living conditions may have immense influence on the production of secondary metabolites. Table II shows the culturability of organisms which are known to produce important secondary metabolites.

Alexandrium spp. (formerly *Gonyaulax* or *Protogonyaulax*) are the most extensively studied dinoflagellates. They can be cultured in enriched seawater media, such as Guillard F and K (Guillard and Keller, 1984). However, the growth rates of the dinoflagellates are much lower than those of bacteria, fungi, and other microalgae. In our laboratory, the average doubling time of most organisms in large culture is normally about 4 days, and never exceeds 2 days. The maximum cell concentrations are also normally low. This low productivity is a serious problem when the culture is aimed at the mass production of certain compounds. In the case of *Alexandrium*, the maximum cell concentration of ca. 25,000 cells/ml is attained after 3–4 weeks, and only a few milligrams of a complex mixture of saxitoxin derivatives can be isolated from a 20-liter culture. The brevetoxin-producing organism *Gymnodinium breve* (*Ptychodiscus brevis*) can be cultured in

Table II. *Dinoflagellates Known to Produce Important Secondary Metabolites,*
and Their Culturability

Organism	Metabolites	Culturability
Alexandrium spp. (*Gonyaulax,* *Protogonyaulax*)	Saxitoxin, gonyautoxin derivatives	Culturable
Amphidinium spp.	Macrolides	Culturable
Dinophysis spp.	Okadaic acid derivatives, polyether macrolides	Unculturable
Gambierdiscus toxicus	Ciguatoxins, maitotoxin	Culturable[a]
Goniodoma spp.	Macrolides	Culturable
Gymnodinium breve	Brevetoxins, hemibrevetoxins	Culturable
Gymnodinium catenatum	Saxitoxin, gonyautoxin derivatives	Culturable
Ostreopsis spp.	Polyether toxins	Culturable
Pyrodnium bahamense var. *compressa*	Saxitoxin, gonyautoxin derivatives	Unculturable[b]

[a]Ciguatoxin was not produced in culture.
[b]The Caribbean species can be cultured.

artificial seawater media such as NH-15 (Wilson and Collier, 1955). Its cell density normally reaches 20,000–30,000 cells/ml after 3 weeks. About 2 mg of brevetoxin B and less than 1 mg of brevetoxin A can be isolated from a 20-liter culture. The culture of primarily benthic organisms, such as *Gambierdiscus toxicus*, reaches only a cell concentration of ~2000 cells/ml in stationary culture (Yasumoto *et al.*, 1979).

Among the unculturable organisms, the most important may be *Dinophysis* spp., which produce okadaic acid derivatives and other interesting polyether and macrolide compounds (Yasumoto *et al.*, 1979; Murakami *et al.*, 1982). They are free-swimming and mostly oceanic, but despite efforts by a number of researchers, no sustainable culture of any *Dinophysis* species has been obtained. The examination of wild cells of *D. acuminata* with an epifluorescence microscope showed the presence of other planktonic ingredients inside the cells (L. Maranda and Y. Shimizu, unpublished). Thus, it is very likely that the organisms engage in phagotrophy although they carry chlorophylls. On the other hand, another type of macrolide-producing organism, *Amphidinium* spp., can be easily cultured in ES medium although they are originally symbiotic in flatworms and other marine invertebrates (Guillard and Keller, 1984).

The second serious problem is the lowered productivity or complete lack of production of targeted compounds in culture. In the case of *Alexandrium* spp., it is speculated that wild cells contain 4–5 times the toxins produced in the culture. *Gambierdiscus toxicus*, the alleged origin of ciguatera fish poisons, fails to produce ciguatoxins in culture (Yasumoto *et al.*, 1979), while Yasumoto's group was able to isolate several ciguatoxin congeners from wild cells (Murata *et al.*,

1990). This again exemplifies the intricate influence of nutrients and metabolic status of the organism on the production of certain secondary metabolites. This subject will be discussed again later.

3. SAXITOXIN AND GONYAUTOXIN DERIVATIVES: SODIUM CHANNEL BLOCKERS

3.1. Isolation and Chemistry

Saxitoxin was first isolated as the toxic principle responsible for paralytic shellfish poisoning (PSP) (Schantz *et al.*, 1957). The phenomenon that the normally edible bivalves sometimes become toxic has been known for a long time in many parts of the world. Sommer and Meyer (1937) suggested that the bloom of a dinoflagellate, *Alexandrium catenellum* (*Gonyaulax catenella*) could be the source of the shellfish toxicity. This is probably the first report to point to a dinoflagellate as an origin of secondary metabolites found in marine invertebrates. Later, the toxin saxitoxin was confirmed in the culture of *A. catenellum* (Schantz *et al.*, 1966). In 1975, the structure of saxitoxin was established by x-ray crystallography (Schantz *et al.*, 1975; Bordener *et al.*, 1975). Meanwhile, Shimizu and co-workers discovered that the PSP of the northeast coast of the United States involves a mixture of toxins, and demonstrated the presence of several new toxins (gonyautoxins and neosaxitoxin) in the cultured causative organism, *A. tamarense* (Shimizu *et al.*, 1976). Since then, several more toxins have been isolated from various biological specimens (Shimizu, 1984, 1988, and references therein). Also, new toxin-producing dinoflagellates have been discovered. In general, most strains of these organisms produce multiple toxins, and few produce saxitoxin as a major component, which makes the dinoflagellates a rather poor source of saxitoxin.

The purification of toxins was first accomplished by separation by gel-filtration chromatography on Bio-gel P-2, followed by ion-exchange chromatography on the weakly acidic resin Bio-rex 70 (Oshima *et al.*, 1977). The method has become the standard procedure for preparative purposes. For analytical purposes, separation on a reverse-phase column and fluorimetric detection after post-column reaction with periodate give excellent results (Sullivan and Iwaoka, 1983).

More than 15 analogs of saxitoxin derivatives have been isolated from dinoflagellates and their structures elucidated (Shimizu, 1984, 1988, and references therein). Structurally, they are divided into two series: saxitoxin and neosaxitoxin (Fig. 1). The neosaxitoxin series has an N-hydroxyl group on N-1, and has quite different physicochemical characteristics from the saxitoxin series. Other structural variations come from the presence of an O-sulfate group and stereochemistry at C-11, an N-sulfate group on the carbamoyl group. Decarbamoyl

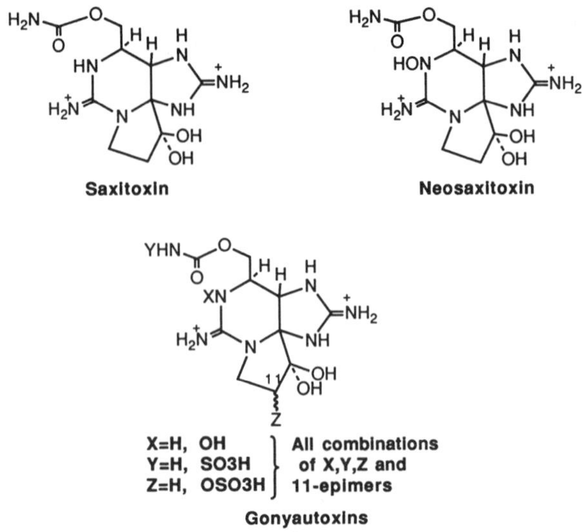

Figure 1. The structures of saxitoxin and gonyautoxin derivatives, sodium channel blockers produced by the dinoflagellates *Alexandrium* and *Pyrodinium* spp.

and 13-deoxydecarbamoyl derivatives have been also recently discovered in a dinoflagellate.

Saxitoxin is a very stable compound in acidic media, but under slightly basic conditions, it undergoes oxidative degradation to aromatized purine derivatives. The N-1-hydroxyl and 11-O-sulfo derivatives are much more unstable in acidic media. The most labile compounds are sulfocarbamoyl derivatives, which are easily hydrolyzed in dilute acid solutions. These differences in chemical nature pose a number of problems to the isolation and purification of the toxins.

3.2. Pharmacology

Saxitoxin and another marine toxin, tetrodotoxin, are important pharmacological tools as selective blockers of sodium channels (Kao, 1972; Narahashi, 1972). Their importance in the study of various agents on neuromuscular systems cannot be overemphasized. The structural variations in the newly discovered compounds have provided crucial information for the understanding of the mechanism of action of the toxins, and, more importantly, the structures of sodium channels or excitable membranes. The analysis of the relationship between the structures and the binding ability of the toxins with channels led to the revision of the previous binding model (Hille, 1975), in which the toxin molecule was postulated to penetrate into the channel and plug the sodium ion path. Newly presented models agree that the toxin does not penetrate inside the channel, but rather binds on the outside of the membrane (Kao and Walker, 1982; Shimizu, 1982; Kao, 1983; Stricharz, 1984). For example, Shimizu (1982) proposed that the toxin binds on the outside of the channel by ion-pairing between the quanidium group and an anionic site and two hydrogen bonds. This model seems to be in agreement with the recent report that the aspartate carboxylate, which is known to be outside of the channel, is the crucial binding site for the toxin (Noda *et al.*, 1989).

4. BREVETOXINS AND CIGUATOXINS: SODIUM CHANNEL ACTIVATORS

4.1. Isolation and Chemistry

Brevetoxins are the toxic principles in the Gulf of Mexico red tide organism which inflicts extensive damage to the marine ecosystem. Because of the organism's very conspicuous toxic effects, its toxic principles have been extensively investigated (Shimizu, 1979; Steidinger and Baden, 1984, and references therein). The isolation and physicochemical properties of the first pure toxin, GB-2 (= *G. breve* toxin 2), were reported at the 4th Food and Drug from the Sea Conference in 1974 (Shimizu *et al.*, 1975). The chemical nature of the compound was speculated as a polycyclic ether (Shimizu, 1979). Later, the identical toxin was reisolated and reported by two other groups (Baden *et al.*, 1979; Risk *et al.*, 1979). In 1980, the x-ray structure of GB-2 was reported under a new name, brevetoxin B (Lin *et al.*, 1981). The structure of the most toxic component, brevetoxin A (formerly GB-1 toxin), was established in 1986 (Shimizu *et al.*, 1986). Several other toxins isolated from the same organism are derivatives of brevetoxin A and brevetoxin B (Fig. 2). In 1989, the structure of a new type of compound, hemibrevetoxin B, was reported, and it is now known there is a series of hemibrevetoxins (Prasad and Shimizu, 1989).

The polyether ring structures of brevetoxins are unprecedented, and are good illustrations that dinoflagellates can be a source of unique bioactive molecules.

Recently the chemistry of brevetoxins has acquired new significance and attention, since the structure of ciguatoxin was found to be closely related to those of brevetoxins. Ciguatoxin is a toxin responsible for ciguatera fish poisoning, which is probably the biggest obstacle in the utilization of fish resources in tropical areas. The chemical structure of the toxin isolated from moray eels has been elucidated, and subsequently the structures of its precursors found in the source organism, a benthic dinoflagellate, *Gambierdiscus toxicus*, have been reported (Murata *et al.*, 1989, 1990) (Fig. 3).

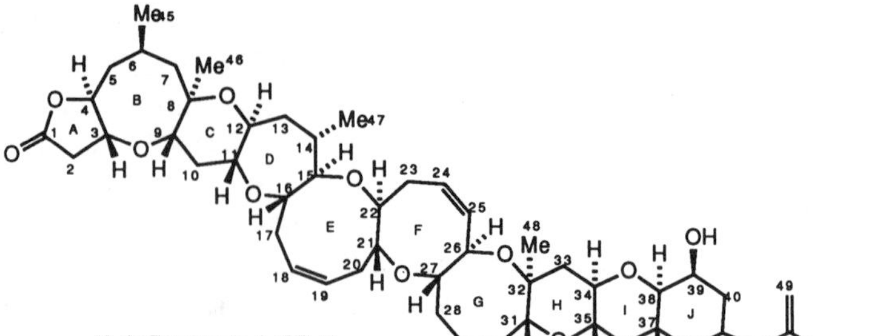

X=O: Brevetoxin A (GB-1)
X=H, OH: GB-7

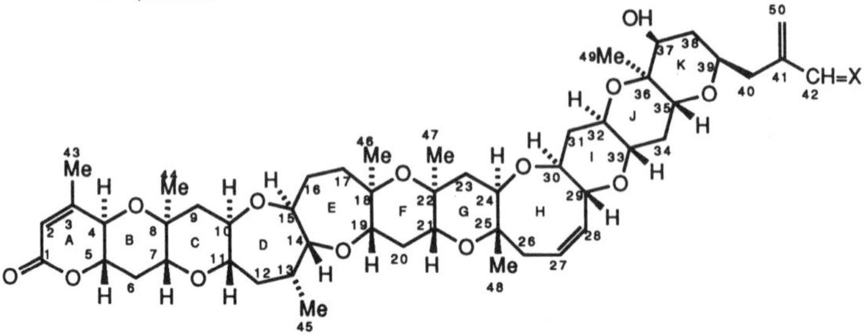

X=O: Brevetoxin B (GB-2)
X=H, OH: GB-3
X=O, 37-OAc: GB-5
X=O, 27, 28β-epoxide: GB-6

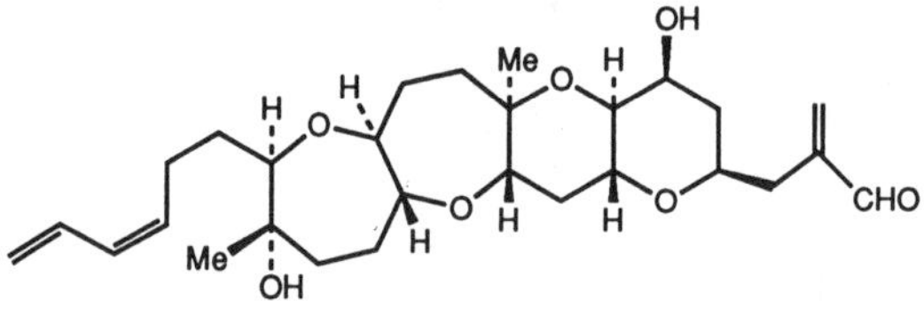

Hemibrevetoxin B

Figure 2. The structures of brevetoxins and hemibrevetoxin, sodium channel activators produced by the dinoflagellate *Gymnodinium breve*.

4.2. Pharmacology

Brevetoxins are selective sodium channel activators, and depolarize excitable membranes (Caterall and Risk, 1981; Huang *et al.*, 1984). Their binding site on sodium channels seems to be different from that of other activators such as scorpion toxin and aconitine (Caterall, 1982). In biomedical research, brevetoxins have quickly become important tools and are widely used in various experiments.

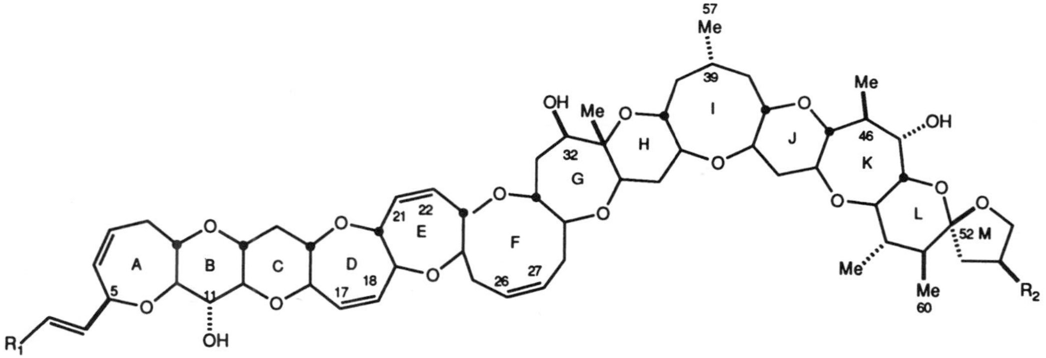

Ciguatoxin: R₁=HOCH₂CH(OH)-; R₂=OH

GT4b: CH₂=CH-; R₂=H

Murata, et al., J. Am. Chem. Soc., **112**, 4380-4386 (1990)

Figure 3. The structure of ciguatoxins derived from the dinoflagellate *Gambierdiscus toxicus* (Murata *et al.*, 1990).

They are also extremely powerful ichthyotoxins and are lethal to guppies at concentrations of 10^{-8} g/ml (Chou and Shimizu, 1982). The toxins have inotropic effects against isolated smooth muscles.

Ciguatoxin is also inotropic and depolarizes excitable membranes (Miyahara *et al.*, 1979), but, judging from the complex symptoms of ciguatera poisoning, including neurological symptoms, the action sites of ciguatoxin are more widespread.

5. OTHER POLYETHER COMPOUNDS PRODUCED BY DINOFLAGELLATES

5.1. Okadaic Acid and Its Derivatives

The most important compounds in this category are okadaic acid and its derivatives. Okadaic acid and its episufide, acanthifolicin, were first isolated from sponges (Tachibana *et al.*, 1981; Schmitz *et al.*, 1981), but they are evidently dinoflagellate metabolites.

Murakami *et al.* (1982) isolated okadaic acid from cultured *Prorocentrum lima*. Okadaic acid and its derivatives *Dinophysis* toxins are toxic principles of so-called diarrheic shellfish poisoning, but their origins were determined to be *Dinophysis* spp. (Yasumoto *et al.*, 1980) (Fig. 4). Okadaic acid is also found in other species (Dickey *et al.*, 1990), and its distribution seems to be widespread.

Okadaic acid and its derivatives are powerful tumor promoters. Their mechanism of action has recently been revealed to be the inhibition of protein phosphatases, especially protein phosphatase-1 and protein phosphatase-2A (Bialojan and Takai, 1988; Haystead *et al.*, 1989). This inhibitory action may lead to a net increase of phosphorylated proteins in cytosols and cause tumor promotion. The compounds have already become very important tools for cancer and cell biology studies, and are marketed at high prices.

A group of toxins, pectinotoxins, isolated from scallops are also polyether derivatives, and very possibly have their origin in the dinoflagellates, *Dinophysis* spp. (Yasumoto *et al.*, 1984).

5.2. Antitumor and Antimicrobial Macrolides

Dinoflagellates produce macrolides, which are very similar to the antibiotics produced by *Streptomyces* and other terrestrial microorganisms.

Antitumor or cytotoxic macrolides, amphidinolides A, B, and C, have been isolated from a symbiotic *Amphidinium* sp. (Kobayashi *et al.*, 1986, 1988; Ishibashi *et al.*, 1987). The dinoflagellate was isolated from *Amphiscolops* sp. and cultured in ES medium. Amphidinolides A, B, and C were obtained in yields of 0.002, 0.001, and 0.001%, respectively. Amphidinolide A has the molecular

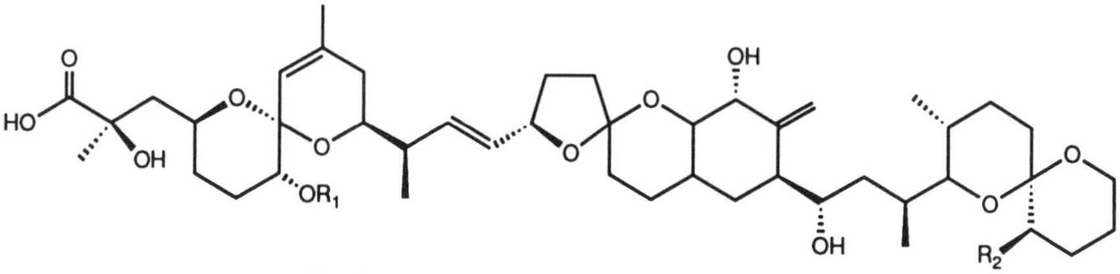

R₁=H, R₂=H: Okadaic acid
R₁=H, R₂=CH₃: Dinophysistoxin-1
R₁=Acyl, R₂=CH₃: Dinophysistoxin-2

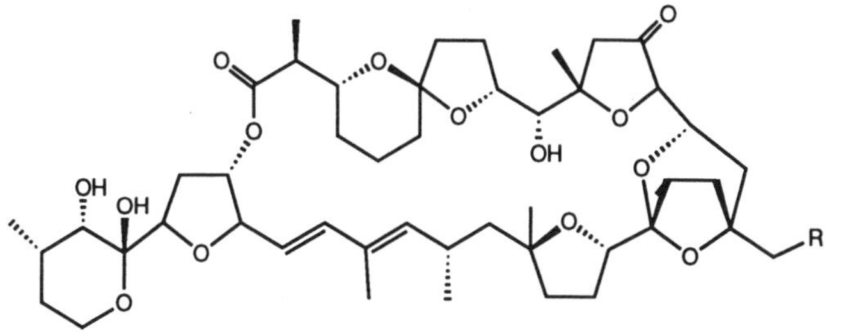

R=OH: Pectinotoxin-1
R=H: Pectinotoxin-2

Figure 4. The structures of okadaic acid and other metabolites produced by the dinoflagellates *Dinophysis* spp.

formula $C_{31}H_{46}O_7$ and a 20-membered lactone ring structure was assigned on the basis of spectroscopic data. Amphidinolide B, $C_{32}H_{50}O_8$; was given a 26-membered lactone structure. Both compounds have epoxide function in the molecules. Amphidinolide C has the formula $C_{41}H_{60}O_9$ and a 25-membered lactone structure. Recently, amphidinolide E, a 19-membered lactone, was isolated from a different species of *Amphidinium* (Kobayashi *et al.*, 1990) (Fig. 5).

Amphibinolide B has the highest antineoplastic activity, LC_{50} 0.14 ng/ml against L1210 mouse leukemia cells, followed by amphidinolide C, 5.8 ng/ml, amphidinolide A, 2.4 µg/ml, and amphidinolide E, 4.8 µg/ml. *In vivo* activity of these compounds has not been reported.

The dinoflagellate *Goniodoma* spp. are known to contain an antifungal agent, goniodomin. From the dinoflagellate *Goniodoma pseudogoniaulax* (*Alexandrium hiranoi*), Murakami *et al.* (1988) isolated goniodomin A, a polyether lactone compound (Fig. 6). The compound was active at a concentration of 0.05 µg against the fungus *Mortierella ramannianus*. The compound can be produced by the cultured organism.

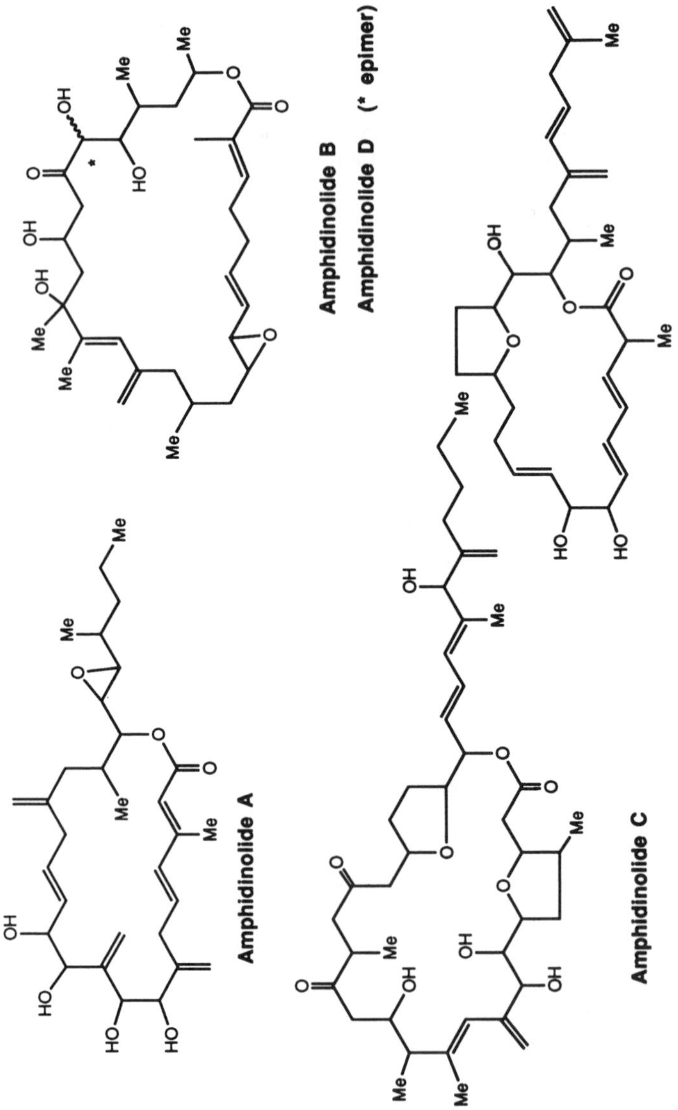

Figure 5. The structures of amphidinolides, antitumor metabolites of the symbiotic dinoflagellates *Amphidinium* spp.

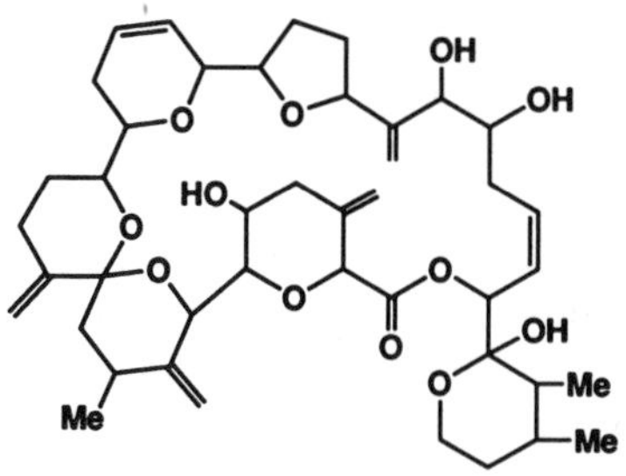

Figure 6. The structure of goniodomin A, an antifungal agent from the dinoflagellate *Goniodoma* sp.

Prorocentrum lima, which produces okadaic acid, also produces a novel toxic nitrogen-containing macrolide, prorocentrolide, in culture (Torigoe *et al.*, 1988). The compound was simply presented as a toxin and no details of its biological activity have been reported, but the molecule seems to be extremely interesting (Fig. 7).

6. MISCELLANEOUS: MAITOTOXIN

Maitotoxin derives its name from the fact that it was first isolated from a tropical fish (local name: maito). Later it was found that the compound is produced by the dinoflagellate *Gambierdiscus toxicus* and transferred to the fish through the food chain (Yasumoto *et al.*, 1976, 1977). In culture, *G. toxicus* produces only

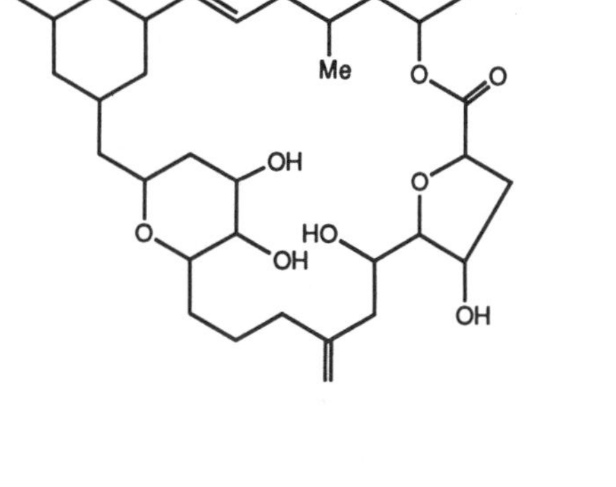

Figure 7. The structure of prorocentrolide, an unusual nitrogen-containing macrolide isolated from the dinoflagellate *Prorocentrum lima*.

maitotoxin, which is water-soluble, and no lipid-soluble toxins are found in the wild cells (Yasumoto *et al.*, 1979).

Maitotoxin has a fairly high molecular weight, and its structure has not been fully elucidated, but it contains a long carbon chain and a sulfate group. It is a powerful calcium channel activator (Takahashi *et al.*, 1982), and is becoming a very important tool in various pharmacological experiments.

7. BIOSYNTHESIS OF DINOFLAGELLATE PRODUCTS AND METABOLISM

As mentioned earlier, the culturing of dinoflagellates and the production of specific compounds in culture are not simple tasks. The understanding of the biochemical pathways operating in the dinoflagellates and the biosynthetic mechanism of the secondary metabolites would be very beneficial for the rational design of the culturing and the production of useful compounds. However, only limited experimental data are available on the biosynthesis of metabolites and metabolic pathways of dinoflagellates.

7.1. Biosynthesis of Saxitoxin Derivatives

The biosynthesis of saxitoxin derivatives in *Alexandrium tamarense* and the toxin-producing cynanobacterium *Aphanizomenon flos-aquae* has been studied by Shimizu and co-workers (Shimizu *et al.*, 1984, 1986, 1989). Experiments using various isotope-labeled precursors showed that the molecule is built from arginine, acetate, and the methyl group from methionine. Some of the detailed mechanisms were also elucidated (Fig. 8).

7.2. Biosynthesis of Brevetoxins

It was shown that the biosynthesis of brevetoxins in *Gymnodinium breve* follows a sequence quite different from that known for polyether compounds produced by *Streptomyces* and fungi (Lee *et al.*, 1986; Chou and Shimizu, 1987). Chou and Shimizu (1987) proposed that dicarboxylic acids are involved in the formation of the basic carbon chain skeletons (Fig. 9).

7.3. Characteristics of Metabolic Pathways

Judging from the above two examples of biosynthesis in dinoflagellates, the metabolic pathways of dinoflagellates seem to be quite different from those seen in the major pathways of terrestrial microorganisms or higher plants. They seem to

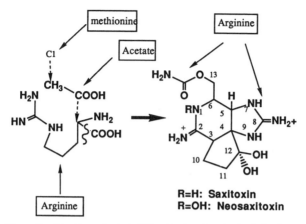

Figure 8. The biosynthetic building blocks of saxitoxin derivatives.

follow alternative paths which are known to serve just as auxiliary routes in other organisms. Based on the analysis of the incorporation patterns of various precursors, it was hypothesized that the key to dinoflagellate metabolism is in the utilization of amino acids, both endogenous and exogenous (Y. Shimizu, unpublished).

For example, in the biosynthesis of saxitoxin, three arginine molecules and one methionine molecules are required. In the case of brevetoxins, the analysis of the labeling pattern of ^{13}C-acetate into the molecules indicates that certain parts of

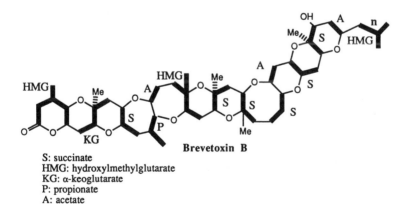

S: succinate
HMG: hydroxylmethylglutarate
KG: α-keoglutarate
P: propionate
A: acetate

Figure 9. The biosynthetic building blocks of brevetoxins as speculated from feeding experiments with *Gymnodinium breve*.

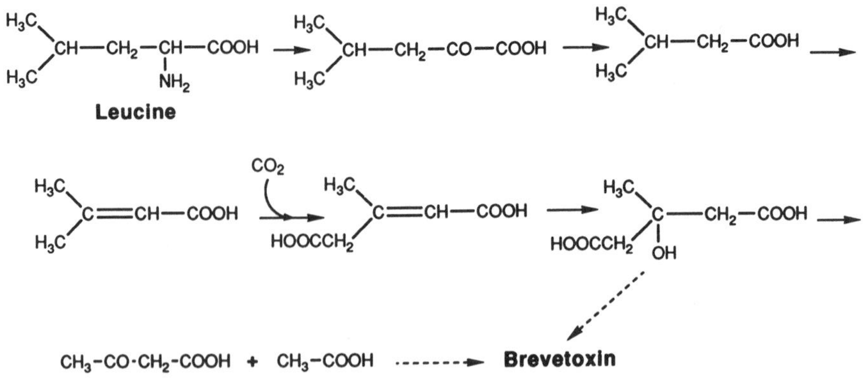

Scheme 1. The speculative fate of the exogenous amino acid leucine in the biosynthesis of brevetoxin in *Gymnodinium breve*.

the molecule come from the degradation products of amino acids (Scheme 1). It was also discovered in feeding experiments that certain amino acids greatly enhance the growth of *the organism*. From these observations, it was speculated that the organism seems to utilize the amino group of exogenous amino acids as a nitrogen source, and incorporate the rest of the carbon skeleton after degradation to small organic acids by its own enzymes or bacteria on the outside of the cell (Scheme 2). Such metabolic pathways may fit well with the dinoflagellate life pattern, which crisscrosses the domains of plants and animals.

8. CONCLUSION

A number of important compounds have been discovered in dinoflagellates, and several of them have already become utilized as indispensable tools in biomedical studies. Also, there is a good prospect of finding therapeutically useful compounds such as antitumor agents in dinoflagellates. The chemistry of dinoflagellates seems to be very promising. On the other hand, the practical production of secondary metabolites in the culture of dinoflagellates involves many problems. There are a number of unanswered questions on such critical subjects as the nutrient requirements of the organisms, the trigger mechanism of the biosynthesis, and the involvement of bacteria in the secondary metabolite production. Many more basic studies of the metabolism and biosynthesis in dinoflagellates will be needed for the practical production of medicinals.

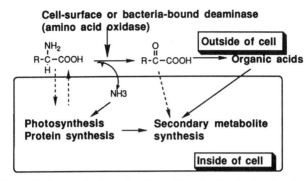

Scheme 2. Schematic view of amino acid utilization in dinoflagellates.

REFERENCES

Baden, D. G., Mende, T. J., and Block, R. E., 1979, Two similar toxins isolated from *Gymnodinium breve*, in: *Toxic Dinoflagellate Blooms* (D. L. Taylor and H. H. Seliger, eds.), Elsevier/North-Holland, New York, pp. 307–334.

Bialojan, C. and Takai, A., 1988, Inhibitory effect of a marine-sponge toxin, okadaic acid, on protein phosphatases, *Biochem. J.* **256**:283–290.

Bordener, J., Thiessen, W. E., Bates, H. A., and Rapoport, H., 1975, The structure of a crystalline derivatives of saxitoxin. The structure of saxitoxin, *J. Am. Chem. Soc.* **97**:6008–6012.

Caterall, W. A., 1982, The emerging molecular view of the sodium channel, *Trends Neurosci.* **5**: 303–306.

Caterall, W A., and Risk, M., 1981, Toxin T_{46} from *Prychodiscus brevis* (formerly *Gymnodinium breve*) enhances activation of voltage sensitive sodium channels by veratridine, *Mol. Pharmacol.* **19**:345–348.

Chou, H. N., and Shimizu, Y., 1982, A new polyether toxin from *Gymnodinium breve* Davis, *Tetrahedron Lett.* **23**:5521–5524.

Chou, H. N., and Shimizu, Y., 1987, Biosynthesis of brevetoxins: Evidence for the mixed origin of the backbone carbon chain and possible involvement of dicarboxylic acids, *J. Am. Chem. Soc.* **109**: 2184–2185.

Dickey, R. W., Borzin, S. C., Faulkner, D. J., Bencsath, F. A., and Andrzejewski, D., 1990, Identification of okadaic acid from a Caribbean dinoflagellate, *Prorocentrum concavum*, *Toxicon* **28**:371–377.

Dodge, D. J., 1984, Dinoflagellate taxonomy, in: *Dinoflagellates* (D. L. Spector, ed.), Academic Press, New York, pp. 17–42.

Gaines, G., and Elbrächter, M., 1987, in: *The Biology of Dinoflagellates* (F. J. R. Taylor ed.), Blackwell, Oxford, pp. 224–281.

Guillard, R. L., and Keller, M. D., 1984, in: *Dinoflagellates* (D. L. Spector ed.), Academic Press, New York, pp. 391–442.

Haystead, T. A. J., Sim, A. T. R., Carling, D., Honnor, R. C., Tsukitani, Y., Cohen, P., and Hardie, D. G., 1989, Effects of the tumour promoter okadaic acid on intracellular protein phosphorylation and metabolism, *Nature* **337**:78–81.

Hille, B., 1975, The receptor for tetrodotoxin and saxitoxin: A structural hypothesis, *Biophys. J.* **15**: 615–619.

Huang, J. M., Wu, C. H., and Baden, D. G., 1984, Depolarizing action of a red tide dinoflagellate brevetoxin on axonal membranes, *J. Pharmacol. Exp. Ther.* **229**:615–621.

Ishibashi, M., Ohizumi, Y., Hamashima, M., Nakamura, H., Hirata, Y., Sasaki, T., and Kobayashi, J., 1987, Amphidinolide-B, a novel macrolide with potent antineoplastic activity from the marine dinoflagellate *Amphidinium* sp., *J. Chem. Soc. Chem. Commun.* **1987**:1127–1129.

Kao, C. Y., 1972, Pharmacology of tetrodotoxin and saxitoxin, *Fed. Proc.* **31**:1117–1123.

Kao, C. Y., 1983, New perspectives as the interaction of tetrodotoxin and saxitoxin with excitable membranes, *Toxicon Suppl.* **3**:211–219.

Kao, C. Y., and Walker, S. E., 1982, Active group of saxitoxin and tetrodotoxin as deduced from actions of saxitoxin analogues on muscle and squid axon, *J. Physiol.* **323**:619–637.

Kobayashi, J., Ishibashi, M., Nakamura, H., Ohizumi, Y., Yamasu, T., Sasaki, T., and Hirata, Y., 1986, Amphidinolide-A, a novel antineoplastic macrolide from the marine dinoflagellate *Amphidinium* sp., *Tetrahedron Lett.* **27**:5755–5758.

Kobayashi, J., Ishibashi, M., Wälchli, M. R., Nakamura, H., Hirata, Y., Sasaki, T., and Ohizumi, Y., 1988, Amphidinolide C: The first 25-membered macrocyclic lactone with potent antineoplastic activity from the cultured dinoflagellate *Amphidinium* sp., *J. Am. Chem. Soc.* **110**:490–494.

Kobayashi, J., Ishibashi, M., Murayama, T., Takamatsu, M., Iwamura, M., Ohizumi, Y., and Sasaki, T., 1990, Amphidinolide E, a novel antileukemic 19-membered macrolide from the culture symbiont dinoflagellate *Amphidinium* sp., *J. Org. Chem.* **55**:3421–3423.

Lee, M. S., Repeta, D. S., Nakanishi, K., and Zagorski, M. G., 1986, Biosynthetic origins and assignments of carbon 13 NMR peaks of brevetoxin B, *J. Am. Chem. Soc.* **108**:7855–7856.

Lin, Y. Y., Risk, M., Ray, S. M., Van Engen, E., Clardy, J., Golik, J., James, J. C., and Nakanishi, K., 1981, Isolation and structure of brevetoxin B from the "red tide" dinoflagellate *Ptychodiscus brevis (Gymnodinium breve)*, *J. Am. Chem. Soc.* **103**:6773–6775.

Miyahara, J. T., Akau, C. K., and Yasumoto, T., 1979, Effects of ciguatoxin and maitotoxin on the isolated guinea pig atria, *Chem. Pathol. Pharmacol.* **25**:177-180.

Murakami, Y., Oshima, Y., and Yasumoto, T., 1982, Identification of okadaic acid as a toxic components of a marine dinoflagellate *Prorocentrum lima*, *Bull. Jpn. Soc. Sci. Fish.* **48**:69–72.

Murakami, M., Makabe, K., Yamaguchi, K., Konosu, S., and Wälchli, M. R., 1988, Goniodomin A, a novel polyether macrolide from the dinoflagellate *Goniodoma pseudogoniaulax*, *Tetrahedron Lett.* **29**:1149–1152.

Murata, M., Legrand, A. M., Ishibashi, Y., Fukui, M., and Yasumoto, T., 1989, Structures of ciguatoxin and its congener, *J. Am. Chem. Soc.* **111**:8929–8931.

Murata, M., Legrand, A. M., Ishibashi, Y., Fukui, M., and Yasumoto, T., 1990, Structures and configurations of ciguatoxin from the moray eel *Gymnothorax javanicus* and its likely precursor from the dinoflagellate *Gambierdiscus toxicus*, *J. Am. Chem. Soc.* **112**:4380–4386.

Narahashi, T., 1972, Mechanism of action of tetrodotoxin and saxitoxin on excitable membranes, *Fed. Proc.* **31**:1124–1132.

Noda, M., Suzuki, H., Numa, S., and Stuhmer, W. A., 1989, Single point mutation confers tetrodotoxin and saxitoxin insensitivity on the sodium channel II, *FEBS Lett.* **259**:213–216.

Oshima, Y., Buckley, L. J., Alam, M., and Shimizu, Y., 1977, Hetrogeneity of paralytic shellfish poisons. Three new toxins from cultured *Gonyaulax tamarensis* cells, *Mya arenaria*, and *Saxidomus giganteus*, *Comp. Biochem. Physiol.* **57c**:31–34.

Prasad, A. V. K., and Shimizu, Y., 1989, The structure of hemibrevetoxin-B: A new type of toxin in the Gulf of Mexico red tide organism, *J. Am. Chem. Soc.* **111**:6476–6477.

Risk, M., Lin, Y. Y., MacFarlane, R. D., Sadagopa-Ramanujam, V. M., Smith, L. L., and Trieff, N. M., 1979, Purification and chemical studies on a major toxin from *Gymnodinium breve*, in: *Toxic*

Dinoflagellate Blooms (D. L. Taylor and H. H. Seliger, eds.), Elsevier/North-Holland, New York, pp. 335–344.

Schantz, E. J., Mold, J. D., Stanger, D. W., Shavel, J., Riel, F. J., Bowden, J. P., Lynch, J. M., Wyler, R. S., Riegel, B. R., and Sommer, H., 1957, A procedure for the isolation and purification of the poison from toxic clams and mussel tissues, *J. Am. Chem. Soc.* **79:**5230–5235.

Schantz, E. J., Lynch, J. M., Vayvada, G., Matsumoto, K., and Ropoport, H., 1966, The purification and characterization of the poison produced by *Gonyaulax catenella* in axenic culture, *Biochemistry* **5:**1191–1195.

Schantz, E. J., Ghazarossian, V. E., Schnoes, H. K., Strong, F. M., Springer, J. P., Pezzanite, J. O,. and Clardy, J., 1975, The structure of saxitoxin, *J. Am. Chem. Soc.* **97:**1238–1239.

Schmitz, F. J., Prasad, R. S., Goppichand, Y., Hossain, M. B., van der Helm, D., and Schmidt, P, 1981, Acanthifolicin, a new episulfide-containing polyether carboxylic acid from extracts of the marine sponge *Pandaros acanthifolium, J. Am. Chem. Soc.* **103:**2467–2469.

Shimizu, Y., 1979, Dinoflagellate toxins, in: *Marine Natural Products, Chemical and Biological Perspectives*, Vol. I (P. J. Scheuer, ed.), Academic Press, New York, pp. 1–42.

Schimizu, Y., 1982, Recent progress in marine toxin research, *Pure Appl. Chem.* **54:**1973–1980.

Shimizu, Y., 1984, Paralytic shellfish poisons, in: *Progress in the Chemistry of Organic Natural Products* (W. Herz, J. Grisebach, and G. W. Kirby eds.), Springer-Verlag, New York, pp. 235–264.

Shimizu, Y., 1986, Toxigenesis and biosynthesis of saxitoxin analogues, *Pure Appl. Chem.* **58:** 257–262.

Shimizu, Y., 1988, The chemistry of paralytic shellfish toxins, in: *Handbook of Natural Toxins*, Vol. 3, *Marine Toxins and Venoms* (A. T. Tu, ed.), Marcel Dekker, New York, pp. 64–85.

Shimizu, Y., Alam, M., Oshima, Y,. and Fallon, W. E., 1975, Presence of four toxins in red tide infested clams and cultured *Gonyaulax tamarensis* cells, *Biochem. Biophys. Res. Commun.* **66:** 731–737.

Shimizu, Y., Alam, M., and Fallon, W. E., 1976, Red-tide toxins, in: *Food-Drugs from the Sea Proceedings 1974* (J. H. Webber and G. D. Ruggieri, eds.), Marine Technology Society, Washington, D.C., pp. 238–251.

Shimizu, Y., Norte, M., Hori, A., Genenah, A., and Kobayashi, M., 1984, Biosynthesis of saxitoxin analogues: The unexpected pathway, *J. Am. Chem. Soc.* **106:**6433–6434.

Shimizu, Y., Chou, H. N., Bando, H., Van Duyne, G., and Clardy, J., 1986, Structure of brevetoxin A (= GB-1 toxin), the most potent toxin in the Florida red tide organism, *Gymnodinium breve* (= *Ptychodiscus brevis*), *J. Am. Chem. Soc.* **108:**514–515.

Shimizu, Y., Gupta, S., and Prasad, A. V. K., 1989, Biosynthesis of dinoflagellate toxins, in: *Toxic Marine Phytoplankton* (E. Graneli, S. Sundstrom, L. Edler, and D. M. Anderson eds.), Elsevier, New York, pp. 62–73.

Sommer, H., and Meyer, K. F., 1937, Paralytic shellfish poisoning, *Arch. Pathol.* **24:**560–598.

Steidinger, K. A., and baden, D. G., 1984, Toxic marine dinoflagellates, in: *Dinoflagellates* (D. L. Spector ed.), Academic Press, New York, pp. 201–261.

Stricharz, G., 1984, Structural determinants of the affinity of saxitoxin for neuronal sodium channels. Electrophysiological studies on peripheral nerve, *J. Gen. Physiol.* **84:**281–305.

Sullivan, J. J., and Iwaoka, W. T., High pressure liquid chromatography determination of the toxins associated with paralytic shellfish poisoning, *J. Assoc. Offic. Anal. Chem.* **66:**297–303.

Tachibana, K., Scheuer, P., Tsukitani, Y., Kikuchi, H., Van Engen, D., Clardy, J., Gopichand, Y., and Schmitz, F. J,. 1981, Okadaic acid, a cytotoxic polyether from two marine sponges of the genus *Halichondria, J. Am. Chem. Soc.* **103:**2469–2471.

Takahashi, M., Ohizumi, Y., and Yasumoto, T., 1982, Maitotoxin, Ca^{2+} channel activator candidate, *J. Biol. Chem.* **257:**7287–7289.

Taylor, F. J. R., 1987, Taxonomy and classification, in: *The Biology of Dinoflagellates* (F. J. R. Taylor, ed.), Blackwell, Oxford, pp. 723–731.

Torigoe, K., Murata, M., Yasumoto, T,. and Iwashita, T., 1988, Prorocentrolide, a toxic nitrogenous macrocycle from a marine dinoflagellate, *Prorocentrum lima, J. Am. Chem. Soc.* **110:**7876–7877.

Wilson, W. B., and Collier, A., 1955, Preliminary notes on the culturing of *Gymnodinium brevis* Davis, *Science* **121:**394–395.

Yasumoto, T., Inoue, A., Bagnis, R., and Garcon, M., 1976, Ecological survey on a dinoflagellate possibly responsible for induction of ciguratera, *Bull. Jpn. Soc. Sci. Fish* **42:**359–365.

Yasumoto, T., Nakajima, I., Bagnis, R., and Adachi, R., 1977, Finding of a dinoflagellate as a likely culprit of ciguatera, *Bull. Jpn. Soc. Sci. Fish.* **43:**1021–1026.

Yasumoto, T., Nakajima, I., Oshima, Y., and Bagnis, R., 1979, A new toxic dinoflagellate found in association with ciguatera, in: *Toxic Dinoflagellate Blooms* (D. L. Taylor and H. Seliger, eds.), Elsevier, New York, pp. 65–70.

Yasumoto, T., Oshima, Y., Sugawara, W., Fukuyo, Y., Oguri, H., Igarashi, T., and Fujita, N., 1980, Identification of *Dinophysis fortii* as the causative organisms of diarrhetic shellfish poisoning, *Bull. Jpn. Soc. Sci. Fish.* **46:**1405–1411.

Yasumoto, T., Murata, M., Oshima, Y., Matsumoto, G. K., and Clardy, J., 1984, Diarrhetic shellfish poisoning, in: *Sea Food Toxins* (E. P., Ragelis ed.), American Chemical Society, Washington, D.C., pp. 207–214.

11

Production of β-Carotene and Vitamins by the Halotolerant Alga Dunaliella

Ami Ben-Amotz

1. DUNALIELLA AND ITS ENVIRONMENT

The chlorophyte *Dunaliella* is classified under the order Volvocales, which includes a variety of ill-defined unicellular species. Members of the genus *Dunaliella* are all motile, ovoid biflagellates with a cell volume ranging from 50 to 1000 μm^3 (Butcher, 1959). The alga contains one large chloroplast with a single pyrenoid and many starch granules in the basal portion of the chloroplast. Like most green algae, *Dunaliella* contains the typical extrachloroplast organelles: membrane-bound nucleus, mitochondria, small vacuoles, Golgi bodies, and eye spots. However, unlike the other green algae, *Dunaliella* lacks a rigid cell wall, and the cell is enclosed by a thin, elastic, plasma membrane covered by a mucus "surface coat." The lack of a rigid cell wall permits rapid cell-volume responses to the extracellular osmotic changes.

    Dunaliella predominates in many aquatic marine habitats and in salt water bodies which contain more than 10% salt. Typical examples are the Dead Sea in Israel, the Pink Lake in Australia, and the Great Salt Lake in the United States.

Ami Ben-Amotz • National Institute of Oceanography, Israel Oceanographic and Limnological Research, Tel-Shikmona, Haifa 31080, Israel.

Marine Biotechnology, Volume 1: Pharmaceutical and Bioactive Natural Products, edited by David H. Attaway and Oskar R. Zaborsky. Plenum Press, New York, 1993.

Dunaliella is probably the most halotolerant eukaryotic organism known, show-ing a remarkable adaptation to a variety of salt concentrations from as low as 0.2% to salt saturation of about 35% (Borowitzka, 1981).

Dunaliella osmoregulates by varying the intracellular concentration of the photosynthetic glycerol in response to the extracellular osmotic pressure. On growth in media containing different salt concentrations, the intracellular glycerol concentration is directly proportional to the extracellular salt concentration and maintains the cell water volume and the required cellular osmotic pressure. Lacking a rigid cell wall, *Dunaliella* shrinks or swells rapidly when exposed to hypertonic and hypotonic conditions, respectively. Within minutes of the osmotic transition, the cell responds by synthesis or elimination of glycerol in a time kinetic process which ceases when the cell volume returns to its original extent. The mechanism of controlling the synthesis and elimination of glycerol in *Dunaliella* is believed to involve a unique "glycerol cycle" activated by a few novel enzymes (Ben-Amotz and Avron, 1981, 1990; Borowitzka and Borowitzka, 1988).

2. β-*CAROTENE PRODUCTION*

Dunaliella is the most enriched β-carotene eukaryotic organism known (Borowitzka and Borowitzka, 1988; Ben-Amotz and Avron, 1990). Under appro-priate cultivation, more than 10% of the dry weight of *Dunaliella* is β-carotene. The accumulation of β-carotene is species-specific and is related physiologically to high light intensity, nitrate limitation, and to most environmental stress conditions such as high salt, high and low temperatures, nutrient deficiencies, etc. β-Carotene in *Dunaliella* accumulates within oily globules in the interthylakoid spaces of the chloroplast and is composed mainly of two stereoisomers: 9-*cis* and all-*trans*. Both the total mount of the accumulated β-carotene and the 9-*cis* to all-*trans* ratio depend on the light absorbed by the cell during one division cycle— the higher the light intensity and the lower the growth rate of the alga, the higher the cellular β-carotene content and the 9-*cis* to all-*trans* ratio. In a series of recent studies, it was shown that the isomerization reaction which eventually causes the accumulation of 9-*cis* β-carotene occurs early in the pathway of carotene synthesis and allows the production of the different β-carotene stereoisomeric intermediates to an equivalent level of the all-*trans* β-carotene (Shaish *et al.*, 1990). The physicochemical properties of 9-*cis* β-carotene differ from those of all-*trans* β-carotene; all-*trans* β-carotene is practically insoluble in oil and is easily crystallized, while 9-*cis* β-carotene is much more soluble in hydrophobic sol-vents, very difficult to crystallize, and generally an oil in its concentrated form. The high proportion of 9-*cis* β-carotene probably accounts for the very high concentration of soluble β-carotene within the lipoidal chloroplastic globule of *Dunaliella*.

The most probable function of the β-carotene globules in *Dunaliella* is to protect the cell against the high-intensity irradiation to which it is exposed in the natural habitat by absorbing energy in the blue region of the spectrum. Strains unable to accumulate β-carotene die when exposed to high irradiation, while the β-carotene-rich *Dunaliella* strains flourish (Shaish *et al.*, 1991).

3. CELL COMPOSITION

Dunaliella is composed of approximately 50% protein, 35% carbohydrate including glycerol, and 8% fat. The composition varies with the growth conditions, with remarkable effects on the glycerol as related to the extracellular salt concentration and on the protein-to-starch ratio by the nitrogen content in the medium (Fabregas *et al.*, 1989). The amino acid composition of *Dunaliella* resembles that of other eukaryotic green algae and higher plants, with a high content of the sulfur amino acids lysine and methionine (Ben-Amotz and Avron, 1989). The carbohydrates include mainly starch of α-1-4-glucosan (Eddy *et al.*, 1958; Eddy, 1965; Finney *et al.*, 1984) and common mono- and disaccharides. The lipid content in *Dunaliella* ranges from 6 to 18% (Tornabene *et al.*, 1980). Most polar lipids in *Dunaliella* are common to other green algae except the zwitterionic polar lipid diacylglycerol-0-(*N,N,N*-trimethyl)-homoserine (DGTS), which was reported to be abundant in the flagellate (Evans *et al.*, 1982). The fatty acids of *Dunaliella* are mostly of 16 and 18 carbons with a minor amount of long-chain fatty acids. In a series of recent studies (Peeler *et al.*, 1989), it was shown that the lipid composition of *D. salina* plasma membranes remains relatively constant in cells grown at varying NaCl concentrations, however, including an increase in DGTS and a decrease in the degree of fatty acid unsaturation on growth at higher salinities, respectively. Sterol analysis of *Dunaliella* reveals the presence of ergosterol and the closely related C29 trienol (Wright, 1979) with other minor sterols as common in other green algae (Wright, 1981).

4. BIOTECHNOLOGY OF DUNALIELLA

Dunaliella is highly suitable for biotechnology. The ability of the halotolerant alga to thrive in media with high salt concentrations allows outdoor cultivation in relatively pure cultures with a low presence of predators and grazers. In addition, its high β-carotene content protects it from the intense solar irradiation in the areas where such cultivation is practical, i.e., arid or desert areas with access to brackish water or seawater. *Dunaliella* is mass cultivated in autotrophic medium at the expense of inorganic nutrients in marine medium. Mass production of *Dunaliella* can be activated by techniques resembling those commonly used for the large-scale production of other algae. However, the commercial production of *Dunaliella* in outdoor cultivation is aimed at maximizing β-carotene production rather than biomass as in *Chlorella* or *Spirulina* culture. As indicated above, β-carotene accumulation in *Dunaliella* increases with increasing light intensity and slower growth rate. However, under these conditions, productivity is rather low due to the slow rate of growth. Thus, in practice, maximal productivity of β-carotene is attained in areas having high light intensity where the growth rate and the nitrogen supply are controlled. The growth rate of the alga is limited synergistically by the salt concentration in the medium and by controlling the medium content of an essential nutrient such as nitrate or sulfate. Theoretical maximal productivity calculations for natural outdoor conditions set an upper limit to the light conversion efficiency of approximately 3%, taking into consideration the photosynthetic machinery, solar irradiation, nutrient supply, temperature, and the design of the bioreactor (Avron, 1989). A light conversion efficiency of 3% is equivalent to 25 g biomass m^{-2} day^{-1}. However, intensive culturing attempts to produce high-β-carotene *Dunaliella* containing about 30 pg β-carotene per cell with a β-carotene to chlorophyll ratio exceeding 7 limit the maximal algal outdoor productivity to 5–10 g *Dunaliella* m^{-2} day^{-1} of dry biomass containing 4% β-carotene.

Intensive cultures of *Dunaliella* are based on open oblong raceways operated with paddle wheels on a culture depth of 10–25 cm and a surface area of 1000–4000 m^2. Intensive culture contains 10–20 mg β-carotene $liter^{-1}$ in a depth of 10–20 cm and with a maximal productivity of β-carotene of 400 mg m^{-2} day^{-1}. The extensive culturing system (Borowitzka *et al.*, 1984) is based on nonmixed, very large ponds of 5000–50,000 m^2 in locations that provide a high yearly average solar irradiation where the β-carotene in *Dunaliella* is controlled by using salt concentrations approaching saturation. The extensively grown *Dunaliella* are usually nonmotile, round, large-volume cells of >1000 μm containing more than 50 pg β-carotene per cell and a very low content of chlorophyll. The culture is based on a β-carotene concentration of 0.1–1.0 mg $liter^{-1}$, which yields a productivity of around 0.05–0.1 g *Dunaliella* m^{-2} day^{-1}, of more than 6% β-carotene, with a maximal productivity of β-carotene of 10 mg m^{-2} day^{-1}. The selection of an intensive versus extensive mode of cultivation is related to different considerations, but most important for the second is the availability of the natural resources solar irradiation, arid land, and salt water.

Recently a superintensive mode of mass culturing *Dunaliella* was introduced (Photo-Bioreactions, Ltd., UK, unpublished). The algae are grown in a small-diameter, plastic, transparent, close-circuit tubing system called "photobioreactors." The small-diameter tube allows a short light path to maximize the cell density under tight physiological growth control. The system is expected to yield a biomass concentration about ten times higher than in the open raceways with expected consistent yield and high β-carotene content.

The wide range of *Dunaliella* biotechnological commercial approaches, from the intensive small-bore plastic tubes through the open raceways to the extensive unmixed, large-body, open ponds, allows the production of the algae at different categories of economics. At present, the question of the most effective cultivating production process is still open for verification and conclusions.

5. PRODUCTS OF DUNALIELLA CULTIVATION

β-carotene is currently the most important product of *Dunaliella* cultivation. It is used as a food-coloring agent and as provitamin A in human food and animal feed. It is added to different commercial products such as cosmetics, vitamin preparations, and pharmaceuticals. The worldwide market is dominated by synthetic all-*trans* β-carotene, a product of complex chemical synthesis developed by researchers at Hoffmann La Roche around 1950 (Isler, 1971). Recent epidemiological and oncological studies suggest that normal to high levels of β-carotene in the body may protect it against cancer (Krinsky, 1989a). It has been suggested that the antioxidant function and the ability to quench various radical species may explain the effect of β-carotene as a chemopreventive agent (Krinsky, 1989b). Therefore, humans and animals fed on a diet high in carotenoid-rich vegetables and fruits and who maintain higher than average levels of serum β-carotene have a lower incidence of several types of cancer.

The interest in a natural source of β-carotene is increasing with the buildup of information relating carotenoids to preventive medicine. Of special interest in this regard is the observation that natural β-carotene as found in *Dunaliella* and in most fruits and vegetables contains a mixture of stereoisomers of β-carotene of much higher fat solubility and lower crystallization property than the fat-insoluble, crystallizable, all-*trans* β-carotene. The antioxygenic requirement for higher absorbed β-carotene in disease prevention may create a new expanded market for the natural β-carotene stereoisomeric mixture. Most products of *Dunaliella* in the commercial market are extracts of β-carotene in edible oil containing between 1.5 and 30% *Dunaliella* β-carotene. The others are dried *Dunaliella* powder in capsules or tablets of about 5% β-carotene used predominantly in the health food market under the label of natural β-carotene. The increasing production of *Dunaliella* and its introduction into margarine and other food products as a natural food coloring may gradually replace the synthetic all-*trans* β-carotene in "all-natural" food items. The natural β-carotene sold in 1990 for approximately US$2000 kg^{-1} at a total world market of around 10 tons.

No other products of *Dunaliella* have reached the market stage based on feasible economics. Nevertheless, the production of *Dunaliella* biomass for β-carotene allows the extraction and purification of many other biochemical agents, including glycerol, enzymes, vitamins, amino acids, fatty acids, and growth regulators.

REFERENCES

Avron, M., 1989, The efficiency of biosolar energy conversion by aquatic photosynthetic organisms, in: *Microbial Mats* (Y. Cohen and E. Rosenberg, eds.), ASM Press, Baltimore, pp. 385–387.

Ben-Amotz, A., and Avron, M., 1981, Glycerol and β-carotene metabolism in the halotolerant alga *Dunaliella*: A model system for biosolar energy conversion, *Trends Biochem. Sci.* **6:** 297–299.

Ben-Amotz, A., and Avron, M., 1989, The biotechnology of mass culturing *Dunaliella* for products of commercial interest, in: *Algal and Cyanobacterial Biotechnology* (R. C. Cresswell, T. A. V. Rees, and S. Shah, eds.), Longman, New York, pp. 91–114.

Ben-Amotz, A., and Avron, M., 1990, The biotechnology of cultivating the halotolerant alga *Dunaliella*, *Trends Biotechnol.* **8:**121–126.

Borowitzka, L. J., 1981, The microflora: Adaptation to life in extremely saline lakes, *Hydrobiologia* **81:**33–46.

Borowitzka, L. J., and Borowitzka, M. A., 1988, *Dunaliella*, in: *Microalgal Biotechnology* (M. A. Borowitzka and L. J. Borowitzka, eds.), Cambridge University Press, Cambridge, pp. 27–58.

Borowitzka, L. J., Borowitzka, M. A., and Moulton, T. P., 1984, The mass culture of *Dunaliella salina* for chemicals: From laboratory to pilot plant, *Hydrobiologia* **116/117:**115–121.

Butcher, R. W., 1959, An introductory account of the smaller algae of British coastal waters, I. Introduction and Chlorophyceae, *Fish. Invest. Ser.* **31:**175–191.

Eddy, B. P., 1965, The suitability of some algae for mass cultivation of food, with special reference to *Dunaliella bioculata*, *J. Exp. Bot.* **7:**372–380.

Eddy, B. P., Fleming, I. D., and Manners, D. J., 1958, Alpha-1:4-glucosans, Part IX. The molecular structure of starch-type polysaccharide from *Dunaliella bioculata*, *J. Chem. Soc.* **28:**2827–2830.

Evans, R. W., Kates, M., and Wood, G., 1982, Identification of diacylglycerol-0-(N,N,N-trimethyl)homoserine in the halotolerant alga *Dunaliella parva*, *Chem. Phys. Lipids* **31:**186–195.

Fabregas, J., Abalde, J., Cabezas, B., and Herrero, C., 1989, Changes in protein, carbohydrates and gross energy in the marine microalga *Dunaliella tertiolecta* (Butcher) by nitrogen concentrations as nitrate, nitrite and urea, *Aquacult. Eng.* **8:**223–239.

Finney, K. F., Pomeranz, Y., and Bruinsma, B. L., 1984, Use of algae *Dunaliella* as a protein supplement in bread, *Cereal Chem.* **61:**402–406.

Isler, O. (ed.), 1971, *Carotenoids*, Birkhauser, Basel.

Krinsky, N. I., 1989a, Antioxidants junctions of carotenoids, *Free Radical Biol. Med.* **7:**617–635.

Krinsky, N. I., 1989b, Carotenoids as chemopreventive agents, *Preventive Med.* **18:**592–602.

Peeler, T. C., Stephenson, M. B., Einspahr, K. J., and Thompson, G. A., 1989, Lipid characterization of an enriched plasma membrane fraction of *Dunaliella salina* grown in media of varying salinity, *Plant Physiol.* **89:**970–976.

Shaish, A., Avron, M., and Ben-Amotz, A., 1990, Effect of inhibitors on the formation of stereoisomers in the biosynthesis of β-carotene in *Dunaliella bardawil*, *Plant Cell Physiol.* **31:** 689–696.

Shaish, A., Ben-Amotz, A., and Avron, M., 1991, Production and selection of high β-carotene mutants of *Dunaliella bardawil*, *J. Phycol.* **27**:652–656.

Tornabene, T. G., Holtzer, G., and Peterson, S. L., 1980, Lipid profile of the halophilic alga *Dunaliella salina*, *Biochem. Biophys. Res. Commun.* **96**:1349–1356.

Wright, J. L. C., 1979, The occurrence of ergosterol and (22E, 24R)-24-ethylcholesterol-5,7,22-trien-3-beta-ol in the unicellular chlorophyte *Dunaliella tertiolecta*, *Can. J. Chem.* **57**:2569–2571.

Wright, J. L. C., 1981, Minor and trace sterols of *Dunaliella tertiolecta*, *Phytochemistry* **20**:2403–2405.

12

Marine Microorganisms: A New Biomedical Resource

William Fenical and Paul R. Jensen

1. INTRODUCTION

Microorganisms have had a profound effect on medicinal science since the discovery that not only are they the cause of infection, but they produce organic substances that can cure infection. The discovery of penicillin in 1929 heralded the era of antibiotics and the realization that microorganisms are a rich source of clinically useful natural products [see Betina (1983) for an outline of the history of antibiotics]. Since that time, between 30,000 and 50,000 natural products have been discovered from microorganisms. Of these substances, more than 10,000 are biologically active and more than 8000 are antibiotics (Berdy, 1989; Betina, 1983). This tremendous rate of discovery is testament to the inherent ability of microorganisms to produce bioactive metabolites and the heavy investment of industry in microbial resources. Due to this investment, over 100 microbial products are in use today as antibiotics, antitumor agents, and agrichemicals. Despite the many discoveries, most antibiotics of microbial origin come from terrestrial bacteria belonging to one taxonomic group, the order Actinomycetales. Although these bacteria continue to be studied extensively, it is clear that the rate of discovering

William Fenical and Paul R. Jensen • Marine Research Division, Scripps Institution of Oceanography, University of California at San Diego, La Jolla, California 92093-0236.

Marine Biotechnology, Volume 1: Pharmaceutical and Bioactive Natural Products, edited by David H. Attaway and Oskar R. Zaborsky. Plenum Press, New York, 1993.

novel metabolites from terrestrial actinomycetes is decreasing and that new sources of bioactive natural products must be explored.

One such resource can be found in marine microorganisms, which are only now beginning to be explored for their potential as producers of biomedically relevant metabolites. Marine microorganisms encompass a complex and diverse assemblage of microscopic life forms and occur throughout the oceans, including environments of extreme pressure, salinity, and temperature. Marine microorganisms have developed unique metabolic and physiological capabilities that not only ensure survival in a great variety of extreme habitats, but also offer the potential for the production of metabolites which would not be observed from terrestrial microorganisms.

Early indications that marine microorganisms represent a resource for biomedically relevant compounds came from the work of Rosenfeld and Zobell (1947) and Grein and Meyers (1958), who showed that marine bacteria produce antimicrobial agents. In addition, it was shown that seawater has bacteriocidal properties, and it was suspected that this was due, in part, to the production of antibiotics by planktonic algae (Steemann-Nielsen, 1955) and by bacteria (Baam *et al.*, 1966). Despite this early evidence, relatively little research has been directed toward the study of natural products from marine microorganisms. Results from preliminary studies are encouraging, and it is now known that marine microorganisms are capable of producing unusual natural products that are not observed from terrestrial sources. Many of these compounds have antibiotic and other biological activities, and it is clear that a greater investment in the development of marine microbiology as a resource for biomedically relevant substances is needed.

In this chapter, we provide a brief overview of the marine microbial world and a summary of the bioactive metabolites that have been discovered from its constituents. We hope that in reading this chapter it becomes clear that marine microorganisms encompass many diverse taxonomic groups, and that only a superficial examination of their potential as a source of bioactive agents has been made. Most of the bioactive compounds discovered come from only a few groups of microorganisms, and as new and unusual marine microorganisms are studied, it is certain that new metabolites will be discovered. On the basis of the scope and diversity of marine microorganisms, they must be considered a major, but largely undeveloped, biomedical resource.

2. MICROBIAL DIVERSITY IN MARINE ENVIRONMENTS

Microorganisms can be defined as all life forms that cannot be seen with the unaided eye. This broad definition encompasses an extensive and diverse assemblage of organisms which exhibit widely different morphological, ecological, and physiological characteristics. Due in part to their small size, microorganisms have been notoriously difficult to classify, and many systems have been adopted since their discovery. Recently, a new classification system based on phylogenetic similarity has been proposed and is gaining wide acceptance (Woese and Fox, 1977). In this system, all cellular life forms are placed into three primary groups— the archaebacteria, the eubacteria, and the eukaryotes. These groups were established through the use of new techniques in molecular biology (e.g., 16S rRNA sequence analysis) which allow conclusions to be made about phylogenetic relationships between microorganisms. These new groupings split the previously unified prokaryotes into the archaebacteria and the eubacteria, two groups that are now believed to be as evolutionarily distinct from each other as they are from the eukaryotes.

Due to the highly complex nature of the marine microbial world, it is a challenge to discuss the entire spectrum of microbes in a comprehensive manner. This discussion can be facilitated, however, by grouping microorganisms according to certain shared characteristics. For the purpose of this review, we have grouped marine microorganisms primarily according to the definitions of Woese and Fox (1977). The constituents of these groups have been further categorized according to their principal mode of obtaining organic carbon and energy (Table I). In addition, the chemoheterotrophic eubacteria, which represent a large and heterogeneous group, have been further divided according to morphological and physiological characteristics.

The methods by which microorganisms obtain carbon and energy are important classification criteria, and therefore a brief description of how these categories are defined is in order. In the most general sense, microorganisms can be divided into two nutritional groups, autotrophs and heterotrophs. Autotrophs are primary producers, generating organic molecules from carbon dioxide. The energy required for carbon dioxide incorporation and for other cellular processes comes from either sunlight (photoautotrophy) or the oxidation of reduced inorganic substances (chemoautotrophy). Sometimes "litho" is inserted after the prefix, e.g., a photolithotrophic or chemolithotrophic autotroph, to clarify that the energy source (electron donor) is inorganic. Heterotrophic microorganisms use preformed organics as a carbon and, in most cases, as an energy source (chemoheterotrophy). Likewise, "organo" is sometimes used, e.g., "chemoorganotrophic heterotroph," to clarify that the energy source is organic.

It should be noted that unlike higher plants and animals, many microorganisms are highly versatile in their nutritional requirements. For example, some chemoheterotrophic bacteria can use light as an auxiliary energy source (photoheterotrophy), and some chemoautotrophic bacteria can utilize organic substrates. Because of the high degree of metabolic diversity found within discrete taxonomic groups of microorganisms, and because some individual microorganisms have the ability to utilize a variety of nutritional strategies depending upon environmental conditions, there are few absolutes in the taxonomic groups listed in Table I. Also,

Table I. Classifications of Marine Microorganisms

Archaebacteria	Autotrophic eubacteria	Chemoheterotrophic eubacteria	Eukaryotes
Chemoautotrophs Methangens Thermoacidophiles[a] Chemoheterotrophs Halophiles	Photoautotrophs Anoxygenic photosynthesis Purple and green photosynthetic bacteria (order Rhodospirillales) Oxygenic photosynthesis Cyanobacteria (order Cyanobacteriales) Prochlorophytes (order Prochlorales) Chemoautotrophs Nitrifying bacteria (family Nitrobacteraceae) Colorless sulfur-oxidizing bacteria[a] Methane-oxidizing bacteria (family Methylococcaceae)	Gram-positive Endospore-forming rods and cocci Non-spore-forming rods Non-spore-forming cocci (family Micrococceae) Actinomycetes (order Actinomycetales) and related organisms Gram-negative Rods and cocci Aerobic (family Pseudomonadaceae) Facultative (family Vibrionaceae) Anaerobic (sulfur-reducing bacteria) Gliding bacteria (orders Cytophagales and Beggiatoales) Spirochaetes (order Spirochaetales) Spiral and curved bacteria (family Spirillaceae) Budding, and/or appendaged bacteria Mycoplasmas (class Mollicutes)[b]	Photoautotrophs Microalgae Chemoheterotrophs Protozoa Flagellates Amoebae Ciliates Fungi Higher fungi Ascomycetes Deuteromycetes Basidiomycetes Lower fungi (class Phycomycetes)

[a]Includes heterotrophic genera.
[b]Only two reports documented.

grouping bacteria by common morphological features, as has been done to some extent for the chemoheterotrophic eubacteria, inherently combines bacterial groups which are nutritionally distinct.

We believe Table I includes all of the major categories of marine microorganisms (except acellular forms, e.g., viruses); however, many of these groups are discussed superficially and throughout this chapter we cite more detailed references for additional information. As general references, we highly recommend Sieburth (1979) for a detailed and comprehensive review of marine microorganisms, Stanier *et al.* (1986) for a general microbiology text, and Atlas and Bartha (1987) for information pertaining to microbial ecology. In addition, the text *Marine Microbial Ecology* by Wood (1965), although outdated, is still a valuable reference, and *Microbial Seascapes* (Sieburth, 1975) offers a fascinating view, by electron microscopy, of many marine microorganisms.

2.1. Archaebacteria

The archaebacteria appear to represent a primitive evolutionary line of prokaryotes. In addition to the phylogenetic differences that separate them from the eubacteria, these bacteria demonstrate unique biochemical features, including the lack of peptidoglycan as a cell wall structural polymer (making them insensitive to beta-lactam antibiotics) and ether-linked rather than ester-linked membrane lipids. The archaebacteria are a heterogeneous group that includes both autotrophs and heterotrophs; however, they all share the characteristic of inhabiting extreme environments. Because this kingdom was recently proposed, no systematic subdivisions have been made. There are, however, three highly distinct groups—the halophiles, the thermoacidophiles, and the methanogens. We refer to Woese *et al.* (1978) for a description of the archaebacteria and Kandler (1982) for recent biochemical studies.

The halophilic archaebacteria require at least 12–15% NaCl to survive and grow well even at concentrations up to saturation. They are red due to a high carotenoid content and dominate in high-salt environments, such as salterns and salt lakes, where they can occur in sufficient concentrations to cause the pink color often associated with these habitats. The halophilic archaebacteria are chemoheterotrophic; however, they can also generate energy from sunlight using a unique mechanism of photophosphorylation. These bacteria are easily cultured on high-salt media, and detailed descriptions of their taxonomy and physiology are available (Rodriguez-Valera, 1988). For a complete description of the microbiology and biogeochemistry of hypersaline environments, see the recent text by Javor (1989).

The thermoacidic archaebacteria are a heterogeneous group defined by their ability to grow at low pH and at high temperatures. Representative thermoacidic bacteria can function at temperatures exceeding 90°C and at a pH of less than 1, in habitats that must be considered the most extreme environments on this planet, including geothermal hot springs. Certain genera, e.g., *Sulfolobus*, can be grown in the lab on organic substrates, but appear to grow in nature as chemoautotrophs via the oxidation of sulfur compounds. This process results in the formation of sulfuric acid and is largely responsible for the acidity of the habitats in which they are found.

The methanogens are strict anaerobes and have the ability to produce methane by the reduction of carbon dioxide and some simple organic materials (acetate, formate, methanol, etc.). These bacteria are widespread in marine habitats, including hypersaline environments, where organic matter is being decomposed through biological processes. Methanogens reduce these materials to methane gas, which rises out of the sediments and becomes available for aerobic processes (e.g., oxidation by methanophilic bacteria). The methanogens and other archaebacteria have evolved unique mechanisms to survive in extreme environments, including the production of heat-stable lipids and enzymes that preserve membrane integrity and cellular function at high temperatures. These and other bacteria inhabiting extreme environments have great potential for biotechnology, and this subject was recently reviewed by DaCosta *et al.* (1988).

2.2. Eubacteria

Most prokaryotes, including the cyanobacteria, are eubacteria. They are almost exclusively unicellular, although some are filamentous or coenocytic, and from a nutritional standpoint are both diverse and versatile. The eubacteria have evolved nutritional strategies to exploit virtually every imaginable resource and, in many cases, can utilize more than one nutritional strategy, depending on environmental conditions. The most recent and comprehensive treatises on the systematics of these bacteria can be found in vols. 1–4 of *Bergey's Manual of Systematic Bacteriology*. These detailed volumes describe the Gram-negative bacteria of general, medicinal, or industrial importance (Vol. 1) (Krieg and Holt, 1984), the Gram-positive bacteria other than actinomycetes (Vol. 2) (Sneath *et al.*, 1986), the archaebacteria, cyanobacteria, and remaining Gram-negative bacteria (Vol. 3) (Staley *et al.*, 1989), and the actinomycetes (Vol. 4) (Williams *et al.*, 1989). In addition, methods for the isolation and culture of virtually all known groups of eubacteria have been described in detail in Volumes I and II of *The Prokaryotes* (Starr *et al.*, 1981).

2.2.1. Photoautotrophic Eubacteria

The photoautotrophic eubacteria can be divided into two groups based on the type of photosynthesis they perform (Table I). The bacteria that do not liberate oxygen during photosynthesis (anoxygenic photosynthesis) are usually referred to as photosynthetic bacteria and are represented by the purple and green bacteria of the order Rhodospirillales. These bacteria can be distinguished from other phototrophs by the presence of bacterial chlorophyll, which differs structurally from the chlorophylls of cyanobacteria, algae, and higher plants, and by their inability to use water as an electron donor in photosynthesis. These bacteria photosynthesize only under anaerobic conditions, using reduced inorganic sulfur compounds or H_2 as an electron donor. Although some photosynthetic bacteria can grow heterotrophically in the dark, their principal mode of nutrition is photosynthesis. Photosynthetic bacteria are common in shallow marine sediments, where they play an important ecological role in the conversion of toxic hydrogen sulfide to less toxic oxidized compounds. Methods for the isolation and culture of these bacteria have been described in detail (Pfennig, 1967; Stanier *et al.*, 1981).

It should be noted that bacteria capable of performing anoxygenic photosynthesis in the presence of oxygen have been described (Shiba *et al.*, 1979). This is unusual in that it was previously believed that bacteria could photosynthesize anoxygenically only under anaerobic conditions. The bacteria capable of doing this are facultative phototrophs, growing heterotrophically using light as an additional energy source, and are common in aerobic marine habitats, where they have been shown to constitute up to 6.3% of the bacterial population. These aerobic photosynthetic bacteria include the marine genus *Erythrobacter* as well as certain methylotrophic species. The state of knowledge regarding these bacteria has been discussed in detail in the recent text by Harashima *et al.* (1989).

The prokaryotes that generate oxygen during photosynthesis (oxygenic photosynthesis) possess the same type of chlorophyll as algae and higher plants (chlorophyll a) and are represented by the cyanobacteria and the Prochlorales. The cyanobacteria (formerly known as blue-green algae) are a diverse group of Gram-negative eubacteria represented by more than 1000 species of filamentous and unicellular phototrophs. The classification of these bacteria has been problematic, as some species have been described using the terms of the Botanical Code, under the division Cyanophyta, while others have been described using the International Code of Nomenclature of Bacteria, under the order Cyanobacteriales.

The cyanobacteria are unique in their ability to fix nitrogen while performing oxygenic photosynthesis, and for this reason, some can grow under conditions of extreme nutrient limitation when supplied only with light, mineral media, carbon dioxide, and molecular nitrogen. Cyanobacteria are common in the marine environment and occur in many lighted habitats, including upper sediment layers and as symbionts in the surface tissues of some invertebrates, for example, sponges (Wilkinson, 1987). It should be noted that some cyanobacteria are also capable of performing anoxygenic photosynthesis; therefore, this is not an absolute criterion for separating them from the photosynthetic bacteria. For a general text on cyanobacteria see Carr and Whitton (1982).

Members of the Prochlorales are uncommon relative to cyanobacteria, and only one marine genus has been described—*Prochloron* (Lewin, 1977). These phototrophs differ from cyanobacteria in that they possess both chlorophylls a and b, and lack bilin pigments. *Prochloron* is most often found as a symbiont in ascidians, but has also been reported in association with other invertebrates and free-living in some freshwater habitats. These bacteria have never been successfully cultured, and therefore little is known of their physiology except for what can be determined from cells isolated from host invertebrates (these cells can be obtained in large quantities). For an extensive review of *Prochloron* see the recent text by Lewin and Cheng (1989).

2.2.2. Chemoautotrophic Eubacteria

The chemoautotrophic eubacteria generate organic molecules from CO_2 and energy from the oxidation of reduced inorganic substrates. These Gram-negative prokaryotes are widely distributed in the marine environment and play a critical role in the recycling of elements that have been reduced through biological and geochemical processes. There are three major groups of chemoautotrophs and these can be distinguished based on the type of substrate oxidized (Table I). Although many of these bacteria were once considered obligate chemoautotrophs, it is now believed that all nitrifying and sulfur-oxidizing bacteria can both assimilate and metabolize organic substances to some degree (Sieburth, 1979).

The nitrifying bacteria oxidize either ammonia to nitrite (generic names beginning with *Nitroso* e.g., *Nitrosococcus*) or nitrite to nitrate (generic names beginning with *Nitro*, e.g., *Nitrobacter*), but never both. This process is extremely important in the nitrogen cycle, since positively charged ammonium ions bind to acidic sediment particles, where they become unavailable for biological processes. The nitrifying bacteria, by regenerating nitrite and nitrate from ammonium, convert nitrogen to a form readily available for other biological processes.

Reduced inorganic sulfur compounds can be oxidized by many taxonomically unrelated bacteria, including the purple and green photosynthetic bacteria. By restricting this group to colorless sulfur bacteria, the photosynthetic bacteria, already discussed with the phototrophic prokaryotes, can be distinguished. The remaining sulfur-oxidizing bacteria represent a diverse group that includes thermophilic archaebacteria, e.g., *Sulfolobus*, and certain of the gliding bacteria. Most information about sulfur-oxidizing bacteria comes from work with small, unicellular species within the Nitrobacteraceae, such as *Thiobacillus*. These bacteria are aerobic, do not deposit intracellular sulfur granules, and are some of the only colorless sulfur-oxidizers within the Nitrobacteraceae that have been successfully cultured. These bacteria are obligate or facultative in their sulfur-oxidizing abilities, are common in marine sediments, and are often tolerant to the acidic conditions resulting from the oxidation of hydrogen sulfide.

The sulfur-oxidizing bacteria of the order Cytophagales are characterized by gliding motility and include large filamentous forms that deposit intracellular sulfur granules. These bacteria are common in hydrogen sulfide-rich marine habitats, where they can form visible mats. They are often microaerophilic (requiring low levels of oxygen), in which case they are found inhabiting the interface between aerobic and anaerobic environments. Because of their precise oxygen requirements, and the difficulties in providing hydrogen sulfide and oxygen simultaneously under controlled conditions, there has been little success in culturing these bacteria (with the exception of a few strains of Beggiatoa). In addition to bacteria assigned to the Cytophagales, there are large-celled species with uncertain taxonomic affiliations, e.g., *Achromatium*, that should be included in this discussion for their apparent chemoautotrophic growth by the oxidation of reduced sulfur compounds.

Recently, high concentrations of sulfur-oxidizing bacteria have been discovered in association with deep-sea hydrothermal vents (Jannasch and Wirsen, 1979). These bacteria are the primary producers in hydrothermal vent habitats and represent the base of a food chain which supports a rich and unusual invertebrate fauna. In addition to free-living forms, sulfur-oxidizing bacteria have been found as symbionts in certain vent-associated invertebrates, most notably vestimentiferan worms, where they can occur in densities of up to 10^9 cells/g wet weight (Cavanaugh *et al.*, 1981). These symbiotic bacteria have not been successfully cultured, and therefore their classification remains uncertain. It is clear, however, that in many cases they are the primary source of organic materials for the host. Since the discovery of chemoautotrophic bacteria as symbionts in vent-associated invertebrates, similar associations have been found in invertebrates inhabiting other hydrogen sulfide-rich habitats. For a review of chemoautotrophic bacteria/invertebrate symbioses see Southward (1987).

The methane-oxidizing bacteria (methanophiles) are a diverse group of Gram-negative prokaryotes defined by their unique ability to utilize methane as a sole carbon and energy source. These bacteria are often included in discussions of chemoautotrophs because of their general inability to utilize organic materials with carbon–carbon bonds. Methanophiles are aerobic and can be found in upper sediment layers where they utilize biologically produced methane as it rises from deeper anaerobic sediments [for a more detailed discussion of bacteria that metabolize C-1 compounds see Crawford and Hanson (1974)]. Methanophiles and other chemoautotrophs have the remarkable ability to utilize many resources that intuitively would not be expected to support life. These organisms are important in nutrient cycling and as primary producers, and their unique metabolic pathways offer potential for the production of novel metabolites.

2.2.3. Chemoheterotrophic Eubacteria

Chemoheterotrophic eubacteria are the most thoroughly studied of the bacteria, in part because many can be easily cultured in nutrient-rich media, where they use preformed organic materials as both carbon and energy sources. They are too large and diverse a group to summarize by principal mode of nutrition alone, and for this reason, they have been subdivided based on differences in cell wall structure (Gram-positive vs. Gram-negative), morphology, and affinity for oxygen (Table I).

2.2.3a. Gram-Positive Bacteria. Differences in the chemical composition of the bacterial cell wall can be distinguished by Gram staining. This test differentiates between two major taxonomic groups, the Gram-positive and the Gram-negative bacteria. From early studies of marine bacteria, it was determined that most seawater bacteria are Gram negative (Zobell and Upham, 1944). Gram-positive bacteria are usually reported as being less than 10% of the total seawater population, and it has been suggested that the Gram-negative cell wall is better adapted for survival in the marine environment (Sieburth, 1979). These reports, however, are based upon seawater samples, and there is evidence that Gram-positive bacteria occur as a higher percentage of the total marine microflora in sediments and on surfaces (Kriss, 1963). Unfortunately, most research on marine bacteria has focused on Gram-negative varieties, and therefore little is known about the distribution and ecological role of Gram-positive bacteria in marine habitats.

The actinomycetes (order Actinomycetales) and related microorganisms are a morphologically diverse group of Gram-positive bacteria ranging from strains with branching filaments that form well-developed mycelia, to unicellular rods and cocci that form typical bacterial colonies. Actinomycetes are common soil bacteria and are best known for their ability to produce unusual secondary metabolites, a characteristic for which they are unrivaled in the microbial world. Many of these secondary metabolites have antibiotic properties, and because of this, the pharmaceutical industry has invested heavily in the study of actinomycetes.

Although it is well documented that actinomycetes can be isolated from marine habitats (Goodfellow and Haynes, 1984), it has been proposed that they are not indigenous marine bacteria. This proposal is based on the assumption that actinomycete growth from marine samples originates from spores which have been washed into the sea, where they remain viable but metabolically dormant. Recent studies have shown that actinomycetes have distributions in marine sediments and requirements of seawater for growth that cannot be explained by the theory that they are metabolically inactive terrestrial bacteria (Jensen *et al.*, 1991). Based on this, it can be concluded that actinomycetes adapted to the marine environment represent a physiologically unique resource of microorganisms for industry. The reports of novel metabolites from marine actinomycetes, many of which will be discussed in this text, support this conclusion.

In addition to actinomycetes and related genera such as *Arthrobacter*, Gram-positive, endospore-forming bacteria of the genera *Bacillus* and *Clostridium* (family Bacillaceae) can also be isolated from marine sediments. Like actinomycetes, these bacteria produce highly resistant spores. Thus, their physiological activity in the marine environment has been questioned. However, *Bacillus* species readily grow in seawater-based media when sufficient nutrients are supplied, and therefore it is reasonable to propose that growth occurs under appropriate conditions in the marine environment. Considering that the genus *Bacillus* has been a productive source of antibiotic and insecticidal materials, it seems appropriate to suggest that, like marine actinomycetes, these Gram-positive bacteria represent an unexplored and potentially rich source of novel metabolites. In addition to the spore-forming Gram-positive bacteria, non-spore forming cocci (family Micrococceae) (Gunn and Colwell, 1983) and rods have been reported from marine environments. Although these bacteria make up a small percentage of the total heterotrophic population, they should not be overlooked as an important component of the marine microflora.

2.2.3b. Gram-Negative Bacteria. The Gram-negative eubacteria represent the largest and most diverse group of marine chemoheterotrophic prokaryotes. We have grouped these bacteria according to morphological characteristics and, within the rods and cocci, according to their affinity for oxygen (Table I). Most Gram-negative marine bacteria do not have distinct morphological features other than being either rod or coccus shaped. These aerobic and facultatively anaerobic bacteria are common in seawater and are best represented by the families Pseudomonadaceae and Vibrionaceae, respectively. They are easily isolated from marine samples, as they grow rapidly on marine agar and can represent a large percentage of the bacterial colonies observed on plates which have been inoculated with marine samples. Although they are rarely distinct morphologically, they include biochemically and ecologically diverse genera such as the bioluminescent *Photobacterium* and the genus *Vibrio*, which can make up a large percentage of the gut microflora of some fish and invertebrates (Leifson *et al.*, 1964). The bacteria in the Pseudomonadaceae, collectively known as "pseudomonads," are extremely common in seawater. It should be noted that these bacteria are not restricted to the genus *Pseudomonas*, but include other genera, e.g., *Xanthomonas* and *Alteromonas* [see P. Baumann and Baumann (1981) for a description of aerobic marine eubacteria].

Most areas of the world's oceans are well oxygenated, and not surprisingly most marine bacteria are aerobic or facultatively anaerobic. There are, however, obligately anaerobic eubacteria that play important roles in the marine environment. One prominent group of anaerobes are the sulfur-reducing bacteria (generic names beginning with *Desulfo*, e.g., *Desulfovibrio*). These Gram-negative bacteria ferment simple organics, and in the process respire anaerobically using oxidized sulfur compounds as a final electron acceptor. The sulfur-reducing bacteria are widely distributed in marine sediments and generate large quantities of hydrogen sulfide that rises out of the sediments and becomes available for oxidation by phototrophic and colorless sulfur-oxidizing bacteria.

Many Gram-negative, heterotrophic eubacteria have distinct morphological features by which they can be grouped. It should be noted that grouping bacteria in this manner can place nutritionally distinct genera together, and therefore not all of the following bacteria are exclusively chemoheterotrophic. Bacteria with distinct morphological features are usually associated with surfaces and are not often found in open ocean waters. They tend to grow slowly relative to pseudomonads and thus are not frequently observed on agar media that have been inoculated with marine samples. Many of these bacteria, however, can be isolated using selective techniques such as those described in Vols. I and II of *The Prokaryotes* (Starr *et al.*, 1981). The groups used in the following discussion of morphologically distinct bacteria are based on those established in *Bergey's Manual of Systematic Bacteriology*, Vol. 1 (Krieg and Holt, 1984) and Vol. 3 (Staley *et al.*, 1989).

The gliding bacteria in the orders Cytophagales and Beggiatoales grow in long, thin filaments and are characterized by their unusual mechanism of motility. Most of these bacteria are chemoheterotrophic; however, they include chemo-autotrophic genera that metabolize hydrogen sulfide (see section on chemo-autotrophs). These common marine bacteria are usually large-celled and relatively slow-growing. They are most frequently found attached to surfaces and can occur in high enough concentrations to form visible mats. These orders include the aerobic genus *Leucothrix*, which is often found in association with crustaceans, and *Beggiatoa*, which is known to form large, visible mats in areas rich in hydrogen sulfide. *Beggiatoa* is one of the few aerobic, large-celled, sulfur-oxidizing bacteria that can be successfully cultured. The myxobacteria (order Myxococcales) are also motile by gliding and, in addition, produce unusual reproductive structures called fruiting bodies. These bacteria are known producers of novel secondary metabolites; however, their occurrence in the marine environment is not well documented. For a recent review of the gliding bacteria and the secondary metabolites they produce see Reichenbach and Hofle (1989).

Spirochaetes are large, coiled bacteria that can be aerobic, facultative, or anaerobic. They belong to the order Spirochaetales and are highly motile, using a unique mechanism of flexing that allows them to move rapidly through viscous liquids where motility by flagellated bacteria is reduced. They are found free-living in the marine environment and as symbionts associated with the crystalline style of certain mollusks (Johnson, 1981). Most mollusks with a crystalline style harbor large numbers of the spirochaete *Cristispira*; however, this genus has never been successfully cultured.

The appendaged or prosthecate bacteria are primarily aquatic and are often found attached to surfaces (Staley *et al.*, 1981). They have complex life cycles that include the formation of cellular extensions such as stalks and hyphae. The functions of these appendages are not entirely understood, although they have been shown to play a role in reproduction and nutrient absorption. These bacteria are adapted to low nutrient concentrations and include such genera as *Caulobacter*. They are not often observed on agar plates; however, they will attach to glass microscope slides which have been placed in seawater (Zobell and Allen, 1935).

The spiral and curved bacteria of the Spirillaceae are Gram-negative rods ranging from comma-shaped to helical coils. They can be distinguished from the spirochaetes in that they move using flagella. These bacteria are common in marine environments and tend to be microaerophilic (preferring low concentrations of oxygen). Included in this group is the genus *Bdellovibrio*, an unusual parasitic bacterium whose life cycle involves penetrating the cell wall of another bacterium, reproducing, and ultimately lysing the host cell.

In addition to the Gram-negative eubacteria with distinct morphological characteristics, there is a group of bacteria, the mollicutes, which are characterized by their lack of a defined cell wall and their resulting pleomorphic shape. Mollicutes are the smallest known living organisms capable of self-reproduction and are best known as parasites of plants, animals, and invertebrates. There is recent evidence that mollicutes (formerly known as mycoplasmas) occur in the marine environment in association with certain invertebrates (Boyle *et al.*, 1987; Zimmer and Woollacott, 1983).

Marine prokaryotes represent a large and heterogeneous group of microorganisms. Because of their tremendous diversity, it is difficult to present a comprehensive description of these bacteria by placing them in generalized groups. The difficulties of this task are compounded by current changes in bacterial systematics. Traditionally, bacteria have been classified based on phenotypic traits. However, new methods in molecular biology, e.g., comparisons of the base composition of DNA (% G + C), are now being used to determine genetic relationships among microorganisms, and this information is rapidly changing the taxonomic position of many bacteria.

Despite the superficial treatment of prokaryotic diversity presented in this chapter, it is clear that many morphologically and biochemically distinct groups of marine bacteria exist. The basic science pertaining to many of these bacteria is not well developed, and only a few groups have been explored as a resource for biomedically relevant natural products.

2.3. Eukaryotes

Eukaryotic microorganisms are more complex and generally larger than prokaryotes. They are distinguished from prokaryotes by the possession of a nuclear envelope and membrane-bound organelles, e.g., mitochondria and chloroplasts.

Eukaryotic microorganisms can be divided into three groups, the microalgae, the protozoa, and the fungi (Table I). The distinctions between these three groups are not always clear, however, as some microorganisms have characteristics of more than one group. From a nutritional perspective, eukaryotic microorganisms obtain nutrients either photoautotrophically, as in the algae (there are no chemoautotrophic eukaryotes), heterotrophically through the engulfing of organic materials (phagocytosis), as in the protozoa, or through absorption, as is characteristic of the fungi. Although many eukaryotic microorganisms can be cultured, they have not been studied extensively as a source of industrially important metabolites.

2.3.1. Photoautotrophic Eukaryotes: Microalgae

Microalgae are the major producers of organic carbon compounds in the sea. They are fundamental to the marine food chain because they convert carbon dioxide into organic materials which can subsequently be utilized by heterotrophic microorganisms, invertebrates, and fish. Microalgae include all unicellular photosynthetic eukaryotes and are classified with the algae primarily according to morphology and the type of photosynthetic pigments they possess. Microalgae occur in 10 of the 12 algal divisions (phyla) as described by Round (1973). They are a diverse group and include the diatoms (Bacillariophyta) and dinoflagellates (Dinophyta), which together make up the bulk of marine phytoplankton (some authors place the dinoflagellates with the flagellated protozoa).

Diatoms are the largest group of microalgae and pelagic forms contribute significantly to primary production in the open ocean, where higher plants are lacking. Their cell wall is composed of silica and forms bivalved frustules that fit one within the other to enclose the cell. The shape of these frustules is an important taxonomic characteristic in the identification of the many thousands of described species. Dinoflagellates are also common in the marine environment and occur both free-living and as symbionts in some marine invertebrates. It is the mutually beneficial symbiosis between the dinoflagellate *Symbiodinium microadriaticum* (zooxanthellae) and scleractinian corals which results in the formation of coral reefs in the tropics. Dinoflagellates are also responsible for red tides and for the production of some marine toxins.

Diatoms and dinoflagellates are just two of the diverse groups of microalgae. Many other divisions, including the exclusively unicellular phyla Euglenophyta and Prasinophyta, are also common in the marine environment. We refer to the text *Phytoflagellates* (Cox, 1980) for more information pertaining to these microorganisms. The microalgae include many thousands of species, many of which have been successfully cultured and utilized for commercial applications. Detailed methods describing the isolation and culture of photosynthetic eukaryotes can be found in the text by Stein (1975).

2.3.2. Chemoheterotrophic Eukaryotes

The chemoheterotrophic eukaryotes are represented in the marine environment by the protozoa and the fungi. They are important ecologically for the conversion of bacterial biomass into more complex life forms, and for the breakdown of certain recalcitrant organic materials. The chemoheterotrophic eukaryotes are common marine microorganisms, and the technology for their mass culture is available. Despite this, they have not been adequately studied for their potential to produce biomedically relevant natural products.

2.3.2a. Protozoa. The phylum Protozoa includes the simplest of unicellular eukaryotic animals. There is some debate over the definition of the protozoa. For the purpose of this review, we will define them as unicellular eukaryotes that obtain nutrients heterotrophically, usually by engulfing food particles. The protozoa can be divided into three groups: the nonphotosynthetic flagellates, the ciliates, and the amoebae. As might be expected, their classification is complex, and we recommend Capriulo (1990) for more precise information.

As the name implies, flagellates are characterized by the possession of flagella. These microorganisms are widespread in the marine environment and constitute an important component of the planktonic biomass. They can be divided into two groups, one of which includes the nonphotosynthetic members of the microalgae, e.g., the phagotrophic euglenids and dinoflagellates. Other flagellates are not related to the microalgae and include the choanoflagellates and the bicoecids. This heterogeneous group includes unicellular and colonial forms that are voracious consumers of bacteria.

The amoeboid protozoa include the naked, amorphous amoebae that are motile and engulf prey by means of cytoplasmic extensions called pseudopodia. Other amoebae have rigid exoskeletons (the loricate amoebae) and include the testaceans, the foraminiferans, and the radiolarians. The tests of ancient loricate amoebae are common in fossil marine muds and have been studied extensively by paleobiologists for information relevant to the evolution of the oceans and their inhabitants.

The ciliated protozoa possess rows of cilia which beat synchronously and are used for locomotion and food collection. The ciliated protozoa are the most homogeneous group of protozoa and are widely distributed in marine environments, with more than 6000 species described. Almost all ciliates have a permanent opening for collecting food, which they do quite efficiently, making them important in the conversion of bacterial biomass into a form that can be consumed by higher animals. These and other protozoa are ecologically important in the marine environment, and we recommend the recent text by Fenchel (1987) for a review of the ecology of the protozoa.

2.3.2b. Fungi. Fungi are heterotrophic eukaryotes which obtain nutrients osmotrophically through the absorption of dissolved organic substances. Most are coenocytic with a filamentous, vegetative growth form known as a mycelium; however, unicellular forms, or yeasts, are also common. Fungi can be divided into two major groups: the higher fungi, which include the classes Ascomycetes, Basidiomycetes, and Deuteromycetes (fungi imperfecti), and the lower fungi, or Phycomycetes.

The Ascomycetes are the largest group of fungi, with over 2000 genera. Most of the described marine fungi belong to this group [for a review of the higher marine fungi see Kohlmeyer and Kohlmeyer (1979)], and they are usually found in shallow waters, in association with decomposing algae and other cellulose-containing materials. The wood-inhabiting species are called lignicolous fungi and are important economically, as they contribute to the deterioration of submerged wooden structures. Some of the higher fungi are also known pathogens of marine algae (Goff and Glasgow, 1980). These pathogens can be problematic in aquaculture and can cause visible galls in natural populations of seaweeds, e.g., the brown alga *Sargassum*. They are also believed to be responsible for the sponge-wasting disease which caused extensive mortalities in commercial sponges in the Bahamas (Galtsoff *et al.*, 1939).

The lower fungi are also common in the marine environment, but there are relatively few studies of these microorganisms. They are best known as parasites and for causing diseases in a variety of marine invertebrates (Sparks, 1985), seagrasses (Short *et al.*, 1987), and algae (Goff and Glasgow, 1980). The phycomycete *Lagenidium callinectes*, for example, is a voracious predator on shrimp larvae and causes extensive problems in commercial shrimp hatcheries. Fungi of the genus *Fusarium* cause black gill syndrome and other crustacean infections that result in serious mortalities in pond-raised penaeid shrimp and cultured lobsters.

Fungi are serious pathogens in the marine environment. They have not been studied extensively as a source of secondary metabolites, and considering that many can be cultured, they represent a resource for novel secondary metabolites. Additional information pertaining to marine fungi can be found in the texts *The Biology of Marine Fungi* (Moss, 1986) and *Recent Advances in Aquatic Mycology* (Jones, 1976).

3. BIOACTIVE METABOLITES FROM MARINE MICROORGANISMS

The potential for marine microorganisms to provide bioactive natural products is immense if judged by the scope and taxonomic diversity of this group. In almost all of the taxonomic classes mentioned in Section 2, there have been indications (or documentation!) of the production of bioactive metabolites. In surveys of selected groups, reports of antibiotic and toxin production have been widespread. In most cases, however, these reports have not been followed by comprehensive chemical studies to isolate and identify the bioactive component present. The antibiotic properties of marine bacteria, first reported by Rosenfeld and Zobell (1947), for example, have been a subject of continuing reports to the present time (e.g., Grein and Myers, 1958; Baam *et al.*, 1966; Doggett, 1968; Gauthier and Flateau, 1976; Hoyt and Sizemore, 1982; Lemos *et al.*, 1985; Meseguer *et al.*, 1986; Nair and Simidu, 1987; Barja *et al.*, 1989; Austin and Billaud, 1990). Various marine bacteria have also been indicated to produce antiviral agents (Gundersen *et al.*, 1967; Fujioka *et al.*, 1980; Toranzo *et al.*, 1982; Katzenelson, 1978). In other microbial classes, early surveys documented numerous cases in which the presence of bioactive metabolites was indicated but not proved (Baslow, 1969). By and large, however, the major studies have involved specific classes of microorganisms, primarily the microalgae, the bacteria, and the marine fungi, which are readily cultured.

3.1. Microalgae

As defined in Section 2, marine microalgae are a large group of phototrophic microorganisms spanning at least ten full plant divisions. These organisms are widely distributed in marine habitats and form the foundation of the marine food web. Although the chemical activities of these organisms as a whole are virtually unknown, selected classes of microalgae have been recognized as being chemically prolific. The dinoflagellates (phylum Dinophyta), in particular, are known producers of "red tide" and other fish toxins in both temperate and tropical waters. Recent investigations have further shown that some dinoflagellates produce polyether antibiotics of unusual structure classes. The reader is referred to the chapter on dinoflagellates in this text by Y. Shimizu for a comprehensive account of the biomedical potential of this group.

Depending upon one's choice of systematics, the blue-green algae or cyanobacteria are phototrophic microorganisms classified by some as microalgae and by others as bacteria. Indeed, the cyanobacteria resemble algae environmentally in that they are a conspicuous component of the phytoplankton. In freshwater environments, toxic cyanobacteria are frequently observed, creating severe dangers for livestock. Recent studies have shown that freshwater cyanobacteria are a significant resource for bioactive molecules, particularly cytotoxic agents with applications in cancer chemotherapy (Moore *et al.*, 1988; Patterson *et al.*, 1991). However, the culture of unicellular marine cyanobacteria appears to be much more difficult, greatly restricting chemical investigations of these organisms.

Many species of microalgae have been reported to produce bioactive metabolites. Steeman-Nielsen (1955) was among the first researchers to report upon the presence of antibiotic substances in planktonic algae. Since that time, several reports (Sieburth, 1964; Duff *et al.*, 1966; Aubert and Gauthier, 1966; Aubert *et al.*, 1966; Antia and Bilinski, 1967) have documented the presence of antibiotics in these microorganisms. In only two cases, however, have the active antibiotics been isolated and defined. Sieburth (1960) showed that the active component from the arctic phytoplankter *Phaeocystis pouchetii* was acrylic acid, a simple metabolite with generalized toxicity against bacteria. Occurring as high as 7% of the dry weight of the alga, acrylic acid was found to be transferred throughout the marine food chain, eventually finding its way to the digestive tracts of antarctic penguins (Sieburth, 1961).

3.2. Chemical Studies of Marine Bacteria

Marine bacteria are by far the best studied chemically of any of the marine microorganisms. This field is growing rapidly and there are frequent additions to the growing literature. The reader is directed to a recent review article by Austin (1989), which gives a good account of marine bacterial metabolites from a microbiological viewpoint.

To the best of our knowledge, the first documented identification of a bioactive marine bacterial metabolite was the highly brominated pyrrole antibiotic **1** (see Fig. 1), isolated by Burkholder and co-workers from a bacterium obtained from the surface of the Caribbean seagrass *Thalassia* (Burkholder *et al.*, 1966). This highly unique metabolite was identified by x-ray crystallographic methods (Lovell, 1966), and is composed of more than 70% bromine by weight. The molecule showed impressive *in vitro* antibiotic properties against Gram-positive bacteria, with minimum inhibitory concentration (MIC) ranging from 0.0063 to 0.2 μg/ml. The antibiotic was not active against Gram-negative bacteria, however, and it proved inactive in whole-animal assays. Although this interesting bacterium was first assigned as *Pseudomonas bromoutilis*, the biochemical characteristics of this isolate were later found to indicate an affinity to the genus *Alteromonas* (Skerman *et al.*, 1980).

Several years later, Andersen *et al.* (1974) isolated a purple-pigmented bacterium which also produced potent antibiotics. The strain, originally defined as a *Chromobacterium* sp., [now revised by L. Baumann *et al.* (1972) to the genus *Alteromonas*] was isolated from seawater samples collected in the North Pacific Ocean gyre. Comprehensive chemical analysis showed that this organism produces several antimicrobial compounds, including the known pyrrole **1**, tetrabromopyrrole (**2**), hexabromo-2,2'-bipyrrole (**3**), and several simple phenolics, including 4-hydroxybenzaldehyde and *n*-propyl-4-hydroxybenzoate. Tetrabromopyrrole (**2**) showed moderate antimicrobial activity *in vitro* against *Staphylo-*

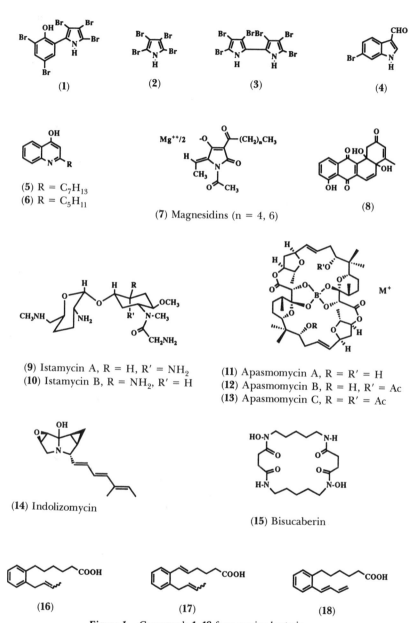

Figure 1. Compounds **1–18** from marine bacteria.

coccus aureus, *Escherichia coli*, *Pseudomonas aeruginosa*, and *Candida albicans*. It was even more active against a group of marine bacteria, and showed autotoxicity against the producing *Chromobacterium* sp. itself.

In a similar investigation, the same group (Wratten *et al.*, 1977) later isolated an antibiotic-producing bacterium from a La Jolla, California, tide pool seawater sample. The yellow strain was identified as a pseudomonad, and careful analysis of its metabolic products showed the production of 6-bromoindole carboxaldehyde (**4**), its debromo analog, and a mixture of 2-*n*-pentyl- and 2-*n*-heptylquinolinol (**5**, **6**), the latter a known antibiotic produced by strains of *P. aeruginosa*. The most potent of these simple antibacterial agents was 2-*n*-pentylquinolinol (**6**), which showed its greatest activity against *S. aureus*. The unique metabolite, 6-bromo-indole carboxaldehyde (**4**), lacked antibiotic properties.

The discovery of this exceptional molecule points strongly to the concept of the genetic uniqueness of marine microorganisms. Like the plants and invertebrates in marine habitats, some bacteria have evolved the ability to concentrate bromide ion from seawater ([Br$-$] = ca. 65 mg/liter in seawater) and, through oxidative pathways, to incorporate this element into organic compounds. This is a common mechanism in marine systems, resulting in bromination of metabolites which possess enhanced bioactivities. The bioactivities reported for these simple brominated metabolites did not appear to provide useful leads in antibacterial chemotherapy.

In another study, a group of antibiotics, the magnesidins (**7**) (see Fig. 1), were isolated as a 1:1 mixture of methylene homologs (four and six methylene groups) as minor pigments of the marine bacterium *Pseudomonas magnesiorubra* (Gandhi *et al.*, 1976; Kohl *et al.*, 1974). *P. magnesiorubra* was originally isolated from the surfaces of the tropical marine green alga *Caulerpa peltata*. These interesting pigments are thought to be oxidation products of prodigiosin, a common tripyrrolic pigment produced by marine as well as terrestrial bacteria.

Since the early 1970s, the Tokyo group at the Institute of Microbial Chemistry (Okazaki and Okami, 1972; Okami and Okazaki, 1972; Okazaki *et al.*, 1975) has been an innovator in the search for new metabolites from marine actinomycetes. Largely working with sediment-derived actinomycetes, Okami and co-workers have investigated a large number of both near-shore and deep-sea habitats. In a series of papers beginning in 1972, several structurally-unique bioactive molecules have been discovered (Okami, 1986). An isolate identified as *Chainia purpurgena* SS-228, found in mud samples obtained from Sagami Bay, was found to produce the benzanthraquinone antibiotic **8** (Kitahara *et al.*, 1975; Okazaki *et al.*, 1975). The antibiotic showed selective inhibition of Gram-positive bacteria, with MIC values of between 1 and 2 μg/ml, and was an active antitumor agent *in vivo* against Ehrlich carcinoma in mice, showing significant life span extension at low doses. In addition, quinone **8** deactivated dopamine-β-hydroxylase (producing 65% inhibition at 0.1 μg/ml). Most interestingly, the production of **8** was only observed in selected seawater media containing the unique Japanese seaweed product "Kobu Cha." Kobu Cha is a dried and pulverized powder produced in Japan from the brown seaweed *Laminaria*. This very interesting observation clearly shows that marine microorganisms have specific nutrient adaptations which relate to their natural nutrient sources in the ocean. It points to the need, in marine microbiology, to develop selective marine isolation and mass culture media which utilize *natural* nutrients and growth factors derived directly from marine sources.

The istamycins A and B (**9** and **10**) are other examples of antibiotics isolated by the Tokyo group from the culture broths of marine actinomycetes (Okami *et al.*, 1979). The istamycins were produced by fermentation of the marine streptomycete *Streptomyces tenjimariensis* SS-939, collected from a shallow-water mud sample from Sagami Bay, Japan. The compounds show strong *in vitro* antibiotic activity against both Gram-negative and Gram-positive bacteria, including those which are known to be resistant to the aminoglycoside antibiotics. Istamycins A and B typically showed MIC values of between 0.10 and 3.0 μg/ml against various species and isolates of *Staphylococcus, Bacillus, Corynebacterium*, and *Escherichia*. The compounds were much less active against *Pseudomonas, Klebsiella*, and *Serratia* species. The istamycins are related to the fortimycins and sporaricins, aminoglycoside antibiotics produced by terrestrial actinomycetes, but appear to be more therapeutically interesting because of their activity against resistant bacteria.

Among the more interesting and unusual compounds isolated from marine microorganisms are the aplasmomycins A–C (**11–13**), also isolated by the Tokyo group. The compounds are produced by a marine actinomycete identified as *Streptomyces griseus* SS-20, isolated from shallow mud samples also from Sagami Bay (Okami *et al.*, 1976; Sato *et al.*, 1978; Hotta *et al.*, 1980). Here, too, this organism only produced the aplasmomycins when cultured with Kobu Cha-containing media, or when grown under conditions (27°C and very low nutrients) which relate to the natural environment of Sagami Bay. The aplasmomycins are antibiotics which inhibit Gram-positive bacteria *in vitro* with MIC values of between 0.8 and 3.0 μg/ml against *S. aureus, Bacillus subtilus, B. anthracis*, and *Corynebacterium smegmatis*. More importantly, aplasmomycin is an effective antimalarial agent *in vivo*, inhibiting *Plasmodium berghei* infection in mice when administered orally. It was on the basis of this potent antiplasmodium activity that aplasmomycin received its name. The structure elucidation of aplasmomycin A was performed by x-ray crystallographic methods. To the surprise of the Tokyo group, the molecule contained an unexpected boron atom in the center of the complex (Nakamura *et al.*, 1977). Although quite remarkable structures, there are precedents for these compounds in the boromycins, similar boron-containing ionophores produced by a terrestrial actinomycete.

The Tokyo group has also experimented with the genetics of some marine actinomycetes. When protoplasts of the istamycin-producing *S. tenjimariensis* SS-939 were fused with the aplasmomycin-producing *S. griseus* SS-20, a new clone, SK2-52, was obtained which produced a new indolizidine antibiotic, indolizomycin (**14**) (Gomi *et al.*, 1984; Yamashita *et al.*, 1985; Okami, 1988).

Screening with a novel macrophage effector bioassay, the Tokyo group has also isolated an interesting new 22-membered ring siderophore, bisucaberin (**15**), from the unicellular marine bacterium *Alteromonas haloplanktis* SB-1123 (Kameyama *et al.*, 1987; Takahashi *et al.*, 1987). Strain SB-1123 was isolated from a deep-sea mud sample collected at -3300 m off the Aomori Prefecture coast. The bacterium required seawater for growth and produced large quantities of bisucaberin (ca. 700 mg/liter) using sardine and cuttlefish powders and maltose as the major carbon sources. Although bacteria derived from deep sediments often require up to 700 atm pressure to allow growth, this organism appears to grow well at surface pressures. The fermentation was conducted for 2 days at 27°C a temperature well above the normally low temperatures recorded for the deep seas. Bisucaberin has the unique biological property of rendering tumor cells susceptible to the cytolytic action of murine peritoneal macrophages. This property has been proposed as a biorational approach to utilizing natural immunological methods as alternatives to potent cytotoxins in cancer chemotherapy. No data have been reported on the whole-animal testing of these interesting compounds.

In other work with unicellular bacteria, Holland *et al.* (1984) have illustrated the production of a series of C_{16} aromatic acids (**16–18**) by the marine bacterium *Alteromonas rubra*. The acids **16–18** showed interesting pharmacological properties in bronchodilator assays and in neuromuscular assays designed to detect relaxant effects. These compounds appear obvious in their biosynthetic origin from fatty acid synthesis.

More recent work from the Tokyo group includes the isolation of a structurally novel monoterpene alkaloid, altemicidin (**19**; see Fig. 2) from a marine strain of *Streptomyces sioyaensis* SA-1758 (Takahashi *et al.*, 1989a, b). This interesting monoterpene alkaloid was detected by screening cultures for toxicity against the common brine shrimp *Artemia salina*. Earlier reports have shown that *Artemia* toxicity translates amazingly well to antitumor activity (Meyer *et al.*, 1982). Indeed, altemicidin showed potent antitumor activity *in vitro* against L1210 murine leukemia and IMC carcinoma cell lines with IC_{50} values of 0.84 and 0.82 µg/ml, respectively. The new compound showed weak antibacterial activity, but was relatively toxic in mice ($LD_{50} = 0.3$ mg/kg iv), probably limiting its chemotherapeutic uses.

In recent studies in our laboratory, we have focused attention on the antitumor-antibiotics produced by deep-sea bacteria and those microorganisms found in tropical marine habitats. Screening for antitumor effects, we isolated a deep-sea bacterium which yielded a series of novel cytotoxic and antiviral macrolides, the macrolactins A-F (**20–25**) (Gustafson *et al.*, 1989). The bacte-

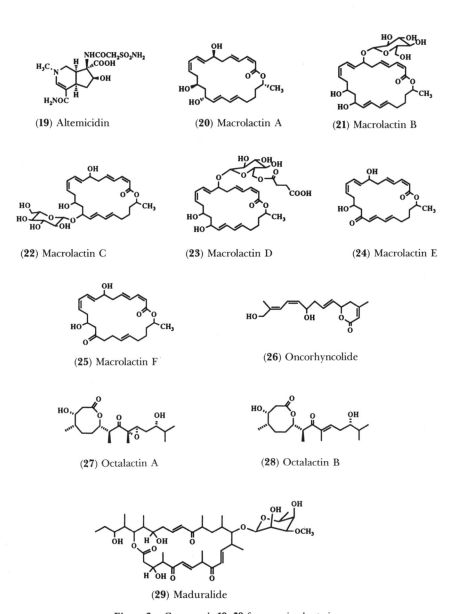

Figure 2. Compounds **19–29** from marine bacteria.

rium, isolate C-237, was Gram-positive, had a requirement for salt, and could not be identified by standard biochemical methods. Under standard fermentation at atmospheric pressure, bacterium C-237 produced the six macrolides and two open-chain hydroxy acids in varying amounts. Macrolactin A (**20**) was produced as the major metabolite (ca. 8 mg/liter) in most of the fermentations. The majority of the biological properties were due to macrolactin A, which showed modest antibacterial activity, but was active against B16-F10 murine melanoma *in vitro* with IC_{50} values of 3.5 μg/ml. More importantly, macrolactin A inhibited several viruses, including herpes simplex (IC_{50} = 5.0 μg/ml) and human immunodeficiency virus, HIV (IC_{50} = 10 μg/ml). The structures of these compounds were determined by combined spectral methods and did not include the absolute stereochemistries at the four chiral centers. In a collaborative effort with Rychnovsky and co-workers (Rychnovsky *et al.*, 1991), we have established the complete stereostructure for macrolactin A as shown in **20**.

High-molecular-weight antitumor agents have also been isolated from marine microorganisms. Umezawa and his co-workers (Umezawa *et al.*, 1983; Okami, 1988) have described the isolation of a new antitumor polysaccharide, marinactan (MACT), on the growth of sarcoma-180 tumors in mice. Marinactan, a heteroglycan largely composed of glucose, mannose, and ficose, is produced by *Flavobacterium uliginosum* MP-55. At daily doses of 10–50 mg/kg in mice, marinactan produced 70–90% inhibition of the growth of this solid tumor.

Additional studies of unicellular marine bacteria include the isolation of several amino acid diketopiperazines from a sponge-derived bacterium of the genus *Micrococcus* (Stierle *et al.*, 1988), the known antibiotic 3-amino-3-deoxy-D-glucose from a deep-sea bacterium (Fusetani *et al.*, 1987), and the interesting lactone oncorhyncolide (**26**) from a seawater-derived bacterium, isolate #157 (Needham *et al.*, 1991). Oncorhyncolide is, biosynthetically, a unique metabolite not seen in prior studies. Current studies of the origin of this metabolite indicate that it could be a terpenoid, or be derived by unique methylation reactions not yet observed in nature.

Although marine actinomycetes have been recognized as major inhabitants of shallow marine waters, their distributions on living surfaces in the ocean remain unclear. We isolated an unidentified streptomycete from the surface of a gorgonian coral (*Pacifigorgia* sp.) from the Gulf of California, Mexico. This isolate (PG-19), when grown in marine media, produced a series of undescribed metabolites possessing cytotoxic and antibiotic properties. Two of the derivatives isolated are the 20-hydroxy derivative of oligomycin-A and the 5-deoxy derivative of enterocin. These compounds possess antibiotic properties similar to their parent molecules. In addition, this interesting organism yielded two totally new cytotoxic agents, octalactins A and B (**27**, **28**), belonging to a new structure class. The compounds are C_{19} ketones possessing rare eight-membered ring lactone functionalities. Octalactin A (**27**) possesses potent *in vitro* cytotoxicity against B16-F10 murine melanoma (IC_{50} = 7.2 × 10^{-3} μg/ml) and HCT-116 human carcinoma (IC_{50} = 0.5 μg/ml) cell lines. Octalactin B is devoid of cytotoxic activity, leading to the conclusion that the epoxy ketone functionality is essential for cytotoxicity. The structures of these compounds were finalized by x-ray studies on octalactin A (Tapiolas *et al.*, 1991), but the absolute stereochemistry was not defined.

Of the marine actinomycetes studied, those of the genus *Streptomyces* have clearly predominated. It must be noted, however, that nonstreptomycetes are common marine actinomycetes (e.g., Jensen *et al.*, 1991) and that their chemical behaviors remain virtually unknown. One example of studies of a marine non-streptomycete is our recent investigation of a sediment-derived actinomycete (CNB-032) found in the shallow sediments of Bodega Bay, California. On the basis of its cell wall constituents and whole-cell sugar composition, this organism keys to a member of the suprageneric group Maduramycetes. Fermentation of this organism resulted in the production of maduralide (**29**), a new member of a rare class of macrolide. Although the biological properties of this macrolide are not yet known, the compound is a member of a rare 24-membered ring lactone group represented previously only by rectilavendomycin (Omura, 1984). The compound was not crystalline, hence x-ray methods could not be applied to the structure determination. The structure assignment, yielding only the planar molecule, was determined by combined spectral methods emphasizing 2D NMR (Pathirana *et al.*, 1991).

The surface of an undescribed jellyfish from the Florida Keys has also recently yielded an actinomycete which produces unique peptide metabolites. This organism was identified as a *Streptomyces* sp. (CNB-091). Extracts of this culture were found to possess significant antibiotic activity. Investigation of the source of this activity led to the isolation of two new bicyclic peptides, salinamides A and B (**30, 31**; see Fig. 3), which possess novel depsipeptide backbones. The structure of salinamide A was derived first by spectral methods and by hydrolysis to yield several chiral amino acids. Subsequently, the crystalline isomer, salinamide B (**31**), was isolated and its structure determined by x-ray methods. Conversion of **30** to **31** with HCl in methanol further linked these molecules, confirming both their relative and absolute stereochemistries (Trischman *et al.*, 1992). The salinamides are antibiotics of moderate potency with selective activity against Gram-positive bacteria. The phenyl ether ring illustrates restricted rotation with all four aromatic protons shown at distinct NMR chemical shifts. This property, and its selective antibiotic activity, have opened the question of its relatedness to the vancomycin antibiotics, which act by binding dipeptide cell wall precursors. Studies are now in progress to determine any mechanistic similarities between vancomycin and the salinamides.

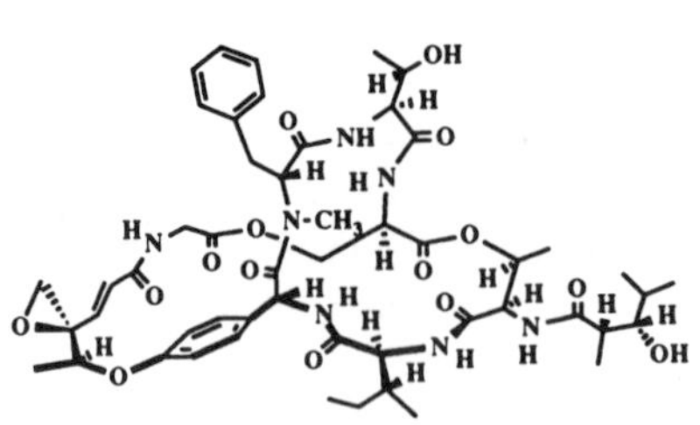

(30) Salinamide A

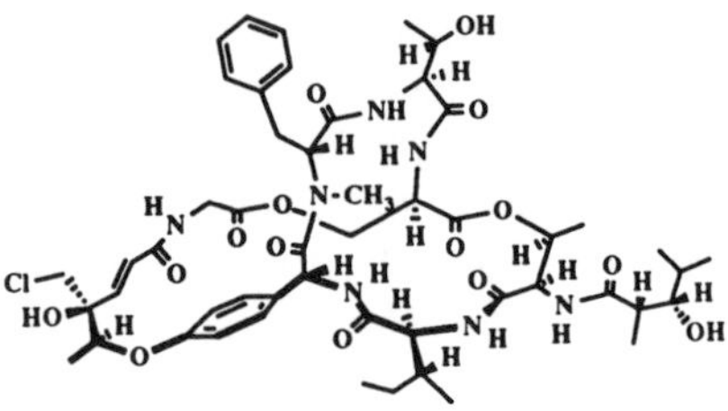

(31) Salinamide B

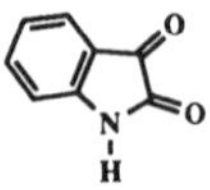

(32) Isatin

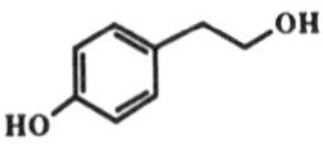

(33) Tyrosol

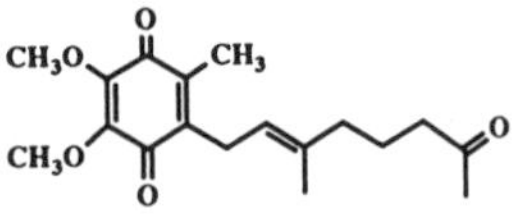

(34)

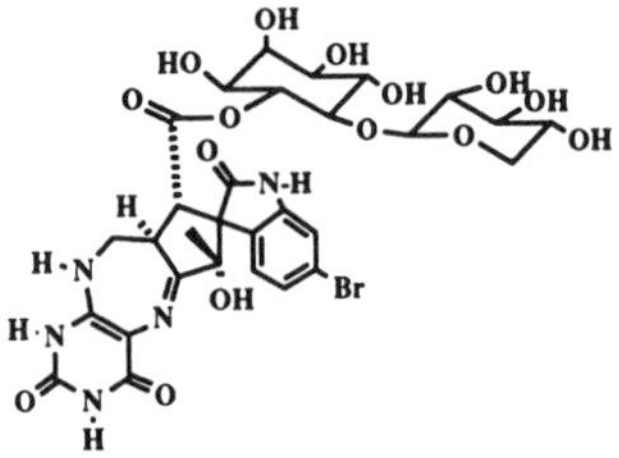

(35) Neosurugatoxin

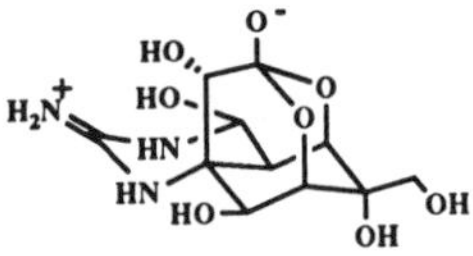

(36) Tetrodotoxin

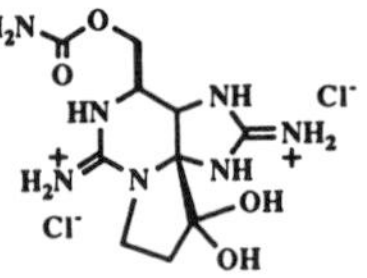

(37) Saxitoxin

Figure 3. Compounds **30–37** from marine bacteria.

3.3. Symbiotic Marine Bacteria—Origins of Marine Toxins

As more evidence is obtained, it is becoming abundantly clear that bacteria form highly specific, symbiotic relationships with marine plants and animals. Our own experience in this area arose from a study of the pathogen resistance of the estuarine shrimp *Palaemon macrodactylus*. We observed (Gil-Turnes *et al.*, 1989) that the eggs of this animal possess significant bacterial epibionts, which, when removed by treatment with antibiotics, leads to the rapid infestation of the eggs by pathogenic fungi, especially *Lagenidium callinectes*. Although there are many plausible mechanisms to explain this protective phenomenon, with *Palaemon* the answer appears to lie in the bacterial production of defensive antifungal agents. Fermentation of the symbiotic bacterium (an *Alteromonas* sp.) led to the isolation of an unexpectedly potent antifungal agent, 2,3-indolinedione, also known as isatin (**32**; see Fig. 3). This compound had been known for years as a synthetic intermediate in the production of indigo dyes. It was not recognized as an antifungal agent, although related indolinediones have shown various biological properties. In a similar study, Gil-Turnes (1988) investigated the American lobster, *Homarus americanus*, and found similar results. In this case, the eggs of *Homarus* were found totally covered with an unidentified unicellular bacterium. In a similar fashion, the cultured bacterium produces very large amounts of tyrosol (2-*p*-hydroxyphenyl ethanol, **33**). At the concentration produced, it appears that this simple phenolic can be effective in controlling pathogenic microorganisms. Gil-Turnes (1988) also studied the tropical filamentous blue-green *Microcoleus lyngbyaceus* from Puerto Rico. Studies of surface-derived bacteria from 65 sampling sites around the island were compared. To our surprise, four specific strains of highly colored bacteria (one purple, one red, and two yellow) were consistently obtained only from the surface of the cyanobacterium. These strains were not isolated from adjacent seawater samples nor from morphologically similar filamentous green algae. Fermentation of the bacteria led to unique results. All four strains produced the same antibiotic material, identified as the quinone **34**. This molecule, produced at levels of 20 mg/liter, possesses significant antifungal and antibacterial activities, and it had never been observed as a natural product. Biosynthetically, this compound appears to be produced by oxidative cleavage of a ubiquinone-type precursor.

These few studies, and considerable additional observations (e.g., Sieburth, 1975, 1979), illustrate that many marine bacteria are adapted to specific surface habitats in marine (and probably all aquatic) environments. Further, it appears that symbiotic relationships have evolved which provide nutrient-rich environments for the bacterium and protection for the host against water-borne pathogens.

In the context of symbiotic bacteria, recent information delineates the true origins of many marine toxins. Evidence of bacterial toxin production was first obtained for the interesting, but not widespread toxin neosurugatoxin (**35**). First found along with prosurugatoxin as an unexpected toxicant in the edible Japanese mollusk, *Babylonia japonica* (Kosuge *et al.*, 1972), this unique toxin was later traced to an unexpected source, a Gram-positive, coryneform bacterium isolated directly from the digestive gland of the animal (Kosuge *et al.*, 1985). It appears that environmental conditions were generated which allowed this bacterium to attain large numbers in *Babylonia*. It does not appear that the bacterium is a true symbiont of the mollusk, since animals from other regions are known to be nontoxic.

The potent marine neurotoxin tetrodotoxin (TTX, **36**) has been known for many years. Until recently, its origin has been considered to be pufferfish of the family Tetraodontidae. In Japan, pufferfish, known as "fugu," are considered a delicacy and great care is taken in preparing and consuming the flesh of these toxic fish. Despite great care and training, a few deaths are reported each year from tetrodotoxin ingestion. The origin of TTX has been a subject of vigorous debate. Although mainly recognized from pufferfish, TTX has been isolated from crabs, an octopus, and even a terrestrial amphibian, all suggesting a microbial source. Recent data have shown conclusively that TTX is produced by numerous unicellular marine bacteria (Yasumoto *et al.*, 1986; Noguchi *et al.*, 1986; Yotsu *et al.*, 1987). These organisms span a diverse range of bacterial groups, with toxin being detected in at least 15 genera of shallow- and deep-sea isolates (Do *et al.*, 1990; Simidu *et al.*, 1990). It is indeed unusual to find the same molecule being produced so widely and by numerous bacterial genera. Perhaps toxin production is linked to a plasmid which can be transferred within marine environments. Continued study of this interesting phenomenon will undoubtedly clarify aspects of the bacterial production of TTX.

The causative toxin involved in "paralytic shellfish poisoning" (PSP) has, for many years, been known to be saxitoxin (STX, **37**) and structurally-related toxins (e.g., the gonyautoxins), metabolites known to be produced mainly by marine dinoflagellates of the genus *Protogonyaulax* from cold-water marine habitats. While this source has been generally accepted, there have been inconsistencies which have caused this concept to be questioned. Toxic shellfish have been observed in areas devoid of *Protogonyaulax*, for example (Ogata *et al.*, 1989). Recently, Kodama *et al.* (1988, 1990) have provided startling evidence that a marine bacterium, identified as a *Moraxella* sp., isolated directly from *P. tamarensis*, is the true producer of STX toxins.

3.4. Chemical Studies of Marine Fungi

Although terrestrial fungi have represented a major biomedical resource (penicillin from *Penicillium*, for example), studies to develop the biomedical potential of marine fungi have been few. The first report which demonstrated that marine fungi may lead to unique metabolites was that involving the small lactone leptosphaerin (**38**; see Fig. 4). Leptosphaerin was isolated from cultures of *Leptosphaeria oraemaris* (Schiehser, 1980) and its structure was determined by x-ray crystallography and through synthesis (Schiehser *et al.*, 1986; Pallenberg and White, 1986). This work appears to have been initiated at Oregon State University in the early 1970s, and aspects of these studies have been summarized

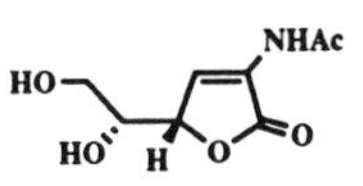

(**38**) Leptosphaerin 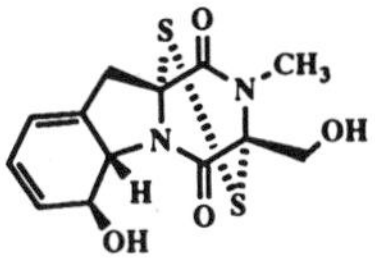   (**39**) Gliotoxin 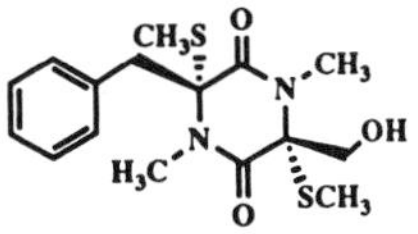   (**40**) Gliovictin

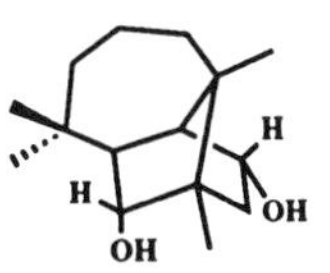

(**41**) Culmorin

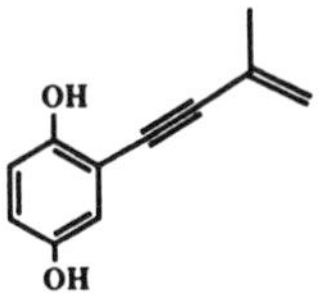

(**42**) Siccayne

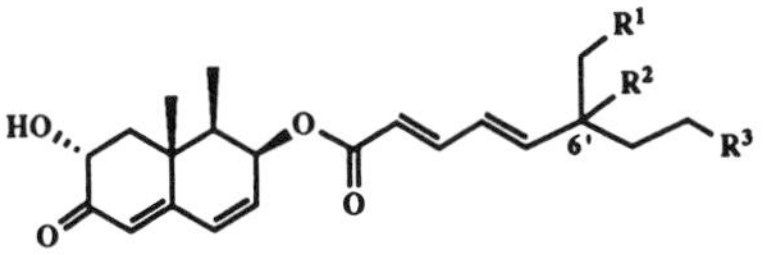

(**43**) Dendryphiellin A, $R^1 = R^2 = H$, $R^3 = OH$
(**44**) Dendryphiellin B, $R^1 = R^3 = H$, $R^2 = OH$
(**45**) Dendryphiellin C, $R^1 = R^2 = R^3 = H$ (6′S)
(**46**) Dendryphiellin D, $R^1 = OH$, $R^2 = R^3 = H$

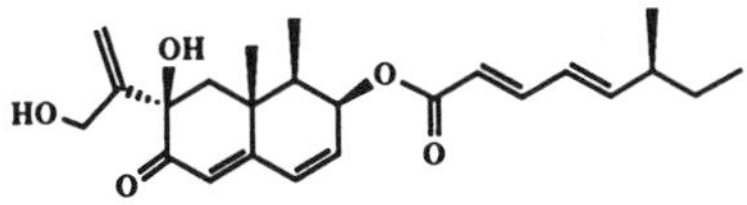

(**47**) Dendryphiellin E

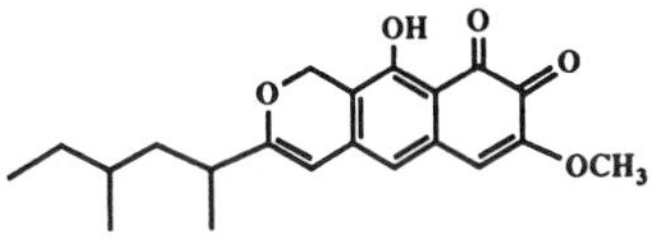

(**48**) Obionin A

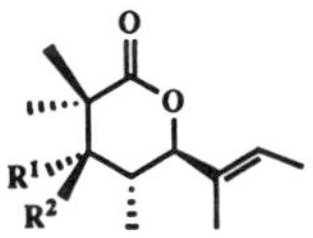

(**49**) Heliascolide A, $R^1 = OH$, $R^2 = H$
(**50**) Heliascolide B, $R^1 = H$, $R^2 = OH$

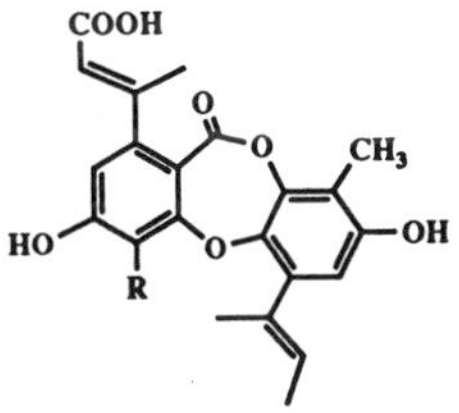

(**51**) Auranticin A, $R = CH_2OH$
(**52**) Auranticin B, $R = CHO$

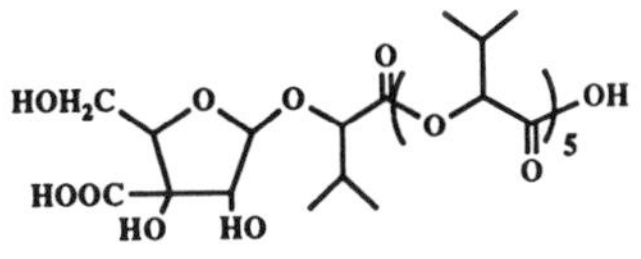

(**53**) Hymenoscyphin A

Figure 4. Compounds from marine fungi.

by Kirk *et al.* (1974). In the late 1970s, Okutani (1977) described the isolation of the common bacterial toxin gliotoxin (**39**) from a marine *Aspergillus* sp. Subsequently, Shin and Fenical (1987) reported the structurally-related metabolite gliovictin (**40**) from the marine fungus *Asteromyces cruciatus*.

One of the most common classes of marine fungi are the lignicolous forms which inhabit submerged wood surfaces. To assess the degree of interspecies chemical competition within this environmental group, Strongman *et al.* (1987) compared the antifungal activities of 27 lignicolous fungi and found four isolates showing inhibitory properties. The source of the antifungal activity in this strain of *Leptosphaeria oraemaris* was traced to the sesquiterpene diol culmorin (**41**), a metabolite known to be produced by several terrestrial fungi. This work suggests that chemical defense is an important adaptation of fungi competing for limited wood substrates.

In a search for antibiotics from Basidiomycetes, Kupka and co-workers detected activity from extracts of *Halocyphina villosa*, one of the four marine Basidiomycetes that have been described (Kupka *et al.*, 1981). Subsequent chemical studies showed the antibiotic properties to be derived from siccayne (**42**), an acetylenic hydroquinone also produced by the terrestrial fungus *Helminthosporium siccans*. Siccayne was weakly antibacterial against a variety of Gram-positive and Gram-negative bacteria, and also showed inhibition of DNA and RNA synthesis *in vitro*.

The first in-depth chemical study of a marine fungus appears to be that reported by the group in Trento, Italy. Fermentation of the marine deuteromycete *Dendryphiella salina* resulted in the production of an interesting Tris-nor sesquiterpenoid, dendryphiellin A (**43**) (Guerriero *et al.*, 1988), as the major metabolite. Closer investigation of the fermentation broths yielded the related metabolites dendryphiellins B-D (**44–46**), which vary only in the positions of hydroxylation on the fatty acid side chains (Guerriero *et al.*, 1989). Finally, the fully intact sesquiterpenoid ester dendryphiellin E (**47**) was isolated, proving that these metabolites are derivatives of the eremophilane sesquiterpene class.

More recent work from the Gloer group at the University of Iowa has led to the isolation of new bioactive compounds from marine fungi. A new polyketide, the ortho-quinone obionin A (**48**), was isolated from the marine fungus *Leptosphaeria obiones* (Poch and Gloer, 1989a). *L. obiones* is a halotolerant ascomycete originally isolated from the surface of the salt marsh grass *Spartina alterniflora*. Obionin A showed central nervous system (CNS) activity in inhibiting binding of the dopamine D-1-selective ligand 3H-SCH 23390 to bovine corpus striatum membrane (IC_{50} = 2.5 µg/ml). Poch and Gloer (1989b) also reported the interesting lactones heliascolides A and B (**49**, **50**) from the marine fungus *Helicascus kanaloanus* (ATCC 18591). This ascomycete was originally isolated from a mangrove environment in Hawaii. The helicascolides, whose structures were determined by combined spectral methods, are epimeric secondary alcohols at the C-3 position.

The most recent paper published from the Iowa group (Poch and Gloer, 1991) reports the structure determination of the auranticins A and B (**51, 52**) from a mangrove isolate of *Prussia aurantaica* (ATCC 14745). Auranticin A shows antibiotic activity against *B. subtilus* and *S. aureus* at concentrations of 5–50 µg/disk. Lastly, in unpublished work, G. K. Poch and J. B. Gloer have isolated an interesting diacid, hymenoscyphin A (**53**), from the salt-marsh fungus *Hymenoscyphus* sp. Hymenoscyphin A shows toxicity against brine shrimp at 10 µg/ml, and consists of an unusual repeating alpha-hydroxy isobutyric acid group. In culture, the production of this metabolite has proved to be unreliable. This is not an uncommon problem encountered in fermentation, and it is expected to be even more problematic with marine organisms whose culture requirements have been largely unstudied.

4. DISCUSSION AND PROSPECTS FOR THE FUTURE

4.1. Marine Microorganisms As a Resource for Bioactive Metabolites

By and large, the pharmaceutical industry has focused on the same groups of microorganisms, mainly the actinomycetes and other Gram-positive isolates, as its primary source for new drug candidates. Historically, this has been an amazingly successful enterprise, with over 100 drugs and animal care products now being produced through fermentation. This success has stimulated current expenditures in excess of $9 billion each year in the search for new bioactive-material-producing microorganisms from traditional terrestrial sources. But, as the return from these traditional sources diminishes, the need to explore entirely new environments for fermentation microorganisms will increase. Indeed, in this regard marine microorganisms represent a massive undeveloped resource.

In this review, data have been presented to show the scope and diversity of microorganisms found in marine environments. On the basis of the few chemical studies reported, and in recognition of the unique compounds which have been isolated, it can be concluded that marine microorganisms could, if effectively explored, represent a major biomedical resource. Given the lack of in-depth knowledge of the nutritional requirements of marine microorganisms, however, the question of how to develop this resource effectively must first be answered.

4.2. Basic Research in Marine Microbiology

In our opinion, it will not be productive to assault marine environments with major biomedical screening programs at this time. Without considerable attention to developing the basic biology of marine microorganisms, explorations for new bioactive metabolites would be limited to those few classes of microorganisms which are readily isolated and grown under "standard" conditions. Unfortunately, we know little about the specific nutrients and growth factors required by most marine microorganisms. Common media constituents such as peptone, simple sugars, etc., are unrealistic marine nutrients, and in the marine environment are apparently replaced with complex carbon sources such as chitin, sulfated polysaccharides, and marine proteins. In addition, we know virtually nothing of the effects of uncommon inorganic elements, such as lithium, silicon, etc., which are also abundant in marine sediments.

It is estimated that less than 5% of the bacteria observed by microscopic methods are found culturable under "standard" conditions. This fact will greatly limit our ability to isolate and culture the majority of the interesting and new bacterial forms present. Most investigations have focused on the Gram-negative bacteria prevalent in seawater, with almost no attention being placed on the Gram-positive groups. We now know that Gram-positive bacteria are major marine organisms commonly associated with surfaces and particles. Because of the severe lack of basic understanding, we are concerned that more emphasis be placed upon the basic biology of marine microorganisms before this resource can be effectively developed.

4.3. The Roles of Industry and Academia

Although there is considerable microbiological expertise in the pharmaceutical industry, few of these investigators have received training in marine microbiology. Even those who have been trained in marine methods have focused their attention on specific groups of microorganisms which are readily culturable and easily controlled. Medical marine microbiology is a new field requiring attention to areas of marine microbiology, new isolation methods, new media development, etc., which are not among the curricula of the major oceanographic institutions. These facts will hinder the immediate development of marine microorganisms as a biomedical resource. However, collaborative programs which combine the biomedical and microbiological expertise of the pharmaceutical industry with the marine microbiological resources available in marine institutions will have a good chance of making significant headway.

REFERENCES

Andersen, R. J., Wolfe, M. S., and Faulkner, D. J., 1974, Autotoxic antibiotic production by a marine *Chromobacterium, Mar. Biol.* 24:281–285.
Antia, N. J., and Bilinski, E., 1967, A bacterial-toxin (lecithinase C) in a marine phytoplanktonic chrysomonad, *Fish. Res. Board Can. J.* 24:201–204.

Atlas, R. M., and Bartha, R., 1987, *Microbial Ecology: Fundamentals and Applications*, 2nd ed., Benjamin/Cummings, Menlo Park, California.

Aubert, M., and Gauthier, M., 1966, Origine et nature des substances antibiotiques présentes dans le mileau marin VII. Note sur l'activité antibactérienne d'une diatomée marine *Chaetoceros teres* (Clève), *Rev. Int. Oceanogr. Méd.* **4**:33–37.

Aubert M., Gauthier, M., and Daniel, S., 1966, Origine et nature des substances antibiotiques présentes dans le mileau marin III. Activité antibactérienne d'une diatomée marine *Asterionella japonica* (Clève), *Rev. Int. Oceanogr. Méd* **1**:35–43.

Austin, B., 1989, Novel pharmaceutical compounds from marine bacteria, *J. Appl. Bacteriol.* **67**:461–470.

Austin, B., and Billaud, A.-C., 1990, Inhibition of the fish pathogen, *Serratia liquefaciens*, by an antibiotic-producing isolate of *Planococcus* recovered from seawater, *J. Fish Dis.* **13**:553–556.

Baam, R. B., Gandhi, N. M., and Freitas, Y. M., 1966, Antibiotic activity of marine microorganisms: The antibacterial spectrum, *Helgol. Wiss. Meeresunters.* **13**:188–191.

Barja, J. L., Lemos, M. L., and Toranzo, A. E., 1989, Purification and characterization of an antibacterial substance produced by a marine *Alteromonas* species, *Antimicrob. Agents Chemother.* **33**:1674–1679.

Baslow, M. H., 1969, *Marine Pharmacology*, Williams and Wilkins, Baltimore, pp. 8–55.

Baumann, L., Baumann, P., Mandel, M., and Allen, R. D., 1972, Taxonomy of aerobic marine eubacteria, *J. Bacteriol.* **110**:402–429.

Baumann, P., and Baumann, L., 1981, The marine Gram-negative eubacteria: Genera *Photobacterium, Beneckea, Alteromonas, Pseudomonas*, and *Alcaligenes*, in: *The Prokaryotes*, Vol. II (M. P. Starr, H. Stolp, H. G. Truper, A. Balows, and H. G. Schlegel, eds.), Springer-Verlag, Berlin, pp. 1302–1331.

Berdy, J., 1982, Search and discovery methods for novel antimicrobials, in: *Bioactive Metabolites from Microorganisms* (M. E. Bushell and U. Grafe, eds.), Elsevier, Amsterdam, pp. 3–25.

Betina, V., 1983, *The Chemistry and Biology of Antibiotics*, Elsevier, Amsterdam.

Boyle, P., Maki, J. S., and Mitchell, R., 1987, Mollicute identified in novel association with aquatic invertebrate, *Curr. Microbiol.* **15**:85–89.

Burkholder, P. R. Pfister, R. M., and Leitz, F. P., 1966, Production of a pyrrole antibiotic by a marine bacterium, *Appl. Microbiol.* **14**:649.

Capriulo, G. M., 1990, *Ecology of Marine Protozoa*, Oxford University Press, Oxford.

Carr, N. G., and Whitton, B. A., 1982, *The Biology of Cyanobacteria*, University of California Press, Berkeley.

Cavanaugh, C. M., Levering, P. R., Maki, J. S., Mitchell, R., and Lidstrom, M. E., 1981, Prokaryotic cells in the hydrothermal vent tube worm *Riftia pachyptila* Jones: Possible chemoautotrophic symbionts, *Science* **213**:340–342.

Cox, E. R., 1980, *Phytoflagellates*, Elsevier, New York.

Crawford, R. L., and Hanson, R. S. (eds.), 1974, *Microbial Growth on C1 Compounds* (Proceedings of the 4th International Symposium), American Society for Microbiology, Washington, D.C.

Da Costa, M. S., Duarte, J. C., and Williams, R. A. D., 1988, *Microbiology of Extreme Environments and its Potential for Biotechnology* (FEMS Symposium No. 49), Elsevier Applied Science, London.

Do, H. K., Kogure, K., and Simidu, U., 1990, Identification of deep-sea-sediment bacteria which produce tetrodotoxin, *Appl. Environ. Microsc.* **56**:1162–1163.

Doggett, R. G., 1968, New anti-*Pseudomonas* agent isolated from a marine *Vibrio*, *J. Bacteriol.* **95**:1972–1973.

Duff, D. C. B., Bruce, D. L., and Antia, N. J., 1966, The antibacterial activity of marine planktonic algae, *Can. J. Microbiol.* **12**:877–884.

Fenchel, T., 1987, *Ecology of Protozoa: The Biology of Free-Living Phagotrophic Protists*. Springer-Verlag, New York.

Fujioka, R. S., Lok, P. C., and Lau, L. S., 1980, Survival of human enteroviruses in the Hawaiian ocean environment, *Appl. Environ. Microbiol.* **39:**1105–1110.

Fusetani, N., Ejima, D., Matsunaga, S., Hashimoto, K., Itagaki, K., Akagi, Y., Taga, N., and Suzuki, K., 1987, 3-Amino-3-deoxy-D-glucose: An antibiotic produced by a deep-sea bacterium, *Experientia* **43:**464–465.

Galtsoff, P. S., Brown, H. H., Smith, C. L., and Smith, F. G. W., 1939, Sponge mortality in the Bahamas, *Nature* **143:**807–808.

Gandhi, N. M., Patell, J. R., Gandhi, J. N., De Souza, J., and Kohl, H., 1976, Prodigiosin metabolites of a marine *Pseudomonas* species, *Mar. Biol.* (Berl.) **34:**233.

Gauthier, M. J., and Flateau, G. N., 1976, Antibacterial activity of marine violet-pigmented *Alteromonas* with special reference to the production of brominated compounds, *Can. J. Microbiol.* **22:**1612–1619.

Gil-Turnes, M. S., 1988, Antimicrobial metabolites produced by epibiotic bacteria: Their role in microbial competition and host defense, Ph.D. Dissertation, University of California-San Diego, La Jolla, California.

Gil-Turnes, M. S., Hay, M. E., and Fenical, W., 1989, Symbiotic marine bacteria chemically defend crustacean embryos from a pathogenic fungus, *Science* **246:**116–118.

Goff, L. J., and Glasgow, J. C., 1980, *Pathogens of Marine Plants*, Center for Coastal Marine Studies, University of California, Santa Cruz, California.

Gomi, S., Ikeda, D., Nakamura, H., Naganawa, H., Yamashita, F., Hotta, K., Kondo, S., Okami, Y., and Umezawa, H., 1984, Isolation and structure of a new antibiotic, indolizimycin, produced by a strain SK2-52 obtained by interspecies fusion treatment, *J. Antibiotics* **37:**1491–1494.

Goodfellow, M., and Haynes, J. A., 1984, Actinomycetes in marine sediments, in: *Biological, Biochemical, and Biomedical Aspects of Actinomycetes* (L. Oritiz-Oritz, L. F. Bojalil, and V. Yakoleff, eds.), Academic Press, Orlando, Florida, pp. 453–472.

Grein, A., and Meyers, S. P., 1958, Growth characteristics and antibiotic production of actinomycetes isolated from littoral sediments and material suspended in sea water, *J. Bacteriol.* **76:**457–463.

Guerriero, A., D'Ambrosio, M., Cuomo, V., Vanzanella, F., and Pietra, F., 1988, Dendrophiellin A, the first fungal trinor-eremophilane. Isolation from the marine deuteromycete *Dendrophiella salina* (Sutherland) Pugh et Nicot, *Helv. Chim. Acta* **71:**57–61.

Guerriero, A., D'Ambrosio, M., Cuomo, V., Vanzanella, F., and Pietra, F., 1989, Novel trinor-eremophilanes (Dendrophiellin B, C and D), eremophilanes (Dendrophiellin E, F and G), and branched C9-carboxylic acids (Dendrophiellic Acid A and B) from the marine deuteromycete *Dendrophiella salina* (Sutherland) Pugh et Nicot, *Helv. Chim. Acta* **72:**438–446.

Gundersen, K. A., Brandeberg, S., Magnussen, S., and Lycke, E., 1967, Characterization of a marine bacterium associated with virus inactivating capacity, *Acta Pathol. Microbiol. Scand.* **71:**274–280.

Gunn, B. A., and Colwell, R. R., 1983, Numerical taxonomy of staphylococci isolated from the marine environment, *Int. J. Syst. Bacteriol.* **33:**751–759.

Gustafson, K., Roman, M., and Fenical, W., 1989, The macrolactins, a novel class of antiviral and cytotoxic macrolides from a deep-sea marine bacterium, *J. Am. Chem. Soc.* **111:**7519–7524.

Harashima, K., Shiba, T., and Murata, N., 1989, *Aerobic Photosynthetic Bacteria*, Springer-Verlag, Berlin.

Holland, G. S., Jamieson, D. D., Reicheldt, J. L., Viset, G., and Wells, R. J., 1984, Three aromatic acids from a marine bacterium, *Chem. Ind.* **1984** (3 December).

Hotta, K., Yoshida, M., Hamada, M., and Okami, Y., 1980, Studies on new aminoglycoside antibiotics, istamycins, from an actinomycete isolated from a marine environment, *J. Antibiot.* **33:**1515.

Hoyt, P. R., and Sizemore, R. K., 1982, Competitive dominance by a bacteriocin-producing *Vibro harveyi* strain, *Appl. Environ. Microbiol.* **44**:653–658.

Jannasch, H. W., and Wirsen, C. O., 1979, Chemosynthetic primary production at East Pacific sea floor spreading centers, *Bioscience* **29**:592–598.

Javor, B., 1989, *Hypersaline Environments: Microbiology and Biogeochemistry*, Springer-Verlag, New York.

Jensen, P., Dwight, R., and Fenical, W., 1991, The distribution of actinomycetes in near-shore tropical marine sediments, *Appl. Environ. Microbiol.* **57**:1102–1108.

Johnson, R. C., 1981, Introduction to the spirocheates, in: *The Prokaryotes*, Vol. I (M. P. Starr, H. Stolp, H. G. Truper, A. Balows, and H. G. Schlegel, eds.), Springer-Verlag, Berlin, pp. 582–591.

Jones, E. B. G., 1974, *Recent Advances in Aquatic Mycology*, Elek, London.

Kameyama, T., Takahashi, A., Kurasawa, S., Ishizuka, M., Okami, Y., Takeuchi, T., and Umezawa, H., 1987, Bisucaberin, a new siderophore, sensitizing tumor cells to macrophage-mediated cytolysis. I Taxonomy of the producing organism, isolation and biological properties, *J. Antibiot.* **40**:1664–1670.

Kandler, E. O., 1982, *Archaebacteria* (Proceedings of the 1st International Workshop on Archaebacteria), Gustav Fisher, New York.

Katzenelson, E., 1978, Survival of viruses, in: *Indicators of Viruses in Food and Water* (G. Berg ed.), University of Michigan Press, Ann Arbor, Michigan, pp. 39–50.

Kirk, P., Catalfomo, P., Blick, J. H., and Constantine, G. H., 1974, Metabolites of higher marine fungi and their possible ecological significance, *Veroeff. Inst. Meeresforsch. Bremerhaven* **1974** (Suppl. 5):509.

Kitahara, T., Naganawa, H., Okazaki, T., Okami, Y., and Umezawa, H., 1975, The structure of SS-229Y, an antibiotic from *Chainia* sp., *J. Antibiot.* **28**:280.

Kohl, H., Bhat, S. V., Patell, J. R., Gandhi, N. M., Nazareth, J. Divekar, P. V., DeSousa, N. J., Berscheid, H. G., and Fehlhaber, H.-W., 1974, Structure of magnesidin, a new magnesium-containing antibiotic from *Pseudomonas magnesiorubra*, *Tetrahedron Lett.* **1974**:983–986.

Kohlmeyer, J., and Kohlmeyer, E., 1979, *Marine Mycology: The Higher Fungi*, Academic Press, New York.

Kodama, M., Ogata, T., and Sato, S., 1988, Bacterial production of saxitoxin, *Agric. Biol. Chem.* **52**: 1075–1077.

Kodama, M., Ogata, T., Sato, S., and Sakamota, S., 1990, Possible association of marine bacteria with paralytic shellfish toxicity in bivalves, *Mar. Ecol. Prog. Ser.* **61**:203–206.

Kosuge, T., Zenda, H., Ochiai, A., Masaki, N., Noguchi, M., Kimura, S., and Narita, H., 1972, Isolation and structure determination of a new marine toxin, surugatoxin from the Japanese ivory shell *Babylonia japonica*, *Tetrahedron Lett.* **1972**:2545–2548.

Kosuge, T., Tsuji, K., Harai, K., and Fukuyama, T., 1985, First evidence of toxin production by bacteria in a marine organism, *Chem. Pharm. Bull.* **33**:3059–3061.

Krieg, N. R., and Holt, J. G., 1984, *Bergey's Manual of Systematic Bacteriology*, Vol. 1, Williams and Wilkins, Baltimore.

Kriss, A. E., 1963, *Marine Microbiology (Deep Sea)*, Wiley, New York.

Kupka, J., Anke, T., Steglich, W., and Zechlin, L., 1981, Antibiotics from basidomycetes XI. The biological activity of siccayne, isolated from the marine fungus *Halocyphina villosa* J and E. Kohlmeyer, *J. Antibiot.* **34**:298–304.

Leifson, E., Cosenza, B. J., Murchelano, R., Cleverdon, R. C., 1964, Motile marine bacteria: I. Techniques, ecology, and general characteristics, *J. Bacteriol.* **87**:652–666.

Lemos, M. L., Toranzo, A. E., and Barja, J. L., 1985, Antibiotic activity of epiphytic bacteria isolated from intertidal seaweeds, *Microb. Ecol.* **11**:149–163.

Lewin, R. A., 1977, *Prochloron*, type genus of the Prochlorophyta, *Phycologia* **16**:217.

Lewin, R. A., and Cheng, L., 1989, *Prochloron: A Microbial Enigma*, Chapman and Hall, New York.

Lovell, F. M., 1966, The structure of a bromine-rich antibiotic, *J. Am. Chem. Soc.* **88**:4510–4511.

Meyer, B. N., Ferrigni, N. R., Putnam, J. E., Jacobsen, L. B., Nichols, D. E., and McLaughlin, J. L., 1982, Brine shrimp: A convenient general bioassay for active plant constituents, *J. Med. Plant Res.* **45**:31–34.

Meseguer, I., Rodriguez-Valera, F., and Ventosa, A., 1986, Antagonistic interactions among halobacteria due to halocin production, *FEBS Lett.* **36**:177–182.

Moore, R. E., Patterson, G. M. L., and Carmichael, W. W., 1988, New pharmaceuticals from cultured blue-green algae, in: *Biomedical Importance of Marine Organisms* (D. G. Fautin, ed.), California Academy of Science, San Francisco, pp. 143–150.

Moss, S. T., 1986, *The Biology of Marine Fungi*, Cambridge University Press, Cambridge.

Nair, S., and Simidu, U., 1987, Distribution and significance of heterotrophic marine bacteria with antibacterial activity, *Appl. Environ. Microbiol.* **53**:2957–2962.

Nakamura, H., Iitaka, Y., Kitahara, T., Okazaki, T., and Okami, Y., 1977, Structure of aplasmomycin, *J. Antibiot.* **30**:714–719.

Needham, J., Andersen, R., and Kelly, M. T., 1991, Oncorhyncolide, a novel metabolite of a bacterium isolated from seawater, *Tetrahedron Lett.* **32**:315–318.

Noguchi, T., Jeon, J.-K., Arakawa, O., Sugita, H., Deguchi, Y., Shida, Y., and Hashimoto, K., 1986, Occurrence of tetrodotoxin and anhydrotetrodotoxin in *Vibrio* sp. isolated from the intestines of a xanthid crab *Atergatis floridus*, *J. Biochem.* **99**:311–314.

Ogata, T., Sato, S., and Kodama, M., 1989, Paralytic shellfish toxins in bivalves which are not associated with dinoflagellates, *Toxicon.* **27**:1241–1244.

Okami, Y., 1986, Marine microorganisms as a source of bioactive agents, *Microb. Ecol.* **12**:65–78.

Okami, Y., 1988, Bioactive metabolites of marine microorganisms, in: *Horizons on Antibiotic Research*, Japan Antibiotic Research Association, Tokyo, pp. 213–227.

Okami, Y., and Okazaki, T., 1972, Studies on marine microorganisms I. Isolation from the Japan Sea, *J. Antibiot.* **25**:456–460.

Okami, Y., Okazaki, T., Kitahara, T., and Umezawa, H., 1976, Studies on marine microorganisms V. A new antibiotic, aplasmomycin, produced by a streptomycete isolated from shallow sea mud, *J. Antibiot.* **29**:1019–1025.

Okami, Y., Hotta, K., Yoshida, M., Ikeda, D., Kondo, S., and Umezawa, H., 1979, New aminoglycoside antibiotics, istamycins A and B, *J. Antibiot.* **32**:964–966.

Okazaki, T., and Okami, Y., 1972, Studies on marine microorganisms II. Actinomycetes in Sagami Bay and their antibiotic substances, *J. Antibiot.* **25**:261–266.

Okazaki, T., Kitahara, T., and Okami, Y., 1975, Studies on marine microorganisms IV. A new antibiotic SS-228Y produced by *Chainia* isolated from shallow sea mud, *J. Antibiot.* **28**:176–184.

Okutani, K., 1977, Biotoxin produced by a strain of *Aspergillus* isolated from marine mud, *Bull. Jpn. Soc. Sci. Fish.* **43**:995.

Omura, S., 1984, *Macrolide Antibiotics*, Academic Press, Orlando, Florida.

Pallenberg, A. J., and White, J. D., 1986, The synthesis and absolute configuration of (+)-leptosphaerin, *Tetrahedron Lett.* **27**:5591–5594.

Pathirana, C., Tapiolas, D. M., Jensen, P. R., Dwight, R., and Fenical, W., 1991, Structure of maduralide: A new 24-membered ring macrolide glycoside produced by a marine bacterium, *Tetrahedron Lett.* **32**(21):2323–2326.

Patterson, G. M. L., Baldwin, C. L., Bolis, C. M., Caplan, F. R., Karuso, H., Larsen, L. K., Levine, I. A., Moore, R. E., Nelson, C. F., Pschappat, K. D., Tuang, G. D., Furusawa, E., Furusawa, S., Norton, T. R., and Raybourne, R. B., 1991, Antineoplastic activity of cultured blue-green algae (Cyanophyta), *J. Phycol.* **27**:530–536.

Pfennig, N., 1967, Photosynthetic bacteria, *Annu. Rev. Microbiol.* **21**:285–324.

Poch, G. K., and Gloer, J. B., 1989a, Helicascolides A and B: New lactones from the marine fungus *Helicascus kanaloanus, J. Nat. Prod.* **52:**257–260.

Poch, G. K., and Gloer, J. B., 1989b, Obionin A: A new polyketide metabolite from the marine fungus *Leptosphaeria obiones, Tetrahedron Lett.* **30:**3483–3486.

Poch, G. K., and Gloer, J. B., 1991, Auranticins A and B: Two new depsidones from a mangrove isolate of the fungus *Preussia aurantiaca, J. Nat. Prod.* **54:**213–217.

Reichenbach, H., and Hofle, G., 1989, The gliding bacteria: A treasury of secondary metabolites, in: *Bioactive Metabolites from Microorganisms* (M. E. Bushell, and U. Grafe, eds.), Elsevier, Amsterdam, pp. 79–100.

Rodriguez-Valera, F., 1988, *Halophilic Bacteria*, Vol. 2, CRC press, Boca Raton, Florida.

Rosenfeld, W. D., and Zobell, C., 1947, Antibiotic production by marine microorganisms, *J. Bacteriol.* **54:**393–398.

Round, F. E., 1973, *The Biology of the Algae*, 2nd ed., St. Martin's Press, New York.

Rychnovsky, D. S., Skalitzky, D. J., Fenical, W., Gustafson, K., and Pathirana, C., 1991, Stereochemistry of the macrolactin family of macrocycles: A degradative study, in: *Abstracts 201st National Meeting of the American Chemical Society*, Atlanta, Georgia, Abstract #187.

Sato, K., Okazaki, T., Maeda, K., and Okami, Y., 1978, New antibiotics, aplasmomycins B and C, *J. Antibiot.* **31:**632.

Schiehser, G. A., 1980, The isolation and structure of leptosphaerin: A metabolite of the marine ascomycete *Leptosphaeria oraemaris*, Ph.D. Dissertation, Oregon State University, Corvallis, Oregon [University Microfilm, Ann Arbor, Michigan QK 623, L46,SC4].

Schiehser, G. A. White, J. D., Matsumoto, G., Pezzanite, J. O., and Clardy, J., 1986, The structure of leptosphaerin, *Tetrahedron Lett.* **27:**5587–5590.

Shiba, T., Simidu, U., and Taga, N., 1979, Distribution of aerobic bacteria which contain bacteriochlorophyll *a, Appl. Environ. Microbiol.* **38:**43–45.

Shin, J., and Fenical, W., 1987, Isolation of gliovictin from the marine deuteromycete *Asteromyces cruciatus, Phytochemistry* **26:**3347.

Short, F. T., Muehlstein, L. K., and Porter, D., 1987, Eelgrass wasting disease: Cause and recurrence of a marine epidemic, *Biol. Bull.* **173:**557–562.

Sieburth, J. M., 1960, Acrylic acid and "antibiotic" principle in *Phaeocystis* blooms in Antarctic waters, *Science* **132:**676–677.

Sieburth, J. M., 1961, Antibiotic properties of acrylic acid, a factor in the gastrointestinal antibiosis of polar marine animals, *J. Bacteriol.* **82:**72–79.

Sieburth, J. M., 1974, Antibacterial substances produced by marine algae, *Dev. Ind. Microbiol.* **5:** 124–134.

Sieburth, J. M., 1975, *Microbial Seascapes*, University Park Press, Baltimore.

Sieburth, J. M., 1979, *Sea Microbes*, Oxford University Press, Oxford.

Simidu, U., Kita-Tsukamoto, K., Yasumoto, T., and Yotsu, M., 1990, Taxonomy of four marine bacterial stains that produce tetrodotoxin, *Int. J. Syst. Bacteriol.* **40:**331–336.

Skerman, V. B. D., McGowan, V., and Sneath, P. H. A., 1980, Approved lists of bacterial names, *Int. J. Syst. Bacteriol.* **30:**225–240.

Sneath, P. H. A., Mair, N. S., Sharpe, M. E., and Holt, J. G., 1986, *Bergey's Manual of Systematic Bacteriology*, Vol. 2, Williams and Wilkins, Baltimore.

Southward, E. C., 1987, Contribution of symbiotic chemoautotrophs to nutrition of benthic invertebrates, in: *Microbes in the Sea* (M. A. Sleigh, ed.), Halstead Press, New York, pp. 83–118.

Sparks, A. K., 1985, *Synopsis of Invertebrate Pathology: Exclusive of Insects*, Elsevier, Amsterdam.

Staley, J. T., Hirsch, P., and Schmidt, J. M., 1981, Introduction to the budding and/or appendaged bacteria, in: *The Prokaryotes*, Vol. I (M. P. Starr, H. Stolp, H. G. Truper, A. Balows, and H. G. Schlegel, eds.), Springer-Verlag, Berlin, pp. 451–455.

Staley, J. T., Bryant, M. P., Pfennig, N., and Holt, J. G., 1989, *Bergey's Manual of Systematic Bacteriology*, Vol. 3, Williams and Wilkins, Baltimore.

Stanier, R. Y., Pfennig, N., and Truper, H. G., 1981, Introduction to the phototrophic prokaryotes, in: *The Prokaryotes*, Vol. I (M. P. Starr, H. Stolp, H. G. Truper, A. Balows, and H. G. Schlegel, eds.), Springer-Verlag, Berlin, pp. 197–211.

Stanier, R. Y., Ingraham, J. L., Wheelis, M. L., and Painter, P. R., 1986, *The Microbial World*, Prentice-Hall, Englewood Cliffs, New Jersey.

Starr, M. P., Stolp, H., Truper, H. G., Balows, A., Schlegel, H. G. (eds.), 1981, *The Prokaryotes*, Vols. I and II, Springer-Verlag, Berlin.

Steemann-Nielsen, E., 1955, The production of antibiotics by plankton algae and its effect upon bacterial activities in the sea, *Deep-Sea Res.* 3(Suppl):281–286.

Stein, J. R., 1975, *Handbook of Phycological Methods: Culture Methods and Growth Measurements*, Cambridge University Press, Cambridge.

Stierle, A. C., Cardellina, J. H., II, and Singleton, F. L., 1988, A marine *Micrococcus* produces metabolites ascribed to the sponge *Tedania ignis*, *Experientia* 44:1021.

Strongman, D. B., Miller, J. D., Calhoun, L., Findlay, J. A., and Whitney, N. J., 1987, The biochemical basis for interference competition among some lignicolous marine fungi, *Bot. Mar.* 30:21–26.

Takahashi, A., Nakamura, H., Kameyama, T., Kurasawa, S., Naganawa, H., Okami, Y., Takeuchi, T., Umezawa, H., and Iitaka, Y., 1987, Bisucaberin, a new siderophore, sensitizing tumor cells to macrophage-mediated cytolysis. II Physico-chemical properties and structure determination, *J. Antibiot.* 40:1671–1676.

Takahashi, A., Kurosawa, S., Ikeda, D., Okami, Y., and Takeuchi, T., 1989a, Altemicidin, a new acaricidal and antitumor substance. I. Taxonomy, fermentation, isolation and physicochemical and biological properties, *J. Antibiot.* 42:1556–1561.

Takahashi, A., Ikeda, D., Nakamura, H., Naganawa, H., Kurasawa, S., Okami, Y., Takeuchi, T., and Iitaka, Y., 1989b, Altemicidin a new acaricidal and antitumor substance. II Structure determination, *J. Antibiot.* 42:1562–1566.

Tapiolas, D. M., Roman, M., Fenical, W., Stout, T. J., and Clardy, J., 1991, Octalactins A and B, cytotoxic eight-membered ring lactones from a marine bacterium, *Streptomyces* sp., *J. Am. Chem. Soc.* 113:4682–4683.

Toranzo, A. E., Barja, J. L., and Hetrick, F. M., 1982, Antiviral activity of antibiotic-producing marine bacteria, *Can. J. Microbiol.* 28:231–238.

Trischman, J., Tapiolas, D. M., Fenical, W., Dwight, R., Jensen, P. R., McKee, T., Ireland, C. R., Stout, T., and Clardy, J., 1992, Structure of the salinamides, novel bicyclic depsipeptides from a surface-derived marine bacterium, *J. Am. Chem. Soc.*, submitted.

Umezawa, H., Okami, Y., Kurasawa, S., Ohnuki. T., Ishizuki, M., Takeuchi, T., Shiio, T., and Yuigari, Y., 1983, Marinactan, antitumor polysaccharide produced by marine bacteria, *J. Antibiot.* 36:471–477.

Wilkinson, C. R., 1987, Significance of microbial symbionts in sponge evolutionary ecology, *Symbiosis* 4:135–146.

Williams, S. T., Sharpe, M. E., and Holt, J. T., 1989, *Bergey's Manual of Systematic Bacteriology*, Vol. 4, Williams and Wilkins, Baltimore.

Woese, C. R., and Fox, G. E., 1977, Phylogenetic structure of the prokaryotic domain: The primary kingdoms, *Proc. Natl. Acad. Sci. USA* 74:5088.

Woese, C. R., Magrum, L. J., and Fox, G. E., 1978, Archaebacteria, *J. Mol. Evol.* 11:245–252.

Wood, E. J. F., 1965, *Marine Microbial Ecology*, Chapman Hall, London.

Wratten, S. J., Wolfe, M. S., Andersen, R. J., and Faulkner, D. J., 1977, Antibiotic metabolites from a marine pseudomonad, *Antimicrob. Agents Chemother.* 11:411.

Yamashita, F., Hotta, K., Kurasawa, S., Okami, Y., and Umezawa, H., 1985, New antibiotic producing actinomycetes, selected by antibiotic resistance as a marker. I. New antibiotic production generated by protoplast fusion treatment between *Streptomyces griseus* and *S. tenjimariensis*, *J. Antibiot.* **38**:58–63.

Yasumoto, T., Yasumura, D., Yotsu, M., Michishita, T., Endo, A., and Kotaki, Y., 1986, Bacterial production of tetrodotoxin and anhydrotetrodotoxin, *Agric. Biol. Chem.* **50**:793–795.

Yotsu, M., Yamazaki, T., Meguro, Y., Endo, A., Murata, M., Naoki, H., and Yasumoto, T., 1987, Production of tetrodotoxin and its derivatives by *Pseudomonas* sp. isolated from the skin of the pufferfish, *Toxicon.* **25**:225–228.

Zimmer, R. L., and Woollacott, R. M., 1983, Mycoplasma-like organisms: Occurrence with larvae and adults of a marine bryozoan, *Science* **220**:208–210.

Zobell, C. E., and Allen, E. C., 1935, The significance of marine bacteria in the fouling of submerged surfaces, *J. Bacteriol.* **29**:239–251.

Zobell, C. E., and Upham, H. C., 1944, A list of marine bacteria including sixty new species, *Bull. Scripps Inst. Oceanogr.* **5**:239–292.

13

Academic Chemistry and the Discovery of Bioactive Marine Natural Products

D. John Faulkner

1. INTRODUCTION

University scientists who study marine natural products have already made some significant contributions toward the discovery of new pharmaceutical agents. Over the past 20 years, their research programs have provided a spectacular array of novel compounds, many of which possess useful biological or pharmaceutical properties. The most important of these have been obtained from sessile marine organisms, such as sponges, tunicates, soft corals, gorgonians, and bryozoans. The association of bioactive compounds with sessile and soft-bodied organisms could be predicted by application of the principles of chemical ecology that were developed in collaboration with marine biologists and ecologists as an integral part of marine natural products chemistry. In recent years, technical advances in analytical instrumentation and methodology have greatly enhanced the productivity of the relatively small number of marine natural products chemists. If they can now take full advantage of the new bioassays made possible by the biotechnol-

D. John Faulkner • Scripps Institution of Oceanography, University of California at San Diego, La Jolla, California 92093-0212.

Marine Biotechnology, Volume 1: Pharmaceutical and Bioactive Natural Products, edited by David H. Attaway and Oskar R. Zaborsky. Plenum Press, New York, 1993.

ogy explosion, there is no doubt that marine natural products chemists will make even greater contributions to drug development in the next decade.

In order to appreciate the rapid development of marine natural products chemistry, one must examine the history of natural products chemistry and its relationship to the pharmaceutical industry. The origins of natural products chemistry can be traced back to the earliest scientific studies of those plants that provided the *materia medica* of the middle ages. Phytochemistry and the related science of pharmacognosy are still dominated by academic researchers, although these fields have been declining in popularity for several decades. Perhaps the most significant change in the field of natural products chemistry came with the discovery of antibiotics and other drugs from microorganisms. The microbial strains that produced the wonder drugs of the 1950s became items of commerce and much of the natural products chemistry of the 1960s was performed by pharmaceutical companies that were eager to maintain the commercial advantages afforded by patent law. The change in emphasis from plant products to a combination of fermentation products and synthetic chemicals as the bases of the pharmaceutical industry led to the decline of natural products chemistry as an academic pursuit, especially in the United States and other countries with strong pharmaceutical industries. Against this background, it is remarkable that marine natural products chemistry should have become so well established during the past 20 years.

2. THE HISTORICAL RECORD

Before examining the current status of academic research to discover new drugs from marine organisms, one must evaluate the historical record. I will not review the research that was performed by those pharmaceutical companies that have specialized in marine natural products research because the public record on which I must rely provides a very incomplete record of their progress.

The origins of the antiviral drug Ara-A (**1**) can be traced back to the serendipitous isolation of the arabinosyl nucleosides spongothymidine (**2**) and spongouridine (**3**) from the sponge *Cryptotethia crypta* by Bergmann and Feeney

(1950, 1951). Scheuer (1986) has traced the steps that led from this chance discovery to the eventual marketing of Ara-A (Vidabarine) in 1980. Rather than repeat Scheuer's excellent account of the key events, I wish to offer a few rather general observations. Since the late Werner Bergmann was primarily interested in sterols, his discovery and structural elucidation of nucleosides from sponges may have been a chance event. It was a timely discovery because there was a great interest in nucleosides and nucleotides at that time. Whether the arabinosyl nucleosides would have been synthesized as part of a routine program to synthesize nucleoside analogues is a matter of speculation, but it must be considered as a likely event considering the research directions of the time. Nonetheless, it is significant that the sponge nucleosides were widely studied in the late 1950s and early 1960s so that they were known to have interesting pharmacological properties prior to the synthesis of first Ara-C and then Ara-A. The role of the academic chemist, while probably not intentional, was to discover the first examples of a new group of compounds, the arabinosyl nucleosides, that are among the few antiviral agents on the market.

The isolation of prostaglandins from the Caribbean gorgonian *Plexaura homomalla* was a startling and valuable discovery in 1969. At that time, the prostaglandins were so scarce and so valuable that a cartoon in a London newspaper depicted bowler-hatted businessmen setting off in their yachts to make their fortunes by hunting the elusive gorgonian in what was described as a "nautical gold rush." By the late 1960s, the prostaglandins were already known to be very important substances in the mediation of a diverse array of physiological functions. However, research on prostaglandins was being severely hampered by inadequate supplies of the active compounds, PGE_2 (4) and $PGF_{2\alpha}$ (5). When

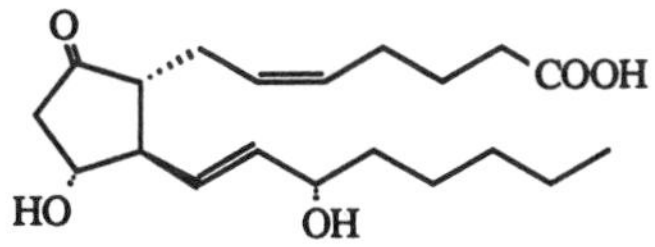

(4) 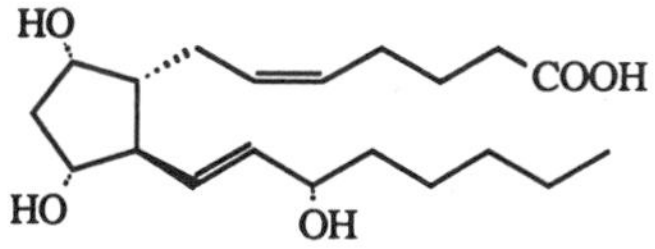   (5) 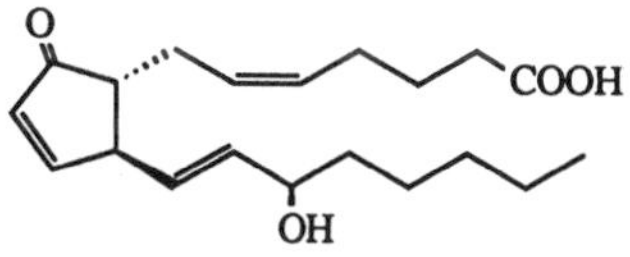   (6) 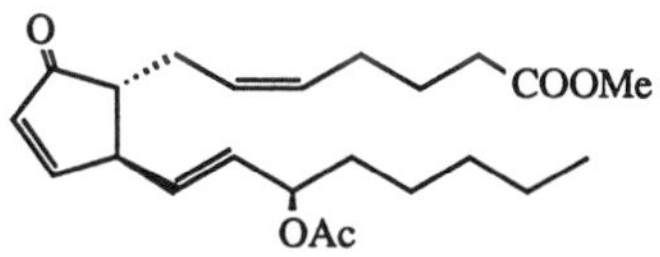   (7)

Weinheimer and Spraggins (1969) reported that *Plexaura homomalla* contained > 1% dry weight of 15-*epi*-PGA_2 (6) and its acetate, methyl ester derivative (7), it seemed that the supply problem might be solved. Although prostaglandins of the 15-*epi*- series were not physiologically active, scientists at the Upjohn Company devised a synthetic route to convert the marine natural products into active prostaglandins (Bundy et al., 1972). Just when it seemed that the ecology of the shallow-water Caribbean reefs might be irrevocably changed by the commercial harvesting of gorgonians, several total syntheses of prostaglandins were reported. These syntheses solved the problem of obtaining reliable supplies of prostaglandins for research and clinical trials. This was by no means the end of academic interest in the production of eicosanoids by coelenterates (Carmely et al., 1980; Kikuchi et al., 1982; Kobayashi et al., 1982; Baker et al., 1985; Iguchi et al., 1985, 1986) and red algae (Gregson et al., 1979; Higgs and Mulheirn, 1981; Nagle and Gerwick, 1990, and references cited therein). The halogenated prostanoids, such as the punaglandins (e.g., punaglandin 1, **8**), from *Telesto riisei* (Baker et al.,

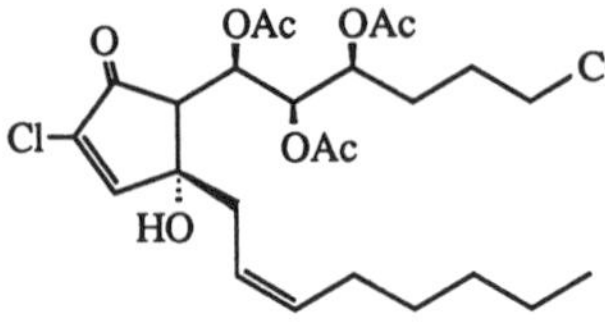

(8) 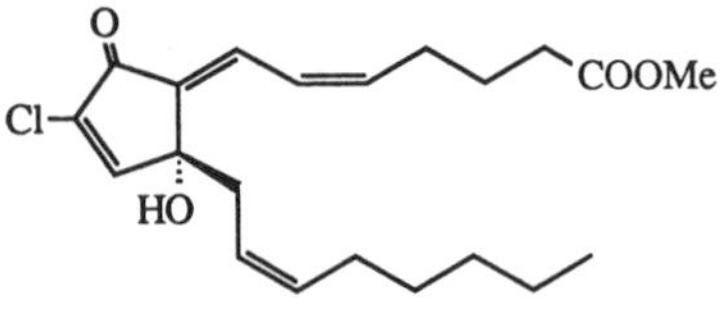   (9)

1985), and related compounds (e.g., chlorovulone 1, **9**) from *Clavularia viridis* (Iguchi et al., 1985, 1986), are being studied for their potential as antitumor agents. The serendipitous discovery of prostanoids in *P. homomalla* did not have a lasting impact on pharmaceutical development, but it clearly demonstrated the pharmaceutical potential of the marine environment.

A collaborative program in marine chemistry and pharmacology was established by the California Sea Grant College Program in 1979 in order to facilitate the discovery of new marine natural products with promising pharmacological properties. Within 10 years, the program has described more than 20 compounds or groups of compounds that could be considered as candidates for drug development, although few have yet attained the stage of commercial development. One of the most notable discoveries was that manoalide (**10**), an abundant metabolite of the sponge *Luffariella variabilis*, possessed useful anti-inflammatory properties. For several years, the chemist's role in the development of manoalide was almost invisible. The collaboration of my group with that of Prof. Robert Jacobs began in 1979, but we did not publish any of our chemical studies until 1987. Our decision to delay publication resulted largely from a desire to obtain the necessary patent

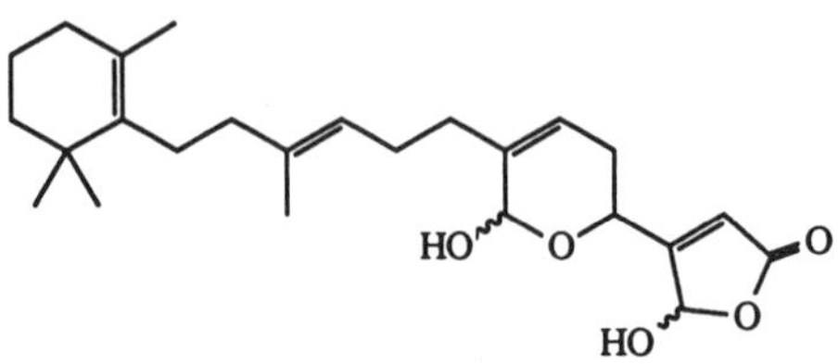

(10)

protection that would enable a pharmaceutical company to invest in the commercial development of manoalide. In retrospect, it is not clear that we made the correct decision.

The manoalide story provides some interesting insights into the role of academic chemists in drug development. In 1980, a few months after we had started our studies, Dr. Jacobs was convinced that manoalide showed unusual promise as an anti-inflammatory and analgesic agent. We therefore started a thorough study of manoalide and its derivatives that eventually formed the basis of a structure-activity study. We were very disappointed when de Silva and Scheuer (1980) published the structure of manoalide a few months later. After Dr. Jacobs had determined that manoalide acts as an inhibitor of the enzyme phospholipase A_2 (Glaser and Jacobs, 1986) and patent applications had been filed, we began the frustrating task of finding a company that might be interested in pursuing the commercial development. The first company to be interested in manoalide was unable to negotiate a mutually beneficial licensing agreement with the University of California and decided instead to build a synthetic program around the structure of manoalide. The university then negotiated an option for a second company to develop manoalide as a topical anti-inflammatory agent. In 1985, we accepted a contract to provide sufficient manoalide to complete phase I of a clinical trial. The collection of the sponge did not present a problem, but the large-scale isolation work proved to be incompatible with the other research being performed in the laboratory. In addition, some specimens of *L. variabilis* contained the isomeric compound luffariellin A (**11**) (Kernan *et al.*, 1987) in place of or admixed with

(**11**)

manoalide (**10**), so that each piece of sponge had to be extracted separately, thereby increasing the complexity of the isolation procedure. Although we were able to overcome all the isolation problems, we could not work on a large enough scale to provide the material on schedule and the isolation work had to be completed elsewhere. This was a very good example of an academic group going too far along the path from discovery to development. We should have accepted responsibility for recollection of the sponge because our experience was invaluable in this regard. We should have declined a contract for large-scale isolation of manoalide because this was incompatible with academic goals and facilities. Our involvement with the development program delayed an academic program that has shed light on the chemistry involved in the interaction of manoalide with phospholipase A_2

(PLA$_2$) (Glaser *et al.*, 1989; B. C. M. Potts *et al.*, 1992). Although the final chapter in the manoalide story has yet to be written, it now appears almost certain that future drug developments will involve synthetic PLA$_2$ inhibitors rather than manoalide and related natural products. Manoalide will continue to be marketed as a molecular probe for phospholipase A$_2$.

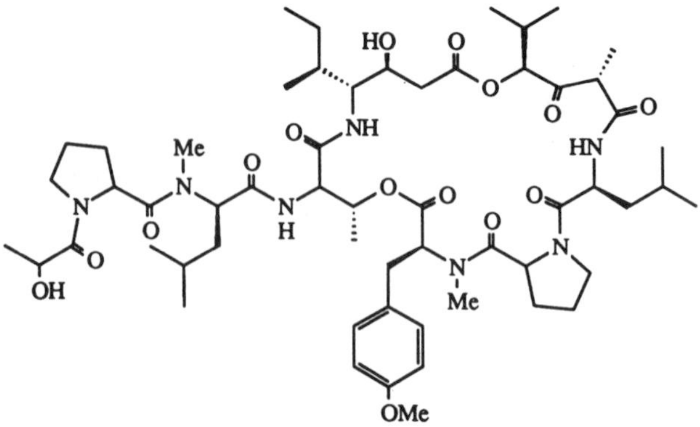

(12)

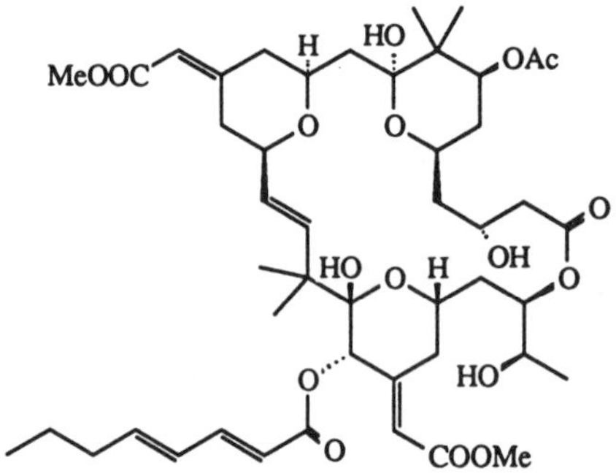

(13)

The commercial development of either didemnin B (12) or bryostatin 1 (13) as antitumor agents will ultimately depend not only on the efficacy of the compounds, but also on their availability (Suffness *et al.*, 1989). Didemnin B has been obtained in reasonable yield from the tunicate *Trididemnum solidum*, but it is not a simple matter to collect the tunicate in large quantities. The compound used in phase II clinical trials was obtained from natural sources, but future supplies might well be synthesized. An account of the development of didemnin B has been presented by Rinehart (1988). In the case of bryostatin 1 (Pettit *et al.*, 1982), the supply of material from natural sources is a more critical problem. The bryostatins are isolated from the bryozoan *Bugula neritina* in low and variable yields. Although there is a synthetic route to a bryostatin (Kageyama *et al.*, 1990), that synthesis is not considered commercially feasible. Collaborative research efforts are underway to culture *B. neritina* and also to determine whether the bryostatins are synthesized by a symbiont that might be cultured. For both didemnin B and bryostatin 1, the National Cancer Institute has acted as a sponsor of the research and the professional staff has provided pivotal assistance to the marine natural products chemists in the development of these agents. At the present time, this level of governmental assistance is unique to the anticancer and anti-AIDS research areas.

There are several more marine natural products that have reached the stage where commercial evaluation of their potential is underway and many others that probably deserve a similar treatment. Among the compounds studied by Dr. Jacobs and his collaborators in the University of California, specific pharmacological activities have been ascribed to the bengamides (Adamczeski *et al.*, 1989, and references cited therein), cymopol and related compounds (Högberg *et al.*, 1976), deacetylerytholide A (Look *et al.*, 1984), elatol and elatone (White and Jacobs, 1981), ircinins I and II (Cimino *et al.*, 1972), jaspamide ($\equiv$ Jasplakinolide) (Zabriskie *et al.*, 1986; Crews *et al.*, 1986), lophotoxin (Culver *et al.*, 1985), mycothiazole (Crews *et al.*, 1988), onchidal (Abramson *et al.*, 1989, and references cited therein), pseudopterolide A (Bandurraga *et al.*, 1982), the pseudopterosins (Look *et al.*, 1986; Roussis *et al.*, 1990), scalaradial (Cimino *et al.*, 1974) and related compounds, sceptrin (Walker *et al.*, 1981), stypoldione (O'Brien *et al.*, 1986, and references cited therein), variabilin (Faulkner, 1973; Jacobs *et al.*, 1981), and zonarone (Fenical *et al.*, 1973). Similar lists of pharmacologically-active compounds have been generated by other academic and industrial groups. These results demonstrate that marine natural products chemistry is a very active and successful field of study.

3. THE CURRENT STATUS OF ACADEMIC RESEARCH

The academic scientists involved in marine natural products research are usually organic chemists who have specialized in structural elucidation. Modern analytical methods have allowed them to elucidate the structures of molecules that a few years ago could only be tackled by x-ray crystallographic methods and they can now enjoy the exhilarating experience of identifying complex molecules using intricate arguments based on interpretation of spectral data. However, no matter how satisfying it may be for the chemists to solve the intellectual problems of structural elucidation, they realize that structural elucidation is only a small part of drug development. In order to help define appropriate roles for academic chemists in the development of pharmaceutical applications for marine natural products, I will examine five important components of a marine chemistry and pharmacology program:

1. Collection and identification of marine organisms.
2. Screening of crude extracts for bioactivity.
3. Isolation and identification of novel natural products.
4. Pharmacological screening of pure compounds.
5. Commercial development of marine natural products.

My analysis of these topics is based on personal experience and anecdotal information. The analysis cannot cover all interactions between academic scientists, government agencies and the pharmaceutical industry. It is presented in the hope that recognition of the different and sometimes conflicting goals of the various players will result in stronger collaborative research programs.

4. COLLECTION AND IDENTIFICATION OF MARINE ORGANISMS

Field research and the collection of marine organisms is one of the most demanding aspects of marine natural products research. Most chemists do not have the time to become expert marine biologists or taxonomists. The best that they can do is to become familiar with the groups of organisms that they wish to study and, more importantly, learn to recognize organisms that have already been studied. It is far better to eliminate frequently studied species in the field than to find only known compounds after extensive chemical studies. The identification of marine organisms is often a problem, not only in the field but also in an absolute sense. A surprisingly large number of marine invertebrates have not been described in the scientific literature and others are the subject of seemingly continuous taxonomic revision. Nonetheless, it is essential that all specimens reported in the marine natural products literature be examined by an appropriate expert and that a properly preserved voucher specimen be deposited in a recognized museum or collection. Marine natural products chemists rarely obtain excellent voucher specimens from their field collections, but they should always try to prepare vouchers that are adequate for taxonomic study. In addition to voucher specimens, the collector must provide the taxonomist with appropriate field collection notes and photographs. The collection notes and photographs are particularly important if the specimens are to be efficiently recollected. Although it may seem a burden to those involved in collecting specimens for chemical studies, it is well worth dedicating extra time for recording, describing, and photographing specimens as they are collected so that each subsequent field trip will be more productive.

Should an academic research group employ professional divers or students for the collection of specimens? This question is generally answered very simply and pragmatically: Most academic research groups cannot afford to employ professional divers. Yet experienced or professional divers, particularly those with marine biology or ecology backgrounds, are frequently the most efficient collectors. Their experience and judgment make any collecting trip a safer proposition, particularly when the use of SCUBA equipment is involved. To allow graduate students the opportunity to collect their own specimens provides a valuable learning experience for the students, but they must be subjected to a rigorous diver training program that prepares them for research rather than sport diving. Perhaps the best compromise is to have an experienced diver accompany small groups of student collectors, so that the collecting trip becomes an extension of the student's diver training course. Students always prefer to study specimens that they have personally collected and will often make an extra effort to isolate minor products from "their" sample.

It is rare for a pharmaceutical company to have an experienced collector on its staff unless it has hired a graduate from an academic marine natural products program. It is even more unusual for a company to have two or three such individuals to make up a collecting team that meets minimum diving safety requirements. In order to have a drug discovery program with a marine natural products component, most companies must either purchase marine organisms or extracts from commercial sources or collaborate with academic institutions to obtain marine specimens. Both strategies have advantages and disadvantages. Companies that purchase organisms have near total control over their discoveries within the limits of their contractual obligations to the supplier, but they frequently find that they have the same samples as all other companies that have purchased samples. Companies that collaborate with academic marine natural products groups will usually obtain a better selection of specimens, extracts, and compounds, but must deal with the conflict between the academic requirement for publication of results and the industrial desire for confidentiality.

The recollection of large quantities of marine organisms to provide the quantities of materials that are required for clinical trials is best left to specialists. Although they can and should provide the expertise needed to make recollection possible, academic groups rarely benefit from participating in the recollection of a single organism. It is better for an academic scientist to provide logistical and scientific advice for a recollection that is organized and financed by an industrial group than to accept a potentially lucrative contract for the recollection of a single organism.

5. SCREENING OF CRUDE EXTRACTS FOR BIOACTIVITY

The goal of a drug discovery program is to obtain the widest possible screening for each crude extract so that no useful compound is overlooked. The bioassays available to pharmaceutical companies are superior in number and scope to those that marine natural products chemists can conduct in their own laboratories. Collaborative research programs between industrial and academic groups are therefore highly desirable.

Although the ultimate goal of a drug discovery program is to isolate compounds that have *in vivo* activity, the use of *in vivo* assays for screening crude extracts is problematic. Most academic groups working in natural products chemistry have neither the expertise nor the funding to perform *in vivo* bioassays. Academic colleagues can rarely afford to perform routine screening for the chemist, since this activity diverts them from the basic research activities on which their academic careers are dependent. The chemist must usually rely on off-campus facilities provided by industrial or government laboratories for *in vivo* screening. However, "mail-order" screening presents some serious disadvantages. Preparing, mailing, and tracking the samples is time-consuming and requires greater attention to detail than is generally anticipated. Also, because the turnaround time for *in vivo* bioassays can exceed 2 months, a complete bioassay-guided fractionation may develop too slowly to satisfy the usual academic schedules. The effectiveness of using *in vivo* assays to screen crude extracts can also be questioned because it involves screening a multicomponent mixture of chemicals against multiple targets in the bioassay organism. It is normally more reliable and efficient to screen crude extracts in *in vitro* assays and reserve the *in vivo* assays for pure compounds.

An *in vitro* bioassay for screening crude extracts should have a very specific target, require very little material, and take little time to perform. Antimicrobial and enzyme inhibition assays are examples of fast, simple assays that can easily be performed by the chemist. A simple *in vitro* bioassay can often be used in place of a more complex and time-consuming *in vivo* assay to track the active compound(s) during a bioassay-guided fractionation and purification.

6. ISOLATION AND IDENTIFICATION OF BIOACTIVE NATURAL PRODUCTS

It is always preferable to perform bioassays and bioassay-guided chemical isolations in the same location. In cases where an off-campus collaborator reports that a crude extract has a potentially important activity, particularly in an *in vivo* bioassay, several ways to succeed in bioassay-guided fractionation are possible, depending on the complexity of the bioassay. If the assay is relatively simple, it can be set up and performed on campus. If this is not possible, the activity profile of the extract may indicate a simpler assay that fortuitously tracks the desired bioactivity. For example, we have successfully tracked cytotoxicity by substituting an *in vitro* assay against *Candida albicans*. This is obviously risky, for it can result in the isolation of compounds with only antifungal activity and no cytotoxicity. Consequently, all fractions must be saved for eventual screening in the event that the "wrong" compound has been targeted. In cases where the bioassay-guided fractionation involves a complex proprietary assay, the isolation should be performed in laboratories associated with the bioassay.

In recent years we have adopted a compromise procedure to investigate crude extracts that show promising activity in off-campus bioassays. All fractionations are followed by thin layer chromatography (TLC) and ^{1}H NMR spectroscopy in order to isolate the major secondary metabolites in the extract. After these metabolites have been identified, all pure compounds and all fractions, often pooled according to their TLC characteristics and/or NMR spectra, are screened in the off-campus bioassay. In most cases the activity is associated with one of the identified compounds. If not, the search must continue by using rigorous bioassay-guided fractionation methods. This procedure is particularly valuable if the structural elucidations can be performed using nondestructive spectroscopic methods that conserve most material for the bioassays.

The identification of novel marine natural products depends on a combination of spectroscopy and chemical conversions. In recent years, the amount of "wet chemistry" in the study of marine natural products has been declining. This is an inevitable result of the increasing sophistication of NMR spectroscopy and the other spectroscopic methods used in structural elucidation. However, the reliability of a structural elucidation is less dependent on the sophistication of the spectroscopic methods than it is on the number of spectroscopic experiments performed and the quality of the data supporting each facet of the structural elucidation. For example, NMR experiments that define the relative locations of two atoms are frequently used to locate the positions of both atoms; this is not a logical interpretation. Rather than defining a structure that can be shown to fit the spectral data, it is best to examine many possible structures and treat each proposed structure as a hypothesis that cannot be proved but can only be disproved by incompatible data. The rigorous application of this philosophy will improve the accuracy of structural elucidation by spectroscopic methods, but it is doubtful that the precision of x-ray crystallography will ever be surpassed.

7. PHARMACOLOGICAL SCREENING OF PURE
COMPOUNDS

Because much of the funding for marine natural products research is derived from biomedical sources, it is the duty of the chemist to obtain the broadest possible pharmacological screening for all new compounds. This is easily arranged because many pharmaceutical companies and academic pharmacologists want to screen new marine natural products. They are particularly interested in compounds having novel structural features, but they will also screen familiar compounds in new assays. Because of the competition between pharmaceutical companies, there is no coordination to prevent duplication of screening. However, it is better that a compound be screened many times than not at all.

When chemists complete the structural elucidations of new marine natural products, they are faced with a dilemma. Should they publish the structures immediately or wait until they have obtained a pharmacological profile of the new metabolite? Since the efficacy of a marine natural products program is dependent on both publication and pharmaceutical utility, there is no inherently correct choice. Those seeking new drugs will favor extensive pharmacological studies before publication. However, the competitive nature of scientific research, particularly as it applies to marine natural products chemistry, tends to force publication of the structural study as soon as it is completed, even though the pharmacological value of the compound might never be known. It is unfortunate that the first scientist to publish a structure receives more credit than the person who publishes a more comprehensive paper that includes the relevant pharmacological properties and structure-activity studies. The best strategy is to perform chemical and pharmacological studies in parallel so that there is minimal delay in presenting an interdisciplinary report.

8. COMMERCIAL DEVELOPMENT OF MARINE NATURAL PRODUCTS

Academic institutions cannot meet all the requirements for commercial development of a new pharmaceutical agent. In fact, some academic scientists, especially in the United States, consider the process of commercial development incompatible with academic research programs. This is not the case in Japan or some European countries where there is extensive collaboration among industry, government, and universities to promote the commercial development of basic research discoveries. In the United States, where interactions between industry and universities are more awkward, there are many compelling reasons for formally separating the drug discovery and drug development programs, but these should not inhibit the free exchange of basic research data between academic and industrial scientists. Academic scientists cannot easily participate in an industrial drug development program, but they should continue to perform the basic studies that expand the realm of their invention.

Even when there is excellent scientific collaboration between academic and industrial scientists, legal wrangling often inhibits interaction between industry and the academic world in the United States, where lawyers adopt an adversarial position even when negotiating a collaboration agreement. This may well explain the higher degree of cooperation between academics and industry in Japan and Europe. Academic scientists should be aware that the legal formalities may present the most difficult hurdle to overcome on the path from drug discovery to commercial development.

In an effort to foster research cooperation between university and industrial researchers, the National Cancer Institute has established a number of National Cooperative Drug Development Groups to perform all aspects of research and development required to transform a bioactive crude extract into a commercial product. Although these programs are in their infancy, they seem to promise a smoother route to drug development by encouraging rapid exchange of information between all parties, by making best use of the skills of each individual in the consortium, and by clearly defining responsibilities in research and development. The cooperative programs combine the individualism of academic research programs that is so important in drug discovery with the experience, size, and financial abilities of industry that are required for drug development. In addition, the industrial partner generally provides a broad range of pharmacological expertise that the academic chemist cannot easily find elsewhere. These programs have yet to deal with conflicts between "academic freedom" and "industrial secrecy," but it is expected that they will be minimized. In any case, it is obviously beneficial to establish the ground rules for the collaborative program as early as possible so that each party understands their responsibilities and the limits of their responsibilities. These cooperative drug development programs are still at an experimental stage, but it is hoped that they will provide a smooth route from discovery to development.

9. CONCLUSIONS

Academic scientists have discovered many interesting marine natural products with interesting pharmacological properties. In most cases, however, synthetic derivatives will be marketed rather than the natural product itself. Pharmaceutical companies need both a strong patent position and reliable supplies of material before making a commitment to take a novel compound down the long and expensive path of drug development. A company can obtain a much stronger patent position on a synthetic derivative based on a marine natural product than on the natural product itself. Although reliable supplies of some marine natural products might be obtained by harvesting of a marine organism, pharmaceutical companies would prefer that the compound be synthesized. However, it may be a very difficult task to synthesize some of the more complex marine natural products, particularly by an economically feasible route. In the future, some complex marine natural products with unique pharmacological properties, particularly those from marine microorganisms, may be good candidates for laboratory culture.

It should be emphasized that one of the most important roles of academic natural products chemistry is the training of graduate students and postdoctoral fellows. They are particularly well prepared for careers in the interdisciplinary environment of the pharmaceutical industry. It is not unusual for a marine natural products chemistry student to be an accomplished diver and field researcher with a basic understanding of marine biology and ecology. They have usually participated in studies involving bioassay-directed isolation and can understand the special requirements of interdisciplinary research involving complex pharmacological assays. Their chemical training provides experience in separation science, practical spectroscopy, and structural elucidation and, more importantly, they are trained to use interdisciplinary approaches to solve both practical and theoretical problems.

It is encouraging to know that a number of pharmaceutical companies are actively evaluating the pharmacological potential of marine extracts and marine natural products and that others have established synthetic programs based on marine natural products discovered by academic laboratories. Those companies that are actively involved in collaborative research projects with academic research groups report a very satisfactory number of lead compounds derived from marine sources. An increase in the number of such collaborations will benefit the U.S. pharmaceutical industry and will allow the academic chemists to concentrate on those areas of research where they excel. Further improvements in the areas of technology transfer and university-industrial relations are needed so that the academic chemist is relieved of the task of "selling" research results to industry. Direct support of academic programs from industrial sources can also be beneficial provided that this can be accomplished without violating the principles of academic freedom and conflict of interest. Future research programs should aspire to foster collaborations that will allow the academic scientists to concentrate on the discovery of novel marine natural products with desirable pharmacological properties and their roles in the biological processes in the ocean.

REFERENCES

Abramson, S. N., Radic, Z., Manker, D. C., Faulkner, D. J., and Taylor, P., 1989, Onchidal: A novel naturally occurring irreversible inhibitor of acetylcholinesterase with a novel mechanism of action, *Mol. Pharmacol.* **36**:349–354.

Adamczeski, M., Quiñoà, E., and Crews, P., 1989, Novel sponge-derived amino acids. 5. Structures, stereochemistry, and synthesis of several new heterocycles, *J. Am. Chem. Soc.* **111**:647–654.

Baker, B. J., Okuda, R. K., Yu, P. T. K., and Scheuer, P. J., 1985, Punaglandins: Halogenated antitumor eicosanoids from the octacoral *Telesto riisei*, *J. Am. Chem. Soc.* **107**:2976–2977.

Bandurraga, M. M., Fenical, W., Donovan, S. F., Clardy, J., 1982, Pseudopterolide, an irregular diterpenoid with unusual cytotoxic properties from the Caribbean sea whip *Pseudopterogorgia acerosa* (Pallas) (Gorgonacea), *J. Am. Chem. Soc.* **104**:6463–6465.

Bergmann, W., and Feeney, R. J., 1950, The isolation of a new thymine pentoside from a sponge, *J. Am. Chem. Soc.* **72**:2809–2810.

Bergmann, W., and Feeney, R. J., 1951, Contributions to the study of marine products. 32. The nucleosides of sponges, *J. Org. Chem.* **16**:981–987.

Bundy, G. L., Schneider, W. P., Lincoln, F. H., and Pike, J. E., 1972, The synthesis of prostaglandins E_2 and $F_{2\alpha}$ from (15*R*)- and (15*S*)-PGA$_2$, *J. Am. Chem. Soc.* **94**:2123–2124.

Carmely, S., Kashman, Y., Loya, Y., and Benayahu, Y., 1980, New prostaglandin (PGF) derivatives from the soft coral *Lobophyton depressum*, *Tetrahedron Lett.* **21**:875–878.

Cimino, G., De Stefano, S., Minale, L., and Fattorusso, E., 1972, Ircinin-1 and -2, linear sesterterpenes from the marine sponge *Ircinia oros*, *Tetrahedron* **28**:333–341.

Cimino, G., De Stefano, S., and Minale, L., 1974, Scalaradial, a third sesterterpene with the tetracarbocyclic skeleton of scalarin, from the sponge *Cacospongia mollior*, *Experientia* **30**:846–847.

Crews, P., Manes, L. V., and Boehler, M., 1986, Jasplakinolide, a cyclodepsipeptide from the marine sponge, *Jaspis* sp., *Tetrahedron Lett.* **27**:2797–2800.

Crews, P., Kakou, Y., and Quiñoà, E., 1988, Mycothiazole, a polyketide heterocycle from a marine sponge, *J. Am. Chem. Soc.* **110**:4365–4368.

Culver, P., Burch, M., Potenza, C., Wasserman, L., Fenical, W., and Taylor, P., 1985, Structure-activity relationships for the irreversible blockade of nicotinic receptor agonist sites by lophotoxin and congeneric diterpene lactones, *Mol. Pharmacol.* **28**:436–444.

De Silva, E. D., and Scheuer, P. J., 1980, Manoalide, an antibiotic sesterterpenoid from the marine sponge *Luffariella variabilis* (Polejaeff), *Tetrahedron Lett.* **21**:1611–1614.

Faulkner, D. J., 1973, Variabilin, a new antibiotic from the sponge *Ircinia variabilis*, *Tetrahedron Lett.* **1973**:3821–3822.

Fenical, W., Sims, J. J., Squatrito, D., Wing, R. M., and Radlick, P., 1973, Zonarol and isozonarol, fungitoxic hydroquinones from the brown seaweed *Dictyopteris zonaroides*, *J. Org. Chem.* **38**:2383–2386.

Glaser, K. B., and Jacobs, R. S., 1986, Molecular pharmacology of manoalide, *Biochem. Pharmacol.* **35**:449–453.

Glaser, K. B., de Carvahlo, M. S., Jacobs, R. S., Kernan, M. R., and Faulkner, D. J., 1989, Manoalide: Structure activity studies and definition of the pharmacocore for phospholipase A_2 inactivation, *Mol. Pharmacol.* **36**:782–788.

Gregson, R. P., Marwood, J. F., and Quinn, R. J., 1979, The occurrence of prostaglandins PGE$_2$ and PGF$_2$ in a plant—The red alga *Gracilaria lichenoides*, *Tetrahedron Lett.* **1979**:4505–4506.

Higgs, M. D., and Mulheirn, L. J., 1981, Hybridalactone, an unusual fatty acid metabolite from the red alga *Laurencia hybrida* (Rhodophyta, Rhodomelaceae), *Tetrahedron* **37**:4259–4262.

Högberg, H.-E., Thomson, R. H., and King, T. J., 1976, The cymopols, a group of prenylated bromohydroquinones from the green calcareous alga *Cymopolia barbata*, *J. Chem. Soc. Perkin Trans. 1*. **1976**:1696–1700.

Iguchi, K., Kaneta, S., Mori, K., Yamada, Y., Honda, A., and Mori, Y., 1985, Chlorovulones, new halogenated marine prostanoids with an antitumor activity from the stolonifer *Clavularia viridis*, *Tetrahedron Lett.* **26**:5787–5790.

Iguchi, K., Kaneta, S., Mori, K., Yamada, Y., Honda, A., and Mori, Y., 1986, Bromovulone 1 and iodovulone 1, unprecedented brominated and iodinated marine prostanoids with antitumor activity isolated from the Japanese stolonifer *Clavularia viridis*, *J. Chem. Soc. Chem. Commun.* **1986**:981–982.

Jacobs, R. S., White, S., and Wilson, L., 1981, Selective compounds derived from marine organisms: Effects on cell division in fertilized sea urchin eggs, *Fed. Proc.* **40**:26–29.

Kageyama, M., Tamura, T., Nantz, M. H., Roberts, J. C., Somfai, P., Whritenour, D. C., and Masamune, S., 1990, Synthesis of bryostatin 7, *J. Am. Chem. Soc.* **112:**7407–7408.

Kernan, M. R., Faulkner, D. J., and Jacobs, R. S., 1987, The luffariellins, novel anti-inflammatory sesterterpenes of chemotaxonomic importance from the marine sponge *Luffariella variabilis*, *J. Org. Chem.* **52:**3081–3083.

Kikuchi, H., Tsukitani, Y., Iguchi, K., and Yamada, Y., 1982, Clavulones, new type of prostanoids from the stolonifer *Clavularia viridis* Quoy and Gaimard, *Tetrahedron Lett.* **23:**5171–5174.

Kobayashi, M., Yasuzawa, T., Yoshihara, M., Akutsu, H., Kyogoku, Y., and Kitagawa, I., 1982, Four new prostanoids: Claviridenone-A, -B, -C and -D from the Okinawan soft coral *Clavularia viridis*, *Tetrahedron Lett.* **23:**5331–5334.

Look, S. A., Fenical, W., Van Engen, D., and Clardy, J., 1984, Erythrolides, unique marine diterpenoids interrelated by a naturally occurring di-π-methane rearrangement, *J. Am. Chem. Soc.* **106:**5026–5027.

Look, S. A., Fenical, W., Jacobs, R. S., and Clardy, J., 1986, The pseudopterosins: Anti-inflammatory and analgesic natural products from the sea whip *Pseudopterogorgia elisabethae*, *Proc. Natl. Acad. Sci. USA* **83:**6238–6240.

O'Brien, E. T., Asai, D. J., Groweiss, A., Lipschutz, B. H., Fenical, W., Jacobs, R. S., and Wilson, L., 1986, Mechanism of action of the marine natural product stypoldione: Evidence for reaction with sulfhydryl groups, *J. Med. Chem.* **29:**1851–1855.

Nagle, D. G., and Gerwick, W. H., 1990, Isolation and structure of constanolactones A and B, new cyclopropyl hydroxy-eicosanoids from the temperate red alga *Constantinea simplex*, *Tetrahedron Lett.* **31:**2995–2998.

Pettit, G. R., Herald, C. L., Doubek, D. L., Herald, D. L., Arnold, E., and Clardy, J., 1982, Isolation and structure of bryostatin 1, *J. Am. Chem. Soc.* **104:**6846–6848.

Potts, B. C. M., Faulkner, D. J., de Carvalho, M.S., Jacobs, R. S., 1992, Chemical mechanism of inactivation of bee venom phospholipase A_2 by the marine natural products manoalide, luffariellolide, and scalaradial, *J. Am. Chem. Soc.* **114:**5093–5100.

Rinehart, K. L., Kishore, V., Bible, K. C., Sakai, R., Sullins, D. W., and Li, K.-M., 1988, Didemnins and tunichlorin: Novel natural products from the marine tunicate *Trididemnum solidum*, *J. Nat. Prod.* **51:**1–21.

Roussis, V., Wu, Z., Fenical, W., Strobel, S. A., Van Duyne, G. D., and Clardy, J., 1990, New antiinflammatory pseudopterosins from the marine octacoral *Pseudopterogorgia elisabethae*, *J. Org. Chem.* **55:**4916–4922.

Scheuer, P. J., 1986, Marine Resources: The Search for New Chemicals from Marine Organisms, Sea Grant Quarterly, University of Hawaii Sea Grant College Program, Vol. 8, No. 4.

Suffness, M., Newman, D. J., and Snader, K., 1989, Discovery and development of antineoplastic agents from natural sources, in: *Bioorganic Marine Chemistry*, Vol. 3 (P. J. Scheuer, ed.), Springer, Berlin, pp. 131–168.

Walker, R. P., Faulkner, D. J., Van Engen, D., and Clardy, J., 1981, Sceptrin, an antimicrobial agent from the sponge *Agelas sceptrum*, *J. Am. Chem. Soc.* **103:**6772–6773.

Weinheimer, A. J., and Spraggins, R. L., 1969, The occurrence of two new prostaglandin derivatives (15-*epi*-PGA$_2$ and its acetate, methyl ester) in the gorgonian *Plexaura homomalla*. Chemistry of coelenterates. XV, *Tetrahedron Lett.* **1969:**5185–5188.

White, S. J., and Jacobs, R. S., 1981, Inhibition of cell division and of microtubule assembly by elatone, a halogenated sesquiterpene, *Mol. Pharmacol.* **20:**614–620.

Zabriskie, T. M., Klocke, J. A., Ireland, C. M., Marcus, A. H., Molinski, T. F., Faulkner, D. J., Xu, C., and Clardy, J. C., 1986, Jaspamide, a modified peptide from a *Jaspis* sponge, with insecticidal and antifungal activity, *J. Am. Chem. Soc.* **108:**3123–3124.

Index

Emboldened page numbers indicate chemical structure diagrams. Figures and tables are represented by *f* and *t*.